How PRODUCTS Are MADE

How PRODUCTS Are MADE

An Illustrated Guide to

Product Manufacturing

Volume 6

Jacqueline L. Longe, Editor

GALE GROUP

Detroit
New York
San Francisco
London
Boston
Woodbridge, CT

How Products Are Made, volume 6

STAFF

Jacqueline L. Longe, *Coordinating Editor*
Chris Jeryan, *Managing Editor*

Stacey L. Blachford, *Associate Editor*
Deirdre S. Blanchfield, Melissa C. McDade, *Assistant Editors*

Mary Beth Trimper, *Manager, Composition and electronic prepress*
Evi Seoud, *Assistant Manager, Composition and electronic prepress*
Nikkita Bankston, *Buyer*

Kenn Zorn, *Product Design Manager*
Michelle DiMercurio, *Art Director*
Bernadette M. Gornie, *Page Designer*
Mike Logusz, *Graphic Artist*

Barbara J. Yarrow, *Manager, Imaging and Multimedia Content*
Robyn V. Young, *Project Manager, Imaging and Multimedia Content*
Dean Dauphinais, *Senior Editor, Imaging and Multimedia Content*
Kelly A. Quin, *Editor, Imaging and Multimedia Content*
Leitha Etheridge-Sims, Mary K. Grimes, David G. Oblender, *Image Catalogers*
Pam A. Reed, *Imaging Coordinator*
Randy Bassett, *Imaging Supervisor*
Robert Duncan, *Senior Imaging Specialist*
Dan Newell, *Imaging Specialist*
Christine O'Bryan, *Graphic Specialist*

Maria Franklin, *Permissions Manager*
Margaret A. Chamberlain, *Permissions Specialist*

ISBN 0-7876-3642-8
ISSN 1072-5091

Printed in the United States of America
10 9 8 7 6 5 4 3 2 1

Contents

Introduction

About the Series

Welcome to *How Products Are Made: An Illustrated Guide to Product Manufacturing*. This series provides information on the manufacture of a variety of items, from everyday household products to heavy machinery to sophisticated electronic equipment. You will find step-by-step descriptions of processes, simple explanations of technical terms and concepts, and clear, easy-to-follow illustrations.

Each volume of *How Products Are Made* covers a broad range of manufacturing areas: food, clothing, electronics, transportation, machinery, instruments, sporting goods, and more. Some are intermediate goods sold to manufacturers of other products, while others are retail goods sold directly to consumers. You will find items made from a variety of materials, including products such as precious metals and minerals that are not "made" so much as they are extracted and refined.

Organization

Every volume in this series is comprised of many individual entries, each covering a single product. Although each entry focuses on the product's manufacturing process, it also provides a wealth of other information: who invented the product or how it has developed, how it works, what materials are used, how it is designed, quality control procedures, byproducts generated during its manufacture, future applications, and books and periodical articles containing more information.

To make it easier for you to find what you're looking for, the entries are broken up into standard sections. Among the sections you will find are the following:

- Background
- History
- Raw Materials
- Design
- The Manufacturing Process
- Quality Control
- Byproducts/Waste
- The Future
- Where To Learn More

Every entry is accompanied by illustrations. Uncomplicated and easy to understand, these illustrations may follow the step-by-step description of the manufacturing process found in the text, highlight a certain aspect of the manufacturing process, or illustrate how the product works.

A cumulative subject index of important terms, processes, materials, and people is found at the end of the book. Bold faced volume and page numbers in the index refer to main entries in the present or previous volumes.

This volume contains essays on 100 products, arranged alphabetically, and 15 special boxed sections, describing interesting historical developments or biographies of individuals related to a product. Photographs are also included. Bold faced terms found in main entries direct the user to the topical essay of the same name.

Contributors/Advisor

The entries in this volume were written by a skilled team of technical writers and engineers, often in cooperation with manufacturers and industry associations. The advisor for this volume was David L. Wells, PhD, CMfgE, a long time member of the Society of Manufacturing Engineers (SME) and Professor and Chair of the Industrial and Manufacturing Engineering Department at North Dakota State University.

Suggestions

Your questions, comments, and suggestions for future products are welcome. Please send all such correspondence to:

How Products Are Made

Gale Group, Inc.

27500 Drake Rd.

Farmington Hills, MI 48331-3535

Contributors

Nancy EV Bryk

Chris Cavette

Michael Cavette

Sandy Delisle

Loretta Hall

Gillian S. Holmes

Mary McNulty

Annette Petruso

Perry Romanowski

Randy Schueller

Laurel M. Sheppard

David L. Wells

Angela Woodward

Acknowledgments

The editor would like to thank the following individuals, companies, and associations for providing assistance with Volume 6 of *How Products Are Made*:

Artificial Heart: Jay Caplan, Thermo Cardiosystems Inc., Woburn, Massachusetts. **Artificial Heart Valve:** Mark Spindler and Brett Demchuk, ATS Medical Inc., Minneapolis, Minnesota. **Backhoe:** Richard L. Hall and Roger Zuehl, Case Corporation, Racine, Wisconsin. **Bagpipes:** Mark Cushing, Cushing Bagpipe Company, Andover, New York. **Basketball:** Steven L. Johnson and Elizabeth Daus, Rawlings Sporting Goods Company, Fenton, Missouri. **Breath Mint:** Jason Ford, American Chicle Group, Warner Lambert; Dr. John Flanyak, Brach and Brock Confections Inc. and Confectionery School, University of Wisconsin; John Lux, Brach and Brock Confections Inc., Chicago, Illinois. **Broom:** Dan Koshnick, Cedar Brand Brooms. **Candy Corn:** John Lux, Brach and Brock Confections Inc., Chicago, Illinois. **Cuckoo Clock:** Rosemary Y. Sharp, Linden Clocks, North Smithfield, Rhode Island. **Dishwasher:** Karl Lanes and Russ Maheras, Amana Appliances, Amana, Iowa. **Footbag:** Wham-O Inc., San Francisco, California. **Gyroscope:** Dick Vaughn, Humphrey Inc., San Diego, California. **Hard Hat:** Robert Mundell, E. D. Bullard Company, Cynthiana , Kentucky. **Harpsichord:** John Phillips and Janine Johnson, John Phillips Harpsichords, Berkeley, California; Fred Palmer, Music Department, California State University, Hayward, California. **Hula Hoop:** Greg Lehr and Scott Masline, Wham-O Inc., San Francisco, California. **Kaleidoscope:** Carolyn Bennett, C. Bennett Scopes Inc., Media, Pennsylvania. **Matryoshka Doll:** Ellana and Yuri Burlan, Russian World, New York, New York. **Pillow:** Mike Rodriguez, Royal Pillow, Miami, Florida. **Popsicle:** Doug Grieve, Wilcoxson's Ice Cream Manufacturing Company Inc., Billings, Montana; Elizabeth Snyder, The Ice Screamers, Warrington, Pennsylvania. **Sailboat:** Scott Flick, Ranger Fiberglass Boat Company, Kent, Washington (a subsidiary of Martini Marine Inc.). **Shoelace:** Tom Goltermann, St. Louis Braid Company. **Silicon:** Hayes Kern, Global Metallurgical Sales Inc., Cleveland, Ohio. **Steel Wool:** Paul Bonn, International Steel Wool Corporation. **Toy Model Kit:** Dean Milano, Lewis Nace, and David Carlock, Revell-Monogram, Morton Grove, Illinois. **Trophy:** Catherine Garcia, PDU Plastic Dress Up Company. **Vacuum Cleaner:** Gloria Howard, Castro Valley Vacuum, Castro Valley, California. **Wood Stain:** Richard Boracko, ZAR Wood Stains. **Wrapping Paper:** Laurie Henrichsen, American Greetings, Cleveland, Ohio. **Xylophone:** Gilberto Serna, Century Mallet Instrument Service, Chicago, Illinois; Fred Palmer, Music Department, California State University at Hayward, California.

Photographs appearing in Volume 6 of *How Products Are Made* were received from the following sources:

AP/Wide World Photos. Reproduced by permission: **Artificial Heart.** Jarvik, Dr. Robert (holding mechanical heart), photograph; **Canal and Lock.** Gorgas, William

Crawford (standing, in military uniform), photograph; **Hockey Puck.** Gretzky, Wayne (handling puck), 1994, photograph by Reed Saxon.

Archive Photos Inc. Reproduced by permission: **Basketball.** Jordan, Michael, Ward, Charlie (Jordan, leaping up over Ward), New York City, 1998, photograph by Ray Stubblebine; **DNA Synthesis.** Mullis, Kary (seated, legs crossed, two cameras filming him), photograph; **Sailboat.** Mighty Mary (America3) with its all-women crew, photograph; **Tunnel.** Two men shake hands at the English Channel Tunnel entrance, 1990, photograph.

The Library of Congress: **Boxing Gloves.** Howard Cosell talking with Muhammad Ali, photograph; **Comic Book.** Schulz, Charles M. (drawing cartoon of Charlie Brown), photograph; **Geodesic Dome.** Fuller, R. Buckminster (at head of classroom, feet resting on bottom rung of stool), photograph; **Saxophone.** Coltrane, John (holding saxophone), photograph; **Skyscraper.** Empire State Building, New York City, New York, photograph; **Thompson Submachine Gun.** Gatling, Richard Gordon, photograph.

National Oceanic and Atmospheric Administration (NOAA): **Storm Shelter.** NOAA Historical Photo Collection. National Oceanic and Atmospheric Administration/ Department of Commerce.

Tony Freeman/Photo Edit. Reproduced by permission: **Bagpipes.** Scottish Americans (bagpipers), photograph.

All line art illustrations in this volume were created by **Electronic Illustrators Group (EIG)** of Morgan Hill, California.

Action Figure

Background

An action figure is a doll-like toy designed to resemble characters from movies or literature. The figures can be articulated to hold a variety of poses and may come equipped with accessories, such as clothing, tools, weapons, and vehicles. Action figures are created by assembling molded plastic parts made based on hand-sculpted prototypes.

History

The term action figure was first used in 1964 by the Hasbro Company's Don Levine to describe their new G.I. Joe toy. Levine preferred the name action figure instead of doll because it was more inviting to young boys. In concept, the original Joe was similar to Mattel's Barbie doll, which had been introduced five years earlier. However, the action figure had better articulation, a feature that made it more appealing because the soldier could be bent into a variety of poses. Furthermore, Joe came equipped with numerous accessories and outfits based on real-life military equipment. G.I. Joe was a huge success and additional characters were added to the line. The franchise has remained a strong seller for almost 40 years.

G.I. Joe was followed in 1966 by Captain Action, which was noteworthy because it was a single figure that could be played with as multiple characters. The figure was sold with costumes and accessories from many famous characters including the Phantom, Captain America, Batman, Superman, and Spiderman. Captain Action was the first figure to combine superhero characters with action figures, a trend that continues today.

In 1977, Twentieth Century Fox gave a toy license to the Kenner company to manufacture action figures based on its new movie *Star Wars*. The success of the movie greatly expanded the toy market and ensured the popularity of licensed action figures. Before *Star Wars*, action figures were typically 8-12 in (20-30 cm) tall, but Kenner designed their figures to be only 3.75 in (9.5 cm) in height. Other manufacturers quickly adopted the smaller figure style. A host of other movie and TV show-based toys soon followed, including *Star Trek*, *Battle Star Galactica*, and *Buck Rogers in the Twenty-fifth Century*.

In 1983, federal regulations prohibiting the creation of children's programming based on toys were lifted. This opened a new era in action figures. The Mattel Company took advantage of this opportunity and created a cartoon series based on their 1981 action figure line called "He-Man and the Masters of the Universe." These toys were extremely successful and sold over 55 million units that year. These figures continued to sell through 1990, generating a total of over $1 billion in revenue. Several other toys that were made into cartoons achieved similar success and, thus, began a long standing practice of linking toys and cartoons.

With the 1984 introduction of the "Transformers" series, action figures reached a new level of this sophistication. Transformers were robots able to transform themselves into other objects, such as fighter jets, tanks, or racecars. Since 1984, the Transformers series has debuted several different generations of toys that continue to be popular.

With the 1994 introduction of a line of characters based on Todd McFarlane's **comic**

Once conceived, a prototype of the proposed action figure is created with wire and clay—an armature. The head and facial features of the action figure are created separately and with more detail.

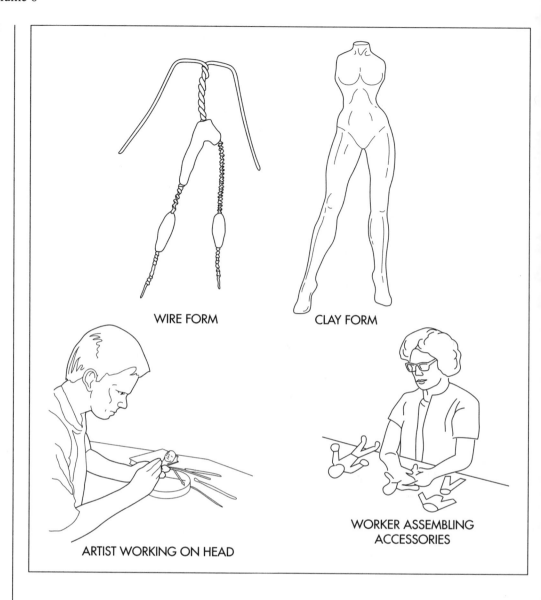

WIRE FORM

CLAY FORM

ARTIST WORKING ON HEAD

WORKER ASSEMBLING ACCESSORIES

book *Spawn*, the action figure industry advanced yet again. These figures were noteworthy because they were made with a much higher degree of detail than any previous toys. This is credited to having the creator of the comic book directly involved with the design of the toys. McFarlane's influence on the development of action figures based upon his comic book has resulted in the detailed toys of today.

Design

Once the character has been selected, the actual design process begins with sketches of the proposed figure. The next step is the creation of a clay prototype. This model is made by bending aluminum wires to form the backbone of the figure, known as an armature. The wire form includes the outline

of the arms and legs posed in the general stance that the figure will assume. The sculptor then adds clay to the armature to give the basic weight and shape that is desired. The clay may be baked slightly during the prototyping process to harden it. Then, the sculptor uses various tools, such as a wire loop, to carve the clay and shape details on the figure.

After creating the basic form, the sculptor may choose to remove the arms and work on them separately for later attachment. This gives the sculptor more control and allows him to produce finer details on the prototype. Working with blunt tools, the sculptor shapes the body with as much detail as is desired. During this process, photo and sketch references are used to ensure the figure is as realistic as possible. Some sculptors may even use human models to guide their design work.

After the general body shape is complete, the sculptor adds the finer details, paying close attention to the eyes, nose, and mouth that give the figure its life-like expression. The designer may attach a rough lump of clay on the main figure as a temporary head while the real head is sculpted on a separate armature. This allows the sculptor to finish the figure's facial expressions independently of the body. At this point, the finished head can be attached to the main armature and joined to the body with additional clay. Once the head is attached, the neck and hair are sculpted to properly fit to the figure. Then, depending on the design of the figure, the costume may be sculpted directly onto the body. However, if a cloth costume or uniform will be added later, the prototype is sculpted without any costume details. During this process, parts of the clay may be covered with aluminum foil to keep it from prematurely drying out. Once everything is completed, the entire figure is baked to harden the clay.

The sculpted prototype is then sent for approval to the manufacturer. Once all design details have been finalized, the prototype is used to make the molds that will form the plastic pieces for the mass-produced figure. The entire sculpting process may take about two weeks, depending on the skill and speed of the sculptor. This process may be repeated several times if revisions must be made to the figure. Several months are typically allowed for this design phase.

Raw Materials

Aluminum wire, modeling clay, and various sculpting tools are used to create the prototype. The actual figure is molded from a plastic resin, such as acrylonitrile butadiene styrene (ABS). This is a harder plastic used to form the main body. Softer plastics, like polypropylene and polyethylene, are commonly used to mold smaller accessory and costume pieces. Various fabrics, such as rayon and nylon, may be used for costume components including body suits, capes, and face masks. As a final decoration, acrylic paints of various colors may be used to decorate the figure. In addition, more elaborate toys may contain miniature electronic components that provide light and sound effects.

The Manufacturing Process

Creating the mold

1 A master mold, or series of molds, are made from the finished clay prototype. These molds are used for mass producing the finished parts for the figure. Approximately two-thirds of the time required to make an action figure is involved in creation and operation of the molds. Patterns based on the prototype are made for each individual piece and sent out for assembly line production. This stage may take several months because the manufacturing plants are typically located overseas, such as Asia.

2 The designers must figure out how to best turn the three dimensional figure into a series of plastic parts. Some figures are simply molded from a single piece of plastic and contain a wire skeleton, allowing limited movement. For figures with greater articulation, each moveable part is molded as a separate component. For example, the *Star Wars* Storm Trooper toys are designed with a hollow chest and solid vinyl limbs. The entire figure is built with swivel joints that allow the limbs to move. Instead of a wire frame, this type of figure contains small gears and washers in the joints of the arms and legs that allow the figure to be bent to any angle and to hold that pose.

Molding the parts

3 Several molding processes can be used to create the plastic parts components. For example, Hasbro originally considered a rotational molding process to give its *Star Wars* figures a smooth, seamless look. However, the size of the pieces caused them to lose some detail and this process was unacceptable. Another problem with rotational molding is that it is a slower process and therefore more expensive for the manufacturer. Instead, Hasbro elected to use an injection molding process. In this process, molten plastic is pumped into a two-piece mold. Pressure is then applied to hold the mold together while the plastic cools and hardens. The mold is then opened and the plastic component is ejected. Each piece of the figure is made in this fashion.

Assembly

4 After all the individual plastic components have been molded, they are assembled to form the finished figures. Instead of glue, some pieces may be bonded together by an ultrasonic sealing method that uses high frequency vibrations to fuse the plastic together. This process provides a tighter seal between pieces and gives a very thin, almost invisible seam. Hasbro uses this process to create figures with a high level of detail without sacrificing manufacturing speed. Glue is used to attach some auxiliary pieces such as chest plates and boots.

Finishing details

5 A figure may be molded entirely from plastic or the figure's clothing may be made of fabric and attached to the body in a separate process. Hasbro makes action figures' costumes from real fabric, just like in the movies. In their *Star Wars* series, Admiral Ack Bar has a vinyl belt insignia, while the character Boba Fett has a canvas-like cape. These pieces are prepared separately from the main figure and added after the figure is complete.

Packaging and shipping

6 While the figure itself is being created, the packaging is being designed as well. The package typically consists of a box with a clear plastic window or a cardboard backing covered with a plastic blister shall. Well-designed packaging graphics can increase the value of the figure and, in fact, some collectors never open figures that they purchase so the artwork stays in mint condition. After the package design is completed, it is mass-produced and sent to the production facility where the figures are made. The figures are inserted in the package, which is then glued or taped shut. The finished units are packed in shipping cases and sent to the United States, usually by boat. From there the toys are distributed to various retailers.

Quality Control

The quality of toy action figures is controlled throughout the manufacturing process. As the molds are created and tested at the production sites, samples are sent back to the manufacturer for approval. Then a series of stringent safety and quality tests must be conducted to ensure the figures are safe for small children and that they will not fall apart during play. A month or two is required to test each figure's articulation many times and to stress test the packaging. Transit tests are used where they actually ship a toy in its box. Additional tests involve dropping the package on the floor and bouncing it around to replicate rough handling it may receive before it arrives on the store shelf.

The Future

Action figures are continually evolving depending upon the latest trends in movies and books. Children's programming remains a major source of inspiration, including popular teen shows such as *Buffy the Vampire Slayer*. A couple of new trends have emerged in the late 1990s. One is a movement toward very realistic, voluptuous female action figures based on comic book characters. Another is the area of sports figures particularly those based on the World Wrestling Federation (WWF).

Other advancements in action figures are coming from new computer technology. Modern action figures can have sound chips and batteries that allow them to play taped sound effects at the press of a button. Similarly this technology can be used to give the figure flashing weapons or eyes that light up. As computer technology advances, action figures will become increasingly interactive and may even be capable of additional functions controlled remotely from a desk top computer.

Where to Learn More

Periodicals

Brady, Mathew. "Toy Making." *Toyfare* (December 1997).

Palmer, Tom, Jr. "Rallying the Troups." *Toyfare* (October 1997).

Rot, Tom. "Rack 'em Up." *Toyfare* (March 1999).

—*Randy Schueller*

Air Freshener

Background

An air freshener is a product designed to mask or remove unpleasant room odors. These products typically deliver fragrance and other odor counteractants into the air. They do so through a variety of product formats, including aerosols, candles, potpourri, and gels. By the late 1990s, sales of air fresheners in the United States had exceeded several hundred million dollars per year. One the most successful new products are Glade Plug-Ins, which use heat generated by electric current to vaporize air-freshening ingredients.

History

Fragrance compounds have been used since antiquity to freshen air and mask odors. For example, the ancient Egyptians were known to use musks and other natural materials to scent their tombs. Over the last 2,000 years a variety of compounds, including numerous spices and floral extracts, have been used for their ability to impart a pleasant aroma. However, it was not until 1948 that the first modern air freshener was introduced. This product, using technology developed by the military to dispense insecticides, was a pressurized spray containing about 1% perfume, 24% alcohol or other solvents, and 75% chlorofluorocarbon (CFC) propellant. This was able to deliver a fine mist of fragrance that remained suspended in the air for a long period of time. This format of the product became the standard in the industry and sales grew tremendously. In the early 1950s, many companies began to add odor-counteractant chemicals to their formulas. These were chemicals that were intended to actually destroy or neutralize offensive odors, as

opposed to simply masking them with fragrance. Perfumery houses showed these active chemicals were capable of reducing a variety of unpleasant odors, such as cigarette smoke, urine and fecal odors, cooking smells, and amine odors typically associated with fish. Compounds used for this purpose included various unsaturated esters, long-chain aldehydes and a few pre-polymers.

Over the next 25 years, aerosol air freshener formulas were modified to improve performance and reduce formula costs. But by the 1970s, the market significantly shifted away from aerosols, due to concerns about destruction of the ozone layer by chlorofluorocarbons (CFCs). While reformulation by the aerosol industry has kept this product form from disappearing completely, alternate air freshener delivery forms have become increasingly popular. In the 1990s, a resurgence in potpourri and candles lead to a host of new air freshening products. For example, Kalib Enterprises Ltd.'s Potpourri, which contains a blend of dry spices and herbs, uses a battery-operated fan to circulate fragrance throughout the room. Arizona Natural Resources Inc.'s Crystal Candle division has introduced candles that kill odors, as well as aromatherapy candles that have specific therapeutic uses.

One of the most innovative, and popular, new formats is Glade Plug-Ins, manufactured by S. C. Johnson of Racine, Wisconsin. Plug-Ins use heat generated by electricity to spread fragrance through the air. It consists of a tiny plastic tray containing a gel-like fragrance concentrate. The consumer simply peels a multilayer barrier film from the top of the tray, leaving a permanent membrane layer that allows the fragrance to diffuse into

By the late 1990s, sales of air fresheners in the United States had exceeded several hundred million dollars per year.

An example of an electrical plug-in air freshener.

the air. The tray is inserted into a warmer unit, which then is plugged into an electrical outlet. As the warmer unit heats up, fragrance permeates at a controlled rate through the film membrane, dispersing into the air.

Design

Plug-Ins consist of a small, plastic tray that holds a gel-like mixture of fragrance. This tray is inserted in a plastic unit equipped with electrical prongs that plug into a standard outlet. The electric current causes a heating element to warm up, vaporizing the volatile fragrance components. The fragrance in the tray lasts several weeks, at which time the consumer simply inserts a fresh tray. The product is designed with a variety of fragrance types to appeal to a wide consumer audience, including Honeysuckle, Mountain Meadow, Country Breeze, Powder Fresh, and Country Garden.

The lid stock of the fragrance tray is specially designed to both hold in the scent and let it out at a controlled rate. Utilizing patented technology, the proprietary laminated film is made with a multilayer barrier and a permeable membrane. S. C. Johnson Wax licensed patented film technology from American National Can that involves a removable barrier and a permanent membrane. The lid material, combined with the proper heat-sealing temperature, pressure, dwell time, and seal design, is designed for easy use by the consumer.

Compatibility between the fragrance formulation and the lid material is key to product performance. During testing, S. C. Johnson researchers further refined the formulations as they learned how they behave with films. Their suppliers also learned a great deal about improving their film technology. The film structure is applied in a one step process during a form-fill-seal operation. An angled piece of the film allows the consumer to easily peel off the outer film/foil barrier layer. The inside membrane, however, remains securely sealed to the tiny tray. As the fragrance is warmed, the membrane allows a continuous and regulated fragrance release. As fragrance is released, the concentrate cracks and dries out, visually signaling the consumer to replace the refill tray. The tray also features a patented ridge down its middle, which is used as a guide for inserting the tray into the warmer unit.

Even the carton that contains the product is unique. This innovative one-piece package went through more than 25 design modifications before fulfilling all marketing and manufacturing requirements. Specifically, it contains a special fifth panel, which secures the warmer unit and fragrance pack during packaging and point of purchase display. The carton features a polyester window that wraps around the side of the folding carton, so the consumer can see the electrical blades of the warmer unit. The carton is designed to show off the warmer unit and its blades, to display the fragrance pack, to easily fit on store shelves, to run on high-volume machinery, and to be reasonably priced.

Raw Materials

The perfume oils used in preparing fragrance concentrate in the air fresheners can be divided into a variety of types. These include aldehydes, which are members of the synthetic fragrance group. When concentrated, aldehydes smell soapy or fatty; however, when mixed in the proper proportions with water, they develop a sweet, flowery smell. Green fragrances are fresh, having an odor similar to cut grass or plant stems and are also usually produced by synthetic perfume oils. However, natural sources of green notes such as galbanum, a tropical resin, or violet leaf oil are also used. Floral notes are some of the oldest and most popular fragrance components. Examples include jasmine-rose complexes blended with lilac and lily of the valley. Herbal-spice fragrance notes are also important components. Lavender, sage, moss, cinnamon, cloves, sandalwood, and cedar are used to provide these notes. Lastly, oriental fragrance notes may be included. These are sweet, heavy, and strong, and are often found in natural animal materials, such as musk. Specific examples of fragrance ingredients used in Plug-Ins include bergamot, bitter orange, lemon, mandarin, caraway, cedar leaf, cloverleaf, cedarwood, geranium, lavender, patchouli, lavandan, rose absolute, and many others. These can be mixed with a variety of synthetic fragrance components, such as aldehydes, ketones, esters, alcohol, terpenes, and so forth. These components are blended together, and mixed with a variety of gelling ingredients. The gel matrix that contains the fragrance can be an organic or inorganic system. These are typically prepared hot, and the fragrance is added as the product cools to preserve the integrity of the fragrance.

The Manufacturing Process

Fragrance pack construction

1 Plug-Ins' fragrance trays are made on form-fill-seal machines at multiple locations, including S. C. Johnson's Waxdale plant. This disposable cartridge is constructed of a transparent, vapor-impermeable polyvinyl film and from a variety of other thermoplastics, such as polyethylene, polypropylene, polystyrene, polyvinyl chloride, and polyvinyl acetate. These thermoplastics are made from different polymers. These manufacturing lines can run either the primary package or the refill package. When running the refill package, a collating/casing machine built by Warren Industries is utilized. This manufacturing arrangement allows the manufacturer to quickly and easily change between different fragrance types. Hand wheels on the carton-assembling machine move the counters to preset numbers when switching between the primary package and the refill carton. This allows changeover between fragrances to be completed in a matter of minutes.

Heater construction

2 The electrical warmer units are produced for S. C. Johnson by Heaters Engineering and Constar Industrial. The housing for the electrical plug apparatus is typically made from a thermoplastic, such as phenol-formaldehyde, epoxy, polyphenylene sulfide, polyphenylene oxide, polycarbonate, and polyimide. Some components may be made from thermoplastic polymers, such as polyethylene, polypropylene, and polyamide. To reduce costs of the unit and keep assembly times to a minimum, the electrical plug housing and the prongs comprise a single structure that is molded together, and the surface of the prongs are given a metallic coating to allow them to conduct electrical current.

Final assembly and packaging

3 The fragrance tray and the heater unit are packaged in a cardboard carton using machinery at Warren Industries. The folding cartons are manually loaded every 20 minutes into magazines on the carton former. Three cartons are erected simultaneously and sent to a Kliklok custom-designed fifth-panel folder. Two flaps on the fifth panel extend over the carton's sides to help support the panel during loading of the fragrance pack and warmer unit.

4 The packing cartons are rotated as the move down the assembly line and a line of hot-melt adhesive is applied to each side.

5 The fragrance packs are loaded into two machines that orient the packs for insertion into the assembled carton. These machines are loaded with fragrance trays every 30 minutes to keep the packaging line operating at its optimal rate. As the carton passes by the feeder bowl, the machines place a fragrance pack in every other carton.

6 The electrical warmer units are then placed into the cartons manually. Shipping cases are erected and fed to the line. Finished packages are manually collated into two rows of six and loaded. Cases are hot-melt sealed, manually palletized, and stretch wrapped.

Quality Control

Quality control occurs at several stages during the manufacturing process. Incoming raw materials used in the fragrance and the gel matrix are assayed to ensure they meet specifications and provide the appropriate odor. The fragrances themselves must comply with regulations established by the California Air Quality Board (CARB), which is responsible for reducing the emission of volatile organic compounds (VOCs), which include the chemicals used in fragrances. By definition, VOCs have vapor pressures of more than 0.1 mm Hg-degrees (68°F or 20°C), or, if the vapor pressure is unknown, contain 12 or fewer atoms of carbon in the molecule. Other states, including New York, have passed similar legislation. These regulations have dramatically impacted the formulation of a variety of air freshener formulations. In addition, plastic components are inspected after molding to ensure they are free from sharp edges and cracks, which could compromise the package integrity.

The Future

The future of Plug-Ins, and other air freshener products, will be determined in part by the consumer product regulatory environment. Just as aerosol sprays were significantly impacted by VOC regulations, similar legislation could effect the fragrance ingredients allowed for use in other air freshener products. Furthermore, advances in packaging and dispersing technology will result in improved products to control room odors. For example, next generation of Plug-In type products are being developed at the time of this writing. These products are designed with a refillable chamber for fragrance oil.

Where to Learn More

Books

Umbach, Wilfried, ed. *Cosmetic and Toiletries.* New York: Ellis Horwood, 1991.

Periodicals

McMath, Robert. "Whether the Cover or Kill, Air Fresheners Smell like Big Business." *Brandweek* 34, no. 8 (February 22, 1993):34.

Packaging 36, no. 3 (March 1991): 40.

—*Randy Schueller*

Angioplasty Balloon

Background

An angioplasty balloon is a medical device that is inserted into a clogged artery and inflated to clear blockage and allow blood to flow. The full medical name for the angioplasty procedure is percutaneous transluminal coronary angioplasty. With extensive use in the United States since 1980, it can relieve angina (chest pain) and prevent heart attacks in people with coronary artery disease. Before angioplasty, bypass surgery was the only option for people with clogged arteries. In bypass surgery, doctors must open the patient's chest to reroute blood vessels to the heart. Angioplasty is less invasive, as the balloon is fed in through the blood vessels, and the chest remains closed. Patient recovery time is also generally faster with angioplasty, than with bypass surgery.

Angioplasty is performed under local anaesthetic, and the patient is kept awake so the doctor can ask if he or she feels any pain during the procedure. The surgeon opens the femoral artery at the top of the leg, and passes a catheter threaded on a thin guidewire into the blood vessel. The catheter, which is a tubular medical device, is about 3 ft (91 cm) long. The surgeon feeds the catheter through the blood vessels into the coronary artery. The catheter releases dye, so its precise position can be seen on a fluoroscope, which is an instrument used for observing the internal structure by means of x ray. When the first catheter is in place at the clogged artery, the surgeon feeds a smaller, balloon-tipped catheter through it. This catheter is about the width of a pencil lead, and the length of the balloon itself corresponds to the length of the affected section of artery—usually less than an inch. The

surgeon guides the balloon-tipped catheter into the narrowed artery. The doctor inflates the balloon for a few seconds. It reaches a diameter of about an eighth of an inch (0.3 cm). If the patient does not feel any pain, then the doctor proceeds to inflate the balloon for a full minute. This clears the arterial blockage, and then the catheters are removed. The patient is treated with prescription drugs to thin the blood and prevent clots, and should recover from the operation within weeks.

In the late 1990s, about 500,000 people a year underwent angioplasty. Medical researchers continued to compare the benefits of angioplasty versus bypass surgery. The principal drawback of angioplasty is that up to half of all patients who undergo the procedure eventually require a repeat procedure. However, new methods are being conducted that prevent some of the scar tissue build-up that can narrow arteries after the procedure.

History

A German physician, Werner Forssmann, was the first known doctor to enter the heart with a catheter. He performed this operation on himself in 1929, when he was 25 years old. Forssmann worked at a small clinic in the town of Eberswald. He was interested in researching a catheter for the heart, but his superior at the clinic forbade him to investigate anything so dangerous. Undeterred, he decided to experiment without his superior's consent. But he did not have access to sterile instruments without the permission of a nurse. Forssmann persuaded a nurse to get him the instruments, convincing her he would use the catheter on her. The pliant woman agreed to let him operate on her. But

In the late 1990s, about 500,000 people a year underwent angioplasty.

The angioplasty balloon is formed through blow molding, a process that uses heated jaws and compressed air to mold and shape the balloon into its correct form.

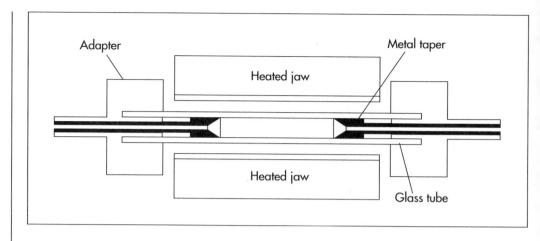

when she was lying on the operating table waiting to be operated on, Forssmann strapped her down so she could not interfere, and then instead performed the operation on himself. He anaesthetized his arm, then slid a catheter 26 in (66 cm) up a vein and into his heart. An x ray verified that the tube was actually inside his heart. In 1956, Forssmann was awarded the Nobel Prize for his work, shared with two other doctors, Andre Frederic Cournand and Dickinson W. Richards, who had extended his ideas.

An Oregon doctor, Charles Dotter, investigated the dilation of narrowed arteries by means of catheters in the 1960s. Dotter opened narrowed leg arteries by passing progressively larger catheters through them. Dotter's work was taken up in Europe, though it received little attention in the United States. A balloon catheter for opening the iliac artery (at the top of the leg) was developed in 1973 by a Dr. Porstmann. Dr. Andreas Gruentzig, working at the University Hospital of Zurich in Switzerland, is credited with performing the first balloon angioplasty to open a clogged coronary artery. Gruentzig worked throughout the 1970s perfecting a balloon catheter that was thin and flexible enough to do the job. In 1977, he performed his first procedure. The patient suffered angina due to a single blocked artery. Gruentzig performed the operation with a team of doctors standing by to do an emergency bypass if the operation failed. But the angioplasty was successful. Gruentzig taught the technique to others, and brought his technology to the United States when he emigrated to Atlanta, Georgia, in 1980. Gruentzig died in a plane crash in 1985, but within 10 years of his introduction

of angioplasty, the procedure was being performed on over 200,000 patients annually. That number rose over the next decade as the technique was refined, and better prescription drugs were found to prevent scarring after the dilation.

Raw Materials

The key requirements of angioplasty balloons are strength and flexibility. A variety of plastics has been used that combine these traits. The first angioplasty balloons in use in Gruentzig's time were made of flexible PVC (polyvinyl chloride). The next generation of balloon technology used a polymer known as cross-linked polyethylene. The materials typically used in the twenty-first century are polyethylene terephthalate (PET) or nylon. PET is the kind of plastic commonly used in plastic soda bottles. It is somewhat stronger than nylon, but nylon is more flexible. So either material is used, depending on the manufacturer's preference. Some angioplasty balloons are coated for lubrication, for abrasion resistance, or to deliver an anticoagulatory drug. In these cases, an additional raw material is required.

The Manufacturing Process

Angioplasty balloons are made by extruding material into a tube shape, and then forming the tube into a balloon through a process known as blow molding.

Extrusion

1 The raw materials for the balloon arrives at the manufacturing facility in granulat-

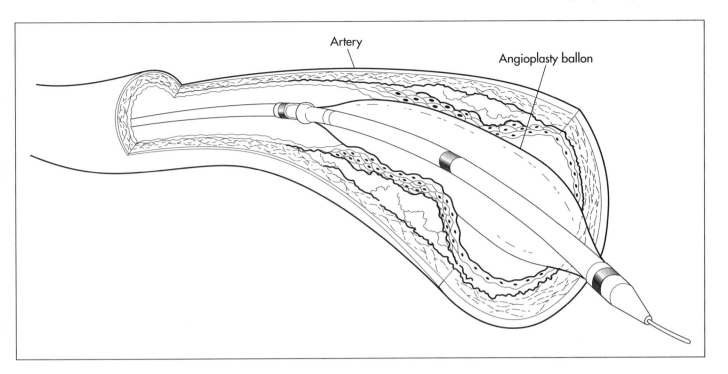

Artery

Angioplasty ballon

ed form. Workers empty the raw materials into a heated, barrel-shaped vat. As the granules melt and liquify, a rotating screw mixes the materials into a homogeneous blend. The liquid plastic is then pumped through an extrusion device. This is a nozzle with one hole cut in it. The liquid comes out the extruder as a long tube. The tube is pulled by a mechanical puller through a cooling bath, which freezes the tubing so that it is solidified. Next, a mechanical cutter chops the tubing to its specified length. At this point, the tubing is called a pre-form.

Balloon forming

2 Next, the balloon is formed through blow molding. Though many tubes are cut, blow molding takes place one piece at a time. A worker inserts the pre-form into a device called a glassform. The manufacturer may have various glassforms that correspond to different finished diameters of the product. Next, one end of the tube is welded shut. The open end is connected to a supply of compressed air. Then two heated jaws close around the part.

3 The compressed air is switched on, and it keeps the pre-form at a constant inner pressure. The heated jaws warm the piece. This warmup time prepares the plastic for the next step. The blow molding process is

controlled by a computer with sensors that determine when the material has reached the optimum temperature for the next step of pressure forming.

4 After the warmup, the computer signals the compressed air machine to switch to a high-pressure mode. The balloon is inflated at this high pressure for a specified amount of time. Shortly after the high-pressure stage begins, the heated jaws stretch the material. Then the formed balloon is cooled, again with compressed air. Now it is ready for removal from the glassform, inspection, and packaging.

Inspection

5 Angioplasty balloons go through a series of inspections for quality control purposes. Balloons are checked both by visual inspection and by machines.

Packaging

6 If the manufacturer makes only angioplasty balloons, and not the catheters that go with the device, the balloons are individually boxed, the boxes are bagged, and then shipped to a larger manufacturer. This manufacturer then assembles an angioplasty kit by wrapping the balloon around a catheter and sterilizing it. First, the balloon is collapsed by a vacuum pump. Then a worker

An angioplasty balloon is inserted into a clogged artery and inflated to clear blockage and allow blood to flow.

glues or heat bonds it to a catheter shaft. The balloon is tested again at this point. Then the worker deflates the balloon again, wraps it around the shaft, places a protective wrapping over it, and sends it to be sterilized, making it ready for hospital use.

Quality Control

Quality control is, of course, extremely important in medical devices. Angioplasty balloons are manufactured one at a time, and typically each piece is inspected once it is formed. A worker inspects the balloon visually for any marked flaws. Then the worker loads the balloon into a machine that tests its wall thickness. Next, the balloon is placed in another machine, which inflates it and checks the internal pressure.

The Food and Drug Administration (FDA) oversees quality control for the medical device industry. A federal study in 1970 revealed thousands of injuries and a significant number of deaths related to medical devices. As a result, in 1976 Congress amended the Food and Drug act to give the FDA authority over medical device manufacturing. The FDA is required to be notified of every medical device before it can be marketed, and manufacturers must prove the device is safe and effective. Because of the importance of quality in angioplasty balloon manufacturing, the companies that make them are generally not aiming for economy of scale, or making a lot of balloons as fast as they can. Instead, the balloons are made in a labor-intensive fashion, one at a time, with step-by-step inspection.

The Future

Angioplasty is simpler and easier on the patient than bypass surgery, the procedure that it, to some extent, replaced. Its biggest drawback is that some 30-50% of patients undergoing the procedure need to repeat it because their arteries clog again. The initial

clogging is known medically as stenosis, and when it happens after angioplasty, it is called restenosis. Most research into angioplasty at the beginning of the twenty-first century concentrates on ways of preventing restenosis. Some angioplasty balloons are coated with prescription drugs, such as heparin, to prevent arterial buildup. Such a drug is routinely given to patients after the procedure, but with coated balloons, the drug can be delivered directly to the affected artery. Some surgeons are also experimenting with a device called a stent, which can be placed in the artery during angioplasty to prevent the vessel clogging again. Stents are small metal tubes that may be either stainless steel or some kind of flexible steel mesh. The newest angioplasty technology is involved with combining the balloon with the stent for the best results for both the patient and the surgeon.

Where to Learn More

Books

Friedman, Steven G. *A History of Vascular Surgery.* Mt. Kisco, NY: Future Publishing, 1989.

Fries, Richard C. *Reliable Design of Medical Devices.* New York: Marecel Dekker, Inc., 1997.

Klaidman, Stephen. *Saving the Heart: the Battle to Conquer Coronary Disease.* Oxford: Oxford University Press, 2000.

Periodicals

Sauerteig, Knut, and Michael Giese. "The Effect of Extrusion and Blow Molding Parameters on Angioplasty Balloon Production." *Medical Plastics and Biomaterials Magazine* (May 1998).

Stone, John. "Balloon Man." *New York Times Magazine* (October 16, 1988): 61.

—Angela Woodward

Artificial Heart

Background

A natural heart has two pumps, each with two chambers. The right atrium pumps oxygen-depleted blood from the body into the right ventricle, which pumps it to the lungs. The left atrium sends aerated blood from the lungs into the left ventricle, which pumps it out to the body. With each heart beat, the two atria contract together, followed by the large ventricles.

Congestive heart failure, which is the steadily declining ability of the heart to pump blood, is one of the leading causes of death. This disease is caused by sudden damage from heart attacks, deterioration from viral infections, valve malfunctions, high blood pressure and other problems. According to the American Heart Association, an estimated five million Americans are living with heart failure and over 400,000 new cases are diagnosed every year. About 50% of all patients die within five years. Heart disease cost the United States health industry about $95 billion in 1998.

Though medication and surgical techniques can help control symptoms, the only cure for heart failure is an organ transplant. In 1998, around 7,700 Americans were on the national heart transplant list but only 30% received transplants. Artificial hearts and pump-assist devices have thus been developed as potential alternatives.

An artificial heart maintains the heart's blood circulation and oxygenation for varying periods of time. The ideal artificial heart must beat 100,000 times every 24 hours without requiring either lubrication nor maintenance and must have a constant power source. It must also pump faster or slower depending on the activity of the patient without causing either infection or blood clots.

The two major types of artificial hearts are the heart-lung machine and the mechanical heart. The first type consists of an oxygenator and a pump and is mainly used to keep blood flowing while the heart is operated on. This machine can only operate for a few hours since the blood becomes damaged after longer times.

A mechanical heart is designed to reduce the total work load of a heart that can no longer work at its normal capacity. These hearts consist of equipment that pulses the blood between heart beats or use an artificial auxiliary ventricle (left ventricle assist device, LVAD) that pumps a portion of the normal cardiac output. Because such devices usually result in complications to the patient, they have generally been used as a temporary replacement until natural hearts can be obtained for transplantation. Worldwide about 4,000 LVADs have been implanted. The market for these devices is estimated at $12 billion per year in the United States.

History

Since the late nineteenth century, scientists have tried to develop a mechanical device that could restore oxygen to the blood and remove excessive carbon dioxide, as well as a pump to temporarily supplant heart action. It took almost 100 years before the first successful heart-lung machine was used on a human being by John H. Gibbon Jr. in 1953. Four years later the first artificial heart (made from plastic) in the western world was implanted inside a dog. The National Heart In-

According to the American Heart Association, an estimated five million Americans are living with heart failure and over 400,000 new cases are diagnosed every year.

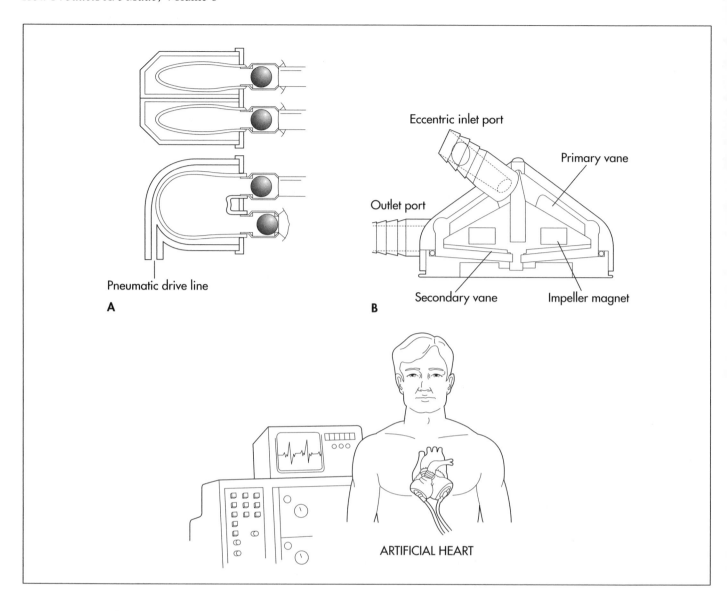

Pneumatic drive line

A

Eccentric inlet port

Outlet port

Primary vane

Secondary vane

Impeller magnet

B

ARTIFICIAL HEART

A. A pneumatic artificial heart. B.
A gyro centrifugal artificial heart.

stitute established the artificial heart program in 1964, leading to the first total artificial heart for human use implanted in 1969.

The emphasis shifted to left ventricular assist systems and blood compatible materials in 1970. During that same year, a LVAD was used successfully. However, blood pump development continued and devices became smaller, lighter, more acceptable, and clinically successful. A number of **polyurethane** and plastic pumps of long-life time were also developed. During the 1980s, the Food and Drug Administration (FDA) imposed more restrictive rules to the Medical Devices Standards Act, leading to higher development costs. Many research groups had to drop out, with only a few remaining today.

Perhaps the most famous scientist is Dr. Robert Jarvik, who invented an artificial heart called the Jarvik-7. This device, made from aluminum and plastic, replaced the two lower chambers of the natural heart and used two rubber diaphragms for the pumping action. An external compressor the size of a refrigerator kept the artificial heart beating. Barney Clark was the first patient to receive this heart. He survived 112 days before physical complications caused by the implant took his life. In 1986, William Schroeder became the second Jarvik-7 recipient, surviving for about 20 months.

The medical community realized that a completely implantable heart could avoid the mobility and infection problems caused by the Jarvik-7. In 1988, the National Institutes of Health began funding development of such hearts and was supporting such a program in 1991 totaling $6 million. Three years later, an electric and battery-powered implantable LVAD became available. In 1999, Charlie Chappis became the first patient ever released from a hospital with such a device. Other artificial hearts of various designs are currently being tested.

Raw Materials

An artificial heart or LVAD is made out of metal, plastic, ceramic, and animal parts. A titanium-aluminum-vanadium alloy is used for the pump and other metal parts because it is biocompatible and has suitable structural properties. The titanium parts are cast at a specialized titanium processor. Except for blood-contacting surfaces, the titanium is machined to a specific finish. Blood-contacting surfaces receive a special coating of titanium microspheres that bond permanently to the surface. With this coating, blood cells adhere to the surface, creating a living lining.

A blood-contacting diaphragm within the pump is made from a special type of polyurethane that is also textured to provide blood cell adherence. Two tubular grafts are made from polyester (which are used to attach the device to the aorta) and the valves are actual heart valves removed from a pig. Other parts that make up the motor are made from titanium or other metals and ceramics.

Design

There are several critical issues when designing a LVAD. Fluid dynamics of the blood flow must be understood so that enough blood is pumped and no blood clots are created. Materials must be chosen that are biocompatible; otherwise the pump could fail. The efficiency of the motor must be optimized so that minimal heat is generated. Because of possible rejection, the total volume and surface area of the entire device should be kept as small as possible. A typical LVAD weighs around 2.4 lb (1,200 gm) and has a volume of 1.4 pints (660 ml).

The Manufacturing Process

1 Most of the components are made to custom specifications by third party manu-

Dr. Robert Jarvik.

Robert Jarvik was born on May 11, 1946, in Midland, Michigan, and raised in Stamford, Connecticut. He entered New York's Syracuse University in 1964, studying architecture and mechanical drawing. After his father developed heart disease, Jarvik switched to pre-medicine. He graduated in 1968 with a bachelor of arts in zoology, but was rejected by medical schools in the United States. He entered the University of Bologna in Italy, but left in 1971 for New York University, earning a master of arts in occupational biomechanics.

Jarvik applied for a job at the University of Utah. The director of the Institute for Biomedical Engineering and Division of Artificial Organs, Willem Kolff, had been developing an artificial heart since the mid-1950s. Jarvik began as his lab assistant, earning his medical degree in 1976.

On December 2, 1982, doctors transplanted the first artificial heart into a human. This plastic and aluminum device, the Jarvik-7, was implanted into Barney Clark, who survived for 112 days after the operation. Several other patients received Jarvik-7 hearts but none lived more than 620 days. The main benefits were that there wouldn't be a wait for a human heart and there was no chance of rejection. The obvious pitfall being patients were forever connected to a compressed air machine via tubes.

The Jarvik-7 was eventually used as a stopgap measure for patients awaiting natural hearts and provided hope that there would not be a wait for transplants. In 1998 Jarvik continued work on a self-contained device to be implanted into a person's diseased heart to make it function correctly.

facturers, including machine shops and printed circuit board manufacturers. The porcine valves are sewn inside the grafts with sutures at a medical device firm that specializes in heart valves.

Once all components are obtained, the LVAD system is assembled and tested, to ensure that each device meets all specifications. Once tested, the LVAD can be sterilized and packaged for shipment.

Forming the polyurethane parts

2 Some artificial heart manufacturers make their own polyurethane parts. One process uses a proprietary liquid solution that is poured on a ceramic mandrel layer by layer. Each layer is heated and dried until the desired thickness is reached. The part is then removed from the mandrel and inspected. Otherwise, a third party manufacturer uses an injection molding or vacuum molding process combined with radio frequency welding.

Assembly

3 Each artificial heart takes several days to put together and test. The assembly process is performed in a clean room to avoid contamination. Each artificial heart consists of up to 50 components that are put together using special adhesives. These adhesives require curing at high temperatures. Several assembly operations happen in parallel, including the assembly of the motor housing and components, the assembly of the percutaneous tube and the attachment of the pusher plates to the polyurethane diaphragm. These subsystems are individually inspected, then final assembly of the complete system occurs. The grafts are assembled separately and attached during operation.

Testing

4 After assembly is completed, each device is tested using special equipment that simulates pressures in the body. All electronic components are tested with electronic test equipment to ensure the proper function of all circuitry.

Sterilization/packaging

5 After the artificial heart is tested and passes, it is sent to an outside service for sterilization. Each device is sealed in plastic trays and returned to the heart manufacturer. It is then packaged in custom suitcases to protect it from contamination and prevent damage.

Quality Control

Most components have already passed inspection before they arrive at the heart manufacturer. Some components are still inspected dimensionally since they require tight tolerances—on the order of millionths of an inch, which requires special measuring tools. To meet FDA regulations, every component (including adhesives) used in the process is controlled by lot and serial number so that tracking problems is possible.

Byproducts/Waste

Scrap titanium is recovered and recycled after remelting and recasting. Otherwise, little waste is produced since most components have passed inspection before leaving the various manufacturers. Other defective parts are discarded. Once a device has been used by a patient, it is sent back to the heart manufacturer for analysis to improve the design.

The Future

Within the next decade, a number of new devices will come on the market. Pennsylvania State University researchers are developing an electromechanical heart powered by radio-frequency energy that is transmitted through the skin. A motor drives push plates, which alternate in pressing against plastic blood-filled sacs to simulate pumping. Patients carry a battery pack during the day and sleep with the device plugged in to an electrical outlet. This artificial heart will be tested in humans by 2001.

Several research groups are developing pumps that circulate blood continuously, rather than using a pumping action, since these pumps are smaller and more efficient. In Australia, Micromedical Industries Limited is developing a continuous-flow rotary blood pump, which is expected to be implanted in a human by 2001. The Ohio State University's cardiology department is developing a plastic pump the size of a hockey puck that is self regulating. This pump is implanted in patients for several weeks until their own heart recovers.

Thermo Cardiosystems, Inc. is also working on a LVAD with a continuous flow rotary pump), expected to be implanted sometime in 2000, and a LVAD with a continuous flow centrifugal pump. The latter is still in an early development phase, but is planned to be the world's first bearingless pump, meaning that it won't have any parts that wear. This is accomplished by magnetically suspending the rotor of the pump. Both these devices will be available with transcutaneous energy transfer, meaning that the devices will be fully implantable.

With fewer donor hearts becoming available, others are also developing an artificial heart that is a permanent replacement. These replacements may be in the form of a left ventricle assist device or a total artificial heart, depending on the patient's physical condition. LVADs are being developed by inventor Robert Jarvik and renowned heart surgeon Michael DeBakey. Total artificial hearts are being jointly developed by the Texas Heart Institute and Abiomed, Inc. in Massachusetts. In Japan, researchers are developing total artificial hearts based on a silicone ball valve system and a centrifugal pump with a bearing system made from alumina ceramic and polyethylene components.

Alternatives to artificial hearts and heart-assist pumps are also under development. For instance, a special clamp has been invented that changes the shape of a diseased heart, which is expected to improve the pumping efficiency by up to 30%. Such a device requires minimal invasive surgery to implant.

Where to Learn More

Periodicals

Bonfield, Tim. "Device to Help Hearts." *Cincinnati Enquirer* (November 7, 1999).

Castor, Tasha. "Ohio State University Cardiology Unit Set to Try Heart Pump." *The Lantern* (May 6, 1999).

"Electric Hearts by 2005." *Popular Mechanics* (March 1997).

Gugliotta, Guy. "Upbeat on Man-Made Hearts: Improved Devices Save Those Too Ill for Transplant." *The Washington Post* (June 28, 1999): A01.

Guy, T. Sloane. "Evolution and Current Status of the Total Artificial Heart: The Search Continues." *ASAIO Journal* (January-February 1998): 28–33.

Hall, Celia. "Thumb-Sized Pump Can Cut Heart Deaths." *The Daily Telegraph* (September 13, 1999): 11.

Hesman, Tina. "Pump Brings New Expectations for Artificial Heart." *Omaha World-Herald* (December 12, 1999).

Hopkins, Elaine. "Device Lets Heart Patient Await Transplant at Home." *Journal Star* (November 30, 1999).

Kinney, David. "Effective Artificial Heart Seems Within Reach." *The Los Angeles Times* (January 23, 2000).

Kolff, William. "Early Years of Artificial Organs at the Cleveland Clinic: Part II: Open Heart Surgery and Artificial Hearts." *ASAIO Journal* (May-June 1998): 123–128.

Kolff, William. "The Need for Easier Manufacturing of Artificial Hearts and Assist Devices and How This Need Can Be Met by the Vacuum Molding Technique." *ASAIO Journal* (January-February 1998): 12–27.

Kunzig, Robert. "The Beat Goes On." *Discover* (January 2000): 33–34.

M2 Communications. "Successful Blood Compatibility Tests for Micromedical's Artificial Heart." *M2 PressWIRE* (March 26, 1999).

Phillips, Winfred. "The Artificial Heart: History and Current Status." *Journal of Biomechanical Engineering* (November 1993): 555–557.

Takami, Y. et al. "Current Progress in the Development of a Totally Implantable Gyro Centrifugal Artificial Heart." *ASAIO Journal* (May-June 1998): 207–211.

Wilson, Steve. "A Life and Death Race Against Time." *Arizona Republic* (November 14, 1999).

Yambe, T. et al. "Development of Total Artificial Heart with Economical and Durability Advantages." *The International Journal of Artificial Organs* (1998): 279–284.

Other

"Progress on Development of an Artificial Heart." http://www.uts.edu.au/new/archives/1999/February/02.html (December 29, 2000).

—*Laurel M. Sheppard*

Artificial Heart Valve

In the United States, more than 80,000 adults undergo surgical procedures to repair or replace damaged heart valves every year.

Background

A heart valve acts as a check valve, opening and closing to control blood flow. This cycle occurs about 40 million times per year or two billion in an average lifetime. Natural valves can develop several problems, either the valve opening becomes narrow or may not close completely. The first condition decreases the pumping efficiency and limits the amount of blood pumped to the body. The second condition can reduce the amount of blood to the rest of the body, as well as result in excess pressure in the lungs, also limiting their efficiency. In the United States, more than 80,000 adults undergo surgical procedures to repair or replace damaged heart valves every year.

Artificial heart valves consist of an orifice, through which blood flows, and a mechanism that closes and opens the orifice. There are two types of artificial heart valves: mechanical devices made from synthetic materials; and biological or tissue valves made from animal or human tissue. In general, biological valves are used for patients who are over 65 or cannot take anticoagulants. Mechanical valves are used for patients that have a mechanical valve in another position, have had a stroke, require double valve replacement, and usually are recommended for those under 40. These type of valves require the patient to take anti-coagulating drugs.

Mechanical valves can be further broken down into three types based on the opening and closing mechanism. These mechanisms are: a reciprocating ball, a tilting disk, or two semicircular hinged leaflets. The first type is based on a ball-in-cage design, which uses a rubber ball that oscillates in a metal cage made from a cobalt-chromium alloy. When the valve opens, blood flows through a primary orifice and a secondary orifice between the ball and housing. About 200,000 of these have been implanted.

The tilting disk valve uses a circular disk retained by wire-like arms that project into the orifice. When the disk opens, the primary orifice is separated into two unequal orifices. About 360,000 of these valves have been implanted. The current design consists of two semicircular leaflets connected to the orifice housing by a hinge mechanism. The leaflets separate during opening, producing three flow areas in the center and on the sides. Over 600,000 bileaflet valves have been implanted.

History

The first recorded surgical operation on a heart valve took place in 1913. Replacement of diseased valves did not take place until 1962, when the first successful biological valves were invented using human tissue from a donor. Ball valves were the first type of mechanical valves and were developed around the same time. Miles Edwards, an electrical engineer who founded a medical device company called American Edwards Laboratories in the 1950s, is credited with co-inventing the first commercially available artificial heart valve. Disk valves became popular in the 1970s after the first successful design was introduced in 1969. The reduced height improved clinical performance. The bileaflet design was first introduced in 1977 and became more popular during the 1980s.

Advances in materials also helped spur the development of mechanical valves. In 1965,

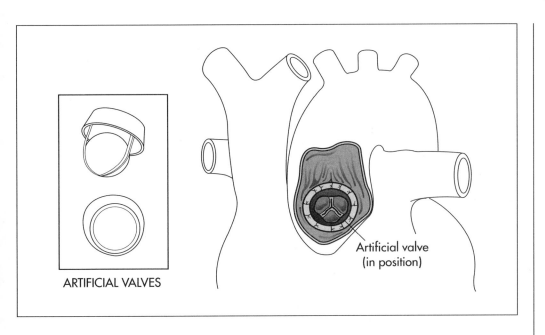

ARTIFICIAL VALVES

Artificial valve
(in position)

scientist Dr. J. C. Bokros from the General Atomic Company was investigating pyrolytic carbon materials for nuclear fuel applications. Because the material's properties were suitable for biomedical applications (durability, blood compatibility), he looked at it for making artificial heart valves. Today, about 90% of all mechanical heart valves implanted have at least one pyrolytic carbon part.

In 1976, medical devices (including prosthetic heart valves) came under the jurisdiction of the Food and Drug Adminstration (FDA). FDA then issued guidelines for Pre-market Approval (PMA) applications for heart valves. In 1993, FDA issued a guidance document based on objective performance criteria. This set the minimum amount of follow-up required for a PMA study at 800 valve-years.

The performance of mechanical valves has been noteworthy. The ball valve, in use for over 30 years, has had only a dozen structural problems that caused no major harm to the patient. The tilt valve had fewer than 1% of failures after 15 years of experience. The most popular type of bileaflet valve only reported several dozen failures to the FDA. However, in early 2000, one valve manufacturer recalled silver-coated valves because of a leaking problem in 2% of patients. In all, there have only been about 50 failures out of the approximately one million valves in service.

Approximately 265,000 prosthetic valves are now implanted worldwide each year, valued at over $700 million. About 60% of these are mechanical valves, with a market value of around $400 million. Over two million mechanical valves have been implanted in patients around the world during the last several decades.

Raw Materials

Most artificial valves are made of titanium, graphite, pyrolytic carbon, and polyester. The titanium is used for the housing or outer ring, graphite coated with pyrolytic carbon is used for the bileaflets, and 100% pyrolytic carbon is used for the inner ring. The pyrolytic carbon is sometimes impregnated with tungsten so that the valve can easily be seen following implantation). The sewing cuff, used to attach the valve to the heart, is made out of double velour polyester.

Titanium is used for its strength and biocompatibility. The outer rings come already fabricated from an outside manufacturer and are made from machined bar stock. Lock rings and wire, used to hold the cuff in place, are also made from titanium. The polyester comes in the form of tubes. All plastic components are deburred by the supplier, which involves removing any bumps from the surface. Occasionally the valve manufacturer may have to deburr some parts.

The pyrolytic carbon coating is produced by depositing gaseous hydrocarbons (usually methane) onto a heated graphite substrate at temperatures of 3,272-4,172°F (1,800-2,300°C) in a chamber. These gases break down into carbon. The inner rings are made from 100% pyrolytic carbon using a fluidized bed process at another manufacturer. This material's atomic microstructure helps resist cracking, making it ductile. However, the processing method can still introduce microcracks that must be detected.

The Manufacturing Process

1 The majority of components are made by a third party, except for the polyester cuffs. These are made by a sewing process that includes various looping, folding, and stitching steps. The manufacturing process therefore consists mainly of various assembly and inspection steps.

Assembly

2 Assembly takes place in a clean room to avoid contamination. The leaflets are attached to the inner rings, which are then placed in the housing or outer ring.

3 While this is being down, the sewing cuffs are being made. A special pressurized heating process is then used to form the cuffs around the valve, which takes place at several hundred degrees. The valves are then mounted into a rotator assembly, which the surgeon uses for implanting.

Sterilization and packaging

4 After the valves are assembled and tested, they are sterilized in a double plastic container. Steam sterilization is used, which involves temperatures up to 270°F (132°C) and times of 15 minutes or more. To make sure the sterilization process has worked, a biological indicator is placed inside. If the indicator shows no growth of bacteria or other viable organisms, the valves and its packaging have been properly sterilized. Each plastic-encased valve is then packaged in a box for shipping.

Quality Control

All components are inspected visually, dimensionally and functionally prior to assembly to make sure they meet specifications. The diameter of each ring is measured and

assigned a size, which is then matched to the appropriate bileaflet to make sure they will fit together. Microscopic analysis using high power magnification is used to check components for scratches. In total, up to 50 inspections are made during the assembly process.

Proof testing is used to determine the structural quality of potentially flawed heart valves. In this method, a valve is loaded to a certain stress level using a special pressurization fixture to see if it will fail at this stress. During the stress test, acoustic emission technology is used to detect minute cracks that might go undetected so that these valves can be rejected. Once the valves are sterilized and packaged, they are inspected to make sure the labels are accurate.

Byproducts/Waste

Due to the stringent quality control procedures, there is little or no waste produced during the assembly process. Any scrap material is recycled if it is feasible. Defective components are returned to the manufacturer. Some chemicals used for cleaning must be disposed of properly following safety regulations.

The Future

Blood clotting is still a problem with mechanical valves and manufacturers continue to improve designs, sometimes using supercomputing modeling tools, as well as surgical procedures. The shape of the orifice is being improved to reduce pressure losses, turbulence and shear stresses. Flow area is maximized by using stronger materials, which minimizes wall thickness. Tapering the sides of the valve pumps blood more efficiently. Operations are also being developed that only require a 3-4 in (8-10 cm) incision instead of 12 in (30 cm). Manufacturing efficiencies will continue to improve.

Researchers are looking at making heart valves out of plastic material that are flexible enough to simulate the opening and closing action. This approach may not require anticoagulation drugs. Others are working on developing artificial heart valves made from a patient's own cells. Experiments have been successful using sheep. Both developments may take decades before they are put in practical use.

Where to Learn More

Periodicals

"Baxter Announces Name of Cardiovascular Spin-Off." *PR Newswire* (January 14, 2000).

Dolven, Ben. "Take Heart." *Far Eastern Economic Review* (November 4, 1999).

Lankford, James. "Assuring Heart Valve Reliability." *Technology Today* (Summer 1999).

"Maker Recalls Heart Valves." *Newsday* (January 25, 2000): A49.

"Medical Carbon Research Institute Announces On-X Prosthetic Heart Valve CE Mark Approval." *Business Wire* (July 24, 1998).

Reed, Stephen. "Sarasota Doctors Trying Out a Better Artificial Heart Valve." *Sarasota Herald Tribune* (June 5, 1998): 1A.

"Researchers Grow Artificial Heart Valves in Sheep." *Reuters Ltd* (November 7, 1999).

Sternberg, Steve. "In Medicine, a Shortage Prevented." *USA Today* (August 3, 1998): 06D.

Other

"Medtronic Announces First Implant." Medtronic, Inc. http://www.medtronic.com (December 29, 2000).

"St. Jude Medical Announces One-Millionth Mechanical Heart Valve Implant." St. Jude Medical, Inc. (April 6, 2000). http://www.prnewswire.com (May 2000).

—*Laurel M. Sheppard*

Baby Carrier

In 1997 alone, over 8,700 children were brought to the emergency room and treated for falls from carriers.

Background

Family members have carried babies in a variety of carriers since the beginning of human civilization. Baby carriers keep the baby close and provide the infant with comfort and security while allowing the carrier some freedom to work and care for other members of the family. The most basic form of baby carrier is the sling. In South America, woven shawls are used to carry babies in front of their mothers. The shawl is anchored over one shoulder, wrapped under the other arm, and tied around the back. The baby is free to nurse and sleep, and is cooled or warmed by the weave and wool of the shawl. In Asia, the sling is tied over one arm but carried lower on the opposite side so the baby rests on the hip. African cultures use the sling in this fashion or with the sling tied in front around the chest with the baby balanced behind. The front-tied sling is also practical for older babies who need to nurse less often and are more settled in their routine.

Other Asian cultures tie a woven cloth around the infant in a complex series of wraps and ties that isolate the child's legs on the mother's sides and allow the baby to move its arms freely against the mother. In cold climates where mothers wear hooded coats or parkas, the hood doubles as a sling that gives the baby room to see around its mother's head or cuddle deep in the hood to nap. Hot climates feature baby carriers of loosely woven airy cloth and even net bags that can be hung from the shoulders or balanced on the head. The cradle boards used by many tribes of Native Americans are ancient forerunners of today's framed backpack that provides rigid strength to carry the child and cloth wrapping for comfort.

Side- and hip-balanced carriers are used by fathers and brothers who hunt and harvest with the baby. Other relatives can baby-sit, thanks to the convenience of baby carriers. The closeness of the baby carrier is not only a convenience but a way of nurturing the whole family.

History

The invention of the cloth infant carrier grew out of one woman's service in the Peace Corps. Ann Moore, a pediatric nurse who worked in Togo, Africa, in the 1960s observed the native women carrying their babies in slings. After she returned to the United States, Moore and her mother, Lucy Aukerman, designed a fabric pouch to resemble these slings but also to be rugged and to have adjustable shoulder and waist supports for the parent. Moore and Aukerman patented the Snugli in 1969.

In 1978, a Nigerian child care professional named Toritse Onuwaje moved to the United States. She was dismayed at the amount of crying American babies did compared to their Nigerian counterparts. She patterned a carrier she called the Baby Wrap on the ukpoma, a cloth carrier worn by members of her native Itshekiri tribe. The women of her tribe work all day with their babies wrapped on their backs, and the babies seldom cry because the mothers' closeness is comforting.

The rigid, molded plastic carrier grew out of the invention of the child safety seat for use in automobiles (and, later, airplanes). The first child safety seats were patented in the late 1970s. In the period from 1982 through 1985, a number of designs of molded plastic carriers were patented with varying configu-

Different types of baby carriers.

rations for carrying a reclining infant. The shape was contoured to support the child's back and seat and incorporated an adjustable handle that could be raised and locked into place while the child is being carried and lowered to lift the baby in and out. Various types of fabrics cushion the child; some carriers use fabric mesh that is suspended in the frame to support the infant, while others are fitted out with padded cloth inserts that fasten in the plastic carrier securely but can be removed for cleaning.

Raw Materials

Slings and cloth carriers are manufactured from fabrics and polyester fiber or foam for stuffing or padding. The fiber is purchased in 500-lb (227-kg) bales. Velcro fasteners, woven fabric webbing for straps, and metal buckles and fasteners for supports around the parent are used to hold cloth carriers in place. Backpack-type carriers can be made mostly of cloth, but a second type consists of an aluminum frame with a cloth support for the baby and padded cloth straps that adjust to the parent.

The rigid or molded infant carrier is made of polypropylene, a medium-impact plastic. It has the advantages of being lightweight,

flexible, and durable. Manufacturers receive the plastic in pellets. Metal components are made of steel by specialized metal fabricators. These parts may include buckles, locking clips, and hardware for adjusting harnesses. Rivets and fabric fasteners may be made of other metals but are also supplied by specialty contractors.

Fabrics used to make infant carriers include durable fabric like denim and nylon mesh and webbing for harnesses. All fabrics have to be specially treated to meet standards for flammability, but they also have to be washable. Chemical cleaning methods and even strong detergents can break down the treatments used to make the material flame-resistant, so these fabrics must be cleaned with water and mild soap. The fabrics are also chosen for their appeal, so color and fashion trends are observed by designers in selecting these materials.

Foam padding and fiber fill are also required to be flame resistant. For manufacturers of molded plastic carriers, most of the cloth liners and pads are sewn by outside suppliers to the specifications of the carrier manufacturers.

All infant carriers bear manufacturers' labels, and printers make paper and adhesive-backed labels according to the manufacturers' requirements and Federal standards for consumer information.

Design

There are three basic designs of infant carriers. The first is the sling of which the Snugli and Baby Wrap are best known. The second is a modification of an aluminum-framed backpack in which the pack portion is fitted to carry a baby and padded straps are provided for the parents' comfort. The third type is the molded plastic carrier that resembles an automobile safety seat with a plastic back, seat, and handle and a padded interior but without the devices needed to secure the carrier in an automobile.

Factors influencing the design of infant carriers are many and relate to how the individual consumer is most comfortable carrying their child. Safety is, of course, the overriding concern. The infant must be comfortably held or restrained in the carrier, a soft carrier must have openings for the child's legs but not so large that the baby will fall out, and a rigid carrier designed to sit on solid surfaces must not tip over. Comfort for the baby is the second leading design factor. All forms of carriers have some type of cloth liner or insert to cushion the baby, and this must be comfortable as well as durable and easily cleaned. And, finally, the person carrying the child must also be comfortable. Cloth infant carriers and backpack-type carriers have to adapt to different body types so the baby can be carried with equal comfort. This is usually accomplished with a set of straps that cross the shoulders and fit around the waist or on the hips. Carriers made of molded plastic must strike the right balance between the weight and strength needed to support the child adequately and light weight for the parent to carry.

These combinations of characteristics have not been easy for manufacturers to produce successfully. Consumer protection groups like the Consumer Product Safety Commission list a number of carriers that have been recalled. The molded plastic carriers have had notable problems such as failure of the handle locks that can cause the carrier to release and tip the baby out. But the cloth carriers have also been made with leg openings that are too large and allow tiny babies to slip through. To the manufacturers' defense, many more problems have been cited related to misuse of the carriers by the adults caring for the infants. Sitting a rigid carrier on a soft surface like a couch permits the baby's movements to tip the carrier over. In 1997 alone, over 8,700 children were brought to the emergency room and treated for falls from carriers. Also, a number of medical reports show that infants are left in one position in the carriers for too long. In the supine or reclining position, the back of the baby's head flattens as it rests against the carrier. Babies held upright in cloth carriers or backpack-type carriers are subject to spinal stress that complicates the natural development of curves in the spine. Sling-type carriers have been found to hold the baby in the best position for its growth.

The Manufacturing Process

Cloth carriers

1 Following initial design, prototypes are sewn by hand and checked for fit on both infants and parents of a range of sizes and weights. When the final design is approved, the pattern on blueprints or computerized drawings is transferred to paper patterns. The patterns are tacked to the fabric with pins and stacked together. A fabric cutter that looks much like a band saw is used to cut through all the layers at once so the pieces for many carriers are produced at the same time.

2 The fabric pieces are stitched together by workers operating industrial-quality sewing machines. Typically, each worker is responsible for one section of the carrier such as the back or the straps. The pieces are stitched together inside out with the seams exposed then turned right side out. Sections that will be padded are left with small openings in the seams where the padding can be added.

3 The sections to be padded are stuffed with polyester fiber or foam padding. The fiber is purchased by the manufacturer in tightly packed bales that are fluffed up with a machine called a picker. The fiber is selected for the packing by the quality of the fibers, and, when it is inserted into the carrier section as padding, it will be stuffed to

the proper density. At the stuffing machine, air pressure blows the fiber into the fabric section. The operator manipulates the section to disperse the fiber throughout the section and to the desired density. The openings through which the stuffing has been added are then stitched together.

4 The sections of the carrier are assembled to create the finished product at another set of sewing machines. Any restraining straps and webbing are stitched securely in place. Hardware and fasteners (if any) are also stitched on or driven into place by special machines, and labels are sewn in place at joins between two fabric sections. A final inspection is performed to check all the seams and fasteners, and the carrier is packed in a plastic bag and boxed. Multiple boxes are packed in shipping cartons for storage or transport.

Backpack-type carriers

1 The cloth sections of the backpack-type carriers are designed, cut, stitched, and padded like the cloth carriers described above.

2 The aluminum frames used to support the carriers are made of tubing formed by aluminum fabricators to the sizes, shapes, and curves specified by the carrier manufacturer. They are received by the carrier manufacturer in bulk lots that are distributed to bins along the assembly line. Assembly line workers pair a cloth carrier to a metal frame and attach it; methods of attachment vary widely among makers but can include Velcro and metal buckles and fasteners. Some frames also have plastic feet or pads where they balance against the parent or so they can be stood on the ground or a solid surface while the infant is placed in the carrier. These feet are inserted into holes in the frame and fastened in place.

3 The frame/carrier assembly is inspected and packaged like the cloth carrier.

Molded plastic carriers

1 Molding of the carrier's plastic shell is the first step in manufacture of this type of carrier. Plastic pellets stored in bulk are melted and injected into specially machined dies or forms for the shell. The dies are the highly detailed product of the design

process, and even the finished texture of the carrier has been designed in the interior of the mold. The shells are extracted from the dies, cooled, trimmed of flashing (excess bits of plastic), and cleaned.

2 At the assembly line, the molded shells and all other parts are distributed to work stations. Each worker is responsible for fitting the carrier with one item or type of item and handing it to the next worker who adds another piece. The padded insert that forms the cushion for the baby is a completed item that has been sewn and stuffed. It is attached to the carrier by Velcro or other fasteners. Alternatively, the nylon mesh used to make a hammock-like support for the infant in some types of carriers is attached to sections of the plastic shell that were designed to trap the mesh and lock it in place. Straps, harnesses, buckles, and labels are added in turn.

3 In the packing department, the infant carriers are individually packed in cartons. The cartons are significant sources of information for consumers and are preprinted with consumer data as well as company information and bright designs. The cartons are stacked and bulk-wrapped in plastic for storage and distribution.

Quality Control

Hands-on assembly is key to the construction of the three types of infant carriers described here. Workers are highly skilled in their own and related parts of manufacture of carriers, and they are rewarded for identifying errors and rejecting imperfect carriers. Molded plastic carriers are nominally the most complex to manufacture; the shell is carefully checked for uniform thickness and strength and compared to a master copy for correctness. Because manufacturers usually make several different lines or models of carriers, each product has a bill of materials listing the part numbers of all the parts making up that model. Assembly workers monitor these, and line managers periodically pull carriers from the assembly line to check the materials, review the overall quality, and test the products.

Byproducts/Waste

The making of infant carriers results in little waste because most of the materials used can be recycled. The molded plastic shells can be reground into pellets that are mixed with new plastic pellets for melting and injection molding. The percentage of reground plastic is kept to a minimum in any single batch of new plastic. Sewn linings of the plastic shells are returned to the supplier and can often be resewn; similarly, cloth carriers can be resewn, assuming the error is not picked up and corrected immediately by the sewing machine operator. Metal parts are also recycled by suppliers.

Safety concerns are important in several aspects of carrier manufacture. Sewing machines, fabric cutters, and power hand tools have automatic emergency shutoffs, safety guards, and other protections. Operation of the injection molding machinery is automated, and workers seldom have reason to approach these machines with their high operating temperatures. Mold operators wear protective gloves. Fabric cutters, sewing machine operators, and those who handle the polyester fiber and foam padding wear masks over their mouths and safety glasses.

The Future

The concept of the infant carrier may be as old as civilization, yet the "modern" baby carrier is less than half a century old. Despite this new or reborn acceptance, the carrier has become an essential part of every baby's layette. This is not surprising because every parent's first priorities are the safety and comfort of that new baby. The alarming aspect of infant carriers may be the number of product recalls. Perhaps this should not be surprising either, because the manufacture of carriers is an attempt for an industry to fit a standard to tiny babies and an even wider variety of sizes and shapes of parents and care givers.

Manufacturers are conscientious about doing their part to keep improving and varying products to prevent child injury and provide security. Responsibility for proper use of infant carriers is beyond the manufacturers' purview, yet they try to design out opportunities for consumer error. Doctors are also adding to the information base by studying

physical changes to babies, particularly flat spots on their skulls and improper spinal development, that may be attributable to infant carriers. Recalls, errors, and skeletal stress sound like negative reasons for a product's future, but all these negatives are strong motivators toward the best possible products for the most babies.

Where to Learn More

Books

Bernhard, Emery. *A Ride on Mother's Back: A Day of Baby Carrying Around the World.* New York: Gulliver Books, Harcourt Brace & Company, 1996.

Periodicals

Karvonen, Karen. "Child Carriers." *Women's Sports and Fitness* 14, no. 5 (July-August 1992): 112.

Quindlen, Anna. "A Mother's Nominees For the Nobel Prize." *New York Times* (September 21, 1986).

Sommars, Jack. "Oh Baby!" *Colorado Business Magazine* 23, no. 3 (March 1996): 16.

Other

Consumer Product Safety Commission. http://www.cpsc.gov (September 18, 2000).

"43 Reasons to Carry Your Baby." *Nurturing Magazine* (1998). http://www.nurturing.ca/carrybaby.htm (January 2001).

The Lemelson-MIT Prize Program: Women's History Month. http://web.mit.edu/invent/www/inventorsR-Z/whm3.html (September 27, 2000).

Loving Attachment. *Choosing a Carrier: Infant Carriers and Spinal Stress.* http://www.lovingattachment.com (September 27, 2000).

The Mayo Clinic. *Misshapen Infant Heads: Not an Epidemic* (May 22, 1999). http://www.mayohealth.org/home?id=CC00010 (September 27, 2000).

—Gillian S. Holmes

Baby Wipes

Background

Baby wipes are disposable cloths used to cleanse the sensitive skin of infants. These cloths are made from non-woven fabrics similar to those used in dryer sheets and are saturated with a solution of gentle cleansing ingredients. Baby wipes are typically sold in plastic tubs that keep the cloths moist and allow for easy dispensing.

History

The technology to create disposable non-woven towelettes was developed in the late 1970s, and the first baby wipe products appeared on the market soon after. Originally, due to the expense of the specialized equipment required to produce these products, major brands like Kimberly-Clark's Huggies and Proctor & Gamble's Pampers dominated the market. As the technology matured and became more affordable, smaller brands began to appear. By the 1990s, many large supermarket chains had their own private label brand of wipes made by contract manufacturers. These private label brands entice consumers with their lower prices and increase profits for the supermarkets.

Baby wipes are sold in the diaper section of supermarkets and generally run from three to five dollars for a 64-count tub. They are important to retailers because they help offset the small profit margins that diaper sales generate. They are merchandised near diapers in the hope that consumers will purchase wipes along with their other infant care products. Wipes are available in different sizes and styles, and a typical store may carry between 10 and 20 different stock keeping units. Total supermarket sales of these pre-moistened towelettes jumped 5% from $251.4 million in 1996 to $263.9 million in 2000.

Design

Baby wipes are designed to be durable enough for heavy duty cleaning tasks, yet still be disposable. The fabric used for the cloths is chosen on the basis of durability, cost, and absorbency. This fabric is then saturated with a cleansing solution designed to be mild yet effective. Packaging is also an important design component and several patents have been granted for containers made specifically for pre-moistened towelettes. These packages are designed to easily dispense single sheets while keeping the towelettes moist until ready for use. Thermo-formed plastic tubs are most commonly used to package wipes in different amounts ranging from a few dozen to several hundred.

Marketers are continually designing new styles, sizes, and formulations of baby wipes. Large-pack refills and attractive graphic labels are some of the recent innovations in the category. One private-label manufacturer uses Jim Henson's Muppet Babies to differentiate its product from competitors. Some products even have character outlines imprinted on the actual wipe. In Canada, premium quality wipes are marketed as having the advantages of being thicker, more absorbent, greater stretchability, hypo-allergenic, alcohol-free, pH-balanced, and/or unscented. Another factor that has impacted baby wipe design is the trend toward natural products. Marketers routinely add a variety of natural ingredients, such as aloe vera and oatmeal, to increase the consumer appeal of their products.

Total supermarket sales of pre-moistened towelettes jumped 5% from $251.4 million in 1996 to $263.9 million in 2000.

An example of a typical box of baby wipes.

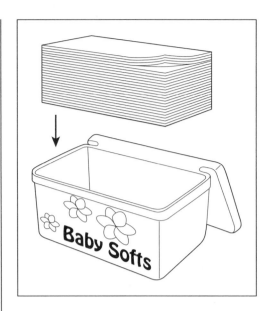

Raw Materials

Non-woven fabric

The material used in baby wipes is a non-woven fabric similar to the type used in diapers and dryer sheets. Traditional fabrics are made by weaving together fibers of silk, cotton, polyester, wool, and similar materials to form an interlocking matrix of loops. Non-woven fabrics, on the other hand, are made by a process that presses a single sheet of material from a mass of separate fibers. Fibers, such as **cotton** and rayon, are used in this process, as well as plastic resins like polyester, polyethylene, and polypropylene.

Cleansing ingredients

Water is the main ingredient and serves as a carrier and diluent for the other ingredients. Baby wipes also contain mild detergents mixed with moisturizing agents, fragrance, and preservatives. The detergents most commonly used are known as amphoteric surfactants, similar to those found in baby shampoos. Sodium diamphoacetate and coco phosphatidyl PG-dimonium chloride are primary surfactants used in wipes. These chemicals don't strip the skin of natural oils and also decrease skin irritation potential. Mildness is a prime consideration given that the wipe solution may be in contact with delicate skin around the anus and genitals.

Humectants such as propylene glycol and glycerine are added to prevent premature drying of the solution and contribute to skin moisturization. In addition, some formulas incorporate oils such as mineral oil, lanolin, or silicones that help to soften skin. Thickeners, such as cellulose derivatives like hydroxymethyl cellulose, control the viscosity of the finished product and keep it the right consistency.

Other ingredients include preservatives, such as methyl and propyl paraben, to ensure the solution does not support microbial growth. Fragrance is usually added to increase consumer appeal and to help overcome body odors, but fragrance-free products are also offered. Featured ingredients may also be added to increase consumer appeal. These include natural ingredients that are known to be kind to the skin such as aloe vera or oatmeal extract.

Packaging components

Packaging used in baby wipes must keep the cloths free from contamination, yet allow for easy dispensing. The package must also prevent the towelettes from drying out. Thermomolded plastic tubs are the packaging choice for most manufacturers. One common design features a hinged lid that allows easy access to the towelettes. These tubs are produced on injection molding equipment by pumping molten polyethylene plastic into a two part mold. Pressure is applied to the mold externally until the plastic cools. When the mold is opened, the plastic tub is ejected and stored until ready for filling.

The Manufacturing Process

Non-woven preparation

There are two primary methods of assembling non-woven fabrics: the wet laid process and the dry laid process.

1 One dry laid process is the "meltblown" method, which is used to make non-woven fabrics from plastic resins. In this method, plastic pellets are melted and then extruded, or forced through tiny holes, by air pressure. As the stream of fibers cools, it condenses to form a sheet. Hot metal rollers are used to flatten the fibers and bond them together.

2 A wet laid process is typically used for softer cloths, like diaper wipes, that use

cotton blends. In this wet process, the fibers are made into liquid slurries with water and other chemicals. The resultant paste is pressed into flat sheets by rollers and then dried to form long rolls of fabric. These rolls are then further processed and slit into narrow widths and then perforated or cut into individual sheets. The finished cloths are classified by their dry weight that is at least 1.4 oz/in^2 (40 g/m^2). Absorbency of the wipes is also an important requirement (quality wipes can absorb between 200% and 600% of their weight in solution).

Formula preparation

3 The ingredients used in the wipe solution are prepared in large batch tanks. Depending on the formula requirements, the tank is charged with the first ingredient which is usually water. The tank may be heated during manufacturing to facilitate blending of powders that must be dissolved or other solids that must be melted. The other the ingredients are added sequentially and mixed until homogenous.

Non-woven treatment

4 Onced prepared, the non-woven cloth is fed from storage rolls onto coating machinery, where the cleansing solution is applied. Several methods can be employed in this process. The cleansing solution can be added by running the fabric through a trough of the solution, or sheets of fabric may be sprayed with the formula from a series of nozzles.

5 Alternatively, individual towelettes may be packaged in sealed foil pouches. In this process, sheets of laminated foil are fed into automated equipment which folds them into a small pouch and heat seals three sides to form an open envelope. Simultaneously, another conveyor line feeds the non-woven cloths into the pouch. A liquid feed mechanism, including conduits extending through the stuffing bars, injects moisturizing liquid into the towelette packet simultaneously with the stuffing of the towelette material.

6 Immediately following this operation, another heat sealer closes the pouch tightly.

Packaging operations

7 The finished cloths are automatically folded, stacked, and transferred to their final package. In one patented method employed by Rockline Inc. of Sheboygan, Wisconsin, the towelettes are folded and stacked so that they can easily be removed one at a time and then the stack is placed in an inner plastic pack. This inner pack is subsequently inserted into an outer tub with a hinged cover.

Quality Control

Each component used in baby wipes must pass a series of quality check points during the manufacturing process. The plastic packaging must be free from mold defects that could cause leakage or improper closure. The non-woven fabric must be uniformly formed and must meet specific tear-strength requirements. Furthermore, prior to manufacture, the cleansing solution must be thoroughly tested. Development chemists evaluate the product to ensure that it is shelf stable and will not undergo any undesirable chemical reactions. They must also test the formula to ensure that it satisfies the requirements for mildness. The most reliable method used to test mildness is known as the Human Repeat Insult Patch Test (HRIPT). In this test an ingredient, or series of ingredients, is applied to human volunteers (usually on the inside of the forearm). The area is then occluded with a patch material and the spot is evaluated by dermatologists or clinicians after a specified time. Any redness or irritation is assigned numerical value and the scores of all the panelists are averaged. A low average score, such as 0 or 1, indicates that the product is essentially non-irritating.

Before ingredients are added to the batch tank, they are assayed to ensure they conform to all relevant specifications. During manufacture, each ingredient is check weighed before it is added to the batch. Then final batch is tested again for basic specifications such as pH, viscosity, and microbial content.

The Future

From a marketing perspective, baby wipes are continually evolving. Supermarkets are planning to boost their declining margins on baby food and national-brand diapers by efficiently promoting private-label baby wipes sales. The market trend is leaning toward larger, more economical size. For example, Huggies recently introduced a 160-count refill package. Smaller travel size packages are also available from some manufacturers. From a technical perspective, as chemists develop new and improved surfactants, future versions of baby wipes will contain milder and more effective cleansing ingredients. Trends in fragrance and featured ingredients will also impact future formulations.

Where to Learn More

Periodicals

Cramp, Beverly. "Scott Worldwide Personal Care and Cleaning's Packaging of Its Baby Fresh Products." *Marketing* (January 19, 1995): 21.

Moore, Amity. "Clean and Mean: Supermarkets are Using a Dual Strategy of Private Label and Price Sensitivity to Beat Mass Merchants in Baby Wipe Rings." *Supermarket News* 47, no. 21 (May 26, 1997): 33.

—*Randy Schueller*

Backhoe

Background

The backhoe is one of the most commonly seen pieces of construction equipment because of its adaptability. Its cousin, the front-end loader, is also a smaller piece of equipment that has a broad bucket like the one on the front of the backhoe for hauling soil, debris, and materials, and lifting them up into trucks. These two machines have some much larger relatives, including the road grader (with a large blade that smoothes soil surfaces), roller compactor (equipped with a heavy roller that compacts soil and asphalt during construction), the bulldozer and crawler tractor (big loaders that move earth by digging, ripping, and blading, with traction from rolling tracks, not tires), the excavator (a track-mounted vehicle with a much larger bucket than the backhoe), and the scraper (with a large bowl in the center of the machine that cuts into the earth and carries the material it has cut in that bowl). More distant members of the vast construction equipment family are cranes, dump trucks, pipe layers, draglines, truck-mounted drills, and shovels.

The key to the power of the backhoe is hydraulic pressure. Hydraulic lines, a reservoir of hydraulic fluid, a pump, and a series of pistons allow the machine's operator to extend its arm and cut through soil with a toothed bucket. The pump exerts pressure on the hydraulic fluid, and operating the levers opens a valve that releases the oil into a piston. The piston expands to lift the arm, swing the bucket, press the bucket into the soil, and lift it out of the excavation. Reversing the valve causes the oil to flow out of the piston and return to the reservoir.

The backhoe's standard equipment is a narrow bucket on the rear end and a loader on the front. The operator effectively makes either device the working end by simply rotating his chair and operating a different set of controls. Typically, if the bucket is being used, the flat front end of the loader is set down on the ground to stabilize the vehicle.

History

The history of heavy excavating machinery began in 1835 when the dipper shovel was invented to excavate hard soil and rock and to load trucks. The dipper shovel was steam-powered and mounted on rails like a train. Rail lines were laid into mines and large excavations so the dipper shovel could move around and load materials into railroad cars or horse-drawn trucks. The dipper shovel had a short boom (lifting arm), a dipper stick (a beam that pivoted out from the boom and gave the shovel its name), and an attached bucket for digging. The dipper shovel was modified in many ways to create the familiar construction equipment of today; the boom was changed, different attachments were added, the weight and balance of the equipment were changed, and the type of tires or tracks were chosen to suit the equipment's primary jobs. Of course, with the invention of gasoline- and diesel-powered vehicles, construction equipment became even more adaptable. Most construction equipment is powered by diesel engines, although electric power, battery power, and propane tanks are used on specialized equipment.

The backhoe is one of the smaller and more versatile descendants of the dipper shovel. The backhoe became an important piece of equipment with the large-scale construction of highways and increased underground

The backhoe's standard equipment is a narrow bucket on the rear end and a loader on the front.

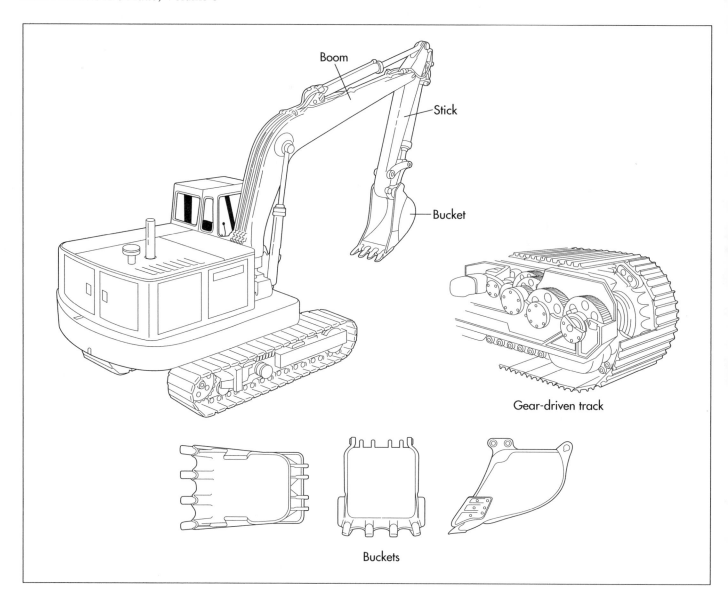

Boom

Stick

Bucket

Gear-driven track

Buckets

A backhoe with highlights of its gear-driven track and bucket.

placement of utilities. Backhoes and trenchers are used to excavate trenches for drainage and utilities. But, from the early 1900s until the late 1950s, the backhoe remained a large piece of equipment, and agricultural tractors were often called into service for smaller, limited access construction projects. Kits were available to adapt the tractors to construction tasks, but sometimes the right connections or attachment points were not provided, and the strains of construction were unsafe for the tractor's design and the operator.

In the late 1950s, a boom in residential development sparked another spurt of changes in backhoe design. Excavation of footings for house foundations, trenching, backfilling (replacing soil in a trench to cover drain-

pipes or utilities), and grading projects required a compact machine capable of a variety of tasks. By 1957, Elton Long, an engineer who had retired from the Case Corporation, reinvented the backhoe in the form of the loader/backhoe that combined two pieces of equipment in one and allowed the agricultural tractor to return to farming. Long's loader/backhoe had rubber tires for mobility and the right swing mechanism and buckets for specialized work. The loader on the opposite end of the machine from the backhoe bucket provided weight and balance when the backhoe was used; likewise, the teeth of the backhoe bucket could be driven into to the ground to provide anchorage as the loader lifted heavy materials. By 1965, other evolutions of the backhoe had created machines exclusively for the construction industry; diesel power, improved hydraulic linkages, four-wheel drive, and other features were added or improved in the 30 years from 1965 to 1995.

By 1995, Case added its L Series loader/backhoes to its product line. The six models in this series have improved hydraulics, more comfortable cabins for the operators, fuel-injection pumps, better cooling efficiency, better access for servicing, improved road performance, improved cycle times (allowing the operator to shift the transmission and accomplish the full cycle of lowering, digging, and raising the bucket), larger fuel tanks, and increased performance of both the backhoe and loader. They range in power from 73 to 99 horsepower (54 to 74 kW), and their loaders are able to lift from about 5,300-7,300 lb (2,400-3,300 kg). The backhoe on the largest L Series machine can excavate to a depth of almost 16 ft (5 m), and the Extendahoe (an adapter that increases the length of the stick) increases that to about 20 ft (6 m).

Raw Materials

Backhoe manufacturers purchase many of its parts as subassemblies, or partially assembled smaller units, that the manufacturer then completes. The manufacture of subassemblies may be done by a number of independent firms that specialize in metal fabrication, hydraulics, or other specialties. The subassemblies that are commonly purchased by backhoe builders include the chassis (body), drive line (the engine, transmission, and front and rear axles), and the loader and the backhoe (the buckets themselves plus the boom, stick, and other attachments). The hydraulic system is supplied as a package including the pump, valves, and hydraulic cylinders. The operator's station may be an open, canopy type or an enclosed cab; these may also be provided by outside suppliers.

Raw materials purchased by the manufacturer and included in many of the subassemblies includes medium-strength alloy steel in the form of thin sheets and as thicker plates that are about 1 in (2.5 cm) thick. The thicker steel plates are used for structural parts of the backhoe, and the thin steel is for housings and cosmetics. Plastics comprise the trim in the interior and around the exterior of the cab, and a temperature-resistant composite plastic is used for the air-cleaner housing. Seals are made of a high-grade elastomeric plastic that can withstand high temperatures and pressures. A lower grade of plastic is formed into the fender and the cab trim. A subcontracted foundry uses ductile iron to cast the backhoe and loader buckets.

Design

By late in the twentieth century, the function, and so the basic design, of the backhoe was clearly defined by the construction industry that uses it. Design improvements continue to be made, but they are in features and performance characteristics, rather than radical design changes. Leading manufacturers like Case Corporation rely on surveys among their customers to collect data leading to design modifications. The company defines the product based on a list of attributes, and these attributes are ranked in importance and in actual performance or delivery by the customers. Case routinely surveys its customers globally to obtain data over the broadest range of operating conditions; it hopes to learn that the design concepts behind its backhoe exceed their customers' expectations.

After collecting survey results and opinions from its customers, Case uses a technique called Quality Function Deployment (QFD) to boil down the input and create a new model or series of models with the requested characteristics. Three or four prototypes of the new design are built, and customers are invited to visit the manufacturing plant for "customer clinics," during which the prototypes are examined and tested. Construction operations are simulated over two or three days, and the customers evaluate the performance of the prototypes and the new features. For example, the operator's comfort may be evaluated in a series of as many as 50 questions and a kind of competition between the earlier design and the prototype. With this detailed input, the manufacturer then performs its own durability, reliability, and other tests and analyzes cost and manufacturability of the redesigned product. Further internal quality evaluations are done before the product is actually launched in the marketplace.

The Manufacturing Process

Receiving and assembling the subassemblies

1 Manufacturing begins at the receiving docks of the factory. Purchased subassemblies and components are unloaded, inventoried, and stored at a number of docks and then directed toward subassembly cells. These cells are a number of work areas where components and subassemblies are put together in more complete units or subassemblies.

2 For example, the components of a canopy-type cab will go to one cell where the steel canopy components are cleaned of oil. One side of the canopy is tack-welded together by a robotic welder and then final welded. While the robot welds, the subassembly operator loads the components for the other side. It is welded in two steps, and the parts of the canopy are welded together, also in a tack and final weld. The completed canopy shell is then loaded on a conveyor to carry it to the next operation.

3 The backhoe casting (poured by an independent foundry) has not been machined. In a subassembly cell, it is machined in a flexible machining center that is computer controlled. Bushings (the bearings that are required at pivot points) are fitted to the machined casting, and the backhoe subassembly is fed through the painting center to the next assembly area. Similarly, the loader arm components are machined, equipped with fittings, and moved through the painting center.

Painting and curing

4 The material handling system consists of conveyors that are both electrically powered and free to move as subassemblies are placed on them or removed. As the structural subassemblies are completed, all are conveyed through the painting center, which has two processes. The pieces are primed using an electrical deposition process that provides a generous protective paint layer. This is called an "e-coat" for the electrical method and is also termed a robust process for its vigorousness. The final coat of paint is applied manually because the painters are able to observe where paint is needed and use their judgment in applying it; that is, manual painting is more flexible than the electrical process. The painted subassemblies are conveyed to curing ovens where the metal is heated to cure the paint.

Hydraulic cylinders and radiator

5 A parallel subassembly is the cylinder factory. The cylinder rods are usually received in precut lengths that have already been internally machined and chromed. Fittings are added in a subassembly area, and the cylinders are processed through their own dedicated paint system where they are e-coated, hand-painted, and oven-cured. The completed cylinders are carried by forklift to the assembly area. The radiator is also a finished assembly, but connections are added so that water lines may be attached to the radiator and, from it, to the engine. Other subassemblies for the cooling, fuel, and lubrication systems are furnished with appropriate connections, attached lines, and sometimes, pumps and valves. Larger components like cylinders and tanks for fluids are added later during the assembly of larger components.

Final assembly

6 All of the subassemblies meet each other in the assembly area. The subassemblies are transported to and delivered at the point-of-use on the assembly line, so there is no lost motion for the assemblers. Each chassis is set on its on assembly cart with its front and rear axles and built up from the deck (bottom or base) to the cab. The cab and canopy are assembled at another area; when each unit is finished, it is carried to the chassis assembly line and attached to the completed chassis.

7 The engine, radiator, transmission, and hydraulic system are mounted onto the chassis. Other systems like the fuel, coolant, and exhaust components are also mounted on the chassis. Hoses and other fittings are attached, and supporting flanges or brackets are added as appropriate. When the cab is in place, the controls are linked to the engine, hydraulics (for moving the buckets), and other systems controlled by the operator. The two buckets—the backhoe and the loader—are the last of the large components

to be put in place, using large pins that fit the inset lugs and bushings. Their hydraulics are fitted, tightened, and tested.

8 The electrical system is the last to be connected; all of the fluid-bearing systems are attached and tested first. Batteries, electrical connections for lighted controls in the cab, and lighting are hooked up. Final body rails and handles are bolted in place, and trim is added.

9 Although all the major parts were pre-painted in the subassembly stage, the finished backhoe makes a last visit to the paint booth for a final coat. Detailing is the last step; decals and warning labels are applied to specific locations based on a template, and each completed backhoe is driven off the assembly line on its own power to the testing area for evaluation.

Quality Control

Quality begins outside the backhoe factory with each of the subcontracted suppliers. They are given product specifications as well as lists of key or critical characteristics (end results) that might not be immediately obvious from the specifications. The suppliers perform their own quality inspections and certify their products with data from those inspections.

As the components are received, they begin a history of documentation, called station control documentation, that travels with them through each stage of manufacture. First, they are logged in and inspected at the receiving dock, then the assemblers inspect them to make sure they meet the specified criteria at each stage of assembly. Each assembler on the line has the authority to reject parts or subassemblies throughout the process. A welder may reject parts for fit or rust, and an assembler can stop the entire assembly line if he or she sees a flaw in materials, subassembly, or appearance.

Independent of the assembly line, the manufacturer also performs random audits. Inspectors may look at components, entire systems, or subassemblies and pull them off the line for inspection. The purposes of these audits are to check the items against specifications, confirm the observations of the assemblers, train assemblers in the finer

points of inspections, and maintain the high standards established by the manufacturer.

A final check is performed on each backhoe. The inspector uses a checklist to validate a set of criteria for the machine's function; for example, there should not be any leaks, torque levels should be appropriate to the pieces, and moving parts should move according to a set of clearly defined motions and limits of motion.

Byproducts/Waste

Backhoe manufacturers do not produce true byproducts, but they make lines with several different models (called derivatives) and accessories. The derivatives are not identical, but they may have a number of features in common to keep costs down and to ease the manufacturing process. The derivatives or models may differ in size, scale, horsepower, or engine displacement. Case's current line of backhoes includes a model that is a loader only. With a three-point hitch and landscaping tools that are manufactured as a separate set of accessories, the loader becomes a loader/landscaper, and its uses are multiplied.

The process for manufacturing backhoes produces little if any waste. Scrap is not generated in the assembly process. In accordance with directives formulated by the U.S. Environmental Protection Agency (EPA) for clean air, paint systems are carefully regulated so they produce little airborne waste. An internal wastewater treatment system treats water that is used to clean materials, product parts, manufacturing equipment, and the factory itself. This internal system discharges into the local city wastewater system, so an external monitor confirms that there are no contaminants in the discharged water. Other materials—principally cardboard packaging and wooden pallets—are reusable or can be recycled.

Safety Concerns

Safety is a primary concern in the factory. Assembly processes are designed to be ergonomic (that is, they allow workers to move without stress or strain), lifting is limited, and safe work zones are built into the assembly line. Overall, the industry establishes and rewards safe work practices, and, through training, workers are constantly made aware of safe work issues. A major manufacturer should have millions of worker hours without any safety-related losses.

The Future

Despite the backhoe's well-established position in the construction industry, there is always room for improvement. Design modifications are driven by customer demand. As of 2000, the two primary areas where customers would like to see more improvements are in the ease of operation and the operator's comfort. The need for simple operation is forced by the fact that there are fewer skilled operators in the marketplace. And operations and reliability are both improving because of the continuing integration of electronics, automation, better engine technology, and on-board diagnostics. It is now up to the manufacturers to cost-effectively incorporate improvements.

The future of the backhoe depends not only on cost-effective design changes but cost consciousness in all aspects of operation including maintenance, durability, fuel efficiency, and resale value. The backhoe is its own best guarantee of a secure future. It is a versatile machine that is becoming even more flexible, thanks to modern technology linked with a proven track record.

Where to Learn More

Books

Adkins, Jan. *Heavy Equipment.* New York: Charles Scribner's Sons, 1980.

Nichols, Herbert L., Jr. *Moving the Earth: The Workbook of Excavation.* Greenwich, CT: North Castle Books, 1976.

Singh, Jagman. *Heavy Construction: Planning Equipment and Methods.* Rotterdam, The Netherlands: A. A. Balkema, 1993.

Other

Case Corporation. http://www.casecorp.com. (December 14, 2000).

—*Gillian S. Holmes*

Bagpipes

Background

The bagpipe is a wind instrument with a number of pipes and a bag. The melody pipe, or chanter, has finger holes that are played to produce the tune. Three other pipes, called drones, have bass and tenor pitches (with one bass and two tenor drones). They are called drones because they produce single notes only that are tuned to the chanter. The piper puffs air by mouth into a blowpipe that fills the bag. The bag is made of animal skin and is held by the player between the side of the chest and arm. The piper's lungs and diaphragm provide air and air pressure to make the reeds vibrate in the chanter and drones to produce one melody and three harmonies with one instrument. When the piper needs to take a breath, squeezing on the bag provides the supplemental air supply to keep the bagpipe playing its continuous sound. The five pipes join the bag at wooden sockets called stocks. In the stock where the mouthpiece is attached to the bag, a leather non-return valve keeps air from escaping back up the pipe. Some bagpipes are heavily ornamented with sterling silver fittings, a velvet or tartan bag cover, and braided silk cords. The colors match those of the Scottish clan (family), military regiment, or other organization to which the piper belongs.

The sound that a bagpipe produces is continuous as the bag is constantly filled by the piper and rhythmically squeezed to feed air to the chanter and drones. To give the effect of detached notes, bagpipe music is written with grace notes that the piper plays rapidly. The range of a set of pipes is limited, so music must be arranged specifically for the bagpipe.

History

Although the familiar bagpipe of the parade band is the Scottish Great Highland bagpipe, bagpipes in many different forms are folk instruments in many cultures around the world. Reputedly, the bagpipe arose in Sumeria or China in about 5,000 B.C., but this has never been substantiated. The oldest references to bagpipes appear in Alexandria, Egypt, in about 100 B.C. The bagpipe may have traveled west through Europe along with spreading populations and the development of individual cultures. Both Roman and Greek writings mention bagpipes in about A.D. 100, and they were known over most of Europe by about the ninth century. The bagpipe probably evolved from a double pipe made of two canes; both were single-reed pipes but one played the tune and the other was the drone. The bags were made of whole skins of goats or sheep (without the hindquarters). More sophisticated instruments had bags that were made of pieces cut from animal skins and stitched together. These types of simple bagpipes are still found on the Arabian and Greek Peninsulas and in North Africa and Eastern Europe.

Illustrations from Geoffrey Chaucer's *Canterbury Tales* show that several of the pilgrims were pipers; Shakespeare also mentions the bagpipe in his play *The Merchant of Venice*. From about the thirteenth through to the sixteenth century, England had many forms of bagpipes with versions for the common folk and more elaborate forms for the royal courts. The popularity of the pipes at court died out around 1560, and the more common forms also lost followers in the south and east.

In Western Europe, the *cornemuse* of France and the *zampogna* of Italy are folk bagpipes

The oldest references to bagpipes appear in Alexandria, Egypt, in about 100 B.C.

Bagpipes.

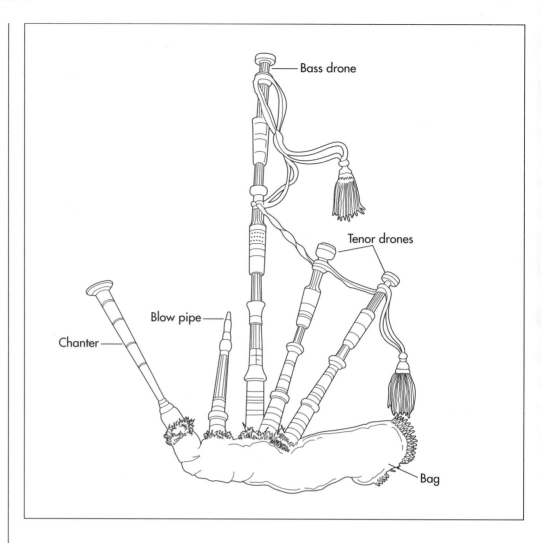

with character. The cornemuse has a chanter and a tenor drone and is blown with an arm-pumped bellows instead of a bag. It is still played today in folk bands or accompanied by a hurdy-gurdy (a three-stringed instrument). The *musette* is also a well-known French bagpipe that became popular while Louis XIV was king. The musette had two chanters and four drones, but all the drones were in a single pipe. The Italian zampogna is played with two hands with a chanter for each hand. The two chanters play melody and harmony, and the instrument also has two drones. All four pipes emerge from a single stock. All of these instruments became popular before 1700.

Although many other varieties of bag- and bellows-blown pipes are part of European musical history, the bagpipe found its real home in the British Isles—primarily in Scotland, Ireland, and northern England—achieving cultural popularity after about 1700 (al-though bagpipes were known long before this time). The French musette may have been the parent of a class of small pipes known as British small pipes, of which the best known is the Northumbrian small pipe that is still played today. The Northumbrian pipe has a cylindrical chanter like a clarinet (rather than a conical one like many other pipes and other wind instruments like the oboe), only seven keys, four single-reed drones that are held in one stock, and a closed bottom end on the chanter. When all the finger holes are covered, the chanter makes no sound, so this fingering is used for staccato (short, rapid) notes and closed phrasing; that is, grace notes are not needed to suggest separate notes. Like the musette, the Northumbrian small pipe dates from about 1700; the chanters for the cornemuse, musette, Northumbrian small pipes, and zampogna all use double reeds.

Another product of about 1700 is the Irish *uilleann* or union pipe, one of the most complicated bagpipes and a bellows-blown instrument. The Irish union pipe has a chanter, three drones, and three companion pipes called regulators. The regulators look like chanters, but they are closed at the bottom and have only four or five keys. The piper plays them like chords with the wrist of the right hand. The chanter itself is articulated by stopping it against the piper's knee. This also pushes the reed to a higher octave, so the Irish pipe has a broader melodic range than other pipes.

The Scottish versions of the bagpipe are the Highland small pipe, the "hydrid union pipe" (also called the Pastoral pipe), the Lowland pipe, the Scottish Border pipe, and the Scottish Great Highland bagpipe. The Highland small pipe was rare early in the twentieth century but is experiencing a rebirth in interest; its small size and soft sound makes it suitable for indoor use. It may be blown by mouth or bellows and has three drones, although they can be tuned differently than drones on other pipes. The hybrid union pipe is also small, has a conical bore, is used in doors, and is able to play two octaves (like the union pipe). The Lowland pipe is bellows-blown, has a cylindrical bore and reeds related to the Northumbrian small pipe, and carries three drones in one stock. It is about half the size of the Scottish Great Highland bagpipe, and, although it went out of fashion in the nineteenth century, it has been revived by makers of antique-type instruments. Finally, the Scottish Border pipe, which is closely related to the Great Highland bagpipe, has a conical bore, is bellows blown, has the drones tied in a common stock, and has toned-down reeds that produce a quieter sound.

The Scottish Great Highland bagpipe is called the *pìob mhór* in Scottish Gaelic. It was used as a martial instrument to inspire troops to battle since the sixteenth century, but, when warring against the English, the Highland clans were accompanied by solo pipers, not bands. Solo pipers also played laments at funerals and folk music for other occasions. The rise of the pipe band did not occur until the rebellious clans were solidly put down, and Scottish regiments were raised under Queen Victoria. Pipe bands quickly became symbols of their regiments and have remained highly visible representatives of Scottish culture to this day.

Participants enjoying a traditional Scottish Highland Games.

Scots enjoy large "gatherings of the clan," which celebrate their heritage and offer opportunities to meet others who share membership in the clan. Most states with a large Scottish and Scotch-Irish population (such as New York and Michigan) have "Highland Games," which feature sports such as "tossing the caber," in which men compete to toss a heavy pole the farthest distance. Bagpipe music is a very important part of this celebration, as it is at any celebration of clan identity. North Carolina, which has one of largest concentrations of people of Scottish descent, hosts the biggest gathering at Grandfather Mountain each July. Campbells mingle with MacGregors and Andersons, while enjoying Scotch whisky and traditional cuisine.

The Highland bagpipe is a large instrument. Five stocks for the three drones (two tenor and one bass), the chanter, and the blowpipe are tied into the bag. The blowpipe is long, so the piper can both play and march with his head erect; the other types of smaller Scottish pipes are often clutched against the chest and require the piper to bend over them slightly to blow into the blowpipe and play the chanter. The drones spread apart like a fan from the piper's left shoulder and out and are held apart by decorative silk cords; the bass drone is the one resting on the piper's shoulder. The two tenor drones are about 16 in (40 cm) long and are tuned to one octave below the chanter. The bass drone is 31.5 in (80 cm) long and is tuned to two octaves below the chanter. The drones are cylindrical bores (like oboes).

The chanter has nine holes including one double-vent hole and eight fingered holes. It is a wide conical bore (like a clarinet) that produces a penetrating, loud sound. Whether this sound is loved or hated, it has migrated with British imperialism, settlement and immigration, and Scottish regiments in wars from the American Revolution through World War II to almost every part of the world. In some places, it has become so popular that it has pushed aside native folk instruments. Piping schools, Scottish Highland Games including pipe band contests, and highly trained manufacturers of Great Highland bagpipes can be found in many countries outside the former British Commonwealth.

Raw Materials

Scottish Great Highland bagpipes dating from the 1700s had pipes made of bog oak. With imperialism and the rise of the "three corner trade" among Africa, America, and Britain, tropical hardwoods became available and have become the woods of choice for constructing pipes. African Blackwood and Brazilian rosewood are ideal for pipes. A brown hardwood called cocus wood is mentioned as a wood for pipes; this was true until the 1920s, but cocus wood is not used now. Many of raw materials used in the manufacture of bagpipes are dictated by the humidity of the region where the bagpipe is to be played. Some tropical hardwoods used to make the chanter and drones, particularly ebony, are ideal for the dampness of the climate in the British Isles but don't work well in the drier parts of the United States. Plastics, particularly acetyl homopolymers, are used by some makers for pipes to avoid the complications of climate.

Bags also require consideration for climate. They must be air-tight and water absorbent. Sheepskin is used in Great Britain, but it is not as durable in drier regions. In the United States elk or cow hide is used, and Australian pipe makers use kangaroo hide. Gortex is a modern material that is sometimes substituted for native hide.

Reeds are the constant in pipe production since the earliest known bagpipes. The water-reed was originally used for pipes as well as the reeds. Today, it is used to make both single and double reeds. Plastics such as polyvinyl chloride (PVC), metals, and **brass** are source materials for reeds for some manufacturers. Ornamentation on bagpipes may have experienced the greatest changes because of concern for preservation

of endangered species. In the 1700s, ivory from elephants, walruses, and narwhals (an Arctic-dwelling whale species) was the most common material for ornamentation because it can be worked and turned into beautiful artwork. Animal horn was also a source. Today, antlers from elk and moose are commonly used as is imitation ivory. Celluloid was an early manmade material to be carved for decoration, but plastics are generally worked now.

The bagpipe maker purchases wood and antler in log form. Plastic is supplied in sheets or rods, and metal for ferrules (bands that are put around the shafts of the pipes to support and strengthen them and caps that protect the pipe ends) is received as metal tubing or castings and may consist of aluminum, brass, nickel, or sterling silver.

Design

The basic design of the Scottish Great Highland bagpipe was established in the 1700s, and its straight, simple lines have been the standard since then. In Victorian times, more combing and beading on the wood came into fashion, and this ornamentation has also become traditional. The pipe maker does have some leeway in the design of the bores of the chanter and drones, but the range of internal dimensions is still limited to maintain the traditional sound. Because each bagpipe is hand-crafted, there are certainly subtle differences among manufacturers. Perhaps the greatest changes in design have been in other families of pipes in which everything old is new again; many pipe makers are reviving antique styles and early forms of bagpipes.

The Manufacturing Process

The wood drones and chanter

1 The pipe maker imports African Blackwood for the chanter and drones in the form of logs. These are sliced into planks and then into squares and are stockpiled for air-drying for a period of three to seven years. Some manufacturers have begun reducing drying times because of the related expense, and there are methods for kiln-drying the wood.

2 When the moisture content of the wood makes it suitable for working, the pipe maker can use a single-flute drill, twist drill, reamer, or gun drill to bore out the cylindrical drones. The single-flute drill makes the cleanest bore, although the carbide-tipped gun drill is a state-of-the-art tool because it uses a coolant hole to blow air or fluid in the bore to clean out chips.

3 The completed bore becomes the center for turning the outside shape of each drone on a lathe. The ferrules (protective metal bands and tips) are press-fit and glued in place, and projecting mounts are threaded on. The drones are finished with applications of wax, oil, lacquer, or varnish. The surface finish depends on the type of wood used, considerations such as humidity, and the pipe maker's style and preferences.

4 The chanter is made like the drones with two major exceptions. First, the bore of the chanter is conical, so it is step-drilled with twist drills then reamed with a single-flute tapered reamer that is 13 in (33 cm) long. The narrow end of the reamer is about 0.13 in (0.32 cm) in diameter and the wide end is approximately 0.87 in (2.22 cm) in diameter. Proper boring of the cone inside the chanter is critical to the tone it will produce. The second process exclusive to the chanter is the drilling or milling of finger holes into the turned bore. After the finger holes are complete, the chanter is surface-finished to match the drones.

5 The stocks are made along with the wood pipes. They are simply straight holes with tie-in grooves at the bottoms. The stocks have to be long enough to accommodate the reeds for the drones. Each stock is equipped with a ferrule at the top to prevent it from splitting.

The bag

6 The bagpipe bags are cut from elk or cow hide; typically, four or five bags can be cut from a single side of a cow hide. The hide is folded, and the bag sides are cut out as a mirror image. The seam is glued with contact adhesive to hold it temporarily until a leather welt can be put in place and the welt and seam are stitched together. The seam and welt are hand stitched with double needles; stitching a single bag takes approximately two hours.

Assembly

7 The 14 pieces comprising a Great Highland bagpipe are assembled by tying. The five stocks are tied into the bag using waxed linen, hemp, or nylon. Some makers use corked joints much like those in clarinets, but they are generally not as popular in pipe manufacture. The chanter and drones are connected to the stocks; only the reeds have to be added to complete the pipes.

8 The final finish is applied to the wood pipes by smoothing them with 80- to 120-grit sandpaper and working up to 400-grit wet sandpaper. Heated oil or wax is then applied by hand using a fine cloth.

If the maker chooses to finish the pipes with lacquer or varnish instead of oil or wax, 220-grit sandpaper is used to smooth the wood before the lacquer or varnish is applied with camel-hair brushes. The lacquer or varnish may be sprayed on in a spray-paint booth.

Reeds

9 The reeds are hand-made from metal tubing and water reed. The chanter takes a double reed that is begun with a brass of copper tube. The tubes may be cylindrical or conical. Two slices of reed are placed against the tube and wrapped in place. Reeds for the drones use tubular lengths of cane or reed instead of slices. On the cane, nodes mark the places where leaves sprouted when the reed was growing. Above a node on the cane, the pipe maker cross-cuts a slice and then makes two parallel cuts perpendicular to the slice. The small tongue made by the three cuts is raised up with the node as a kind of brace at its base. The tongue is about one-quarter to one-half of the diameter of the reed. As air passes through it, this tongue will vibrate to produce its tone. The opposite end of the cane length is tapered and attached to the drone. If modern materials are used, a plastic tube is used for the drone reed with a separate piece of plastic for the tongue. Insertion of the reeds in the pipes completes the bagpipes.

Byproducts/Waste

Dust from the wood used to manufacture bagpipes is highly toxic, and the pipe maker must wear a respirator, not a dust mask, to keep from inhaling the wood dust. Most of the natural products used in making bagpipes are biodegradable. Plastic waste results in very small amounts and is disposed in a landfill. Thinners and other organic compounds are used with lacquer or varnish finishes; but these are usually stored in small quantities with little waste. The primary hazard in bagpipe manufacture is to the pipe maker who must protect himself from the dust hazard and must also wear hearing protection because he works closely with noisy machinery.

Quality Control

Quality control is a constant issue in the production of bagpipes. The pipe maker crafts each bagpipe individually and so is monitoring his own work until the product is complete. Tolerances in boring and turning the pipes are tiny; the sound will suffer if these are not strictly observed. The internal dimension is critical and can only err by plus or minus 0.0005 in (0.013 mm). The exterior diameter can only err by plus or minus 0.1 in (0.25 mm). These tolerances are perhaps the greatest single issue in the quality manufacture of bagpipes. The pipe maker's reputation rests on his ability to create uniformly excellent bagpipes in appearance and more importantly in sound quality.

The Future

Interest in the bagpipe is growing steadily especially in the United States and Canada, which are the two largest markets in the world. The demand for well-made instruments has been steady for a number of years, but the number of bagpipers is growing now. Master pipe-maker Mark Cushing credits the interest in the pipes to two factors. Ethnic interest is prompting people to study pipe-playing because of its connection to their family history. Still more players are attracted by the sound of the pipes and the strong feelings they stir. No matter what the basis for their interest might be, these pipers are encouraged by the many pipe band associations throughout the United States and Canada that provide lessons, encouragement, and a ready audience. Thanks to the swirl of the kilt and the skirl of the bagpipe, pipe makers anticipate a lasting and loving future for their artistry.

Where to Learn More

Books

Baines, Anthony. *Bagpipes.* University of Oxford, England: Oxford University Press, 1999.

Baines, Anthony. *The Oxford Companion to Musical Instruments.* New York: Oxford University Press, 1992.

Cannon, Roderick D. *The Highland Bagpipe & Its Music.* New York: John Donald, 2000.

Collinson, Francis. *The Bagpipe: The History of a Musical Instrument.* Boston, MA: Routledge & Kegan Paul, 1975.

Dearling, Robert, ed. *The Illustrated Encyclopedia of Musical Instruments.* New York: Schirmer Books, 1996.

Sadie, Stanley, ed. *New Grove Dictionary of Musical Instruments.* London: Macmillan Press, 1984.

Other

Cushing Bagpipe Company. http://www.lightlink.com/mcushing (January 2001).

J. Dunbar Bagpipe Maker Ltd. http://www.dunbarbagpipes.com (January 2001).

K. Pettigrew Bagpipes. http://www.bagpipes.co.uk (January 2001).

MacLellan Bagpipes. http://www.highlandpipemaker.com (January 2001).

The Bagpipe Web. http://www.bagpiper.com (January 2001).

The Piping Center, Glasgow, Scotland. http://www.thepipingcentre.co.uk (January 2001).

Uileann Pipes. http://www.uileannpipes.com (January 2001).

—*Gillian S. Holmes*

Baking Powder

Baking powder is a solid mixture that is used as a chemical leavening agent in baked goods. It can be composed of a number of materials, but usually contains baking soda (sodium bicarbonate, $NaHCO_3$), cream of tartar (potassium bitartrate, $C_4H_5KO_6$), and cornstarch. (A base, an acid, and a filler respectively.) Baking powder is made by generating these solids, combining them in unique proportions, and then transferring them to packaging. First developed in the mid 1800s, baking powder formulations have changed little since.

Background

To modify the final characteristics of baked goods, leavening agents such as baking powder or yeast are added to recipes. A leavening agent is a material that releases carbon dioxide (CO_2) under certain conditions. This creates gas bubbles in the dough making it expand. When the product is baked, air pockets are created resulting in food that is light and crispy. Baking powder is generally preferred to yeast because it produces bubbles much faster. Yeast leavened dough takes anywhere from two to three hours to rise. Baking powder dough takes about 15 minutes.

Baking powder is a white solid that typically has three components, including an acid, a base, and a filler. When water is added to the baking powder, the dry base and acid dissolve into a solution. In this form, the compounds react to produce carbon dioxide bubbles, however, the amount of carbon dioxide produced by this reaction varies. Baking powder determines the final texture of the food and can affect the flavor, moisture, and overall palatability.

History

The development of baking powders began with the discovery of carbonate materials. One of the first carbonates was potash (potassium carbonate, K_2CO_3), a material that was extracted from wood ashes. During the eighteenth century, potash production had become a major commercial industry. American colonies exported huge amounts to England where it was used by glass factories and soap manufacturers.

Potash's usefulness to the baking industry was discovered during the 1760s. Prior to this time bakers had to hand knead dough for long periods to get the proper amount of air mixed throughout. For recipes which called for sourdough, pearlash (concentrated potash) was added to offset the sour taste. By chance, bakers found that these types of dough rose quickly. Evidently, the pearlash reacted with the natural acids in the sourdough to produce carbon dioxide gas. This discovery revolutionized the baking industry.

Over time, wood sources became scarce in England and other sources of carbonates were sought. In 1783 the French Academy of Sciences ran a contest for inventors who could develop a process for converting salt (sodium chloride, NaCl) to soda ash (sodium carbonate, Na_2CO_3). This contest was won by Nicolas LeBlanc in 1791. In his process, salt was reacted with sulfuric acid, coal, and limestone to produce soda ash. The soda ash was tried by bakeries as a leavening agent and found to be equivalent to potash. Baking soda was soon after extracted from soda ash and used to sooth stomach acids. The superior leavening properties of this material were discovered by American bakeries by the 1830s. It released

In 1834, Dr. Austin Church developed a process for making baking soda from soda ash. This product is still sold today under the Arm & Hammer name.

The Solvay ammonia process.

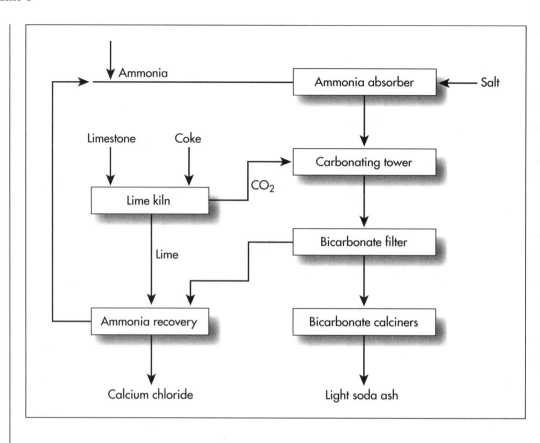

gas quicker and the aftertaste was not as bitter as soda ash.

Another important development in America was the development of potassium bicarbonate ($CHKO_3$) by Natha Read in 1788. He suspended lumps of pearlash over fermenting molasses. This converted the potassium carbonate into potassium bicarbonate. Unfortunately, this process resulted in a less dependable leavening agent when compared to that manufactured in Europe. In 1834, Dr. Austin Church developed a different process for making baking soda from soda ash. This product is still sold today under the Arm & Hammer name.

During the 1860s, various companies introduced other ingredients in their baking soda formulas and sold them as baking powders. These ingredients behaved in a more controlled way in recipes. Over time, different carbonate and acid mixtures have been sold as baking powders. Today, sodium bicarbonate and tartaric acid mixtures remain the most popular.

Raw Materials

As suggested, the primary components of a baking powder are a dry acid, base, and filler. Each of these materials can have a significant impact on the texture and taste of the finished product.

The most common dry base used in baking powders is baking soda, also called sodium bicarbonate. It is a water soluble white crystalline material, and produces carbon dioxide gas by itself when heated above 122°F (50°C). In addition to its use in baking, it is also used in the production of effervescent salts in medicine to prevent excess stomach acidity and in various types of fire extinguishers.

The type of acid used in a baking powder formula is more varied. The first baking powders used cream of tartar, a powdered acid. It was quick reacting and had to be put in the oven quickly or the gas would be spent. This material was perfect for products like pancakes or muffins. Today, there are four major acids used in commercial baking powders including monocalcium phosphate ($CaHO_4P$), sodium acid pyrophosphate ($H_2Na_2O_7P_2$), sodium aluminum phosphate (H_3O_4P), and sodium aluminum sulfate ($NaAlO_8S_2$). Monocalcium phosphate is a fast reacting acid which produces a large amount of gas within three minutes of its addition to baking soda. This is about twice the speed of other acids. Sodium acid pyrophosphate is a slower reacting acid and is used in refrigerated biscuit dough recipes. Sodium aluminum phosphate and sodium aluminum sulfate are also slow reacting acids which generate gas when heated. While these compounds are used, most bakers prefer aluminum-free baking powders due to the unpleasant flavor the aluminum can cause in the baked good.

The third major component of baking powders is an inert filler. The most common of these is cornstarch. The cornstarch has three purposes. First, it helps keep the product dry and easily flowing. Without it, containers of baking powder could bind up and form one large mass. Second, it keeps the acids and bases separated and prevents them from reacting during storage. Finally, it adds bulk to the powder to make it easier to measure and standardize.

Design

While a variety of baking powders are available, all of them meet basic standards and generate almost identical amounts of carbon dioxide. The basic difference between all types is the reaction time. There are two categories of baking powders: single acting and double acting.

Single-acting baking powders immediately produce most of their gas when mixed with a liquid. They are classified by the type of acid they utilize. Those that contain cream of tartar and tartaric acid ($C_4H_6O_6$) create gas rapidly when mixed with baking soda and a liquid. These batters must be cooked quickly or they will go flat. Slower single-acting baking powders are phosphate baking powders that contain either calcium phosphate ($Ca_3O_8P_2$) or disodium pyrophosphate ($H_2Na_2O_7P_2$). Aluminum sulfate ($Al_2O_{12}S_3$) powders react more slowly at room temperature but give a bitter taste to the batter.

Most commercial baking powders are double-acting. These means that initially a small amount of gas is released when it is mixed with a liquid. The primary generation of gas occurs when the batter is heated during cooking. These types of powders allow a batter to be left in an unbaked condition for long periods of time. Often double-acting

baking powders have two acids, one which reacts immediately and one that reacts when heated.

A less often used third type of baking powder is baker's ammonia. It results in a light, airy product but can impart an ammonia flavor if not used properly. It is best used in the production of flat cookies, helping to dissipate the ammonia odor during cooking.

The Manufacturing Process

Baking powder is made in a batch process and involves production of the component raw materials, blending, and packaging.

Production of raw materials

1 The manufacture of baking powder begins with the production of sodium carbonate. Known as the Solvay ammonia process, it was first developed in 1861. In this process ammonia and carbon dioxide are passed through a saltwater ($NaClH_2$) solution in an absorption tower. This results in a compound called ammonium bicarbonate (CH_5NO_3) which reacts with the salt to produce crude sodium bicarbonate crystals and ammonium chloride (ClH_4N).

2 The bicarbonate crystals are filtered out using vacuum filters or centrifuges. They are then washed with water to remove any residual chloride. The resulting solid is then conveyed to the calcining operation. Here, the material is heated and reacted with carbon dioxide to produce sodium carbonate, or soda ash.

3 The soda ash is dissolved, carbonated, and cooled which results in crystallized sodium bicarbonate. This solid bicarbonate material is of a purer concentrate than the intermediate bicarbonate formed earlier in the process. It is then laid out on driers to remove most of the moisture. The product is passed through metal screens to produce the desired particle size and filled into drums for storage.

4 The solid acid for many baking powders is tartaric acid. This material is made using potassium hydrogen tartrate, which is a waste product from wine making. The potassium hydrogen tartrate is first purified and converted to calcium tartrate. Using sul-

furic acid, the calcium tartrate is hydrolyzed to produce calcium sulfate and tartaric acid. These materials are then separated and the resulting tartaric acid is purified and dried.

Blending the powders

5 The sodium bicarbonate, tartaric acid, and cornstarch are transferred to a blending area. Compounders pour the appropriate amount of each solid into mixing containers. These mixers have large, stainless steel blades that thoroughly combine the powders into a single, homogeneous blend. This material is then transferred through vacuum tubing to the filling machine.

Filling and packing

6 The baking powder is placed in a covered hopper and dispensed into the desired package. Baking powders are packaged in a variety of ways depending on the manufacturer. For home use it is typically sold in a 4 or 10 oz (113 or 264g) can. Restaurants can get baking powder in 5 or 10 lb (2.3 or 4.5 kg) metal cans. Industrial bakeries buy it in 50 or 100 lb (23 or 45 kg) fiber cartons. Filling is typically performed by a rapid, carousel filler which forces a specific amount of baking powder into the package which is then sealed. The sealed containers are placed into cardboard boxes and stacked on pallets. The pallets are transferred to trucks or railroad cars and shipped to local grocery stores or commercial bakeries.

Quality Control

To ensure the quality of each batch of baking powder manufacturers monitor the product at each stage of production. The starting raw materials are subjected to various physical and chemical tests to determine if they meet previously determined specifications.

Some of the characteristics that are tested include pH, appearance, and density. The finished product is also tested. Typically, the particle size is checked as are the microbiological characteristics of the powder.

The Future

While baking powders have changed little over the last 100 years, manufacturers are always looking for new ways to make a greater profit. The baking powders of the future may be blended with different ingredients to enhance flavor. They may also be specially formulated for specific types of batter to accentuate characteristics such as gas evolution speed, residual flavor, or blending ease. Certainly, in the future manufacturers will find less expensive production methods.

Where to Learn More

Books

Ciullo, Peter. *Baking Soda Bonanza.* New York: Harper Perennial, 1995.

Kirk-Othmer. *Encyclopedia of Chemical Technology.* Vol. 1. New York: John Wiley & Sons, 1992.

Macrae, R. et al. ed. *Encyclopedia of Food Science, Food Technology and Nutrition.* San Diego: Academic Press, 1993.

Periodicals

"Is Your Baking Powder Still Potent?" *Diabetes Forecast* 50, no. 10 (October 1997).

Lutzow, Susan. "Baking-powder Power." *Journal Information Bakery Production and Marketing* (November 1996).

—*Perry Romanowski*

Basketball

Background

Basketball can make a true claim to being the only major sport that is an American invention. From high school to the professional level, basketball attracts a large following for live games as well as television coverage of events like the National Collegiate Athletic Association (NCAA) annual tournament and the National Basketball Association (NBA) and Women's National Basketball Association (WNBA) playoffs. And it has also made American heroes out of its player and coach legends like Michael Jordan, Larry Bird, Earvin "Magic" Johnson, Sheryl Swoopes, and other great players.

At the heart of the game is the playing space and the equipment. The space is a rectangular, indoor court. The principal pieces of equipment are the two elevated baskets, one at each end (in the long direction) of the court, and the basketball itself. The ball is spherical in shape and is inflated. Basketballs range in size from 28.5-30 in (72-76 cm) in circumference, and in weight from 18-22 oz (510-624 g). For players below the high school level, a smaller ball is used, but the ball in men's games measures 29.5-30 in (75-76 cm) in circumference, and a women's ball is 28.5-29 in (72-74 cm) in circumference. The covering of the ball is leather, rubber, composition, or synthetic, although leather covers only are dictated by rules for college play, unless the teams agree otherwise. Orange is the regulation color. At all levels of play, the home team provides the ball.

Inflation of the ball is based on the height of the ball's bounce. Inside the covering or casing, a rubber bladder holds air. The ball must be inflated to a pressure sufficient to make it rebound to a height (measured to the top of the ball) of 49-54 in (1.2-1.4 m) when it is dropped on a solid wooden floor from a starting height of 6 ft (1.80 m) measured from the bottom of the ball. The factory must test the balls, and the air pressure that makes the ball legal in keeping with the bounce test is stamped on the ball. During the intensity of high school and college tourneys and the professional playoffs, this inflated sphere commands considerable attention.

History

Basketball is one of few sports with a known date of birth. On December 1, 1891, in Springfield, Massachusetts, James Naismith hung two half-bushel peach baskets at the opposite ends of a gymnasium and outlined 13 rules based on five principles to his students at the International Training School of the Young Men's Christian Association (YMCA), which later became Springfield College. Naismith (1861–1939) was a physical education teacher who was seeking a team sport with limited physical contact but a lot of running, jumping, shooting, and the hand-eye coordination required in handling a ball. The peach baskets he hung as goals gave the sport the name of basketball. His students were excited about the game, and Christmas vacation gave them the chance to tell their friends and people at their local YMCAs about the game. The association leaders wrote to Naismith asking for copies of the rules, and they were published in the *Triangle*, the school newspaper, on January 15, 1892.

Naismith's five basic principles center on the ball, which was described as "large, light, and handled with the hands." Players

The basketball Wilt Chamberlain used to score 100 points in a game was sold in the 1990s for $551,844.

A typical basketball is 30-31 in (75-78 cm) in circumference.

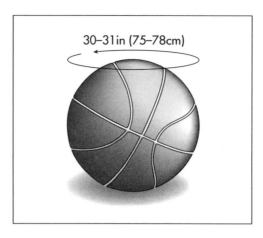

30–31 in (75–78cm)

could not move the ball by running alone, and none of the players was restricted against handling the ball. The playing area was also open to all players, but there was to be no physical contact between players; the ball was the objective. To score, the ball had to be shot through a horizontal, elevated goal. The team with the most points at the end of an allotted time period wins.

Early in the history of basketball, the local YMCAs provided the gymnasiums, and membership in the organization grew rapidly. The size of the local gym dictated the number of players; smaller gyms used five players on a side, and the larger gyms allowed seven to nine. The team size became generally established as five in 1895, and, in 1897, this was made formal in the rules. The YMCA lost interest in supporting the game because 10-20 basketball players monopolized a gymnasium previously used by many more in a variety of activities. YMCA membership dropped, and basketball enthusiasts played in local halls. This led to the building of basketball gymnasiums at schools and colleges and also to the formation of professional leagues.

Although basketball was born in the United States, five of Naismith's original players were Canadians, and the game spread to Canada immediately. It was played in France by 1893; England in 1894; Australia, China, and India between 1895 and 1900; and Japan in 1900.

From 1891 through 1893, a soccer ball was used to play basketball. The first basketball was manufactured in 1894. It was 32 in (81 cm) in circumference, or about 4 in (10 cm) larger than a soccer ball. The dedicated bas-

ketball was made of laced leather and weighed less than 20 oz (567 g). The first molded ball that eliminated the need for laces was introduced in 1948; its construction and size of 30 in (76 cm) were ruled official in 1949.

The rule-setters came from several groups early in the 1900s. Colleges and universities established their rules committees in 1905, the YMCA and the Amateur Athletic Union (AAU) created a set of rules jointly, state militia groups abided by a shared set of rules, and there were two professional sets of rules. A Joint Rules Committee for colleges, the AAU, and the YMCA was created in 1915, and, under the name the National Basketball Committee (NBC) made rules for amateur play until 1979. In that year, the National Federation of State High School Associations began governing the sport at the high school level, and the NCAA Rules Committee assumed rule-making responsibilities for junior colleges, colleges, and the Armed Forces, with a similar committee holding jurisdiction over women's basketball.

Until World War II, basketball became increasingly popular in the United States especially at the high school and college levels. After World War II, its popularity grew around the world. In the 1980s, interest in the game truly exploded because of television exposure. Broadcast of the NCAA Championship Games began in 1963, and, by the 1980s, cable television was carrying regular season college games and even high school championships in some states. Players like Bill Russell, Wilt Chamberlain, and Lew Alcindor (Kareem Abdul-Jabbar) became nationally famous at the college level and carried their fans along in their professional basketball careers. The women's game changed radically in 1971 when separate rules for women were modified to more closely resemble the men's game. Television interest followed the women as well with broadcast of NCAA championship tourneys beginning in the early 1980s and the formation of the WNBA in 1997.

Internationally, Italy has probably become the leading basketball nation outside of the United States, with national, corporate, and professional teams. The Olympics boosts basketball internationally and has also spurred the women's game by recognizing it

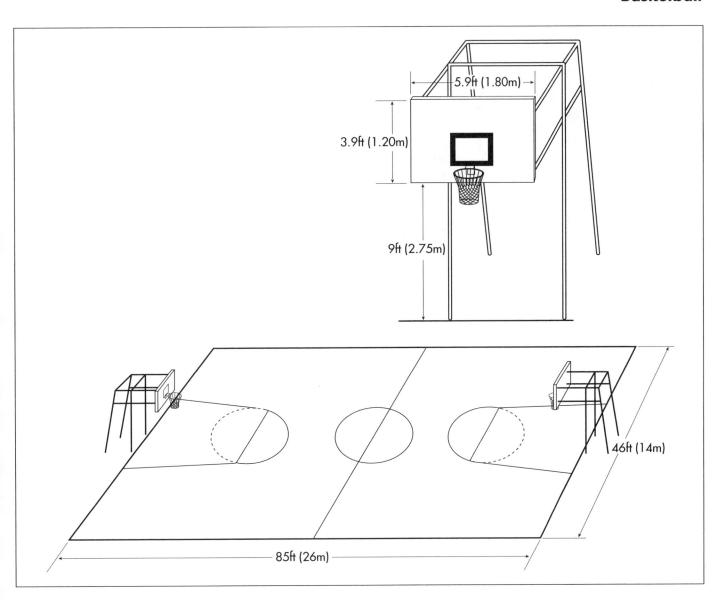

5.9ft (1.80m)

3.9ft (1.20m)

9ft (2.75m)

46ft (14m)

85ft (26m)

as an Olympic event in 1976. Again, television coverage of the Olympics has been exceptionally important in drawing attention to international teams.

The first professional men's basketball league in the United States was the National Basketball League (NBL), which debuted in 1898. Players were paid on a per-game basis, and this league and others were hurt by the poor quality of games and the ever-changing players on a team. After the Great Depression, a new NBL was organized in 1937, and the Basketball Association of America was organized in 1946. The two leagues came to agree that players had to be assigned to teams on a contract basis and that high standards had to govern the game; under these premises, the two joined to form the National Basketball Association (NBA) in 1949. A rival American Basketball Association (ABA) was inaugurated in 1967 and challenged the NBA for college talent and market share for almost ten years. In 1976, this league disbanded, but four of its teams remained as NBA teams. Unification came just in time for major television support. Several women's professional leagues were attempted and failed, including the Women's Professional Basketball League (WBL) and the Women's World Basketball Association, before the WNBA debuted in 1997 with the support of the NBA.

Raw Materials

The outside covering of a basketball is made of synthetic rubber, rubber, composition, or

Michael Jordan.

Michael Jordan was born February 17, 1963. Accepting a basketball scholarship to the University of North Carolina, he became the second Tarheel freshman to start every game. Jordan was named Atlantic Coast Conference (ACC) Rookie of the Year and won the National Collegiate Athletic Association (NCAA) championship in 1982. He led the ACC in scoring and was named college player of the year in 1983 and 1984. Jordan left North Carolina after his junior year and was drafted by the Chicago Bulls as the third overall pick of the 1984 draft.

A broken foot sidelined Jordan for 64 games during the 1985–1986 season. He returned, scoring 49 points against the Boston Celtics in the first game of the playoffs and 63 in the second—an NBA record. During the 1986–1987 season Jordan became the first player since Wilt Chamberlain to score 3,000 points in a season. The Bulls won the 1991–1993 NBA titles. In 1994 Jordan joined the Chicago White Sox minor league baseball team, returning to the Bulls for the remaining 1994–1995 season. In the 1995–1996 season, the team finished 72-10, another NBA record. The Bulls went on to win their fourth NBA title in 1996, fifth in 1997, and sixth in 1998 where Jordan claimed his sixth NBA finals MVP award.

Jordan participated in the 1984 and 1982 Summer Olympics, earning gold medals for the United States. He was named 1985s Rookie of the Year, 1988s Defensive Player of the Year, NBA MVP five times, has a career record for the highest scoring average of 28.5 ppg, played in 11 All-Star games (starting in 10, missing one due to injury), and named All-Star MVP three times. Jordan retired January 13, 1999.

leather. The inside consists of a bladder (the balloon-like structure that holds air) and the carcass. The bladder is made of butyl rubber, and the carcass consists of treads of nylon or polyester. Preprinted decals are used to label the ball, or foil is used to imprint label information. Zinc and copper plates are used in a press to either affix the decals or imprint the foil.

Design

The actual configuration of most basketballs is dictated by the rules or standards of the type of game in which the ball will be used. NBA, WNBA, and other professional leagues have specified dimensions for regulation balls, as described above, and even the imprinted information is specified. Amateur sports bodies have also developed rules and specifications, and there are specialized basketballs made for junior players (younger than high-school age), intermediate players (high-school age), and for indoor, outdoor, or combination play. Promotional basketballs that are much smaller in diameter are also made as souvenirs of many events such as the NCAA Championships.

Basketball designers are always trying to improve the product and build a better basketball. Inventor Marvin Palmquist created the "Hole-in-One" basketball to improve a player's grip; the ball has dimples, much like a golf ball, and can be easily palmed Michael Jordan-style by players with smaller-than-Jordan hands. Even the most skilled NBA star copes with sweaty palms, and this obstacle is addressed in another modification consisting of microscopic holes in the surface, which is made of absorbent **polyurethane**. This is the same material that forms the grip on a tennis racket, but it has been strengthened to withstand the abrasion of bouncing on a wooden basketball court. It absorbs moisture to keep the ball's hide less slippery.

Still other inventors feel the size of the ball is a disadvantage to proper handling and have suggested increasing the circumference from 30 to 36 in (76 to 91.4 cm), resulting in an increase in diameter from 9.6 to 11.5 in (24.4 to 29.2 cm). The so-called Bigball still fits through a regulation hoop and has been used in training sessions by both college and NBA teams. The Bigball must be shot with a higher arc to fall through the hoop, and, after practicing with the larger basketball, the regulation ball seems easier to handle.

The Manufacturing Process

Forming the bladder

1 The making of a basketball begins with the interior bladder. Black butyl rubber in bulk form (and including recycled rubber) is melted in the hopper of a press that feeds it out in a continuous sheet that is 12 in (30.5 cm) wide and 0.5 in (1.3 cm) thick. A guillotine-like cutter cuts the long strip into sheets that are 18 in (45.7 cm) long, and they are stacked up. A hand-controlled machine selects the sheets one at a time and, using a punch press, punches a 1-in-diameter (2.54-cm-diameter) hole that will hold the air tube for inflating the bladder.

2 The sheets are carried on a sheet elevator or conveyor to an assembly line where the air tube is inserted by hand. A heated melding device bonds it to the sheet, which is folded into quarters. Another punch press stamps out a rounded edge and, at the same time, binds the edges to make the seams of the bladder. This bladder is not perfectly shaped.

3 The odd-shaped bladder is taken to a vulcanizing machine. Vulcanization is a process for heating rubber under pressure that improves its properties by making it more flexible, more durable, and stronger. In the vulcanizer, the bladder is inflated. Heating by vulcanization uniformly seals the rubber so it will hold air. Completed bladders are stored in a holding chamber for 24 hours. This quality control measure tests their ability to hold air; those that deflate are recycled.

Shaping the carcass

4 The bladders that withstand the 24-hour inflation test are conveyed from the holding chamber to the twining or winding department. They make this journey suspended from a conveyor system by their air tubes. Machines loaded with spools of either polyester or nylon thread or string wrap multiple strands at a time around each bladder; this is the same process used to make the inside of a golf ball. The irregularly shaped bladders now begin to take on a better, more rounded shape as the precisely controlled threads build and shape the balls. The quality of the thread and the number of strands determine the cost and quality of the ball. The typical street-quality basketball has a carcass made of multiple wraps of three strands of polyester thread. The balls used by professional teams have carcasses constructed of nylon thread that is wrapped using four strands of thread. The same overhead conveyors continue carrying the carcass-encased bladders by their air tubes to the next step in the process where the carcasses and covers will meet.

Crafting the covers of the balls

5 Meanwhile, the exteriors or covers of the balls have been in production as the bladders and carcasses have taken shape. On 60-inch-long (152-cm-long) tables, colored rubber is unrolled from a continuous roll. The smooth rubber does not have pebbling (small bumps) that characterizes the surface of a finished basketball so that the outlines for the panels can be clearly marked on the rubber. A silk screen is moved along a series of metal markers that are guides marking the length of the rubber sheet needed for each ball. The silk screen operator moves the screen by hand and imprints the outlines of the six panels making up the ball. Only one color is used at a time, and, depending on the design, multiple silk screenings may be needed to color the six panels with all the colors on the ball.

6 A hand-operated punch press—equipped with specially designed and tooled dies—punches the rubber outlines to create six separate panels per ball. The same die has a hole that is punched in one of the six panels to make an opening for the air tube. The excess rubber surrounding the panels is lifted off the line and deposited in a bin for recycling.

7 The assembly worker picks up the six panels for a single ball in a specific order and carries them to the vulcanizer. The interior of the vulcanizer for this process is different from the one for the bladders. It is form-fitted to hold the six panels, to create the channels between the panels, and to add any embossed information. The assembler fits the panels individually into specified sections in the vulcanizer. A bladder/carcass is taken off the overhead conveyor, covered with a coating of glue, and placed inside the chamber of the vulcanizer that is lined with

the cover panels. When the ball emerges from the vulcanizer, most of its surface is still smooth (there are no bumps, called pebbling), but the channels and any embossing are formed into the surface.

8 Decals and foil decoration and information (if any) are applied by hand with small heat presses after the smooth ball is retrieved from the vulcanizer. Each ball is carefully inspected for gaps between the panels. These can occur, but each gap is filled during this inspection with a small piece of rubber that is hand-cut to fit the gap. The ball then is fitted into another vulcanizer that unifies the finished surface, blending in any gap fillers, and is specially molded to form the surface pebbling. The vulcanized balls are stored again for 24 hours in a second test to make sure they hold air.

Synthetic laminated covers and leather covers

9 The covers for basketballs that are made of synthetic laminated rubber or leather are also made in panels that are die-cut like the rubber panels. The synthetic laminated panels are shaved or trimmed along the edges, fitted and glued together by hand, and laminated to the carcass to create channels. They are also embossed by a heating process and decals are added. Any glue traces around the edges are removed, and any imperfect panels are replaced in the final inspection of synthetic laminated covers. Leather covers are made of full-grain, genuine leather and are stitched with heavy-duty machines; instead of indented, formed channels, the stitching forms the channels in leather balls. They are printed by silk screening and foil stamping, and their inspection includes a review of the uniformity and color of the leather.

Final testing, inspecting, and packing

10 Balls that pass the second 24-hour air pressure test are "bounce tested" to meet the regulation for inflation pressure that results in each ball bouncing a prescribed height. Balls that pass the bounce test are numbered to show the production run, and the decals and other artwork are inspected and touched up by hand as needed. Each

completed ball is inspected again. The inspector removes the production run tag, and the ball is deflated so it can be easily packed and shipped. Each flattened ball is packed in a polyethylene bag, and the bagged balls are boxed for bulk shipment to the distributor. The distributor also inspects the balls when they are received and is responsible for reinflating them to the correct pressure and packaging them in display boxes for sale. The display boxes may also be packed in bulk for distribution to retailers.

Byproducts/Waste

No byproducts result from the manufacture of basketballs, but most makers have a variety of lines and may also make balls for other sports. Waste is limited. Dies for cutting panels of rubber, synthetic laminate, and leather are carefully designed to space the panels closely and limit the material used. This is especially critical for leather because of the cost; some leather waste is inevitable, though, because leather is a natural material and has irregularities in color, thickness, and surface. All rubber materials can be recycled, and they represent the bulk of material used in making a basketball.

Quality Control

Throughout the manufacturing process, inspections occur regularly to make sure the finished basketball will hold air and to correct any surface variations. Machines like punch presses, dies, vulcanizers, and printing tools are carefully designed initially to maximize use of materials and to create perfect pieces. The assembly process includes many steps that are performed by hand, and the assemblers are trained to watch for imperfections and reject unsuitable products. Inspections and tests also include weight-control testing of the completed carcasses and the panels, regardless of material. Whenever the completed products are stored for any length of time, they are randomly inspected for appearance, size, inflation, and any wobble.

Some distributors have special tests for products bearing their name. For example, Rawlings Sporting Goods Company tests the basketballs they produce for the NCAA Tournament with a unique "Slam Machine" that simulates the workout a ball will get in four games in just five minutes. The machine works by propelling the ball down a chute between two wooden wheels that launch it at about 30 mph (48 kph) toward a backboard that is angled to direct the ball back to the chute. Rawlings also uses this machine to test new designs, materials, glues, and other changes.

The Future

Basketball sales have escalated dramatically with the sport's popularity. Figures from 1998 show that 3.6 million balls were sold in the United States alone for a total of about $60 million. Given the record number of television viewers for the 1999–2000 NBA Championships, many parents and children are likely to purchase basketballs to test their own slam-dunking skills. Participation in the sport and sale of basketballs shows no sign of slowing down.

Another aspect of the worldwide popularity of basketball is that it has sharpened collectors' enthusiasm for souvenir balls, autographed balls, and those from key moments of the great players' games. An example with a high price tag is the basketball Wilt Chamberlain used to score 100 points in a game; it was sold in the 1990s for $551,844.

Where to Learn More

Books

The Diagram Group. *The Rule Book: The Authoritative, Up-to-Date, Illustrated Guide to the Regulations, History, and Object of All Major Sports.* New York: St. Martin's Press, 1983.

Jacobs, A. G., ed. *Basketball Rules in Pictures.* New York: Perigee Books, 1966.

Periodicals

Feldman, Jay. "A Hole New Ball Game." *Sports Illustrated* 18, no. 26 (December 26, 1994): 102.

Jaffe, Michael. "For Better Shooting, Think Big: A Team of Ohio Entrepreneurs Insists that Their Oversized Basketball Will Improve Your Touch." *Sports Illustrated* 74, no. 15 (April 22, 1991): 5.

Mooney, Loren. "Get a Grip." *Sports Illustrated* (November 30, 1998): 16.

Tooley, Jo Ann. "On a Roll." *U.S. News & World Report* 107, no. 8 (August 21, 1989): 66.

Other

Rawlings Sporting Goods Co., Inc. http://www.rawlings.com. (December 14, 2000).

—*Gillian S. Holmes*

Boomerang

The record for maximum time aloft (MTA) is two minutes, 59.94 seconds; the unofficial MTA record, which was witnessed but not thrown in a sanctioned competition, is an astounding 17 minutes, six seconds.

Background

A boomerang is an aerodynamically shaped object designed to fly efficiently through the air when thrown by hand. The term usually refers to an object made to follow a circular flight path that returns it to the thrower. (Some sources describe all aerodynamic "throwing sticks" as boomerangs, separating them into "returning" and "nonreturning" categories.) Traditional designs are V-shaped, but newer versions may have irregular shapes or more than two arms.

Two design components give the boomerang the capability of circular flight. One is the arrangement of the arms, and the other is the airfoil profile shape that allows the arms into wings. During flight, the boomerang spins rapidly (about 10 revolutions per second). The wing profiles create the same lift effect that makes airplanes fly. In addition, the spinning motion creates gyroscopic precession, which pulls the boomerang into a circular path. A similar effect can be seen with a spinning top: if the top's axis is not quite vertical, the upper portion of the toy travels in a circle around the axis.

For a successful flight, the boomerang must also be thrown correctly. It should be held near the end of one wing with the top (curved) surface facing the thrower's body. The boomerang should be almost vertical, with the thrower holding the lowest arm. Depending on wind conditions and the design of the particular boomerang, the upper portion may be inclined up to 30° outward. After drawing the arm backward, the thrower hurls the boomerang with an overhand motion, much as a pitcher would throw a baseball. At the moment of release, the thrower adds a snap of the wrist, as if cracking a whip. The release angle should be between horizontal and 15° above. If there is any breeze, the boomerang must be thrown between 30° and 90° to the right of the oncoming wind (or to the left for a left-handed thrower; right- and left-handed boomerangs are mirror images of each other).

As the boomerang flies forward, it begins to slowly roll over (counterclockwise for a right-hander), ultimately flying horizontally with its flat side down. When it returns to the thrower, it is caught safely at chest height, trapping it between open hands in a sandwich fashion.

Recreational throwers simply enjoy playing a solitaire game of catch. Those who are interested in competition can choose from a variety of events, including precision (returning as close as possible to the launch point) and endurance (making the most catches in a five-minute period). The international record for distance traveled before returning to the thrower is 780 ft (238 m), in a flight that lasted nearly 22 seconds. The record for maximum time aloft (MTA) is two minutes, 59.94 seconds; the unofficial MTA record, which was witnessed but not thrown in a sanctioned competition, is an astounding 17 minutes, six seconds.

History

Boomerangs developed as a refinement of carved throwing sticks (kylies) that were used as weapons, primarily for hunting. The oldest kylie found to date is one formed from a mammoth tusk. Discovered in Poland in 1987, its age has been carbon-dated at about 20,300 years. This 2-ft (60-cm) long, 2-lb (0.9 kg), gently curved implement was probably thrown to kill reindeer. A plastic replica of it has been found to travel an average of 90 ft (27 m) when thrown, although throwing it into a headwind increases its range to an average 123 ft (38 m).

Kylies were used by prehistoric people in all parts of the world. Usually made of wood, they were banana shaped; both faces of each arm were carved into curved, airfoil surfaces. When thrown, they traveled parallel to the ground for distances up to 650 ft (200 m), spinning furiously toward their target. Typically 3 ft (0.9 m) long and weighing 5-10 lb (2.3-4.6 kg), they were effective hunting tools.

There is some evidence that boomerangs were developed in several cultural groups. For example, a boomerang-shaped object found in Germany was made of ash wood. Carbon-dated to an age of 2,400-2,800 years, it is preserved enough to allow archaeologists to reconstruct its entire shape. The replica has been thrown left-handed to produce a complete boomerang trajectory; however, the wing profiles were less than optimal, making it difficult to throw successfully. Evidence suggests that boomerangs may also have been developed in Egypt and India.

In all areas except Australia, hunters devised spears for throwing and bows for shooting arrows, and they stopped using kylies. The Australian aborigines, however, continued to hunt with throwing sticks. Experimenting with designs, the residents of the eastern and southern parts of that continent developed boomerangs, which they used primarily for sport. At major tribal gatherings, they held competitions based on such qualities as the precision of the return and the speed and quality of the flight. Boomerangs were not thrown at animals of prey, although they were sometimes thrown as decoys to lure birds into a net.

The oldest boomerang found in Australia dates to about 14,000 years ago. The origin of the word is uncertain, although it may derive from the cry "boom-my-row" ("return, stick") that British colonizers heard Dharuk tribesmen shout when throwing the instruments in 1788. The traditional method for making a boomerang was to select an appropriately curved piece of wood, usually from the section where the tree trunk joins a large root, making an angle of 95-110°. If necessary, the angle between the wings was adjusted by heating the boomerang over a fire and bending it. The aerodynamic profiles were carved from the wood with an axe, smoothed with a flint, and polished with sand. Designs might be carved into the surfaces, either for decoration or to improve the flight characteristics. The wood was sealed with fish oil or paint.

Boomerangs remained a relatively obscure curiosity until about 1970. A workshop about how to make and throw boomerangs was presented by the Smithsonian Institution in Washington, DC. It sparked great enthusiasm for the sport, and the Smithsonian began sponsoring annual tournaments on the National Mall. The first international championship tournament was held in 1981.

Raw Materials

A wide variety of materials can be used to make a boomerang. Wood remains one of the most popular because it produces good results, is relatively inexpensive, and is easy to work. Generally preferred is aircraft-grade Finnish or Baltic birch plywood, which is laminated from very thin layers of wood. A 0.2-in (5-mm) thick sheet will have between five to 10 layers. To protect the wood from moisture, it is usually sealed with a **polyurethane** coating.

Among the synthetic materials that may be used for boomerangs are polypropylene, acrylonitrile-butadiene-styrene (ABS) plastic, fiberglass, carbon fiber, linen-phenolic resin laminates, and Kevlar para-aramid fiber. Toy boomerangs made from urethane foam or cardboard can be used indoors.

Design

Because proper spinning action is crucial to the performance of a boomerang, there are some fundamental concepts that must be followed during the design process. However, those basic elements leave a great deal of room for creativity, and boomerang makers frequently experiment with innovative shapes.

When creating a new pattern, the designer marks a point in the middle of a sheet of

A boomerang.

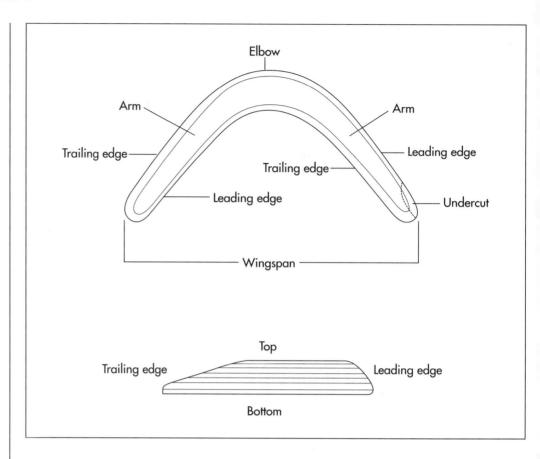

sturdy paper. This first guide mark denotes the center of gravity of the boomerang. As the designer continues to sketch the boomerang, he or she must take care to balance the shape around the center of gravity point. The other important consideration is that the centerlines of each wing of the boomerang must point generally toward the center of gravity (i.e., within 10° in either direction). So the second set of guide marks the designer makes on the paper are the centerlines of the wings. Within the limitations set by the guide marks, the designer can then sketch as basic or whimsical design as desired.

After the design is completely drawn, the designer cuts it out from the sheet of paper. By hanging the pattern successively from the end of each wing, the designer can verify that the planned center of gravity has been adequately preserved. This will be true if the centerline of each wing hangs within 30° of vertical.

Serious designers who seek more precise methods can use more sophisticated techniques, including computer-aided design software.

The Manufacturing Process

The following description focuses on making a V-shaped plywood boomerang. Synthetic materials are worked in a similar manner, but some generate dangerous dust or fumes when being cut or sanded. In this case, dust masks and protective clothing are essential.

Forming

1 The plywood sheet is checked for flatness. If it is not completely flat, it is oriented so the concave side will correspond to the top surface of the boomerang. This significantly increases the boomerang's strength, and it raises the wingtips slightly for better aerodynamics.

2 The pattern is placed on the plywood so that the wood grain runs across from the tip of one arm of the boomerang to the tip of the other arm. The outline of the pattern is traced on the plywood with a pencil.

3 A scroll saw, jigsaw, coping saw, bandsaw, or fret saw is used to cut the

boomerang shape out of the plywood. This basic cutout is called the blank.

4 As an alternative to cutting a single-piece blank, two separate wings can be cut, allowing a section of overlap where they will be joined. Using a router, half the thickness of this overlap section is cut away from each wing. The overlapping sections are joined with wood glue and clamped together until the joint hardens.

5 An outline is drawn on the top of the blank to show the areas to be shaped for the leading and trailing edges of the wings.

6 The profiles of the wings are shaped with a belt sander or by hand with a rasp or a plane. The top of the leading edge of each wing is decreased at a 45° angle, while the rear of the wing is angled down to leave a 0.04-0.08 in (1-2 mm) thick trailing edge. The bottom face of the leading edge is cut back slightly. The tips of the wings are shaped down to the same thickness as the trailing edge. The various layers of the plywood serve as contour lines that help the worker achieve uniform slopes.

7 A shallow section may also be cut out from the bottom surface of each wing. For example, this might consist of a 2-in (5-cm) long strip near the wing tip and behind the leading edge.

Finishing

8 Using progressively finer (80-250 grit) sandpaper, the surface of the boomerang is smoothed carefully with an orbital sander or by hand.

9 After spraying the surface with sanding sealer, fine **steel wool** is used to further smooth the surface. A coat of paint or **wood stain** is followed by one or more coats of clear polyurethane finish.

Tuning

10 The boomerang is thrown several times to test its flight capabilities. Several types of adjustments may be made to tune the boomerang for better performance. For example, the wing profiles might be adjusted by additional sanding.

11 Another tuning technique is to bend the wings, raising their tips about 0.12 in (3 mm) above the plane of the vertex; this is called giving the boomerang a positive dihedral. It may be necessary to heat the boomerang to make it flexible enough for bending and to make the adjustment permanent. This can be done with steam or even in a microwave oven.

12 Twisting the wings to raise or lower the leading edges can also affect the boomerang's performance.

13 Other tuning techniques include drilling holes through the wings, cutting slots in the leading edges of the wings, or drilling shallow holes into the underside of the wings and inserting lead or **brass** plugs to add weight.

Quality Control

Throughout the manufacturing process, the quality of the boomerang is periodically checked. Any unevenness in the boomerang such as unequal sides or bumps will take away from the aerodynamic design. Boomerangs are sanctioned by comities such as the World Boomerang Association and the United States Boomerang Association (USBA). These groups set the standards and rules that any boomerang competitions must adhere to such as safety, skill, and timing.

The Future

Boomerang innovations can be developed in two areas: materials and design. As new materials are developed that are strong, durable, and lightweight, boomerang makers will try using them individually or in combination. For example, the boomerang that flew for more than 17 minutes consisted of a two-layer outer shell of carbon fibers and Kevlar; the shell was filled with epoxy resin mixed with phenolic microballoons.

Two recent innovations suggest ways that designs can be modified to improve aerodynamics. One involved making the upper and lower surfaces of the trailing edges of a boomerang's wings slightly concave. Normally, these surfaces are flat or slightly convex. This design was used for the boomerang that set the current world record for distance. In the other example, a boomerang's wing tips were cut at an angle that made them perpendicular to imaginary lines leading to the center of rotation. Usually, the wing tips are perpendicular to the centerline of the wings. This modification was created by the holder of the unofficial MTA record.

Where to Learn More

Books

Mason, Bernard Sterling. *Boomerangs: How to Make and Throw Them.* Mineola, NY: Dover, 1985.

Periodicals

Drollette, Daniel. "Field Notes: Return to Sender." *The Sciences* (May/June, 1998): 16- 19.

Lane, Marke. "Classic Boomerang." *Workbench* (June/July 1996): 36.

Valenti, Michael. "The Return of the Boomerang." *Mechanical Engineering* (December 1993): 68.

Other

Amateur Boomsmith. http://www.uku.fi/~hniskane/workmain.html (October 27, 2000).

—*Loretta Hall*

Boxing Gloves

Background

Fist fighting has existed as a form of entertainment since the early days of human civilization. Some form of the sport appeared as long as 6,000 years ago in present-day Ethiopia. From there it spread across the ancient world. Throughout the sport's history, segments of society deemed that it was too brutal and have lobbied to restrict or ban it altogether. Partly in deference to those efforts and partly in recognition of the frailty of the human body, practitioners and promoters have developed defenses for use in the sport. The oldest and most little changed of these has been the boxing glove.

History

Boxing was first put on the Olympic program in 688 B.C., and it was there that one of the earliest records of hand protection appears. Olympic fighters wrapped their hands and wrists in leather strips. Initially, the leather was used as protection. Later, the leather was hardened, making these early gloves into weapons. The Romans called these strips cestus and added iron or **brass** studs. Sometimes a large spike called the myrmex was also attached; both instruments could kill an opponent.

It is generally acknowledged that the inventor of the modern boxing glove was an English champion fighter named Jack Broughton. Broughton fought, as did all boxers of his day, with bare knuckles. Broughton developed his gloves—known as mufflers—so that the gentry could practice boxing at the gymnasium without inflicting serious damage. The gloves were reserved for such uses; all public contests were still fought with bare fists. In 1743, Broughton codified the first modern rules of boxing. Strangely, his rules make no mention of gloves. Then in 1867, John Graham Chambers, a member of London's Amateur Athletic Club, published the Marquis of Queensberry rules. Line eight of the rules reads, "The gloves to be fair-sized boxing gloves of the best quality, and new." (The rules also mention that no shoes with springs are to be used.) The rules were gradually adopted for amateur competition, and the use of thinly padded or skintight leather mitts became more widespread. Still, most public and professional bouts were fought with bare knuckles.

American fighter John L. Sullivan is said to have been one of first to popularize the wearing of gloves in public bouts. Sullivan reigned as World Heavyweight Champion from 1882–1892, but many historians do not consider him to be the first modern champion as all the fights in which he won his title were waged under the old Prize Ring rules, which did not require gloves. Ironically, Sullivan did wear gloves in his last fight, in which he lost to the first champion under the Marquis of Queensberry Rules, James "Gentleman Jim" Corbett.

Raw Materials

The skin of a boxing glove is top grain tanned leather, most often cowhide or goatskin because of their durability and flexibility. Lesser-quality gloves will be made from vinyl, but most sanctioning bodies—amateur and professional—require leather gloves. Some manufacturers line their gloves with another layer of leather, but the majority use nylon taffeta. Gloves are stitched with nylon thread and padding is of high-density **polyurethane**,

Boxing was first put on the Olympic program in 688 B.C.

59

A pair of boxing gloves.

Latex, or polyvinyl chloride (PVC) foam delivered in sheet form. Historically, **cotton** batting has been used as padding and many manufacturers still use this material to pad some portion of their models. Some manufacturers also use horsehair.

Design

The primary design consideration involves the glove's padding. In order for a padding material to be effective, it must absorb energy by compressing. The more it compresses, the more energy it absorbs. If a material compresses too much, it ceases to be useful because it becomes simply a thin layer of dense material. Partly because of this, different weight classes require gloves of different weights. A glove's weight is changed by adding or removing layers of padding. If the same glove weight was required for all weight classes, blows thrown by the largest and heaviest boxers would compress the padding beyond its useful range, while blows thrown by the lightest boxers would barely compress the material at all. In addition, many materials that offer excellent energy absorption also display a characteristic known as memory. Once compressed, these materials maintain their deformed state for an extended period of time so that the initial blow with a glove offers normal protection, but subsequent blows are virtually unpadded.

Other design criteria stem from rules and regulations of the various sanctioning bodies. For example, USA Boxing, which regulates much of the amateur competition in the United States and sanctions all Olympic-style competition in the United States, re-

quires that all gloves either be thumbless or have the thumb compartment attached to the body of the glove so that boxers cannot jab each other in the eye. In addition, gloves used for international competition, such as the Olympics, must have a portion of the leather covering the knuckle area dyed white for scoring purposes.

The Manufacturing Process

Paterns and cutting

All boxing gloves are cut, assembled, stitched, stuffed, and finished by hand. The manufacture of a glove begins with a pattern of the individual pieces. While every manufacturer has a different pattern, the basic pieces are the palm, which is cut with a slit down its middle that will eventually form the closure section of the glove; the knuckle area, which is always made from a single piece of leather to avoid seams; the thumb, which is made from two halves; the cuff, which is cut as a wide strip; and a thin strip that will be folded over and sewn onto the edge of the cuff and the closure area to finish the glove. The knuckle piece is cut to be larger than its finished size so that space is left for stuffing.

1 Leather arrives from the tannery in large pieces and is laid out on large cutting tables. The patterns are placed on the leather and arranged to make the most efficient use of that piece. The patterns are then traced onto the leather and the pieces are cut with large scissors. Meanwhile, similar patterns are traced onto the lining material and those pieces are cut. Pieces are made to line the palm, the thumb, the cuff, and the knuckle area.

Assembly and stitching

2 The leather shell of a boxing glove is first sewn together inside out. Stitching is often done on an industrial sewing machine with some of the smaller pieces and finish work being completed by hand. Many of the higher quality gloves are stitched entirely by hand, and double stitching is used throughout all quality gloves.

3 The oversized knuckle piece is stitched to the palm piece. The two pieces are fitted over a buck to assure the correct shape

and the seam is gathered so that the knuckle piece balloons slightly. Gathering the seam also causes the glove to take on its trademark clenched fist shape.

4 Then, the liner pieces are stitched onto this assembled section and the palm is stuffed with padding. The liner is left open at the bottom of the glove, where the cuff will be attached. On many models, the back half of the thumb piece is cut as part of the knuckle piece, and the inner half is sewn onto the knuckle and palm pieces. On others, the thumb is stitched together separately; its lining is attached, and its padding is stuffed. The assembled thumb piece is then stitched onto the glove.

Stuffing the glove

5 The entire glove assembly is now turned right side out. As it is more economical for manufacturers to purchase padding material in standard sheet form, the padding for the knuckle area is made by layering sheets of the material and then cutting it to the desired shape. This also allows glove makers to use one standardized thickness of padding for many glove weights and specifications rather than purchasing or manufacturing a different molded piece for every glove model.

6 The pattern for the glove being made is traced onto the padding material and it is cut. Depending on the manufacturer, pattern pieces may be cut in mass beforehand and kept in stock for assembly.

7 The cut pieces are layered to the specified thickness and are stuffed into the pocket between the knuckle area and its lining.

Finishing the glove

8 The last piece to be stitched to the glove is the cuff. The cuff and its lining are stitched together, and the piece is stuffed. The ends of this assembly are not stitched together as the piece will eventually form part of the gloves closure area.

9 The assembly is stitched to the open end of the glove piece, closing off all the open pockets and sealing the glove's padding.

Muhammad Ali.

Muhammad Ali was born Cassius Marcellus Clay on January 17, 1942, in Louisville, Kentucky. By 17, Ali had won six Kentucky Golden Gloves tournaments in lightweight, welterweight, and heavyweight categories. In 1959 and 1960 he won the Light Heavyweight National Golden Gloves and the National Amateur Athletic Union (AAU) tournaments. In 1960, Ali won the Olympic gold medal for the United States under the light heavyweight category. In 1964 he became heavyweight champion and converted to Islam, renouncing his "slave name" for Muhammad Ali.

On April 28, 1967, Ali refused induction into the U.S. Army on religious grounds. The World Boxing Association (WBA) stripped his title and he was banned from fighting. Joe Frazier was awarded the title, and Ali was sentenced to five years in prison for draft evasion and fined $10,000. In 1970 the U.S. Supreme Court reversed the conviction on a technicality, and the NAACP won its suit proving denial of his boxing license violated Ali's constitutional rights. Ali fought Frazier in 1971 and lost his first professional defeat. In 1974 Ali defeated Frazier but George Foreman now held the title. Ali reclaimed it in Kinshasa, Zaire, billed as the "Rumble in the Jungle."

Ali lost a title defense in 1978 to Leon Spinks, defeating Spinks in a rematch. On June 26, 1979, Ali retired as champion with a professional record of 59 victories and three defeats, but returned in 1980 to fight Larry Holmes for the World Boxing Council (WBC) title. (Holmes won with a technical knockout.) In 1981 Ali boxed professionally for the last time, fighting and losing to Trevor Berbick. In 1977 he was advised to quit boxing because of slowed reflexes, and in 1984 was diagnosed with Parkinson's syndrome. Ali was inducted into the International Boxing Hall of Fame in 1990.

10 If the glove is to be closed with laces, a template is laid over the opening now formed on the glove's underside by the slit in the palm and the open ends of the cuff, and laces holes are punched with an

awl. Each hole is strengthened with stitching, and the entire lace area is finished with several rows of stitching.

11 If the gloves are to be closed with hook and loop material, the loop side is sewn onto the outside face of the cuff, and the hook side is stitched onto the cuff's opposite edge.

12 A single thin strip of leather is folded over the open edge of the cuff and the lace area and is stitched in place to finish the glove. The maker's label and any required sanctioning body labels are sewn onto the back of the cuff and the finished gloves are packaged for shipping.

Quality Control

Virtually every country and state has a boxing commission that regulates professional bouts. Every one of these commissions has its own rules and regulations governing the conduct and equipment of a boxing match. Most amateur competitions in the United States are governed by USA Boxing or Golden Gloves, and each of these bodies specify particular requirements for gloves used in their bouts. What most gloves used today have in common is that they have been tested by the Wayne State University Sports Biomechanics Department in Detroit, Michigan. The University tests a boxing glove by fitting it onto a maple block in the approximate shape of a human fist. The block is attached to a hydraulic ram that can be fired at predetermined rates of acceleration. The gloved block is fired at a biometric human form (a test dummy) that has been fitted with sensors that measure impact. The impact readings for various accelerations are translated onto a scale called a severity index and gloves must fall within a certain range to be acceptable.

The Future

The most surprising aspect of boxing gloves is how little they have changed. The first gloves were leather mitts with little or no padding. Today's gloves have added padding to a greater or lesser degree but not much else. Boxing in general seems to be highly resistant to both change and regulation. For over a hundred years, fighters resisted wearing gloves at all. And since then, they have thwarted most efforts at innovation. The movement to remove the thumbs from gloves, for example, has only succeeded in a few arenas. Gloves have become more heavily padded in recent years and the padding materials themselves have grown more resilient, but many experts insist that this simply allows fighters to punch harder and inflict more damage.

Where to Learn More

Books

Blewett, Bert. *The A-Z of World Boxing: An Authoritative and Entertaining Compendium of the Fight Game from Its Origins to the Present Day.* New Jersey: Parkwest Publications, 1997.

Myler, Patrick. *Gentleman Jim Corbett: The Truth Behind a Boxing Legend.* Robson Books, Ltd., 1999.

Ward, Nathan. *The Total Sports Illustrated Book of Boxing.* Total Sports, 1999.

Other

Hickock Sports.com. http://www.hickoksports.com (April 1, 2000).

International Boxing Hall of Fame. http://www.ibhof.com (May 30, 2000).

USA Boxing. http://www.usaboxing.org (March 16, 2000).

—*Michael Cavette*

Brass

Background

Brass is a metal composed primarily of copper and zinc. Copper is the main component, and brass is usually classified as a copper alloy. The color of brass varies from a dark reddish brown to a light silvery yellow depending on the amount of zinc present; the more zinc, the lighter the color. Brass is stronger and harder than copper, but not as strong or hard as steel. It is easy to form into various shapes, a good conductor of heat, and generally resistant to corrosion from salt water. Because of these properties, brass is used to make pipes and tubes, weather-stripping and other architectural trim pieces, screws, radiators, musical instruments, and cartridge casings for firearms.

History

Ancient metalworkers in the area now known as Syria or eastern Turkey knew how to melt copper with tin to make a metal called bronze as early as 3000 B.C. Sometimes they also made brass without knowing it, because tin and zinc ore deposits are sometimes found together, and the two materials have similar colors and properties.

By about 20 B.C.–A.D. 20, metalworkers around the Mediterranean Sea were able to distinguish zinc ores from those containing tin and began blending zinc with copper to make brass coins and other items. Most of the zinc was derived by heating a mineral known as calamine, which contains various zinc compounds. Starting in about 300 A.D., the brass metalworking industry flourished in what is now Germany and The Netherlands.

Although these early metalworkers could recognize the difference between zinc ore and tin ore, they still didn't understand that zinc was a metal. It wasn't until 1746 that a German scientist named Andreas Sigismund Marggraf (1709–1782) identified zinc and determined its properties. The process for combining metallic copper and zinc to make brass was patented in England in 1781.

The first metal cartridge casings for firearms were introduced in 1852. Although several different metals were tried, brass was the most successful because of it's ability to expand and seal the breech under pressure when the cartridge was first fired, then contract immediately to allow the empty cartridge casing to be extracted from the firearm. This property led to the development of rapid-fire automatic weapons.

Raw Materials

The main component of brass is copper. The amount of copper varies between 55% and 95% by weight depending on the type of brass and its intended use. Brasses containing a high percentage of copper are made from electrically refined copper that is at least 99.3% pure to minimize the amount of other materials. Brasses containing a lower percentage of copper can also be made from electrically refined copper, but are more commonly made from less-expensive recycled copper alloy scrap. When recycled scrap is used, the percentages of copper and other materials in the scrap must be known so that the manufacturer can adjust the amounts of materials to be added in order to achieve the desired brass composition.

The second component of brass is zinc. The amount of zinc varies between 5% and 40% by weight depending on the type of brass.

Brass is easy to form into various shapes, a good conductor of heat, and generally resistant to corrosion from salt water. Because of these properties, brass is used to make pipes and tubes, weather-stripping and other architectural trim pieces, screws, radiators, musical instruments, and cartridge casings for firearms.

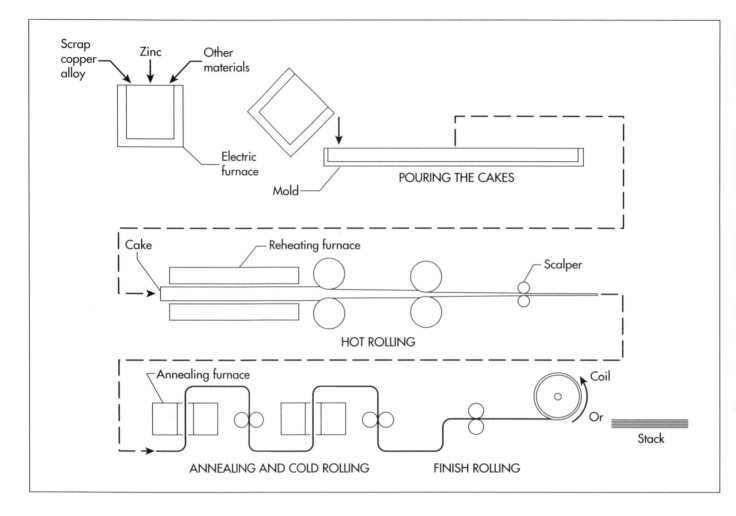

Scrap copper alloy   Zinc   Other materials

Electric furnace

Mold

POURING THE CAKES

Cake   Reheating furnace   Scalper

HOT ROLLING

Annealing furnace   Coil

Or

Stack

ANNEALING AND COLD ROLLING   FINISH ROLLING

A diagram depicting typical manufacturing steps in brass production.

Brasses with a higher percentages of zinc are stronger and harder, but they are also more difficult to form and have less corrosion resistance. The zinc used to make brass is a commercial grade sometimes known as spelter.

Some brasses also contain small percentages of other materials to improve certain characteristics. Up to 3.8% by weight of lead may be added to improve machinability. The addition of tin improves corrosion resistance. Iron makes the brass harder and makes the internal grain structure smaller so that the metal can be shaped by repeated impacts in a process called forging. Arsenic and antimony are sometimes added to brasses that contain more than 20% zinc in order to inhibit corrosion. Other materials that may be used in very small amounts are manganese, **silicon**, and phosphorus.

Design

The traditional names for various types of brass usually reflected either the color of the material or the intended use. For example, red brass contained 15% zinc and had a reddish color, while yellow brass contained 35% zinc and had a yellowish color. Cartridge brass contained 30% zinc and was used to make cartridges for firearms. Naval brasses had up to 39.7% zinc and were used in various applications on ships.

Unfortunately, scattered among the traditional brass names were a number of misnomers. Brass with 10% zinc was called commercial bronze, even though it did not contain any tin and was not a bronze. Brass with 40% zinc and 3.8% lead was called architectural bronze, even though it was actually a leaded brass.

As a result of these sometimes confusing names, brasses in the United States are now designated by the Unified Numbering System for metals and alloys. This system uses a letter—in this case the letter "C" for copper, because brass is a copper alloy—followed by five digits. Brasses whose chemical composition makes them suitable for being formed into the final product by mechanical methods, such as rolling or forging, are called wrought brasses, and the first digit of their designation is 1 through 7. Brasses whose chemical composition makes them suitable for being formed into the final product by pouring molten metal into a mold are called cast brasses, and the first digit of their designation is 8 or 9.

The Manufacturing Process

The manufacturing process used to produce brass involves combining the appropriate raw materials into a molten metal, which is allowed to solidify. The shape and properties of the solidified metal are then altered through a series of carefully controlled operations to produce the desired brass stock.

Brass stock is available in a variety of forms including plate, sheet, strip, foil, rod, bar, wire, and billet depending on the final application. For example, brass screws are cut from lengths of rod. The zigzag fins used in some vehicle radiators are bent from strip. Pipes and tubes are formed by extruding, or squeezing rectangular billets of hot brass through a shaped opening, called a die, to form long, hollow cylinders.

The differences between plate, sheet, strip, and foil are the overall size and thickness of the materials. Plate is a large, flat, rectangular piece of brass with a thickness greater than about 0.2 in. (5 mm)—like a piece of plywood used in building construction. Sheet usually has the same overall size as plate, but is thinner. Strip is made from sheet that has been cut into long, narrow pieces. Foil is like strip, only much thinner. Some brass foil can be as thin as 0.0005 in (0.013 mm).

The actual manufacturing process depends on the desired shape and properties of the brass stock, as well as the particular machinery and practices used in different brass plants. Here is a typical manufacturing process used to produce brass sheet and strip.

Melting

1 The appropriate amount of suitable copper alloy scrap is weighed and transferred into an electric furnace where it is melted at about 1,920°F (1,050°C). After adjusting for the amount of zinc in the scrap alloy, an appropriate amount of zinc is added after the copper melts. A small amount of additional zinc, about 50% of the total zinc required, may be added to compensate for any zinc that vaporizes during the melting operation. If any other materials are required for the particular brass formulation, they are also added if they were not present in the copper scrap.

2 The molten metal is poured into molds about 8 in x 18 in x 10 ft (20 cm x 46 cm x 3 m) and allowed to solidify into slabs called cakes. In some operations, the melting and pouring are done semi-continuously to produce very long slabs.

3 When the cakes are cool enough to be moved, they are dumped out of the molds and moved to the rolling area where they are stored.

Hot rolling

4 The cakes are placed in a furnace and are reheated until they reach the desired temperature. The temperature depends on the final shape and properties of the brass stock.

5 The heated cakes are then fed through a series of opposing steel rollers which reduce the thickness of the brass step-by-step to about 0.5 in (13 mm) or less. At the same time, the width of the brass increases. This process is sometimes called breakdown rolling.

6 The brass, which is now much cooler, passes through a milling machine called a scalper. This machine cuts a thin layer off the outer faces of the brass to remove any oxides which may have formed on the surfaces as a result of the hot metal's exposure to the air.

Annealing and cold rolling

7 As the brass is hot rolled it gets harder and more difficult to work. It also loses its ductility, or ability to be stretched further. Before the brass can be rolled further, it must first be heated to relieve some of its hardness and make it more ductile. This process is called annealing. The annealing temperatures and times vary according to

the brass composition and desired properties. Larger pieces of hot-rolled brass may be placed in a sealed furnace and annealed together in a batch. Smaller pieces may be placed on a metal belt conveyor and fed continuously through a furnace with airtight seals at each end. In either method, the atmosphere inside the furnace is filled with a neutral gas like nitrogen to prevent the brass from reacting with oxygen and forming undesirable oxides on its surface.

8 The annealed pieces of brass are then fed through another series of rollers to further reduce their thickness to about 0.1 in (2.5 mm). This process is called cold rolling because the temperature of the brass is much lower than the temperature during hot rolling. Cold rolling deforms the internal structure of the brass, or grain, and increases its strength and hardness. The more the thickness is reduced, the stronger and harder the material becomes. The cold-rolling mills are designed to minimize deflection across the width of the rollers in order to produce brass sheets with near-uniform thickness.

9 Steps 7 and 8 may be repeated many times to achieve the desired thickness, strength, and degree of hardness. In some plants, the pieces of brass are connected together into one long, continuous sheet and are fed through a series of annealing furnaces and rolling mills arranged in a vertical serpentine pattern.

10 At this point, the wide sheets may be slit into narrower sections to produce brass strip. The strip may then be given an acid bath and rinse to clean it.

Finish rolling

11 The sheets may be given a final cold rolling to tighten the tolerances on the thickness or to produce a very smooth surface finish. They are then cut to size, stacked or coiled depending on their thickness and intended use, and sent to the warehouse for distribution.

12 The strip may also be given a final finish rolling before it is cut to length, coiled, and sent to the warehouse.

Quality Control

During production, brass is subject to constant evaluation and control of the materials and processes used to form specific brass stock. The chemical compositions of the raw materials are checked and adjusted before melting. The heating and cooling times and temperatures are specified and monitored. The thickness of the sheet and strip are measured at each step. Finally, samples of the finished product are tested for hardness, strength, dimensions, and other factors to ensure they meet the required specifications.

The Future

Brass has a combination of strength, corrosion resistance, and formability that will continue to make it a useful material for many applications in the foreseeable future. Brass also has an advantage over other materials in that most products made from brass are recycled or reused, rather than being discarded in a landfill, which will help ensure a continued supply for many years.

Where to Learn More

Books

Brady, George S., Henry R. Clauser, and John A. Vaccari. "Brass." In *Materials Handbook, 14th ed.* New York: McGraw-Hill, 1997.

Hornbostel, Caleb. "Brass." In *Construction Materials: Types, Uses, and Applications.* New York: John Wiley and Sons, 1991.

Kroschwitz, Jacqueline I., and Mary Howe-Grant, eds. "Copper Alloys." In *Encyclopedia of Chemical Technology, 4th ed.* New York: John Wiley and Sons, Inc., 1993.

Other

Metalworld. http://www.metalworld.com (June 19, 2000).

—*Chris Cavette*

Breath Mint

Background

Aromatic herbs have been used throughout history in a number of ways; fragrant soaps, pomanders, bath-water fresheners, potpourri, sachets, incense, scented candles, and natural herbs to sweeten sour breath are common in most cultures and popular today. Aromatic herbs have the advantage of driving away insects, and the mint family has an especially excellent reputation for keeping pests away from people and other plants. It is often grown among other plants, like members of the cabbage family. Spearmint is grown most commonly, but peppermint, and apple, lemon, and pineapple mint are familiar occupants of many gardens.

Mints are perennials that spread rapidly and grow quickly. The mint family is called Labiatae and includes about 160 genera, of which *Mentha* includes the true mints. Spearmint and peppermint are grown extensively in a surprisingly robust health mint industry that produces mainly oil. Over 70,000 acres (28,328 hectares) of farmland in Indiana, Michigan, Washington, and Oregon produce these two types of mint for a wide range of commercial uses.

For natural solutions to the problem of maintaining sweet breath, a small piece of nutmeg or angelica root can be chewed, or a piece of the herb called mace can be placed in the mouth for several minutes. Obviously, the most common herbal breath mints are the mints. A leaf or two from any of the commonly grown mint plants, including peppermint, can be eaten to freshen the breath and aid digestion. Many references about herbs provide recipes for making toothpaste and **mouthwash** from peppermint and other herbs and natural ingredients that avoid the detergents and sugar found in commercial products.

Parsley, fennel, watercress, alfalfa, and pulverized nettle leaves all contain chlorophyll that is used in many commercial breath fresheners like Clorets. Chewing a bud of clove immediately relieves bad breath (especially after eating garlic-laden foods) and aids digestion, as does clove chewing gum.

History

Over 4,000 years ago, people sucked on whole cloves to cleanse their breath. Clove—rather than mint—is probably the oldest and most common herb used for fresh breath. The Pharisees collected tithes in mint and other sweet-smelling herbs, and the Hebrews and Christians spread it on the floors of synagogues and churches as a symbol of cleanliness and hospitality. In the Middle Ages, anise seed was chewed slowly as a breath freshener (and to cover up odors from liquor consumption). Cardamon seeds are also natural breath sweeteners that have been chewed both in the Orient and in Europe since ancient times. In colonial North America, settlers discovered that small bits of calamus (sweet sedge) root and pieces of dried orris root (the root of the Florentine iris) had similar sweetening effects.

In modern times, breath mints were a logical progression from hard candies and chewing gum—it was gum that really launched breath mints as a separate market segment. Hard candies, made from boiling sugar to a hard rolling boil, have been made over the kitchen fire since ancient times; commercially, hard mints like peppermints and glacier mints (clear candies) were made in Victorian Eng-

Over 70,000 acres (28,328 hectares) of farmland in Indiana, Michigan, Washington, and Oregon produce spearmint and peppermint for a wide range of commercial uses.

Tablet-shaped mints are made in a rotary tablet press, consisting of four punch and dies that move along belts. Large rollers continuously move the belts and the punches and dies are pushed up and pulled down by adjustable cams.

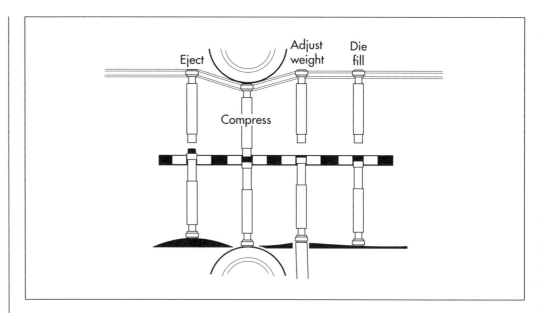

land, on the European continent, and in the United States. But the candy market was as volatile in the 1800s as it is today, and manufacturers have always searched for something new. In 1869, Thomas Adams, an inventor from New York, stumbled on the idea of replacing paraffin wax that was used like chewing gum with chicle, a rubbery fluid produced by some trees. The pelletized chicle was sold in boxes, and new flavors of the chewing gum were introduced over the next 100 years. Many of these flavors had breath- and health-enhancing properties; examples are pepsin (a digestive aid) in Beeman's gum, sassafras and licorice, cloves, Dentyne (the first gum aimed at dental hygiene), Sen-Sen and chlorophyll, cinnamon, and many varieties of mint.

In the 1950s, American Chicle introduced Certs. The need for a mint dedicated to fresh breath had been identified in consumer research. Toothpaste and mouthwash were simply not convenient or portable, and candy mints had no proven association with fresh breath. Certs combined both candy and a breath freshener in a small package. The breath-freshening ingredient was "Retsyn," a mixture of copper gluconate and cottonseed oil that was trademarked by American Chicle. A sugar-free product was introduced in 1982 and reformulated in 1987 with NutraSweet; 1988 retail sales of both the sugar and sugar-free versions of Certs topped $190 million. Also in 1988, the company introduced Sugar Free Certs Mini-Mints, and this began the mini-mint fad that

still has a strong grip on the breath-mint market in 2000. Ten years later, in 1998, the breath-freshener market grew by 13% in one year, while all other gum and candy expanded by only 2.3%, according to the National Confectioners Association.

Breath mints have not been without controversy. The Nutrition Labeling and Education Act requires that every food product have a serving size and the equivalent number of calorie stated on the package. Breath mints posed a significant problem for the Food and Drug Administration (FDA), which took six years to study how to classify serving size and associated calories. Hard candies have a single serving of 0.5oz (15 g), and the FDA initially lumped breath mints in the hard candy category; but 0.5oz (15 g) equals a whole packet of breath mints. The FDA then decided 0.07oz (2 g) (about the same as a single Certs or Breath Saver) would be a suitable single serving. The mini-mint producers objected. Lawyers for Tic Tac claimed that this single serving would be five of their mints. The argument shifted back and forth until the FDA finally ruled that a single serving should equal a single unit—one mint—regardless of size. Effectiveness of breath-freshening was left to be determined by the consumer.

Raw Materials

The bulk of all breath mints is some form of sugar or a sugar substitute. Sugar is present as sucrose or dextrose, and the most common sugar substitutes are sorbitol or zilotrol, (chemists classify both of these as polyhedric alcohols). Binders hold the particles of the other dry ingredients together. Common binders are corn syrup, natural gums like gum arabic, and gelatin. Tablet-type mints also include lubricants that help the dry materials flow through the press, prevent them from sticking to the machinery faces, and pop out of the dies after they have been compressed. Magnesium stearate, stearic acid, and calcium stearate are lubricants in many candies and breath mints. Some mints also include disintegrants that help the mint dissolve, disperse, and become absorbed during digestion. Natural or artificial flavors and the breath-freshener complete the raw materials. Most companies closely guard the secret ingredients of the breath-freshener itself. The outside of the mint may look polished; carnuba wax is the secret to the gloss.

Design

Breath mints come in three basic configurations. The mini-mints like Tic Tac and Blitz mints are tiny tablets or compressed mints that have a hard outer shell made by a process called panning. The design of the mint itself is tiny with a compressed center that is easy to manufacture and an attractive shell with a bright color, cute shape, or shiny finish. They are packaged in small, convenient, eye-catching boxes; several of these have won international design awards, showing that packaging is just as important in the design process as the contents. Roll mints like Breath Savers and Certs are shaped like Life Savers but the holes in the centers may be filled. These mints are hard; the filled center is usually brightly colored. They are also made by the tableting or compressed process. The packages are convenient rolls and have the familiarity of Life Savers.

Other hard mints include square-shaped Velamints from West Germany, which are packaged in squared tubes, and Altoids from England in their collectible mini-tins; these mints are made from an extruded dough. Mints with soft centers are the third class; they can be made in a batch process or by panning. Mentos from Holland and Vikings from Denmark are the leaders in this field.

These mints are larger in diameter than the mini-mints, and they have a hard outer shell with a chewy center. Recent additions to the mint marketplace include Testamints (with Bible verses on the tins) and Web Fuel (tins shaped like computer mice bearing Internet web site addresses).

Many of the leading mints manufactured in Europe were originally aimed at the children's candy market, but this market has very rapid shifts as children's interests change. When sales began to decline among children in Europe, the manufacturers shifted their attention to the adult market in America and the preoccupation with sweet breath. Some of the designs stayed the same despite the shift in market; the bright oranges, reds, and greens of some of these mints themselves and especially their packaging are holdovers from targeting European children.

The adult market in the United States for candy generally has grown steadily since about 1980. By a simple shift in emphasis, Ferrero, the European chocolate company that makes Tic Tacs, was able to keep its eye-catching colors and packaging but market the mint as a "1.5-calorie breath mint." The low calories, breath protection, and cute presentation appeals particularly to young women; from 1980 to 1990, Tic Tac's market share rose from 2% to 12% with this change in advertising approach.

The Manufacturing Process

Tableted or compressed mints

Tableted candies and mints are an offshoot of the pharmaceutical industry that makes pills. The same accuracy that produces just the right dose in a lozenge or tablet for medicinal purposes also makes mints efficiently with the right distribution of flavor and breath freshener. Tablets are made with rotary presses that use a rotating die table and compression rollers to turn out as many as 10,000 mints per minute.

1 The ingredients for mints arrive at the factory in powder form. They are granulated in a mixing and bonding method that helps them flow through a tablet press. The process involves pulverizing (pounding)

them to a fine consistency, mixing (most often in a dry process, although wet mixing can be used), compacting the ingredients, sizing the finished grains (sorting out the coarse particles), mixing the ingredients, and flowing them into the tableting machine. The moisture content is controlled throughout the process (whether it is wet or dry), and the granules are dried on bed dryers (flat systems) or rotary dryers. Mixing—one of the last steps—is the process in which flavors and active ingredients like breath fresheners are added for the most uniform distribution. Lubricants are mixed last so they coat all the other ingredients well.

2 The prepared, granulated, mixed ingredients are conveyed to the tableting machine. While this sounds straightforward, the conveyors cannot have any bends or turns that might sort the materials, and temperature and moisture have to be strictly controlled along the route. Some ingredients, particularly the lubricants, begin to separate from the other ingredients if conditions aren't correct. Some flavors like grape react with sugars if there is too much moisture in the air and begin to turn brown; gelatin also browns if conditions are too warm or dry.

3 The rotary tablet press consists of four punches that move along an upper belt paired to four dies that move along a bottom belt. The belts themselves are continuously turned around large rollers. The punches and dies are pushed up toward each other and pulled down by adjustable cams. As the ingredients enter the rotary tablet press, the granulated ingredients are channeled into a feeder (the upper punch) that fills a die seated in the bottom of the pair. The cavity of the die has to be filled with the volume of granulated ingredients. The second stage of the press adjusts the weight and scrapes excess material off the top of the die. In the third, compression stage, the cams drive the upper punch and the lower die together, compressing the ingredient into a tablet. The punch and die have been designed to have the shape of the breath mint and possibly its name cut into it, so the compressed result has the identity of the mint firmly stamped in place. In the fourth step, the lower cam pushes out (ejects or extrudes) the stamped mint and the upper part pushes it out of the press where it is collected in a bin.

4 The bin funnels the compressed mint tablets to the next process. This may be panning or packaging.

Extruded or batch process mints

5 Some mints are made much like hard candies, and are cooked as a batch that flows in a continuous process that shapes and sizes the batch ingredients into the shape of the mint. In a dry, uncooked process, a dry dough with a sugar base is made. The batch or the dry dough is funneled through a roller with a general shape much like a pointed **ice cream cone** but with the opening shaped to the desired form of the candy, perhaps a triangle, a diamond, or a barrel. Either the cooked batch or the extruded dry dough is forced through this roller, and each candy length is cut as the ingredient emerges. Extruded dough mints can be recognized by their irregular surface.

The panning process

6 Panning is not usually used to make an entire candy or mint but to give a mint a finished coating. Hollow globe-shaped pans with a hole in one side are made of copper and are rotated much like small cement mixers so the hole stays angled upward. Mints made by compression, batching, or extrusion are placed in relatively small quantities in the pans. Sugar, flavors, and colors are added; as the pan rotates, a hard shell of the sugar forms on the outside of the mint. In the same process or another panning operation, wax or a polishing agent may be put in the pan with the mints to give them an attractive luster. The rotation of the pan can also help develop the finished shape of the mint; the oval shape of many of the minimints is created during panning.

Packaging

7 When the mints are finished by the processes described above, they are carried to packaging machines to be wrapped. Usually, they are carried a short distance on conveyors, and inspectors watch the passing flow of mints and pick out broken or imperfect examples. Depending on the type of packing, the mints may be simply channeled into a funnel that deposits them in small boxes or tins. If they are wrapped in tubes of paper, they are vibrated and gently pressed

into line and wrapped with the preprinted packaging. The packages must also be carefully designed to protect the product; an inner paper of foil, a foil/wax paper laminate, or odor-free plastic is needed for roll mints. Cellophane wraps over paper, tin, or plastic novelty boxes may be needed to keep the mints inside isolated from air and moisture.

The plastic novelty packs were made possible by the uniformity of compressed candies and mints. Pez-Haas originated the famous Pez dispenser because the square candies fit so neatly. Similarly, high-speed wrapping of roll-type candies was pioneered by the Life-Saver Company.

Quality Control

Quality control of breath mints begins with chemistry when the food scientists who devise a new mint select the combination of ingredients, processes, and machinery that can produce the desired product. Throughout the process, few hands touch the operation but many eyes watch. The machines are maintained and cleaned with great attention to detail, not only for health and safety but because some materials are abrasive to the expensive dies that form the mints. Rooms for various processes have controls for temperature, light, and humidity; dust is also carefully controlled because the very fine dust generated from pulverizing ingredients can actually explode in the right combination of conditions. Skilled observers watch the high-speed presses, the turning pans, and the sorting and packaging machines. Mints are rejected for the slightest flaws, yet there is little waste because they can be reground and mixed back into the powders to form later batches.

The Future

Breath mints seem to have a secure place in American life because of the importance of appearance, good health, and cleanliness in our modern lifestyle. Mints, however, are part of the volatile candy and confection market and frequently change face to match the latest trends. The breath-mint manufacturers smell out the latest trends by testing their ideas—from new packaging to actual mint samples—among focus groups of actual consumers. If the focus groups endorse the product, a mini test market, like a small town, will be given samples to assess wider appeal. If the new breath mint passes that test, large-scale sales and marketing campaigns are launched. The explosive sales of mini mints show both the present and the future of the breath mint; clever, eye-catching packaging that makes breath mints portable and trendy have helped and will help consumers and manufacturers alike breathe easily.

Where to Learn More

Books

Bremness, Lesley. *The Complete Book of Herbs*. New York: Viking Studio Books, 1988.

Castleman, Michael. *The Healing Herbs: The Ultimate Guide to the Curative Powers of Nature's Medicines*. New York: Bantam Books, 1991.

Hylton, William H., ed. *The Rodale Herb Book: How to Use, Grow, and Buy Nature's Miracle Plants*. Emmaus, PA: Rodale Press Book Division, 1976.

Kowalchik, Claire, and William H. Hylton, eds. *Rodale's Illustrated Encyclopedia of Herbs*. Emmaus, PA: Rodale Press Book Division, 1976.

Thomas, Lalitha. *10 Essential Herbs*. Prescott, AZ: Hohm Press, 1996.

Periodicals

Edmondson, Brad. "But Candy is Dandy." *American Demographics* 9, no. 5 (May 1987): 22.

Gordon, Raymond. "Invaders in Small Packages." *Forbes* 135 (April 22, 1985): 122.

Poniewozik, James. "It's Not Just a Breath Mint, It's a Web Portal." *Fortune* 138, no. 4 (August 17, 1998): 40.

Skrzycki, Cindy. "Sizing Up Breath Mints Leaves a Bad Taste in Some Mouths." *Washington Post* (January 16, 1998).

Other

American Chicle Group. *Pioneering the Confection Industry*. 1990.

Enrique Bernat F.S.A., manufacturer of Smint. http://www.smint.com (December 22, 2000).

Ragold, Inc., manufacturer of Velamints and Dilbert Mints. http://www.ragold.com (December 22, 2000).

—*Gillian S. Holmes*

Broom

Background

Brooms have been used for centuries to sweep up, in, and around the home and workplace. They may be made from a variety of materials, both man-made and natural. Man-made bristles are generally of extruded plastic and metal handles. Natural-material brooms may be constructed of a variety of materials, including brush, but generally include stiff grasses such as broomcorn and/or sotol fiber. Broomcorn brooms have been made for at least 200 years and are considered superior brooms. Plastic brooms merely move dirt around, however, broomcorn stalks actually absorb dirt and dust, wear extremely well, and are moisture-resistant. Broomcorn brooms are the most expensive of the manufactured brooms.

Broomcorn is actually a variety of upright grass of the species sorghum referred to as *Sorghum vulgare,* or *S. bicolor variety technicum,* belonging to the family Gramineae and cultivated for its stiff stems. Broom bristles are derived when these stiff, tasseled branches—that bear seeds on the ends—are harvested and dried. The seeds are edible, starchy, and high in carbohydrates. They can be used for human consumption (in cereals) or for animal feed. The tasseled stalks, used in the manufacture of brooms, can grow 2-8 ft (0.61-2.4 m) tall. Sorghum is especially valued in hot and arid climates due to its resistance to drought.

Mexico grows and processes most of the broomcorn and sotol fiber used in American broom production. Sotol fiber, a yucca fiber, is sometimes used on the inside of the broom and is wrapped with more expensive broomcorn, thus lowering the price of the natural-bristle broom.

The production of broomcorn brooms is still largely a craft production with a single operator working quickly at a machine, making brooms by hand. There have been some changes in the manufacture of broomcorn brooms within the last several decades, but those changes have been very minor. Essentially, the handcraft has changed little since mid-twentieth century.

Brooms were often used in matrimony rituals to symbolize a union.. Enslaved African-Americans married one another in a civil ceremony referred to as "jumping the broom" in which the couple would literally jump over a broom to signify matrimony. Today, African-Americans occasionally recreate this custom by jumping over a broom at weddings, using specially handmade and decorated brooms for this purpose. These brooms then become a centerpiece within the new household.

History

Ashes and dirt were moved around and out of the house using bundled branches and brush for centuries. Native grasses were dried and bundled together, often decoratively woven at the top or tied tightly with yarn or fabric to keep the brooms together. Southerners have used native sweet grass and other grasses for their long stalks with tasseled ends for broom bristle. The course of American broom history was altered in the late eighteenth century, when some say that in 1797 Levi Dickenson, a farmer from Hadley, Massachusetts, used a bundle of tasseled sorghum grass (also called broomcorn) to make a broom for his wife. It is likely these early broomcorn brooms were simply lashed or woven together, resulting in the fact that they often fell apart. Other experi-

Brooms were often used in matrimony rituals to symbolize a union.

Examples of brooms.

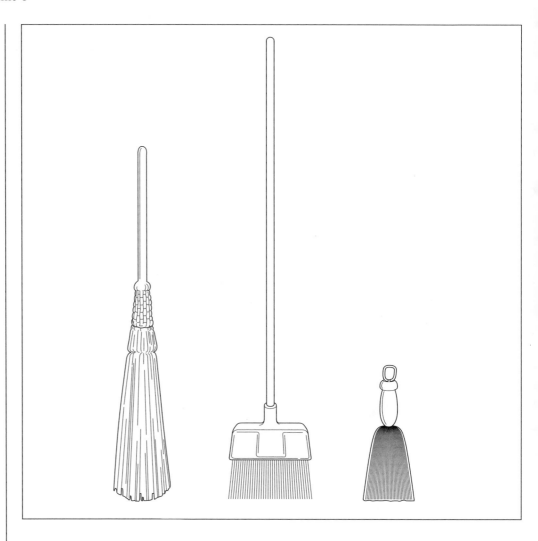

ments with attaching the circular bundles of broomcorn led to wooden handles. By about 1810, wooden handles with holes drilled into them were used to lash the broomcorn to the handle using wooden pegs.

Whether Levi Dickenson was the first American to use sorghum to make brooms is in contention. However, nearly all acknowledge that the United Society of Believers, familiarly called the Shakers, quickly moved into the broom-making business about 1798 by growing broomcorn and making brooms. The Shakers' Watervliet, New York, community took the lead in manufacturing brooms, although nearly all the Shaker communities constructed and sold them throughout the century. The Shakers are credited with inventing the flat broom. They recorded that Theodore Bates of Watervliet examined the circular bundled broom and determined that flat brooms would move dust and dirt more efficiently.

The bundles were put into a vice, flattened, and sewn in place.

The Shakers led the way in improving the broomcorn broom. They appear to be the first to find that wire more effectively secured the broomcorn to the wooden handle rather than tying or weaving. They developed treadle machinery to wind broomcorn around the handle while securing it tightly. They developed special vices to flatten the broom for sewing into the requisite flat shape. Still other machinery was devised to quickly separate the seeds of the broomcorn from the tassel bristles. Using foot-powered machinery, the Shakers could make two dozen brooms per person per day—quite a feat for the early nineteenth century.

Today, the machinery is electrically powered. However, in even the largest American broom factory, the production of broomcorn brooms is still remarkably a hand craft.

(One factory foreman in a large broomcorn factory says he can pick up a broom and tell who amongst his staff made it because each one is made according to the skills and preferences of the maker.) A single machine and operator sits at a machine and constructs a broom. The machines, and the methods, have not changed in over 40 years.

The most significant development in the history of the product resulted from the North American Free Trade Agreement (NAFTA) in 1994 when tariffs were lifted from broomcorn brooms imported from Mexico. Cheaper than American brooms (labor is cheaper and broomcorn is grown there in huge quantities), the Mexican-made broom importation obliterated many American broomcorn manufacturers. American broomcorn manufacturers pressed for more restrictive tariffs, but such tariffs were overruled. Today, there are only about 15 broomcorn manufacturers left in the United States.

Raw Materials

The material used is broomcorn, which is shipped bundled in large bales. The bundles are grouped according to the length of the grass and color. Sotol fiber from the yucca plant may be used in cheaper broomcorn brooms. White metal wire, of approximately 18 gauge, is used to secure the broomcorn and sotol to the handle. A small nail or two is used to secure the wire to the handle. The handles are generally of wood. Sometimes American hardwood is used, but more likely the wood used is ramin wood, an imported wood of dense, heavy, coarse grain. Thick twine is used to sew the brooms flat using a sewing machine. Finally, water is necessary in that the broomcorn must be wetted completely in order to be worked.

The Manufacturing Process

It is important to note that brooms made from broomcorn are made at a station, using a single piece of machinery. Using this machine, brooms are largely still assembled by hand. The process described below is used by the largest manufacturer of brooms and the factory uses about 28 makers to produce 6,000 brooms per day.

1 The raw material for the broom, the broomcorn, comes into the factory already processed and bundled. The bundles are sorted by length and are sorted by the color of the fiber. Bundles are grouped together in a bale weighing about 120 lb (54 kg). Broomcorn must be wet in order to be worked effectively and must be quickly dunked in water before being delivered to the operator. Each bale is lifted with a crane and submerged in a tank of water for 10 seconds. The bundles are then removed from the water using the crane.

2 Workers break apart the wet bales and separate the smaller bundles within the bales. The bundles are placed on racks and rolled to operators who sit at broom-manufacturing equipment.

3 An operator sits at a broom-making machine and has the broomcorn and solid handles there to work. An individual handle is picked up by the operator. The operator inserts a metal wire into a hole drilled near the bottom of the handle. Then, the *insides* are first applied to the broom. In this process, the lowest-grade grass is pressed around the wooden handle, forming the center of the broom. This thick bundle of grass is secured tightly to the handle using the wire attached to the handle through the hole.

4 Then, the shoulders and sides of the broom are given shape as smaller bundles of lesser grade grasses are placed along each side of the center bundle of grass. This *side corn* is secured to the central bundle of grass using more tinned wire that is wrapped by hand tightly around the side corn as well as the central body of grass.

5 Next, the grass is cut off in a straight line just above the wire by the operator using a knife.

6 Over this foundation of lower-grade broomcorn or other grasses is now added the outside of the broom, or the broomcorn we see when we look at a broom. The *hurl*, the best grade of broomcorn used in a broom, is attached to the broom. It is laid atop the center section and shoulders, completely covering it. The hurl is physically attached to the broom using the same piece of white metal wire used earlier in the process.

7 The final construction step is referred to as the run down. The operator runs the wire that secures the hurl down to the handle and nails it off, thus securing the cut end to the wooden handle. The grasses and broomcorn are now completely secured to the broom.

8 The brooms are now constructed but are not finished yet. In order to complete the broom, the broomcorn must be dried out completely. The brooms are moved by rack or palette into a very large drying room that is thermostatically controlled. Depending on the weather, the brooms are left in this large, hot room for five to six hours. When instruments inside the room indicate that no more moisture is being released from the brooms, the heat kicks off and the broomcorn has completely dried.

9 The brooms are now seeded, meaning that cylinders roll vertically over the broomcorn, thus removing all the seeds and small pieces of broomcorn not secured to the handle that will fall out quickly upon use.

10 The seeded brooms are taken to sewing machine operators who run the brooms through a heavy-duty sewing machine with two needles that is threaded with thick twine. The brooms are put through the machine and the broom is flattened and its shape is maintained through the double, triple, or quadruple rows of sewing (depending on the machine and company) that holds the grasses tightly. It takes about 45 seconds to sew the brooms into a flat shape.

11 The brooms are moved by cart to final finishing, where they are trimmed across the bottom so they are even, packaged, and sent for distribution.

Quality Control

Broomcorn is carefully graded so that the manufacturer understands the quality of the product that is shipped in the bale. Broomcorn is categorized by length and by color, with the brown-red broomcorn considered inferior. Inferior broomcorn may be used on the inside of the broom close to the handle and the operator ensures that the inferior product remains out of sight. Machinery

must be in good shape as well. Each individual broom-making machine is maintained, and the craftsman at each machine knows instantly when the machine is amiss. Other machinery such as the hydrostatic dryers or the seed removers are carefully monitored to ensure they perform efficiently. However, it is the broom makers themselves who are the key to monitoring quality of broomcorn brooms. Because the manufacture is completed using one operator per station who works from beginning to end on the product, he or she is sees and handles the product for nearly the entire process (except the sewing process). Each operator can tell whether the product has gone awry and can set aside such a broom so that it will not make it to a retail outlet.

Byproducts/Waste

There is little waste as the processing of the broomcorn and sotol occurs in Mexico. However, there are still seeds to remove and the shorter grasses captured within a bundle of grass is hauled away immediately. These grasses and seeds can be a fire hazard and do not stay long in the factory. (Broomcorn is a bit difficult to catch afire but once it begins burning it is difficult to stop.)

The Future

The broom model has changed little over the past 200 years. Today, there are brooms with synthetic fibers that attract dirt and dust. There are also brooms made of finer, polypropylene fibers with angled bristles. Smaller whisk brooms are also available as are brooms with easy to grip rubber handles. The Internet has brought broomcorn brooms to every home with easy ordering and delivery.

Where to Learn More

Books

Andrews, Edward Deming. *The Community Industry of the Shakers.* Albany, NY: The University of the State of New York, 1993.

Nylander, Jane. *Our Own Snug Fireside.* New Haven: Yale University Press, 1994.

Other

Broom Shop. http://www.broomshop.com (January 2001).

Organization of American States Website: NAFTA Dispute Settlement: Broomcorn Brooms. http:www.sie.oas.org (January 2001).

R.E. Caddy & Company, Inc. http://www.recaddy.com (January 2001).

—*Nancy E.V. Bryk*

Canal and Lock

The first canal designed to accommodate seaworthy ships was built by the Pharaoh Sesosteris I 4,000 years ago, linking the Red Sea to the Nile River and thus to the Mediterranean Sea. Remnants of this canal, which was alternately neglected and rebuilt half a dozen times, suggest that it was about 150 ft (46 m) wide, 16 ft (5 m) deep, and 60 mi (97 km) long.

Background

A canal is a man-made waterway. Canals are built for a variety of uses including irrigation, land drainage, urban water supply, hydroelectric power generation, and transportation of cargo and people. Navigation canals may be shallow facilities designed for barge traffic, or they may be deep enough to accommodate ocean-going ships.

To conserve water and to facilitate two-way travel, canals are built level. If there is a difference in elevation between the ends of a canal, the channel is built as a series of level sections linked by locks. A lock is a rectangular chamber with gates at both ends; with both gates closed, the water level within the lock can be adjusted to match the canal water level on either side. Thus, a vessel entering the lock can be raised or lowered in order to enter the next level canal section. Alternative types of locks are occasionally built to traverse large elevation changes; the entire lock chamber, containing boats floating on water, is hoisted vertically or moved up an inclined section of rail line.

In the United States, about 15% of the intercity freight (measured as a combination of distance and weight) is carried by water, on either artificial canals or navigable rivers. (Straightening or deepening a river to accommodate vessels of a certain desired size is called canalizing the waterway.) In 1997, 1.2 billion tons (1.1 billion metric tons) of cargo was transported on United States waterways; by 2020, the annual tonnage is expected to be twice that amount. Transportation of cargo by barge costs about half as much as moving it by rail and about one-eighth as trucking it.

The U.S. Army Corps of Engineers oversees some 25,000 mi (40,234 km) of commercially navigable waterways and about 240 lock chambers. Half of these locks are more than 50 years old, and some are too small for the large collections of barges that are commonly linked and moved by a single modern tugboat. With the expected increase in waterborne cargo transportation, it will be necessary to refurbish, replace, and enlarge many locks. Canal construction, which includes not only transportation waterways but also channels designed for environmental purposes, will also continue.

History

Canals

Reportedly, canals were built in Egypt as much as 6,000 years ago. The first canal designed to accommodate seaworthy ships was built by the Pharaoh Sesosteris I 4,000 years ago, linking the Red Sea to the Nile River and thus to the Mediterranean Sea. Remnants of this canal, which was alternately neglected and rebuilt half a dozen times, suggest that it was about 150 ft (46 m) wide, 16 ft (5 m) deep, and 60 mi (97 km) long.

The oldest canal still in use was begun in China in the fourth century B.C. for the purposes of collecting grain taxes and transporting troops. Rebuilt and enlarged about A.D. 600, the channel became known as the Grand Canal; its latest extension, completed about A.D. 1280, brought its length to 1,114 mi (1,795 km). It remains the longest navigable canal in the world.

For centuries, canals were dug using only hand tools. While building the Canal du Midi between 1665 and 1681 to link the Atlantic

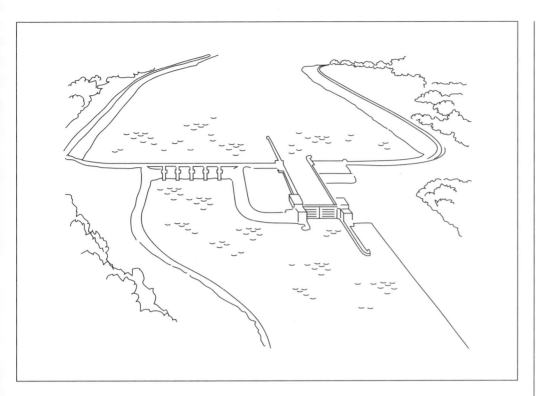

Ocean and the Mediterranean Sea across southern France, gunpowder was first used to blast a **tunnel**. The invention of steam-powered machinery at the end of the seventeenth century helped fuel the "canal eras" in Europe and North America in the eighteenth and nineteenth centuries during which thousands of miles of canals were built. Horse-drawn plows continued to be used especially for final trimming of the channel to desired slopes and depths. Full mechanization of canal construction was achieved in 1946, when an American company built the first canal trimming and lining machines.

Locks

Flash locks, the first attempts to carry boats over difficult elevation changes on rivers or canals, date from the third century B.C. Greek historian Diodorus Siculus, writing 200 years later, described how Ptolemy II had improved the Nile-to-Red Sea canal by building a type of lock. Consisting of a single gate, flash locks carried boats downstream on a rush of water; boats headed upstream could be pulled forward while floating on the torrent released by opening the gate.

Flash locks were hazardous, and they used large amounts of water. A significant im-

provement came in A.D. 984, when the first double-gate lock was built on China's Grand Canal. Also called a pound lock (because it impounds water) this was the predecessor of modern conventional locks. Its gates were panels that lifted vertically. A similar lock with vertical-lift gates built in the Netherlands in 1373 also controlled its water level by partially opening either the upstream or the downstream gate. Pound lock operation was greatly improved in 1485 when an Italian lock was built with smaller, valve-controlled openings in the gates.

Leonardo da Vinci invented the miter gate in 1480. Two gates, each more than half as wide as the lock, swing on vertical hinges. In the open position, they are flush with the lock's walls. In the closed position, they meet in a V pointed upstream so that the higher water level presses against them to promote a tight seal.

The base of the upstream gate in a lock is higher than the base of the downstream gate. If the lock was very deep, water filling the chamber through an opening in the upper gate would create turbulence or even swamp a boat in the chamber. This problem was solved in France during the seventeenth century, when valve-controlled water channels for filling and emptying a lock were built

A lock is a rectangular chamber with gates at both ends; with both gates closed, the water level within the lock can be adjusted to match the canal water level on either side. Thus, a vessel entering the lock can be raised or lowered in order to enter the next level canal section.

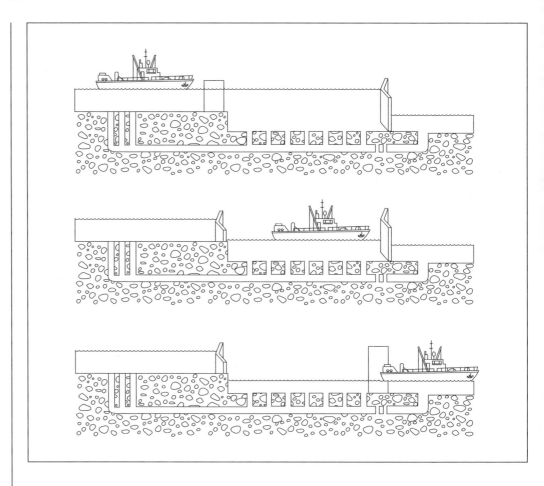

into the lower portion of the chamber's stone lining.

Until the early nineteenth century, lock gates were made of wood, and lock chambers were lined with wood, stone, bricks, or turf. In 1827, cast iron was first used to build both a lock and its gates in Cheshire, England. Following the development of the Bessemer process for mass production of steel, that material was utilized for construction of locks and gates, both as primary elements and as reinforcing bars for concrete.

Raw Materials

Waterproof linings keep a canal's water from seeping into the ground. For many years, the best choice was puddle, a mixture of sand, clay, and water that dried to a waterproof state. Modern materials and additives that are more durable include concrete, fly ash, bentonite, bituminous materials, and plastic sheeting.

Locks are usually made of concrete, occasionally lined with steel. If construction of the lock exposes bedrock, the floor need not be lined. The gates are made by welding together steel plates and reinforcement beams. The vertical edges of the gates are fitted with effective sealing materials such as white oak. In 1999, a French company developed lock gates made of glass-fiber reinforced plastic laminate mounted in stainless steel frames.

Design

Early canals followed the most level surface route possible because large-scale earthmoving was so difficult and expensive. Better excavation equipment and lock construction capabilities permit the construction of shorter, more direct canal routes. Because of geographic obstacles, sections of some canals are built in tunnels or on aqueducts (water-carrying bridges).

It is important to fill and empty a lock chamber while producing minimal water turbulence. Modern designs place sluices in the gate sills or in the chamber's floor or walls. A lock may also be equipped with a submerged bubbler that releases air below the gate closure area; the resulting gentle turbulence keeps the area clear of debris that might prevent the gate from sealing properly.

Various gate designs are available for modern locks, and different types may be used on the upstream and downstream ends of a single lock. Miter gates are one of the most popular choices. Another common choice is the Tainter gate, a curved plate that rotates vertically. In this efficient design, which is used for valves in water-transfer culverts as well as main lock gates, water pressure actually assists in the gate's rotation. Flat gates that slide up, sideways, or down into the lock floor are other options, as are flat gates that are hinged at the bottom and curved gates that rotate horizontally into wall recesses.

The Manufacturing Process

Canal

1 A thorough survey of the canal route must be made, not only for correct alignment but also for accurate depths of cuts and fills.

2 Primary excavation is done with earth-moving equipment like bulldozers and excavators.

3 A crawler-track mounted trimming machine is used to remove the final 12-18 in (30-46 cm) of soil to create the desired wall slopes and flat bed for the canal. Excavated soil is carried by conveyor and loaded onto trucks that will haul it away.

4 A layer of highly permeable material may be spread over the walls and floor to promote drainage of groundwater under the canal.

5 Grids of reinforcing steel are built and lifted into place on the canal walls and floor. Concrete blocks hold the grids above the surface so concrete can flow under them.

6 Concrete is mixed, vibrated to remove air pockets, and applied with a slipform machine, encasing the reinforcing steel. Slipforming means pouring concrete between the canal surface and a form (mold)

William Crawford Gorgas (1854–1920).

William C. Gorgas was born October 3, 1854, near Mobile, Alabama. In 1875 Gorgas received a bachelor of arts degree from the University of the South. Desiring a military career he decided to enter the Army with a medical degree. After graduating from Bellevue Medical College in New York and interning at Bellevue Hospital, Gorgas was appointed to the Medical Corps of the U.S. Army in June 1880. Tours of duty followed in Texas and North Dakota, with nearly 10 years at Fort Barrancas in Florida—Gorgas was assigned this yellow fever area because he had previously had the disease and was immune.

After Havana, Cuba, was occupied by American troops in 1898, Gorgas took charge of a yellow fever camp at Siboney, soon becoming chief sanitary officer of Havana. Acting on information that a certain mosquito carried yellow fever, Gorgas quickly destroyed that mosquito's breeding ground, ridding the city of yellow fever.

In 1904, work commenced on the Panama Canal. Gorgas went to the Canal Zone to take charge of sanitation, succeeding in Panama and Colón. Gorgas came to be generally regarded as the world's foremost sanitary expert, and a number of foreign governments and international commissions sought his aid. His book *Sanitation in Panama* quickly became a classic in the public health field. In 1914 he was appointed surgeon general of the Army, serving in that capacity until his retirement four years later. Gorgas died on July 3, 1920, and is buried in Arlington National Cemetery.

that slowly moves forward; the concrete hardens quickly enough to retain its shape by the time the form moves out of contact with it. Specialized machines place transverse and longitudinal expansion joints in advance of the slipform machine.

Lock

7 A temporary cofferdam is built around the proposed lock site. Steel sheet piles are driven into the ground to form a series of adjacent, vertical cells that extend above the waterline. The cells are filled with sand. Water is pumped out of the enclosed space to create a dry construction area. The lock site is excavated. If necessary, piles are driven into the ground to support the lock structure.

8 Wooden forms are built to shape the floor and walls of the lock. Space for culverts and valve chambers is included within the forms, as are slots for the gate hinges and recesses to contain open gates. Additional forms are built to shape the approach walls that will guide vessels into the lock.

9 After reinforcing steel cages are constructed in the forms, concrete is poured. When the concrete has cured (hardened), the forms are removed.

10 Control valves for filling and emptying the lock are installed, along with hydraulic and mechanical equipment for operating the valves and gates.

11 Gates are prefabricated and shipped to the site. Very large gates may be shipped in sections, which are welded together as they are installed in the lock.

12 Accessories such as guardrails, mooring posts, and escape ladders are installed on the lock walls.

Byproducts/Waste

Disposing of the excavated material is one of the challenges of canal and lock construction. It may be used to construct embankments or be spread over the surrounding countryside where it is carefully landscaped for erosion control as well as appearance. Construction of the Divide Cut, a 29-mi (46-km) canal on the Tennessee-Tombigbee Waterway, in the early 1980s required the disposal of 150 million cubic yards (115 million cubic meters) of dirt.

During lock construction, excavated material can be used to fill cofferdam cells. When the cofferdam is removed, the material may be used to fill in behind the riverbank side of the lock wall.

Lock operation uses large amounts of water; for example, it takes 3.5 million gal (13 million L) of water to fill a 600-ft (180-m) long, 110-ft (34-m) wide lock with a 7-ft (2.1-m) lift. On some canals and rivers, water supplies are limited and conservation is important. Rather than being released downstream as the lock is emptied, some of the water may be diverted to a side pond, where it can be stored and used to help fill the chamber for its next operation.

The Future

Because of the significant need to recondition or replace locks as well as build new facilities, the U.S. Army Corps of Engineers sponsors and conducts research on waterway construction. Scale models of proposed designs can be tested at its Waterways Experiment Station. Through its Innovations for Navigation Projects (INP) Research Program, it supports the development of new ways to build, repair, and operate canals and locks. Recent topics included improvements in underwater concrete and grout placement and development of low-density, high-strength concrete for building modular lock sections that can be installed without construction of a cofferdam. Other current research topics involve protection devices for lock walls and gates from barge impacts, improved designs for water intake and discharge systems, better equipment and control devices for operating gates, and techniques for enlarging (rather than replacing) existing locks that have inadequate capacity.

Where to Learn More

Periodicals

"Pakistan Project Requires World's Largest Canal Machines." *Engineering News-Record Magazine* (December 20, 1999): B-3.

Other

"The ABC's of Lock Building." Trent-Severn Waterway. http://collections.ic.gc.ca/ waterway/rg_eng_i/abc.htm (May 16, 2000).

"Canal Machinery." Guntert and Zimmerman. http://www.guntert.com/Canal/Canal. htm (May 26, 2000).

International Canal Monuments. http://www. icomos.org (May 17, 2000).

"Lock Construction Project." U.S. Army Corps of Engineers. http://www.lrl.usace.army.mil/olmsted/cd_pics/960517ad.htm (May 26, 2000).

Tennessee Tombigbee Waterway. http://www.tenntom.org (May 10, 2000).

—*Loretta Hall*

Candy Corn

One candy company alone estimates it produces 4.3 billion pieces a year and that Americans eat about 20 million lb (9.1 million kg), or about 8.3 billion kernels.

Background

Candy corn are the small pieces of triangular candy made primarily from corn syrup, honey and sugar (it is usually fat-free) and is traditionally colored in a specific pattern of three stripes. It is recognized by the white tip, orange in the center, and yellow at the widest end.

Candy corn has been remarkably popular in America for well over a century. The amount of these sweet kernels consumed in a single year is staggering. One candy company alone estimates it produces 4.3 billion pieces a year and that Americans eat about 20 million lb (9.1 million kg), or about 8.3 billion kernels.

Candy corn is considered a "mellow creme," a candy that has virtually no oils or fats in it but has a marshmallow flavor. It is also of a variety of candies that are made using the process confectioners refer to as starch casting, in which the candy is formed in a machine using cornstarch as the molding agent. This particular kind of cooked candy must set for at least a day before it may be packaged for sale.

History

Candy corn has been around for over a century. Some believe that it was homemade at one point, but others think that it has largely been mass-produced since its invention in the late nineteenth century. No one is quite sure who invented these little morsels. However, it is believed that Americans knew of the candies in the 1880s. By the turn of the century, the Goelitz Candy Company in Illinois, run by German immigrants, was making significant quantities of the confection.

In the early twentieth century, candy corn became the company's single best seller. Goelitz is now not the sole candy corn producer in the country, nor is it the largest. No matter who makes candy corn, the process has remained largely unchanged. Originally, candy slurry was cooked as a fondant and poured, from large ladles or buckets, into separate triangular-shaped molds. Each color was poured into the mold separately. The colorful kernels were allowed to dry before packaging. Now, however, the process is entirely mechanized.

Major recent changes have included the broadening of the use of candy corn to other times of the year. Thus, these candy kernels take on color combinations suited to the time of year they are slated for consumption—red, pink, and white for Valentine's Day, and green and white for St. Patrick's Day, for example. Green and red candy corn is referred to as reindeer corn and is becoming a popular Christmas treat.

Raw Materials

Candy corn ingredients vary with the manufacturer. However, the most important ingredients for the production of candy corn include corn syrup and sugar. Gelatin and soy protein are added to produce a firm-bodied candy. Also often used in the production of candy corn is salt, honey, artificial flavors and colors, and a confectioner's glaze of oil and wax that gives the candy a sheen. Cornstarch, essentially a corn flour, is an extremely important part of the molding process as the candy is injected or squirted into molds made of damp cornstarch. However, the cornstarch is simply the molding agent and does not become part of the candy itself.

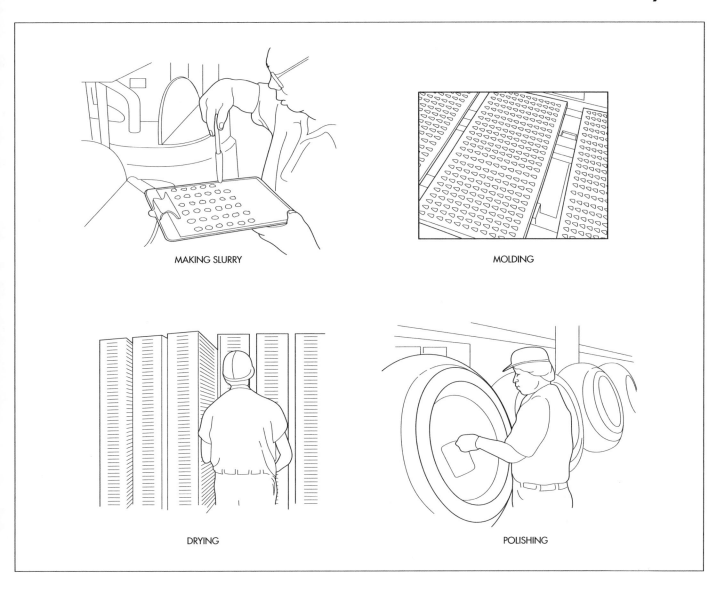

MAKING SLURRY

MOLDING

DRYING

POLISHING

The Manufacturing Process

Candy corn is manufactured using a process referred to by confectioners as starch casting. In this process, the shape of a candy or a candy center is formed by making impressions in a powder called cornstarch. The filling of each of these separate impressions is filled with liquid candy. Starch is an effective material because it easily retains specific shapes. Cornstarch also helps to remove moisture from the candy as it dries. Much of the candy making process, including the starch casting, occurs within a special candy-making machine called a Mogul. The candy is made and then must be left to dry. Thus, it is at least a 24-hour process from the beginning to end of the production process. Occasionally, it may be as long as

48 hours, depending on the moisture of the batch of candy and ambient humidity within the factory.

1 A stacker puts trays into the Mogul via conveyor belts. The Mogul puts cornstarch into these plastic trays. This cornstarch is specially treated so that it adheres to itself and is conducive to molding. Then a leveler, made of plastic, brushes over the top of the trays and levels off the cornstarch that had been placed within the trays.

2 The plastic trays advance to the dye within the Mogul. Here, several hundred triangles leave hundreds of impressions in the cornstarch in these trays. The liquid candy is prepared in three separate colors in three separate candy kitchens in large kettles in which the sugar, corn syrup, salt, honey,

A candy-making machine called a Mogul deposits the slurry in corn starch molds. Each colored section of candy corn is deposited at different stages—first, the white tip;then, the orange stripe; and finally, the yellow stripe at the wide end.

soy protein, gelatin, and flavors are beaten together with mixers. This slurry is cooked to specific temperatures and is then delivered to the depositors separately so the colors are not inadvertently mixed.

3 Next, the Mogul is ready to deposit liquid candy into each cornstarch impression. Each separate cornstarch impression has a nozzle into which liquid candy is loaded and deposited. Thus, if there are 300 triangular molds, there are 300 nozzles above the tray ready to squirt in the liquid slurry. The candy is deposited in three phases. First, the nozzle injects the white tip into the top end of each triangle. Then, it deposits the center, or orange stripe against the white tip. Finally, the depositor finishes off the candy corn with the yellow stripe at the wide end.

4 The filled trays of wet candy corn are now conveyed to the end of the Mogul and automatically stacked. When there are enough trays stacked together, they are moved from the Mogul and sent to the "dry rooms" to set.

5 These wet candy corns must have the moisture removed from them in order to be separated easily from the molds. Thus, the trays sit in the dry rooms between 24-36 hours, depending on the batch of candy and the weather or humidity surrounding the factory. The longer the candy sits in these rooms, the drier they become. Moisture levels are tested to ensure the candy corn includes just the right amount of water.

6 The candy corn, still in the trays, is conveyed back to the Mogul. This machine then turns the trays completely over, dumping the candy and the cornstarch out of the trays. Then, the candy and the cornstarch are separated from one another. The candy is sent to the next processing point. However, the cornstarch is transferred to a sifter that removes any bits of candy corn, then is sent to the drying drum, which removes the moisture from the starch. The corn starch is sifted a final time, then is sent back to the Mogul for use in molding other candies.

7 The candy corn is rather dull-looking and lackluster at this point and must be polished to give it an appealing sheen. The candy corn is conveyed to hollow globe-shaped vessels called polishing pans into which confectioner's glaze (made of oil and wax) is added. The candy is then sent to large revolving drums. As the drums revolve, the pieces of candy corn gently rub against each other, so each piece is polished as it tumbles, much like a rock tumbler.

8 After the candy is polished to a bright sheen, the candy is ready for packaging. It is removed from the polishing apparatus and sent to the packaging machines, which automatically weigh the candy, put it in bags, seal each bag, and put the candy into cases that may be shipped.

Quality Control

Everyone within a candy company is responsible for quality control. There is, of course, the human eye and human hand that looks at the product after it is dumped from the trays before polishing and discards the misshapen pieces. Next, machinery is meticulously maintained because if one part breaks down, the entire mechanized process is completely set back. The Mogul, a large and complex starch casting machine, receives great care and inspection.

All materials are visually inspected before accepted for production, but they are also examined using microtests by microbiologists within the company. They are carefully checked for dangerous contaminants and health hazards such as *E. coli*, salmonella, and staph. The liquid candy slurry is also checked for several properties. The color of each of the three candy batches must be just right because batches that are not of the correct color are rejected for depositing within the corn starch molds. The slurry is also checked for density, weight, and viscosity so that the candy moves to the depositor easily, is deposited easily, and will set correctly. Similarly, the melting point of the sugar is monitored. Finally, the moisture of the drying candy corn is carefully monitored so that the candy is neither too hard nor too soft.

Byproducts/Waste

There is virtually no unused material or scrap left after the production of candy corn. Most significant, the cornstarch is completely sifted out separately from the product, dried, and sifted again so that it may be quickly reused. Generally, any candy slurries that are not of the correct color are easily corrected. Candy corn that is somehow misshapen or considered inferior may be melted down and reused. Of course, any candy that falls on the floor is never reused.

Where to Learn More

Other

Herman Goelitz, Inc. "Jelly Belly Factory Tour." (2000). http://www.jellybelly.com (December 2000).

Herman Goelitz Inc. "The Goelitz Family: Candy Corn & Jelly Belly." German American Corner (1996-2000). http://www.german heritage.com/biographies/atol/goelitz.html (December 2000).

Inglis, Margaret. "Candy Corn, Your Friend and Mine." *CuisineNet Café* (October 22, 1997). http://www.cuisinenet.com/cafe/you_ gonna_eat_that/1997/00008-1.shtml (December 2000).

National Confectioners' Association. http://www.candyusa.org (December 2000).

—Nancy E.V. Bryk

Caramel

Milk is the essential ingredient that distinguishes a caramel from a hard candy. It is the milk solids that chemically change to produce the caramel.

Background

Caramel is often eaten as little brown, sweet, buttery nuggets wrapped in cellophane, but it is also delicious in candy bars and on top of fresh popcorn. The best caramels are sweet and just a bit chewy. Caramels can, in fact, have a variety of textures. Caramel manufacturers use the term "short" to characterize a caramel that is too soft (perhaps too moist) or "long" for a caramel that is quite chewy. Caramels are, in some ways, rather similar to other candies in that the basis for candy is generally sugar, corn syrup, and water. However, caramels vary in an important way in that they also contain milk and fat. While hard candies are plastic or malleable at high temperature, but glass-like (clear and easily cracked) when cooled, caramels are plastic at both high temperature and room temperature. Caramels are softer because they have been cooked to a lower temperature than hard candies (to approximately 245°F [118°C], or the firm ball stage) and contain more moisture. Because of this soft texture, caramel may be extruded at lower temperatures, inserted into a mold, and put into a variety of other candies or candy bars to add flavor, binding, and texture.

What makes a caramel a caramel? The action of the heat on the milk solids, in conjunction with the sugar ingredients, imparts a typical caramel flavor to these sweets. Essentially, the entire batch of candy undergoes a chemical reaction referred to by chemists as the Maillard reaction. In a conventional caramelization process, the sugar syrups are cooked to the proper moisture level, added to the fat and milk, heated, and then allowed to caramelize (develop the characteristic flavor and brown color) in a browning kettle. The confectioner can watch the chemical reaction take place in the kettle as the batch turns from a milky white color to rich brown. The nose can smell the slight burning of the milk solids, too—and a pleasant odor it is. If cooked even further, to about 290°F (143°C), the mixture essentially becomes toffee, a hard-crack caramelized candy.

There is no question that chocolate is a wonderful ingredient in candies, but what would a Snickers bar, a caramel apple, or Milk Duds be without caramel? If not used in a bar, the caramel batch may be poured into a pan, scored, and cut apart in squares for plain consumption. Vanilla caramels, the type most frequently eaten, are flavored with vanilla; chocolate caramels have a bit of chocolate added to the batch, turning it a deep brown color. However, maple caramels, those with molasses and brown sugar, and cream caramels are other delicious varieties. Most Americans have only tasted mass-produced caramels. However, many small confectioneries are springing up that manufacture caramels in fairly small quantities, making them gourmet treats. Caramels are fairly easily made at home as well.

History

It is difficult to know when humans first craved the sugar that gave them that extra bit of energy and satisfied their sweet tooth cravings. Many believe that the earliest sweet treat was honey—simple to acquire and needs no processing. The Arabs and the Chinese prepared candies of fruits and nuts dipped in honey. But during the Middle Ages, refined sugar of any kind was very expensive and a rare treat. Even in the New

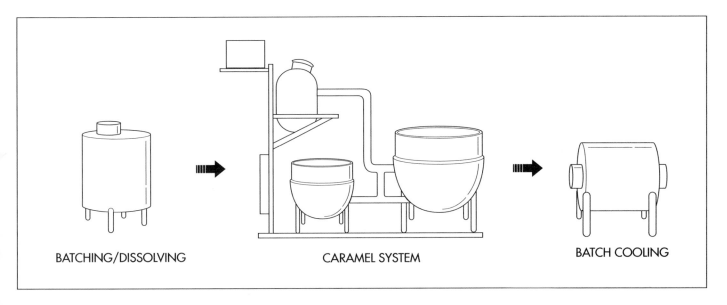

BATCHING/DISSOLVING   CARAMEL SYSTEM   BATCH COOLING

During the caramel production, ingredients are batched, machine mixed, cooked steadily, cooled, extruded, and formed into small caramel squares.

World sugar was an expensive commodity, and refined sugar was purchased in a cone or a loaf and pieces were carefully cut off with sugar nips. But by about 1650, Americans were boiling water and sugar and making hard candies in deep kettles in the fireplace. Surely someone added fat and milk to these concoctions and made caramels.

By the early nineteenth century, Americans used sugar beet juice to make new candies. Still, hard candies were the primary confections. By the mid-1800s, there were nearly 400 American candy manufacturers that were producing primarily the hard candies often sold in general stores—they were cheap to make, easy to transport, and did not spoil easily. Caramels were made at these small confectioneries as well. In fact, Milton Hershey began his chocolate empire not with chocolate, but with caramel. Hershey was born in 1857 in Pennsylvania, and rather than become a printer, he founded a candy-making business in Lancaster, Pennsylvania. By 1886, he had founded the Lancaster Caramel Company, surely utilizing traditional recipes that were found in many regional cookbooks. He learned about chocolate-making because he sought new coatings for his famous caramels. Other long-lived candy companies were founded upon caramels, including Goetze's Candy Company, which began in 1895 and is still going strong making cream caramels, among other things.

Originally, the production of caramels occurred using copper pots over direct gas flames, watched carefully by a master confectioner who used a candy thermometer to monitor the temperature, and poured out the cooling caramel batch onto a marble slab or a water-cooled table and scored it into squares. The heavy, deep candy kettles (that some gourmet caramel-producers still swear by) have given rise to batch cookers with vacuum systems for quick cooling of the caramel syrup that run with little assistance from a machine operator. Brach's Confections, Inc. is among the largest caramel manufacturers in the country, with caramels being a staple of their output.

Raw Materials

The raw materials vary with the manufacturer and type of caramel under production. However, the most frequently made caramel, the vanilla caramel, contains many ingredients if it is mass-produced. The ingredients include milk, sometimes sweetened condensed milk, corn syrup, sugar, oil, whey, calcium carbonate, salt, flavor, butter, another type of fat such as vegetable oil, molasses, and corn starch. Milk is essential to distinguish the caramel from a hard candy, and it is the milk solids that change chemically to produce the caramel. Corn syrup lends additional sweetness to the candy batch but also keeps the mixture from becoming grainy, which would indicate there is too much sugar in the batch (graininess will ruin a batch of caramels). Corn syrup also lends body to the slurry. At least one fat is added to the mixture as well. Butter is often the only fat added by gourmet caramel-makers as it provides superior taste, but this proves to be very expensive for mass-production. So other fats are added along with a fairly small amount of butter. As maple caramels or other flavored caramels are produced, the ingredients vary accordingly.

The Manufacturing Process

There are a number of different caramel-making systems used for the mass-production of these candies. The process described below utilizes one of the many different systems for this purpose. The process is essentially the same, however—the batch is machine mixed, cooked steadily, cooled, extruded, and formed into small squares.

1 All of the ingredients listed above are automatically batched and weighed using a batching and dissolving machine made expressly for the manufacture of caramels. The liquid and the dry ingredients are loaded into the machine. Then, the ingredients are weighed with great precision by computer in the upper weigh tank. The ingredients are mixed by propellers in this upper weigh tank.

2 After blending, this milky slurry drops automatically to the lower mixing and dissolving tank. Steam heat brings the mixture up to a pre-determined temperature. In the mix tank, gear-driven agitation equipment dissolves the ingredients thoroughly. Surface scrapers skim along the bottom and sides where burned protein solids have a tendency to accumulate. These burned solids are redistributed and mixed back into the slurry to ensure that the whole mixture is a homogeneous batch.

3 The heated mixture is then sent to the heated surge tank. An operator's command transfers the batch into a stainless steel scraped surface heat exchanger for the final evaporation. Here a small variable-speed gear motor drives a scraping system within the evaporator. The syrup is forced through a small space that is jacketed with steam, thus forcing evaporation within the mixture.

4 The batch has now had much of its moisture removed and is thickening. It is gravity-fed into a steam-jacketed caramelizing tank where the caramelizing is ensured by exposing the batch to steam. The caramelizing mixture is re-circulated from the bottom to the top of the tank, with incoming syrup mixing with the caramelizing slurry, promoting homogeneity of the product. A discharge valve directs the mixture into the next processing machine, the cooling wheel.

5 This caramel candy, now at about 240°F (116°C), has to be cooled. There are many ways to cool the caramel, including moving it into cool rooms and running it through cooling tunnels. The system described above utilizes as cooling wheel. The caramel is water-cooled on the outside surface of a large wheel that is four feet wide and twelve feet in diameter. The caramel is laid in a film about 1/8-inch (3.2-mm) thick on this wheel. The wheel completes a half-turn and the caramel comes off the wheel, becoming solid and of a consistency so that the candy may be cut and packaged.

6 A batch roller takes the caramel film and shapes it into a rope. The rope is then shaped and sized into the thickness of a finished caramel. Caramels are not molded; instead, they are shaped by being cut from the thick rope. As the caramels are cut they are automatically individually wrapped. From there, the caramels may weighed and placed in a sealed bag and packed into cartons for shipping. If caramels are cooked to just the correct temperature, they can be shipped easily in any type of weather and will hold their shape. If they are undercooked just by a few degrees, they may do poorly after packaging and become too soft.

Quality Control

The machinery involved in the process of candy making is automated. The making of caramels requires precise measurements of ingredients, since too much sugar makes the candy grainy (the sugar does not entirely dissolve in the liquid) and makes it an inferior product. If there is too much moisture in the product, the caramel will be too gooey in warm weather. Too little moisture and cooked at too high a heat, and a "long" or chewy caramel is the result. So, the machinery must be very carefully checked and calibrated for accuracy in the mixing and weighing of materials. Temperature controls, too, must be extraordinarily accurate, since just a few degrees can affect the consistency of caramels. Human operators on the floor use their eyes and hands in order to maintain quality. Master caramel-makers are essential to the production of gourmet caramels, made in smaller batches of 30-50 lb (14-23 kg) at a time. Their experience can detect any slight variation that may result in an inferior batch just by the look, smell, and feel of the batch.

As with all food manufacture, the quality of all consumable ingredients must be checked for quality. Corn syrup must be of the high quality needed for this candy manufacture. All other ingredients must be tested for quality as represented by the suppliers.

Where to Learn More

Books

Rombauer, Irma S., and Marion Rombauer Becker. *The Joy of Cooking*. Indianapolis: Bobbs-Merrill, l953.

Other

Hershey Foods Corporation. http://www. hersheys.com (December 13, 2000).

National Confectioners Association. http:// www.candyusa.org (December 13, 2000).

—*Nancy E.V. Bryk*

Cherries

Background

Cherries may be either deliciously sweet and deep brown-red, or quite tart and bright red. The two most common are the sweet cherry, *Prunus avium L.*, and the sour (often referred to by growers as the pie or tart) cherry *Prunus cerasus L..* Sour cherries have a lower sugar content and a higher acid content than its sweet counterpart. Not surprisingly, sour cherries are slightly less caloric than sweet cherries, containing about 60 calories per 3.5 oz (100 g) portion compared to 80 calories for sweet. Cherries are high in vitamin C, carbohydrates, and water, and include trace amounts of fiber, protein, vitamin A, vitamin B_1 (thiamin), vitamin B_2 (riboflavin), niacin, calcium, phosphorus, iron, and potassium.

Cherries are found in the wild and have been domesticated for centuries. There is a myriad of cherry types, resulting from new varieties and hybrids developed for hardiness and flavor. This fruit is found in Asia, Europe, and North America, with Iran, Turkey, United States, Germany, and Italy leading in the production of cherries. Together, 10 countries produce over 1.1 million short tons (over one million metric tons) of cherries annually.

Cherry trees offer products other than the fruit itself. The lovely, fragrant cherry blossoms are a rite of spring and are actually a tourist draw in places such as Washington, DC, and Door County, Wisconsin. In addition, parts of the tree itself have long been used for medicinal purposes. The bark, leaves, and seeds of the cherry trees contain cyanogenic glycosides—poisons that are lethal if ingested by children or animals. Native Americans and others use the leaves and carefully prepare teas with them for the treatment of colds or coughs. Others have experimented with cherry stalk tea in the treatment of kidney diseases. The cherry has also been associated with virginity from ancient times to the present day. The association may be derived from the fact that the red colored fruit that encircles a small seed symbolizes the uterus of Maya, the virgin mother of Buddha, who was offered fruit and succor by a holy cherry tree while she was pregnant.

History

The sweet cherry originated in the area between the Black and Caspian Seas in Asia Minor. It is likely that bird feces carried it to Europe prior to human civilization. Greeks probably cultivated the fruit first. Romans cultivated the fruit as it was essential to the diet of the Roman Legionnaires (their use likely spread the fruit throughout Western Europe). It is believed that English Colonists brought the fruit to the New World prior to 1630, but they do not seem to have flourished in the eastern United States. Spanish Missionaries brought sweet cherries to California, and varieties were brought west by pioneers and fur traders as well. Sour cherries also are native to Asia Minor, and were brought over to the New World by settlers rather early as well.

Today, the United States probably produces more tart cherries than sweet because the former are easier to grow. They are simply less fussy and are affected less by bad weather. Thus, they flourish in greater numbers. Now, cherry growers are able to purchase a variety of cherry types that best suit the soil and climate in which they operate.

Cherries are mechanically harvested using a tree shaker that knocks the fruit from the tree. Once picked, cherries are quickly processed and sent to market or to be frozen, canned, or used in other products.

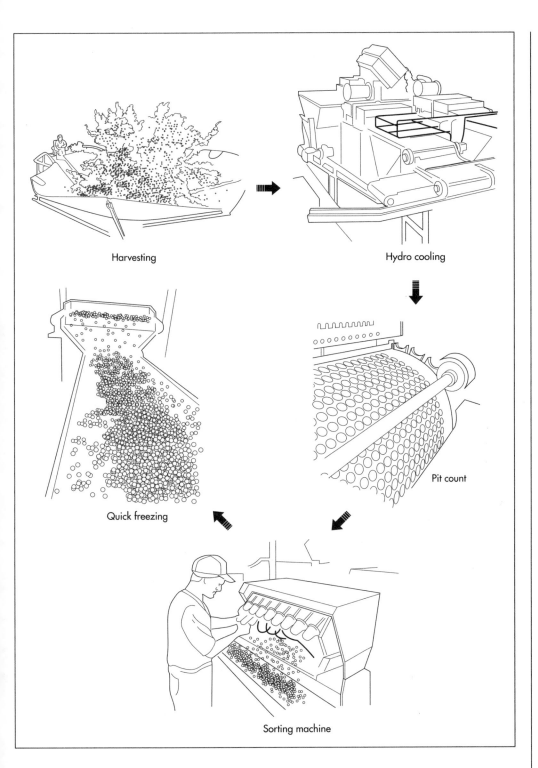

Harvesting

Hydro cooling

Pit count

Quick freezing

Sorting machine

New cultivars (cultivated varieties) of both sweet and sour cherries are being developed that are hardier than older varieties; German varieties are proving to be extraordinarily successful for cultivation in this country.

Raw Materials

Generally, cherries flourish in deep, well-drained, loamy soils. Cherries require cooler climes rather than hot ones because they must be chilled for about 1,000 hours annually. The cherry trees bloom relatively late in spring, so frost is less of a hazard for this stone fruit than others such as peaches or apricots. However, too much frost late in the spring may adversely affect cherry production. The clime must be one that does not have excessive rain during harvest since too much rain at that time can cause the fruit (particularly sweet cherries) to crack. Tart cherries are a bit easier to cultivate and are more tolerant of frost as well as humid, rainy weather. The relative ease with which tart cherries are grown may be one reason why so many are grown in the United States.

Trees of good stock are also necessary for successful cultivation of cherries. It is imperative to acquire stock through tree nurseries that are suited for the soil and climate of the grower's region. Bees, however, ensure that the cherry trees flower and ultimately produce fruit, and are an extremely important ingredient in the cultivation of cherries. Bees are usually brought into the tree orchard in the spring as the flowers first bloom in order to distribute pollen so that the fruit blossoms. Bee hives are generally rented by cherry growers each year. It remains imperative that fertilizers are applied to domesticated cherry trees via foliar (leaf-applied) feedings. Pesticides and fungicides are applied before harvest to deter diseases and pests.

The Production Process

Soil preparation

Different varieties of cherry trees flourish in slightly different soils. Generally, cherries prefer a moderate pH of 6 or 7. Most orchard owners periodically test the soil to ensure the pH is near that mark and may add special fertilizers to treat the soil. Extensive use of fertilizers may encourage vigorous growth but may retard blooming and fruit bearing, so cultivators must carefully assess their use of fertilizers.

1 Root stocks are carefully chosen by cherry growers for their lineage and compatibility with the soil and climate of the orchard. Lineage, as one grower puts it, means that the stock is from healthy, dependable trees from reputable fruit nurseries. There is much contention about the most dependable root stocks for both sweet and tart cherries. A new root stock from Germany (a significant source for cherries) named Gisela allows production from dwarf trees with high yield efficiency and fairly early production.

2 Some varieties of cherry trees, particularly those of the Pacific Northwest, do not naturally produce many branches. Thus, the center of the tree may be dense with a central limb. It is therefore essential for growers to prune the trees regularly so that all the flowers (and ultimately, fruits) receive the amount of sunlight and air circulation required for fruit production. This pruning may be done prior to harvest, after harvest, or at both times. Some growers are experimenting with ways to encourage branching (which still may require pruning). These trees must be carefully maintained. It takes five or six years for sweet cherry trees to produce fruit, with maximum yields obtained at about that time. Sweet cherry trees produce fruit for up to 30 years. Tart cherry trees produce fruit after about three years, and produce fruit for 20 to 25 years.

Fungicides and insecticides

The schedule for applying fungicides and insecticides may vary from orchard to orchard. Some growers apply the first fungicides at floral bloom in spring to prevents leaf spot. Insecticides to keep off bore worms and/or other insecticides may be applied every two weeks or so until harvest.

3 Bees must pollinate the flowers. Just as the trees begin to blossom, cherry growers let bees loose in order to distribute the pollen so that fruit will blossom. The flower must be pollinated in order for the tree to bear fruit. Bees may be set in alternate rows to ensure pollination. Generally, 25-50% of flowers must set fruit each year in order for the crop to be commercially viable.

4 It is approximately two months from flower to fruit. As the fruit ripens, growers hope for no frost and just the right amount of rain—too much rain will crack and damage the sweet cherries. Maturity is gauged by a variety of means, and may vary by grower. Traditionally, color has been a key indicator. Growers are increasingly moving toward determining fruit removal force—the easier it is to remove the fruit, the more mature it is. This maturity is measured by a pull gauge that pulls the fruit from its pedicel. Just before harvest, some growers who use tree-shakers to shake the cherries off their stems apply a spray that makes it easier for the cherries to drop off the tree.

5 Both sweet and tart cherries intended for processing are shaken from trees when ripe. Tree trunks are shaken by a machine that forces the cherries from the tree; it takes just five seconds to drop the fruit from the tree using a shaker. The fruit drops onto a cloth or plastic cover so that it can be easily gathered. (Tree shaking is an ordeal for the grower as well as the tree—the machines are very expensive and if the shaking is done incorrectly, the machine vibrations may damage trees, particularly young trees.) However, sweet cherries that are to be consumed fresh are laboriously hand-picked and carefully boxed for prompt sale.

6 Cherries are now ready to be processed into consumer or retail produce items. Cherries that are to be processed (canned, dried, or frozen) are quite delicate and are easily bruised. They also have a short shelf life, so they must be processed immediately. Tart cherries shaken from trees are immediately plunged into cold water and conveyed to processing plants, where they are washed, de-stemmed, pitted, and packed for freezing within hours of harvesting. Sweet cherries picked for fresh consumption may be hydro-cooled or dumped into cold water by pickers, then packed in shallow flats after being sorted based upon their size and color. Sweet cherries are then immediately shipped out, since their shelf life is just two weeks. They are still prone to brown rot and a variety of molds during this time.

7 For many growers, there is little preparation of the trees for the winter. After harvesting, another spray is applied to foliage to prevent harmful leaf spot. Pruning of limbs and branches often happens after harvesting as well. Other than that, the trees are left unprotected. Cherry trees, like most trees, prefer a fall that gradually gets colder rather than one that is very warm and then very cold suddenly. The gradual cooling of the tree is called "hardening off" so that the tree is eased into the cold weather.

Quality Control

Cultivating a commercially viable cherry crop has many components. First, the soil pH and nutrients must be tested frequently (generally by a state university extension service) so that foliar fertilizers meet the requirements of the trees. Generally, growers keep a record of these soil tests. Second, the grower must understand the climate and soil types well enough to choose root stock that will flourish in that area. Third, pesticides or insecticides must be very carefully mixed and applied according to U.S. Environmental Protection Agency (EPA) standards, recommendations of state university extension services, and the product label. Fourth, pollination of the cherry blossoms is absolutely imperative; if there are few bees in the area, growers must rent bees for this purpose. Fifth, the trees must be carefully shaken during harvesting (if the cherries are to be harvested) so that the tree is not irreparably damaged. Finally, vigilant pruning and assessing the amount of air and sunlight densely packed trees receive is imperative for large yields.

The Future

Perhaps the biggest issue looming for the cherry industry, which is fiercely independent and highly competitive, will be federal regulation of the crop (as other crop-growers are weaning themselves from these regulations). Tart cherry crops have been particularly problematic in the last several years. A bumper crop of tart cherries has resulted in exceedingly low crop prices (tart cherries are less affected by the vagaries of weather than sweet cherries and can be harvested in huge quantities). Several years ago the market was so saturated with tart cherries in Michigan that some growers were receiving five cents a pound for the crop, far below the twenty-cents per pound needed to break even. Federal regulations could establish the amount of cherries that may be offered for sale at market. Excess cherries may be frozen or stored, or given to charity. Some growers are trying to find ways to utilize these tart cherries in ingenious ways. A Michigan cherry grower recently combined lean ground meat with tart cherry pulp, resulting in a lean and tasty meat that appealed to the health-conscious. Others have turned to gourmet foods such as dried cherries, yogurt-covered cherries, or have developed specialty cereals in order to utilize the abundance of tart cherries.

Other issues involve the land upon which the cherries are grown. The cultivation of cherries is very labor-intensive and subject to the weather. Equipment is expensive, too; a cherry shaker alone may cost $175,000. Younger generations increasingly are unwilling to manage the family cherry orchard, realizing that much hard work may not even pay off in profits. Even established cherry growers are wondering if the work is worth the prices and uncertainty. In addition, many of these orchards are located in lush, lovely areas, and taxes on the prime parcels of land are putting some of the growers out of business. Families are deciding that it is not worth running the business, and are selling orchards that will be plowed under to make way for new housing.

Where to Learn More

Periodicals

Flesher, John. "State Cherry Growers Plot Strategies to Resurrect Their Troubled Industry." *Detroit News* (January 2, 1996).

Herzog, Karen. "Times, Taxes Shake Smaller Growers Out of Business in Door County." *Milwaukee Journal Sentinel* (August 15,1999).

Other

California Cherry Advisory Board. http://www.calcherry.com (December 2000).

Cherry Marketing Information. Growers' Info. http://www.cherrymkt.org/growers/growers.html (March 2000).

—*Nancy E.V. Bryk*

Chess

Background

Chess is a classic two person board game. It is played with specially designed pieces on a square board made up of 64 alternating light and dark squares arranged in eight rows and columns. First appearing around A.D. 600, the game steadily evolved into the modern game known today. The earliest methods of production involved carving the board and pieces out of wood or stone. Today, a variety of common modern manufacturing methods such as injection molding and lithographic printing are employed to mass produce thousands of games.

The objective of the modern chess game is to force the opponent's most important piece, the king, into checkmate. This is a position in which the king cannot be moved to avoid capture. The player with the white pieces begins the game by moving a piece to another square following the rules that govern piece movement. The players alternate moves until one player is either checkmated, resigns, or there is a draw. Thousands of books have been published relating to the strategies during the three key stages of chess, including the opening, the middle game, and the end game.

History

While the exact time and place of chess's origin is debated, most scholars believe it was developed sometime around the sixth century A.D. It is a descendant of a game called *chaturanga*, which was commonly played in India during that time. (Chaturanga is derived from a much older Chinese game.) The name chaturanga is a Sanskrit word that refers to the four divisions of the Indian army, including elephants, chariots, cavalry, and infantry. These pieces became the basis for the four types of pieces in the game. Two of the key similarities between chess and chaturanga is that different pieces have different powers and victory is based on what happens to the king.

During subsequent years, chaturanga spread throughout the Middle East, Asia, and Europe. Chaturanga was introduced to China around A.D. 750 and then to Korea and Japan by the eleventh century. In each of these places, it took on different characteristics. For example, Chinese chess has nine files and 10 ranks. It also has a boundary between the fifth and sixth ranks, which makes it a slower game than the Western version. In Persia, the game was called *shtranj* and it was in this form that it was introduced to Western Europe when the Moors invaded Spain. By the tenth century, the game was commonly played throughout Europe and Russia..

Shtranj caught the interest of philosophers, kings, poets, and other nobility, and eventually became known as the "royal game." The best players wrote down the moves of each of their games. This practice eventually led to the development of puzzles in which the solver had to find solutions, like finding checkmate in a specific number of moves. During the fifteenth century some significant rule changes were made. For example, castling was introduced, as was the initial two-square pawn advance. One of the most important changes was the transformation of the counselor piece into the queen, the strongest chess piece. These improvements helped make the game popular throughout Europe. Some of the best players during this time—Ruy Lopez and Damiano—put together chess instruction books

First appearing around A.D. 600, chess steadily evolved into the modern game known today.

Chess pieces and a chessboard
set up.

Pawn   Rook   Knight   Bishop   Queen   King

that also helped to make the game more widely accepted.

The rules and piece design steadily evolved, reaching the current standard during the early nineteenth century. In the twentieth century, chess experienced a tremendous growth in interest resulting in the development of various chess organizations and the crowning of a world champion. The first computer chess program was introduced in 1960. Steady improvements in technologies and algorithms led to the 1996 defeat of the world champion, Garry Kasparov, by a computer called Deep Blue.

Design

Historically, the game's pieces have been both simple and highly decorated. Prior to A.D. 600, the pieces were plain. These were replaced by detailed sets depicting royalty, warriors, and animals. From the ninth to the twelfth centuries, Islamic rules prohibiting the depiction of living creatures resulted in basic pieces made from clay or stone. This change is actually thought to have increased interest in the game at the time because it made sets more widely available and was less distracting to the players. When the game spread to Europe and Russia, highly ornate sets were fashionable.

The standard set for modern chess pieces was introduced by Nathaniel Cook in 1835. His set was patented in 1849 and endorsed by the leading player of the day, Howard Staunton. Staunton's promotion of the set as the standard led to it being known as the Staunton pattern. Today, only Staunton sets are allowed in official international competitions.

A typical chess set has 32 pieces. These are broken down into two sets of 16 pieces each. In each set there are eight pawns, two rooks, two knights, two bishops, one queen, and one king. The different pieces are distinguished by their appearance. The designs vary from simple plastic shapes to intricate, hand-carved statues. While piece size varies depending on the specific set, the tallest piece is typically the king, followed closely in height by the queen. The shortest, least notable pieces are the pawns. The rook has varied considerably over the years, being represented as a ship, castle turret, or a warrior in a chariot.

The chess board is square and made up of 64 alternating light and dark squares arranged in eight rows and columns. The vertical columns extending from one player to the other are known as files. The opposite rows are called ranks.

An important aspect of large scale chess piece manufacture is the process of designing the mold. A mold is a cavity machined from steel. When liquid plastic or molten metal is injected into the mold, it takes on the inverse of the mold's shape when it cools. This results in a finished piece. The mold cavity is highly polished because any flaw can result in a flawed final piece. For making chess game pieces, a two part mold can be used. To make the piece, the two mold sections are joined together and injected with the base raw material. The mold is then opened and the piece drops out. Special release agents and a tapered design help make the parts easier to remove. When molds are designed they are made slightly larger to compensate for the fact that plastic shrinks while it cools.

Raw Materials

Chess sets have been made with a number of raw materials over the years. Materials as diverse as ivory, glass, wood, clay, pewter, stone, and various metals have been used. Today, the most widely available chess sets are made of plastic. Plastic is a mixture of high molecular weight polymers and various fillers. For a plastic to be suitable in chess-piece manufacture it must be easily colored and heat stable, and have good impact strength. The most often used plastics are thermoset plastics such as polymethyl methacrylate (PMMA).

Polymers found in plastics are typically colorless, so colorants are added to make the chess pieces look more appealing. Colorants include soluble dyes or comminuted pigments. Titanium dioxide can be used for white colored pieces. For more ornamental sets, other inorganic materials such as iron oxides can be used to produce yellow, red, black, brown, and tan pieces.

Various filler materials are added to the plastics to produce durable, high quality pieces. For manufacturing ease, plasticizers are often added to the plastic. Plasticizers are nonvolatile solvents that increase the flexibility of the polymer. To improve the overall properties of the plastic, reinforcement materials such as fiberglass may be added. Other additives include ultraviolet (UV) protectors, heat stabilizers, antioxidants, and manufacturing aids.

The Manufacturing Process

The basic steps involved in the creation of a chess game include creating the mold for the pieces, producing the pieces, producing the board, and final assembly. The following manufacturing procedure represents a method that mass producers of the game might use. Some shops still make their sets by hand, a time consuming process that involves carving the pieces from the raw material.

Making the pieces

1 In the earliest phase of manufacture, designs for the chess set pieces are drawn out on a board and used as a guide in making the molds. Pieces are then handmade, typically starting by making the general outline of the piece with a wire frame. Clay is then molded around the frame and shaped to look exactly like the desired piece.

2 When the clay model hardens, a plaster mold of it is produced. From this mold, a steel die (or mold) is then machined, which will allow the exact duplication of the clay model. In some cases, a set of steel molds are connected together so that the whole set of chess pieces can be made in a single injection molding step.

3 With the steel molds made, plastic pellets are transformed into chess game pieces using injection molding. In this process, pellets are put into a hopper connected to the injection molding machine. They are forced through a high-pressure

screw and melted. The screw is turned, forcing the melted plastic through a nozzle and into the mold. Just before the plastic is injected, the two halves of the mold are brought together to form the shape of the chess piece. Inside the mold, the plastic is held under pressure for a set amount of time and then allowed to cool. As it cools, the plastic hardens, the mold is opened, and the chess game piece is ejected. The mold then closes again and the process begins again.

Making the board

4 The construction of the board depends on the starting raw material. Wood and stone sets are cut or carved to specifications. For mass produced sets, the main raw material for the board is cardboard. The cardboard is first cut in a square to the exact dimensions desired, and then it is printed. The printing process involves a printing press fitted with plates. When the press is turned on, the plate passes under a roller and gets coated with water. An ink roller is passed over the plate and ink attaches to the plate in specific printable spots.

5 Ink is transferred from the plate to a rubber roller. The rubber roller is passed over the cardboard, which causes a transfer of ink. The cardboard is then passed to the next roller assembly where the next color is added by a similar process. The ink is specially formulated so that it dries before it enters the next roller assembly. This process of wetting, inking, and printing allows for continuous manufacture of printed chess boards. After all the printing is done, a special clear polymer coating may be applied to protect it and give it a glossy look.

Final assembly

6 To finish production of a game set, all the different components are brought to the packaging area. The exact package depends on the final design, however, in most cases the pieces are put into a box along with the board. During this stage, instruction sheets or other booklets are also put in the box. It is then taken by conveyor to a shrink-wrapping machine..

7 On the shrink-wrap machine, the box is loosely wrapped in a thin plastic film. It is then passed through a heating device that shrinks the film and wraps the box tightly. The boxes are then put into cases and stacked on pallets. They are transferred to trucks that deliver them to local sales outlets.

Quality Control

The quality of the chess game parts are checked during each phase of manufacture. Line inspectors check the plastic parts to ensure they meet size, shape, and consistency specifications. The primary test method is typically visual inspection. When a damaged plastic part is found, it is set aside to be melted again and reformed into a new chess game piece.

The Future

The future of chess sets is likely to involve the improvement of computerized chess sets. Currently, many manufacturers produce single person, computerized games that allow the player to compete against a computer. In the years to come, these computer chess games are likely to become more sophisticated, challenging even the best players in the world. In addition to the current game, variations have been developed. Future chess sets may involve multiple levels in which pieces will be able to attack not only forward and backward, but also up and down. New board shapes have already been introduced making it possible for up to four players to be involved in a game at once.

Where to Learn More

Books

Carraher, C. E., and R. B. Seymour. *Polymer Chemistry.* 5th ed., revised. Undergraduate Chemistry Series. New York: Marcel Dekker, Inc., 2000.

Chabot, J. F. *The Development of Plastics Processing: Machinery and Methods.* Society of Plastics Engineers Monographs. New York: John Wiley & Sons, Inc., 1992.

Goichberg, Bill, et al. *U.S. Chess Federation Official Rules of Chess.* New York: David McKay Co., 1993.

Golombek, Harry. *Chess: A History.* New York: Putnam Publishing Group, 1976.

Kirk-Othmer Encyclopedia of Chemical Technology. New York: John Wiley & Sons, Inc., 1992.

Levy, David, and Monty Newborn. *How Computers Play Chess.* New York: W.H. Freeman & Co., 1990.

—Perry Romanowski

Clothes Iron

Background

A clothes iron is a household appliance used to press the wrinkles out of and creases into clothes. When the iron is turned on, the consumer moves it over an item of clothing on an ironing board. The combination of heat and pressure removes wrinkles.

Irons have evolved over hundreds of years from simple objects made of metal (though they were sometimes made of glass or other materials) that were often heavy and hard to use. Before heated dryers were invented, irons served another purpose as well. Hot irons killed parasites and bacteria in clothing, and eliminated mildew. Most modern irons are made of metal and plastic, and have many features such as steam, temperature controls, and automatic shutoff. Steam provides an additional means for removing wrinkles from clothing.

History

Though objects have been used for thousands of years to remove wrinkles and/or press clothing, for much of that time only the wealthy had their clothes so treated. Because the use of such implements was hard and laborious, only the rich could afford to employ people (usually slaves or servants) to do the work. In about 400 B.C., Greeks used a goffering iron to create pleats on linen robes. The goffering iron was a rolling pin-like round bar that was heated before use.

Empire-era Romans had several tools similar to the modern iron. One was a hand mangle. This flat metal paddle or mallet was used to hit clothes. The wrinkles were removed by the beating. Another implement was a *prelum*. This was made of wood and not unlike a wine press. Two flat heavy boards were put between a turnscrew, also made of wood. Linen was placed between the boards and the increasing pressure applied by the turnscrew created pressure to press the fabric.

The ancient Chinese also had several primitive types of irons, including the pan iron. The pan iron looked rather like a large ice cream scoop. This iron had an open compartment with a flat bottom and a handle. The compartment held hot coal or sand, which heated the bottom of the pan iron. It was moved across clothing to remove wrinkles.

By about the tenth century A.D., Vikings from Scandinavia had early irons made of glass. The Vikings used what was called a linen smoother to iron pleats. The mushroom-shaped smoother was held near steam to warm up, and was rubbed across fabric.

What contemporary consumers would recognize as an iron first appeared in Europe by the 1300s. The flatiron was comprised of a flat piece of iron with a metal handle attached. To heat the iron, it was held over or in a fire until it was hot. When a garment was pressed with the flatiron, it was picked up with a padded holder. A thin cloth was placed between the garment and the iron so that soot would not be transferred from implement to the finished garment. The flatiron was used until it was too cool to do its job. Many people owned several flatirons so they could heat one or more while one was being used.

In approximately the fifteenth century, an improvement over the flatiron was introduced. The hot box (also known as the box iron or slug iron) was made of a hollow

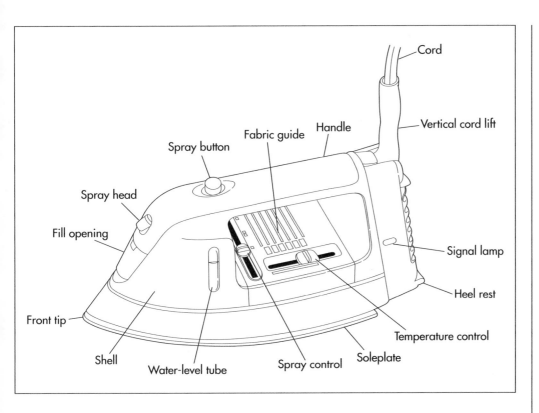

Labels in figure: Cord, Vertical cord lift, Handle, Fabric guide, Spray button, Spray head, Fill opening, Signal lamp, Heel rest, Temperature control, Soleplate, Spray control, Water-level tube, Shell, Front tip

metal box with a smooth bottom and a handle. Inside, hot coals, bricks, slugs (heated metal inserts) or some other heating element were placed. This eliminated the need for an extra cloth between clothing and iron because the iron did not get the clothes dirty. Both the flatiron and hot box were used for several hundred years.

Many innovations in iron technology came in the nineteenth century. When cast iron was invented in the early part of the nineteenth century, some of the problems with flatirons were solved. With the advent of cast iron stoves, flatirons could be heated on top of them, which was much cleaner than a fire. By the 1820s, cast iron was also used to make flatirons. These irons were called sad irons because they were heavy, weighing about 15 lb (5.6 kg), and hard to move.

Like flatirons, sad irons were heated on the stovetop, but they sometimes heated unevenly. The handle also heated up, which posed problems for users. American Mary Potts solved these predicaments in 1870. She made a cardboard base and filled it with plaster of Paris. This was placed around the iron's body and kept it cooler for more even heating. Potts also devised a detachable wooden handle that was **spring** loaded for

the sad iron. Because wood does not hold heat in the same way that iron does, the person using the iron would not be burned.

After gas became available in American homes in the late 1800s, gas irons came into existence. The earliest were patented in 1874. Homes had individual gas lines into them, and the gas iron was hooked up to the gas line by a pipe. The iron contained a burner to which the gas flowed. When the burner was lit with a match, the iron heated up. The iron was very hot and gas sometimes leaked, but the gas irons were lighter than sad irons. Other fueled irons soon followed. These irons were heated with oil, gasoline, paraffin, and other fuels.

The electric iron was invented in the 1880s when electricity became widely available in homes. The first electric iron was patented by Henry W. Seeley in 1882. His iron was hooked up to an electrical source by detachable wires. The electricity stimulated the iron's internal coils. But Seeley's iron, like many early electric irons, did not have electric cords. The irons were heated on a stand. One big problem with Seeley's iron was that it heated very slowly on the stand, and cooled quickly while in use. This iron had to be reheated frequently.

By the turn of the century, iron technology had progressed considerably and irons became more common in American households. In 1903, irons with electric cords directly attached to the iron were being sold. Earl Richardson invented a sole plate (the bottom part of the iron that is made of metal and does the actual pressing) that improved how and where sole plates were heated for better ironing. His iron had more heat in the tip than in the center and was known as the Hotpoint.

In the 1920s, Joseph Myers improved the iron and cord by adding an automatic heat control made of pure silver. Thermostats soon became a standard feature. The first cordless irons were introduced in 1922, though they did not catch on. (The first successful cordless irons were sold in 1984).

In 1926, the steam iron was introduced by the Eldec Company. Steam made it easier to smooth dry stiff fabrics. Previously the user sprinkled water on dry clothing, or clothing had to be ironed when damp. The steam irons employ a water tank that allows heated water vapor to be created and applied through small holes on the sole plate. Steam irons did not become popular until the 1940s.

Edward Schreyer conquered the problem of rusting sole plates in 1938. He developed an aluminum alloy that would not rust or leak. Irons that could vary between steam or dry were introduced in the 1950s. The first iron with automatic shut off was introduced in 1984.

Contemporary irons have nonstick coating on the sole plate, an innovation that was introduced in 1995. Most featured bodies made of plastic and more holes on the sole plate to allow steam to come through. A whip holds the cord out of the way during use. In 1996, about 13-14 million irons with a variety of features were sold in the United States.

Raw Materials

Irons are made primarily of plastic and metal (aluminum and steel). The materials often come to the factory in the form of plastic resins, aluminum ingots, and steel sheets. The metal is used to make the sole plate, thermostat and other internal mechanisms. Plastics are used to make the exterior and handle, as well as the water tank. Certain components, like the spring for the thermostat, cord, plug, and related connections are usually outsourced by iron companies.

The Manufacturing Process

First, each sub-assembly of the iron is produced, most often on separate, automated production lines. Then the iron is assembled.

Sole plate

1 The sole plate is cast of molten aluminum. Part of the mold creates the holes that are essential in a steam iron. Heated metal is inserted into a mold under pressure, cooled, and released.

2 The cooled sole plate is treated in one of three ways. It is polished, coated with a non-stick PTFE material, or covered with another metal. Such metals include stainless steel.

To complete one or more of these processes, the sole plates are put on a large automated carousel, which rotates through each step.

3 To polish the plate, an automated belt sander uses bands of abrasive to polish and buff the plate. The finish required determines which grade of abrasive is used.

4 An automated spray-painting machine applies non-stick coating. After application, the sole plate is baked in an automated industrial process.

5 To coat with another metal, the external metal cover is created by an automated stamp press. The resulting cover is either pressed or riveted onto the sole plate, through a smaller machine press.

Thermostat

6 In an injection mold, a small metal post is cast.

7 A spring is mounted onto the metal post. This spring is a bimetallic switch made of two different metals with divergent linear thermal coefficients bonded together. The spring actually controls the iron's temperature.

8 Power contacts are attached to the end of the spring, which let the electricity through so the iron can be heated. This whole process is generally automated.

Water tank

9 In a two-part injection mold, heated plastic is inserted to make upper and lower sections of the tank. Several openings on the tank are created as part of the mold.

10 The mold is put under pressure, cooled, and released as a one-part tank.

11 Other parts for the tank (pump, internal chamber, piston, buttons, and other parts) are created by similar injection molding processes.

12 On an automated assembly line, the parts are put together, with each of the other parts put onto the water tank.

Housing

13 In an injection mold, heated plastic is inserted into a mold under pressure, cooled, and released.

Handle

14 In an injection mold, heated plastic is inserted into a mold under pressure, cooled, and released.

Assembly

When all the parts are manufactured, the iron is assembled on an automated assembly line.

15 The sole plate is the first part on the assembly line. The thermostat is either screwed onto the plate, or welded to the plate by a robot.

16 To the sole plate-thermostat subassembly, the water tank is put in place. It is secured on an automated line with screws or other industrial fasteners.

17 The handle and body are attached over the sole plate, thermostat and water tank, and fixed by screws. Sometimes this process is automated, but it also can be done manually.

18 The electrical cord is the last piece to be added. Sometimes this process is automated, but it also can be done manually.

19 After an automated testing process, the irons are inspected by hand.

20 Completed irons are packed into individual boxes with instructions and other documents by hand. (For some companies, this is an automated process.)

21 The individual boxes are placed in shipping cartons or master packs for distribution to warehouses.

Quality Control

Before the manufacturing process begins, all the raw materials are sample checked for consistency. As each subassembly of the iron is manufactured, the pieces are checked for correct functionality.

During the assembly process, an iron is removed from the production line and taken apart by an independent department to look for errors. Any mistakes result in the whole batch of irons being checked and corrected if necessary. After the product is assembled, a worker also checks the iron for electrical functionality and water integrity.

Byproducts/Waste

Any excess metal or plastic from the molding process is reused in the process, if possible. Anything that is unable to be used is recycled.

The Future

The most obvious improvements on the iron probably will be to the sole plate. Better coatings will probably be invented that are more resistant to damage from zippers or other protrusions on garments and reduce drag over fabric. These improvements might be in the form of better alloys or better nonstick coatings. Internal mechanisms that better control heat and steam also will continue to evolve.

Where to Learn More

Books

Alpine, Elaine Marie. *Irons*. Minneapolis, MN: Carolrhoda Books, 1998.

Walkley, Christina, and Vanda Foster. *Crinolines and Crimping Irons: Victorian Clothes: How They Were Cleaned and Cared For.* London: Peter Owen, 1978.

—*Annette Petruso*

Cognac

Background

Cognac, a type of brandy, is considered to be one of the finest, if not the finest, of the spirits. It is made from white grapes grown in the Charante region of France, and is named after the town of Cognac in the French region of Charante. Cognac is sometimes called "burnt wine" (from the Dutch word *brandewijn* because the wine is subjected to a double heating.

Although cognac is primarily enjoyed as a beverage, it is also used in cooking for sauces, marinades, fruit preserves, and chocolates. Cognac is splashed over dishes for flamed presentations, in marinades, fruit preserves, and chocolates.

History

Necessity, as the mother of invention, certainly describes the development of cognac. In the seventeenth century, the town of Cognac in the French region of Charante was an exporter of salt and wine. The wine was particularly popular with the Dutch and English merchants who visited the region. They would often distill the wine so that the ship voyage home would not affect the quality of the spirit.

A vintner named Chevalier de la Croix-Marrons is the first person known to heat wine and then send it back through the still again, thus creating "burnt wine." The wine was then stored in oak barrels. It would be diluted upon arrival. However, the merchants found that the distilled wine had improved with age and by its contact with the wood.

In the eighteenth century, two men whose names would become synonymous with cognac, each separately built successful distilleries that manufactured cognac. Jean Martell, a French former smuggler, arrived in Cognac and built a distillery on the Charent River. In 1765, James Hennessy, an Irishman who served in the French navy, also set up shop on the river as Hennessy Connelly and Company. The following year, Hennessy's company received its first order from the American colonies. Soon after, cognac was also exported to the Far East.

The name "cognac" was not affixed to the distilled wine until about 1783. At about that time, the French government developed rules for labeling, classifying the cognac by it smoothness. V.S. (Very Superior) is aged at least two and one-half years. V.S.O.P.(Very Superior Old Pale), or Reserve, is aged in wood at least four years. X.O. (Extra Old, Napoleon, or Extra) is that which has been aged at least five years. These are bare minimums. Most houses age their cognacs for twice the minimum required.

At first, warehouses were built on the river primarily for ease of transport. Therefore, the cellars were damp. This proved beneficial to the cognac because the dampness reduced its strength but not its volume. A dry cellar produces a harsher brandy. Even today, distillers try to build warehouses near rivers, or they keep their cellars humidified.

Since early times, the distillation has been carried out in a large copper pot still, called an alembic, topped with a long "swan's neck". By French law, the stills are limited to small capacities in order to ensure a slow and precise distillation. French law also defines the distillation period. It begins in November and ceases by March 31.

It takes nearly half a century for alcohol content of distilled brandy to decrease to a drinkable 40%.

Since early times, the distillation of wine into cognac has been carried out in a large copper pot still, called an alembic, topped with a long "swan's neck". By French law, the stills are limited to small capacities in order to ensure a slow and precise distillation.

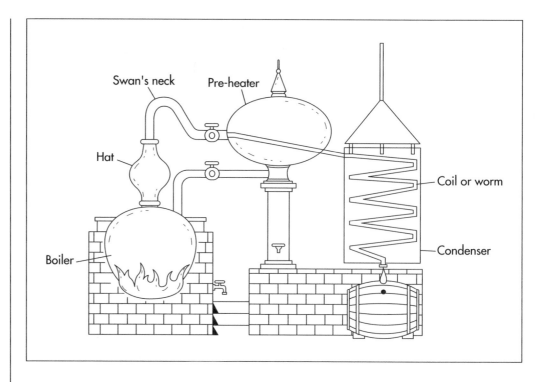

One aspect of the distillation process that has changed is the method of heat. At first, wood was used, then coal. In the present day, **natural gas** provides the heat source.

Raw Materials

The ideal grapes for distilling cognac are Colombard and Saint-Emilion. Blanche, Folle Juirancon, Monfis, and Sauvignon are also used. They are grown in six specific subdivisions, or *crus*, in the delimited region of Charante established by the French government in 1909. Each cru produces a distinctive flavor. Grande Champagne, the area around the towns of Cognac and Seconzac, yields the most delicate and fragrant brandy. Grapes in the Petite Champagne, which surrounds Grande Champagne on the southwest and east, are faster to mature and less subtle in taste. In the hills north of Grande Champagne is the Borderies. Grapes grown here produce a rounder and softer taste. Brandy made from grapes of the remaining three areas, Fins Bois, Bons Bois, and Bois Ordinaires, are used primarily to flavor other brandies.

The grapes are harvested in the winter when they are fruity and have the potential to produce juice that is 8-9% alcohol. Grapes with less than 8% alcohol are too pale to produce the desired aroma; those with more than 9% possess an inadequate concentration.

The wooden barrels, or casks, in which the cognac ages are an essential element in the process. Tannin and vanillin present in the wood lend their properties to the cognac. One-hundred-year-old trees from the forests of Limousin and Tronçais are the primary types used. Limousin wood is extremely rich in tannin and accelerates maturation. Tronçais wood releases its tannin at a much slower rate. A side effect of this process is the blackening of the walls and roofs of the warehouses caused by the growth of fungus.

Most of the major cognac producers control the manufacture of their casks. After the wood is split, it is stacked and seasoned in the open air for a minimum of three years. In the cooperage room of the distillery, the cooper shapes the wood into barrels. Metal bands hold the planks of wood together so that glue and nails, which would affect the flavor of the cognac, are not necessary.

It takes nearly half a century for alcohol content of the distilled brandy to decrease to a drinkable 40%. To facilitate the maturation process, distilled water and other varieties of brandy are blended in to the cognac.

Caramel may be added to even out color variations. Sugar syrup may be added to sweeten and enrich less mature brandies.

The Manufacturing Process

Pressing the grapes

1 The grapes are pressed and the juice is allowed to ferment naturally. No sugar or sulfure dioxide is added.

First distillation

2 As soon as the wine has fermented, it is poured into pot stills enclosed in brick kilns. Each still holds approximately 660 gallons, or the equivalent of 3,000 bottles. The kilns are heated to a temperature range between 173°F (78.3°C) and 212°F (100°C) until the alcohol vaporizes and separates from the rest of the liquid.

3 The vapors are collected in the cowl and the swan's neck of the still. They then pass into the serpentine–like condenser coil. The condensed liquid, called "broullis," is reduced one–third from the original amount and measures about 30% alcohol by volume.

Second distillation

4 The broullis is heated a second time in a process known as "bonnechauffe." This is a exacting process because the distiller has to decide at what moment to isolate what is known as the "heart" of the liquid, to separate it from the "head" and "tails." The head portion is too high in alcohol content while the tail is lacking in substance. These portions are redistilled several times and used in blending.

5 The remaining liquid is the clear "eaude vie." It has been reduced by an additional one–third and is 70% alcohol by volume. This significant reduction in volume means that the distillation of cognac is a costly operation. It takes 9 liters of wine to make one liter of cognac. The amount of spirit that is lost to vaporization (known as "the angels' share") can equal more than 20 million bottles annually.

Casking the distilled brandy

6 The eau de vie is piped into oak casks. The casks are housed in large damp warehouses, or cooperages. The vintage, cru, and date are marked on each cask with chalk.

Generally speaking, the brandy is first stored in newer casks for periods between one and two years. The amount of time is dependent on the level of tannin that is desired. Tannin is strongest in new oak, so the brandy must possess enough character to absorb large amounts of tannin.

Aging and blending the cognac

7 The brandy is progressively moved into older casks, once again dependent on the finished product desired. The cooper presides over the transfer, tasting the brandy once a year to determine if it should be moved to another cask. Most of the tasting is done by the nose; very little cognac is actually orally tasted.

8 Cognacs over varying vintages and from different crus are blended to create the quality desired. This blending occurs over a minimum of one year with samples taken annually. Representative samples of each blending are kept for at least a decade. Each cognac house has a house style that is consistent year after year. Often this style is passed down from family member to family member as a memory of taste and aroma.

Bottling the cognac

9 The bottles in which cognac is packaged are a source of great pride to each company. Many of the bottles are handcrafted of crystal. They are often sealed with wax and draped with satin ornaments.

The cognac is bottled on an a slow moving assembly line at the rate of about 5,000 bottles per hour. Each bottle is inspected and hand–dried. After bottling, the cognac is either packed for shipping or stored for future shipments.

Where to Learn More

Books

Coyle, L. Patrick. *The World Encyclopedia of Food*. New York: Facts on File, 1982.

Lang, Jenifer Harvey, ed. *Larousse Gastronomique*. New York: Crown, 1998.

Robinson, Jancis, ed. *The Oxford Companion to Wine*. Oxford: Oxford University Press, 1994.

Periodicals

Gugino, Sam. "High Spirited." *The Wine Spectator* (January 31, 1998).

Other

"Blending Tools." http://le-cognac.com. (January 17, 2000).

"The Cognac of Madame Raymond Ragnaud and Her Children." http://le-cognac. com/raymond_ragnaud/ragnaud3_us.html. (January 12, 2000).

"Thomas Hine & Co. Handcrafted Cognac." http://www.le-cognac.com/hine/hine_ charte_us.html. (December 1999).

—*Mary McNulty*

Coir

Background

What is commonly called a coconut, as found in grocery stores, is actually only the single seed of a fruit of the coconut palm tree (*Cocos nucifera*). Before being shipped to market, the seed is stripped of an external leathery skin and a 2-3 in (5-8 cm) thick intermediate layer of fibrous pulp. Fibers recovered from that pulp are called coir. The fibers range from sturdy strands suitable for brush bristles to filaments that can be spun into coarse, durable yarn. In the United States, the most popular uses for coir are bristly door mats, agricultural twine, and geotextiles (blankets that are laid on bare soil to control erosion and promote the growth of protective ground covers).

Although coconut palms grow throughout the world's tropical regions, the vast majority of the commercially produced coir comes from India and Sri Lanka. Coconuts are primarily a food crop. In India, which produces about one-fourth of the world's 55 billion coconuts each year, only 15% of the husk fibers are actually recovered for use. India annually produces about 309,000 short tons (280,000 metric tons) of coir fiber.

Coir fibers are categorized in two ways. One distinction is based on whether they are recovered from ripe or immature coconut husks. The husks of fully ripened coconuts yield brown coir. Strong and highly resistant to abrasion, its method of processing also protects it from the damaging ultraviolet component of sunlight. Dark brown in color, it is used primarily in brushes, floor mats, and upholstery padding. On the other hand, white coir comes from the husks of coconuts harvested shortly before they ripen. Actually light brown or white in color, this fiber is softer and less strong than brown coir. It is usually spun into yarn, which may be woven into mats or twisted into twine or rope.

The other method of categorization is based on fiber length. Both brown and white coir consist of fibers ranging in length from 4-12 in (10-30 cm). Those that are at least 8 in (20 cm) long are called bristle fiber. Shorter fibers, which are also finer in texture, are called mattress fiber. A 10-oz (300-g) coconut husk yields about 3 oz (80 g) of fiber, one-third of which is bristle fiber.

The only natural fiber resistant to salt water, coir is used to make nets for shellfish harvesting and ropes for marine applications. Highly resistant to abrasion, coir fibers are used to make durable floor mats and brushes. Strong and nearly impervious to the weather, coir twine is the material hops growers in the United States prefer for tying their vines to supports. Coir is becoming a popular choice for making geotextiles because of its durability, eventual biodegradability, ability to hold water, and hairy texture (which helps it cling to seeds and soil).

History

Palm trees belong to one of the world's oldest plant families, and coconut palms have been cultivated for at least 4,000 years. In Sanskrit, the precursor of the modern languages of Hindi and Urdu, the coconut palm was called "the tree that provides all the necessities of life." In fact, it is one of the world's most useful trees, providing food, drink, fibers, fuel, and building material. Coconut fruits are very hardy; they can even float in the ocean for great distances and still remain viable. Indigenous to Southeast Asia, the coconut palm spread throughout

Palm trees belong to one of the world's oldest plant families, and coconut palms have been cultivated for at least 4,000 years.

The inside layers of the fruit from a coconut palm.

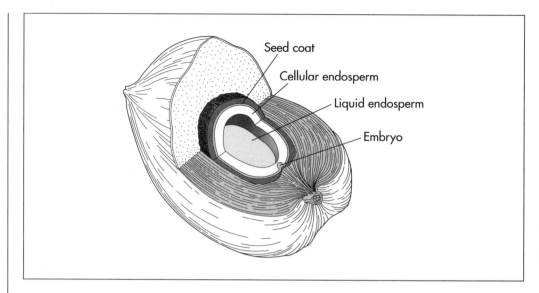

Seed coat

Cellular endosperm

Liquid endosperm

Embryo

the tropics either naturally or by human travel and trade.

About A.D. 60, a Greek sailor wrote about a coconut-producing East African village, probably on the coast of present-day Tanzania, whose boats were made of planks sewn together with fibers. By the eleventh century, Arab traders (whose route stretched from China to Madagascar off the southeastern shore of Africa) were teaching residents of what are now Sri Lanka and India how to extract and process coconut fibers. During the thirteenth century, Marco Polo—while visiting the port of Hormuz on the Persian Gulf—discovered that the masterful Arab seamen built their ships without nails, sewing them together with coconut fiber. In China, Polo found that the Chinese had been using coconut fiber for 500 years.

Halfway around the world, coir also played a significant role in the exploration of Micronesia and Polynesia, where the product is commonly called sennit. For example, early settlers of Hawaii arrived from the Marquesas Islands around the fifth century in a large, double-hulled canoe lashed together with coconut fiber. In fact, sennit lashings were the primary mechanism for connecting pieces to construct boats, buildings, weapons, and tools until European explorers brought iron nails to the region in the late eighteenth century.

Coir production changed little until efforts to mechanize it began in the middle of the twentieth century. In India, a defibering machine was invented in 1950. Coir processing is an important economic activity in India,

where it provides jobs for more than 500,000 people. Because mechanization would eliminate a significant number of those jobs, it is being introduced gradually. In 1980, the primary producing countries of India and Sri Lanka began an ongoing effort to identify and correct technological limitations on coir production.

Raw Materials

Coconut palms flower monthly. Because it takes a year for the fruit to ripen, a tree always contains fruits at 12 stages of maturity. Harvesting usually take place on a 45-60 day cycle, with each tree yielding 50-100 coconuts per year.

Fresh water is used to process brown coir, while sea water and fresh water are both used in the production of white coir. In 2000, researchers announced that adding a broth containing a certain combination of 10 anaerobic (living without oxygen) bacteria to salt water can dramatically hasten the fiber extraction process without seriously degrading product quality.

In Europe and Asia, brown coir mats may be sprayed with latex rubber for use as padding in mattresses or automobile upholstery.

The Manufacturing Process

Harvesting and husking

1 Coconuts that have ripened and fallen from the tree may simply be picked up off

DEHUSKING

EXTRACTING

DRYING

BUNDLING

SPINNING

The outer layers covering the coconut seed are processed and spun into fibers commonly known as coir.

the ground. Coconuts still clinging to the 40-100 ft (12-30 m) tall trees are harvested by human climbers. If the climber picks the fruit by hand, he can harvest fruits from about 25 trees in a day. If the climber uses a bamboo pole with a knife attached to the end to reach through the treetop vegetation and cut selected coconuts loose, he can harvest 250 trees per day. (A third harvesting technique, in which trained monkeys climb trees to pick ripe coconuts, is used only in countries that produce little commercial coir.)

2 Ripe coconuts are husked immediately, but unripe coconuts may be seasoned for a month by spreading them in a single layer on the ground and keeping them dry. To remove the fruit from the seed, the coconut is impaled on a steel-tipped spike to split the husk. The pulp layer is easily peeled off. A skilled husker can manually split and peel about 2,000 coconuts per day. Modern husking machines can process 2,000 coconuts per hour.

Retting

Retting is a curing process during which the husks are kept in an environment that encourages the action of naturally occurring microbes. This action partially decomposes the husk's pulp, allowing it to be separated into coir fibers and a residue called coir pith. Freshwater retting is used for fully ripe coconut husks, and saltwater retting is used for green husks.

3 For freshwater retting, ripe husks are buried in pits dug along riverbanks, immersed in water-filled concrete tanks, or suspended by nets in a river and weighted to keep them submerged. The husks typically soak at least six months.

4 For saltwater retting, green husks are soaked in seawater or artificially salinated fresh water. Often this is accomplished by placing them in pits along riverbanks near the ocean, where tidal action alternately covers them with sea water and rinses them with river water. Saltwater retting usually takes eight to 10 months, although adding the proper bacteria to the water can shorten the retting period to a few days.

5 Mechanical techniques have recently been developed to hasten or eliminate retting. Ripe husks can be processed in crushing machines after being retted for only seven to 10 days. Immature husks can be dry milled without any retting. After passing through the crushing machine, these green husks need only be dampened with water or soaked one to two days before proceeding to the defibering step. Dry milling produces only mattress fiber.

Defibering

6 Traditionally, workers beat the retted pulp with wooden mallets to separate the fibers from the pith and the outer skin. In recent years, motorized machines have been developed with flat beater arms operating inside steel drums. Separation of the bristle fibers is accomplished by hand or in a machine consisting of a rotating drum fitted with steel spikes.

7 Separation of the mattress fibers from the pith is completed by washing the residue from the defibering process and combing through it by hand or tumbling it in a perforated drum or sieve. (Saltwater retting produces only mattress fibers.)

8 The clean fibers are spread loosely on the ground to dry in the sun.

Finishing

9 Bristle fibers that will not immediately be further processed are rolled and tied into loose bundles for storage or shipment. More mechanized producers may use a hydraulic press to create compact bales.

10 Similarly, mattress fibers may simply be baled with a hydraulic press. However, if more processing is desired, the fibers are combed with mechanical or manual carding tools, then loosely twisted into a thick yarn (wick), and wound into bundles. Later, the wick can be re-spun into a finer yarn. Techniques vary from simple hand spinning to use of a hand-operated spinning wheel or a fully automated spinning machine.

11 Depending on its intended final use, the yarn may be shipped to customers, or multiple strands may be twisted into twine and bundled for shipment. Both traditional manual techniques and newer mechanical methods are used to braid twine into rope and to weave yarn into mats or nets.

12 For some uses, such as upholstery padding, bristle fiber is loosely spun into yarn and allowed to rest. Then the fibers, which have become curly, are separated. These fibers are lightly felted into mats that are sprayed with latex rubber, dried, and vulcanized (heat treated with sulphur).

Byproducts/Waste

By weight, coir fibers account for about one-third of the coconut pulp. The other two-thirds, the coir pith (also known as coir dust), has generally been considered a useless waste material. Although it is biodegradable, it takes 20 years to decompose. Millions of tons sit in huge piles in India and Sri Lanka. During the last half of the 1980s, researchers successfully developed processes to transform coir pith into a mulching, soil treatment, and hydroponic (without soil) growth medium that is used as an alternative to such materials as peat moss and **vermiculite**. Before being compressed into briquettes for sale, the coir pith is partially decomposed through the action of certain microbes and fungi. An Australian company has also recently begun turning coir pith into an absorbent product used to remediate oil spills.

The retting process used in coir fiber production generates significant water pollution. Among the major organic pollutants are pectin, pectosan, fat, tannin, toxic polyphenols, and several types of bacteria including salmonella. Scientists are experimenting with treatment options, and at least one coir manufacturing company claims to be treating its effluent water.

The Future

As improved technology increases production, industry groups and governmental agencies are actively promoting new uses for coir fiber. Geotextiles is one promising area. The Indian state of Kerala designated 2000 as Coir Geotextiles Year, which it observed by increasing marketing efforts and supporting research to improve production. The annual world demand for geotextiles is 1.2 billion square yards (1 billion square meters) and growing. Although natural fibers account for only 5% of that, the proportion is expected to increase as more users turn away from nonbiodegradable synthetics.

Another new product under development is an alternative to plywood that is made by impregnating a coir mat with phenol formaldehyde resin and curing it under heat and pressure.

Where to Learn More

Books

Fremond, Yan. *The Coconut Palm.* Berne: International Potash Institute, 1968.

Other

Kew Royal Botanic Gardens. http://www.rbgkew.org.uk/ksheets/coir.html (October 2000).

Punchihewa, P.G., and R.N. Arancon. "Coconut." In *Post-Harvest Operations Compendium.* Food and Agriculture Organization of the United Nations (FAO). http://www.fao.org/inpho/compend/text/ch15.htm (November 2000).

Sudhira, H.S., and Ann Jacob. &ldqo;Reuse of By-Products in Coir Industry: A Case Study." Internet Conference on Material Flow Analysis of Integrated Bio- Systems (March-October 2000). http://www.ias.unu.edu/proceedings/icibs/ic-mfa/jacob/paper.html (October 2000).

—*Loretta Hall*

Comic Book

Background

A comic book portrays a story through a series of sequential illustrations that incorporate short bits of text containing dialogue, sounds, or narratives. The story may be humorous, or it may present a world of adventure, mystery, or fantasy. Most comic books are printed on a regular basis and have one or more central characters who appear in each issue. A particular story may be told in a single issue, or it may continue from one issue to the next over a period of time. The artistic style of a comic book is often attributed to a single artist, although most comics are produced by a team of artists and writers working together.

History

The use of sequential illustrations to tell a story dates to prehistoric times when early humans painted series of images on rocks and cave walls. Egyptian hieroglyphics are another form of sequential illustrations that tell a story.

Hand-drawn illustrations appeared regularly in newspapers and magazines starting in the 1800s. Many of them used humorous or unflattering portrayals of well-known people and were the origin of modern cartoons and comics.

The first newspaper comic strip in the United States was Richard Felton Outcault's "The Yellow Kid," which appeared in the *Hearst New York American* on February 16, 1896. It was published in the Sunday supplement to the paper and was quickly joined by other comic strips.

By the 1910s, the Sunday comics were so popular that newspapers would occasionally publish small books containing reprints of past strips, which they would distribute to promote the paper and gain new readers. Soon, other publishing companies were assembling comic strips from several papers and selling them to merchants to be given away as premiums. In 1934, Eastern Color Printing Company decided to sell these books directly to the public for 10 cents each. American News, which controlled distribution to newsstands throughout the country, initially refused to handle the books, so Eastern Color took them to chain stores and quickly sold 35,000 copies. Faced with this astounding success, American News reconsidered and ordered 250,000 copies of *Famous Funnies No. 1* from Eastern Color. It went on sale in July 1934 and became the first regularly published comic book to be sold at a newsstand.

During the late 1930s, many of the now-famous superheroes made their first appearances in comic books, and comic book sales soared as good triumphed over evil. By the early 1950s, however, readers grew tired of superheroes, and some comic book publishers turned instead to lurid crime and horror stories with graphic illustrations. Some people felt this material was unsuitable for children, and the comic book industry came under public criticism and federal investigation in 1954. In response, many comic book publishers banded together and issued the Standards of Comics Code Authority, which defined appropriate material for comics.

Comic books enjoyed a resurgence of interest during the 1980s, when fresh new artists created a whole new cast of heroes and heroines. Today, comic books are as popular as ever, and the comic book industry is a million-dol-

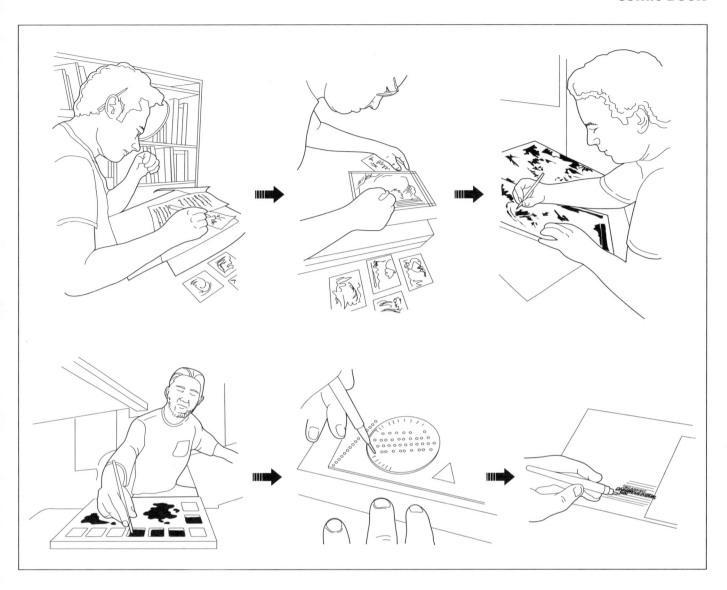

lar business that includes movies, television series, toys, costumes, and many other items.

Raw Materials

During the preparation of a comic book, a variety of art materials may be used to create the original hand-drawn page masters and color guides. These materials include various sizes, weights, and finishes of paper, as well as several different drawing mediums including pencils, inks, markers, and paints. After the master pages have been scanned and colored on a computer, the computer uses the color guides as a reference to generate four pieces of plastic film that are used in the printing process.

The actual comic book itself is printed on a variety of papers using four colored inks—cyan (pronounced SIGH-ann, a shade of blue), magenta, yellow, and black. These four inks are printed in an interlocking pattern of tiny dots, which our eyes perceive as various colors. The printed comic pages are then bound together with staples or glue to form a comic book.

Design

Because each new issue of a comic book requires new artwork, the design process is part of the manufacturing process. The exception is when a new comic title or series is first introduced. That design process involves the same creative and artistic abilities required to produce any new work of art and may include idea generation, preparation of sketches, and the development of a series of refinements before the final characters and themes emerge.

Creating a comic book is a detailed process that includes drafting the plot, designing thumbnail sketches and then the original drawings, and finally adding color and lettering.

Charles Schulz.

Charles Schulz was born in Minneapolis, Minnesota, on November 26, 1922. After World War II, Schulz freelanced for a Catholic magazine and taught in the correspondence school, renamed the Art Instruction Institute. His work appeared in the *Saturday Evening Post*, and eventually he created a cartoon entitled "Li'l Folks."

The United Feature Syndicate of New York proposed publication of Schulz' "Li'l Folks," but it was renamed "Peanuts" by the company. In 1950 the cartoon made its debut in seven newspapers. Within a year the strip appeared in 35 papers, and by 1956 in over 100. In 1955 and 1964, Schulz received the Reuben award from the National Cartoonists Society. By 1965 "Peanuts" appeared in over 2,300 newspapers and the classic cartoon "A Charlie Brown Christmas," produced by Bill Melendez and Lee Mendelson, won a Peabody and an Emmy award.

Schulz also received the Yale Humor Award in 1956, and the School Bell and National Education Association awards in 1960; plus honorary degrees from Anderson College in 1963 and St. Mary's College of California in 1969. A "Charles M. Schulz Award" honoring aspiring comic artists was created by the United Feature Syndicate in 1980. The year 1990 marked the 40th anniversary of "Peanuts" and the Smithsonian Institution featured an exhibit titled, "This Is Your Childhood, Charlie Brown...Children in American Culture, 1945–1970." By the late 1990s the syndicated strip ran in over 2,000 newspapers throughout the world. Schulz died on February 12, 2000, the night before his last original "Peanuts" strip ran announcing his retirement.

The final product of the initial design process may be a prototype comic book known as an "ashcan," a term that was first used in the 1930s when comic book publishers sought to protect new titles by copyrighting them. Rather than take the time to develop new characters or plots to go with the new title, a publisher simply took pages from a previous comic book and pasted the new title on the cover. Once the publisher was granted a copyright, the pasted-up prototype was often thrown in the ashcan—a metal container used to dispose of ashes from the stove or fireplace and commonly found in many households and businesses of that era.

The concept of the ashcan was given a more modern meaning in 1984 when one comic book creator produced a limited number of black and white prototype comics for his friends and staff. In more recent times, several publishers have released small runs of ashcans in a variety of sizes and colors as promotional items for the full-production versions.

The Manufacturing Process

Comic book publishers may be small, independent operations that produce a single comic book title on an irregular basis, or they may be large, well-established companies that produce several comic book titles every month. The manufacturing process varies depending on the size of the operation and the equipment available. Here is a typical sequence of operations that a medium-sized company would use to produce a comic book.

Writing

1 Although most people think of a comic book as a series of pictures, it is the written plot that gives the story its direction. The writer and artist discuss the proposed story and exchange ideas. At this stage, they may use a number of formal or informal techniques for developing ideas. They may make notes on small index cards arranged on a table or they may outline the flow of the story on a display board. During the course of their discussion, they decide on the situations, locations, characters, and other details of the story. This helps define the overall plot from beginning to end.

2 Because most comics have a fixed number of pages, the writer and artist must

then decide how to break up the story to fit each page. They discuss which scenes and dialogue are critical to keep the story flowing and how the characters and action should be depicted to have the greatest impact. Sometimes they follow general industry practices, which define such things as the optimal number of action scenes per page or the amount of dialogue per word balloon, but other times they rely on their own personal style.

3 Once the story has been refined, the writer creates a script. This includes general descriptions of the scenes and characters in the order they appear, the accompanying dialogue or descriptive text, and general instructions to the artist. The result is very much like a script written for a movie or play.

Drawing

4 The artist reads the script and makes a rough sketch of each page, called a thumbnail. The thumbnail helps the artist decide how each scene should be depicted, and how the different scenes should be arranged on the page. Some artists sketch each scene on a small piece of adhesive-backed note paper and then move them around on a larger piece of paper to achieve the desired effect.

5 Using the thumbnail as a guide, the artist begins drawing each page in pencil. Some artists like to work on standard 8.5 x 11 in (22 x 28 cm) white paper and then photoenlarge the pencil drawings onto 11 x 17 in (28 x 43 cm) illustration boards before inking the final copies; others make their pencil drawings directly on the larger boards. The artist usually starts drawing the main elements of each scene with a hard pencil that makes very light lines. When all the main elements are in place, the artist considers the overall effect and makes any changes before proceeding.

6 The artist then darkens the main elements with a softer pencil and adds the backgrounds and other details. Areas for the dialogue balloons, sound effects, and narrative boxes are blocked out in blue pencil to distinguish them from the illustrations.

7 At this point, an editor may review the pencil drawings and make changes.

Sometimes the editor may ask the artist to redraw a portion of a scene to correct an error or clarify an item. In other cases, the editor may have to shorten the dialogue or narrative to fit in the space left by the artist.

8 When the pencil drawings are complete, they are enlarged onto 11 x 17 in (28 x 43 cm) illustration boards if they were drawn on smaller paper. They are then sent to the inker. The inker's job is much more than just tracing over the pencilled lines of the artist with black ink. It involves the selection of line widths, adding shadows, visually separating the foreground from the background, and creating special effects like splatter or wash to give the illustrations texture. The inker uses a variety of pens and brushes to produce a finished black and white page. Many inkers have their own unique style that adds to the artist's original drawings.

9 The final step in the drawing process is adding the lettering for the dialogue, sound effects, and narratives that appear in the script. This can be done using hand lettering, adhesive labels, or computer-generated digital type. The letterer selects a typeface that not only conveys the actual words or sounds, but also conveys the action or emphasis of the scene with its size, style, and placement.

Coloring

10 The finished pages, including the front and back covers, are sent to the colorists who add the colors and prepare the four-color separation films required for printing. The original artwork is first photocopied and then scanned into a computer. The photocopy is hand-colored using colored markers, pencils, and paints to become a guide when coloring the pages on the computer. The scanned copy becomes an electronic file that forms a digital outline of the page to be colored.

11 With the color guide as a reference, the colorist begins to add colors to the digital outlines of each page starting with the backgrounds and working forward. This is done using a custom software package that allows the colorist to trace the outline of any part of the image with the cursor, and then apply and blend colors to that area to

match the color guide or to achieve a special effect. For many colors, the computer already has the information on file. For example, if one character always wears the same clothes, information about the colors of that character's boots, mask, or cape are stored in the computer to ensure they look the same from one issue of the comic book to another.

12 As the colorist selects and applies each color, the computer automatically assigns a code to it. This code is used to identify the four color components that make up that particular color—cyan, magenta, yellow, and black. When these four colors are printed in an interlocking pattern of tiny dots, our eyes perceive them as hundreds of different colors, even though there are really only four colors of ink on the page. The color variations depend on the concentration of each of the four color components. Thus a particular shade of red may have the code M80Y87, for example, which represents 80% magenta and 87% yellow.

13 When all the pages have been colored, a proof copy of the entire comic book is printed from the computer for final review and approval. The computer then prints a piece of plastic film for each of the four component colors on each page. Each piece of film has hundreds of thousands of tiny dots to represent the location and concentration of that color component across the page.

Printing

14 The individual pages are arranged so they will appear in the proper order when the comic book is assembled. Usually, two or more pages are printed on each side of a single sheet of paper. For example, page 2 might be printed on the left half of a sheet and page 23 would be printed on the right half. On the other side of the sheet, page 24 would be printed on the left and page 1 would be printed on the right. On the next sheet, pages 4 and 21 would be printed on one side, and 22 and 3 would be printed on the other. And so on. When the sheets of paper are stacked on top of each other and folded in the middle, the pages appear in the proper order. On some printing presses, as many as eight pages can be printed on each side of a large sheet, then cut and folded as required.

15 The plastic films for the four colors on each page are used to produce four aluminum printing plates. A bright light is projected through each film and onto the plate, which is coated with a chemical that is sensitive to light. Where there are dots on the film, they block the light and the chemical remains on the plate. Where there are no dots, the light passes through the film and burns away the chemical. This process is repeated for all of the pages that appear on each side of a single sheet (see Step 14).

16 The plate for the first color on the front side of the sheet is fastened around a circular drum in the printing press, and the plate for the back side is fastened around another drum below it. When the press is turned on, water flows over the rotating plates, while rollers with colored ink press against them. Where the chemical dots remain on the plates, the ink sticks; where the chemical has been burned away, the ink washes off and doesn't stick. The sheets of paper are fed between the rotating plates, and the front and back (top and bottom) sides are printed at the same time.

17 This process is repeated for each of the four colors. In some presses, a long roll of paper is fed between four sets of rollers, and all four colors are printed in a single pass through the press. The printed sheets or the roll of paper are then cut to the proper size, stacked, folded, and stapled or glued to form the finished comic book.

The Future

The future of comic books looks as dynamic as some of its superhero characters. Comic books offer a visual portal into a world of humor, action, and adventure that can stimulate a reader's imagination.

Where to Learn More

Books

Alvarez, Tom. *How to Create Action, Fantasy, and Adventure Comics.* Cincinnati, OH: North Lights Books, 1996.

Periodicals

Allstetter, Rob. "Fire Drill." *Wizard* (September 1996): 48-51.

Grant, Paul J. "Brush Off." *Wizard* (August 1995): 52-54, 56.

Grant, Paul J. "Letter Perfect." *Wizard* (February 1996): 44-47.

Tierney, Matt. "Separation Anxiety." *Wizard* (January 1996): 40-43.

White, Paul. "In the Can." *Wizard* (February 1994): 86-89.

Other

Comic Art and Graffix Gallery. http://www.comic-art.com (September 18, 2000).

Comic Book Fonts. http://www.comicbook-fonts.com (September 30, 2000).

The Comic Page. http://www.dereksantos.com/comicpage (September 30, 2000).

International Museum of Cartoon Art. http://www.cartoon.org (September 18, 2000).

Words and Pictures Museum. http://www.wordsandpictures.org (September 18, 2000).

—*Chris Cavette*

Corkscrew

In just over 100 years following the first patent, more than 350 corkscrews patents were granted in England and some 250 were awarded in the United States.

Background

Uncorking a bottle of wine presents a challenge. There is no way to grip the cork, which is completely recessed in the bottle's neck. Furthermore, pulling the cork out requires a force of 25-100 lb (110-450 N), depending on whether the cork is moist or dry (from the bottle being stored on its side or upright, respectively).

A corkscrew is an implement designed to mechanically remove the cork. It consists of a handle and a helical or screw-like blade that is commonly called a worm. (A corkscrew without a center post is actually a helix, with all the coils having the same radius; however, it is often called a spiral, even though that term technically means a curve with a constantly changing radius.) Screwing the worm through the center of the cork allows the corkscrew to grip the cork internally. Pulling on the handle draws both the corkscrew and the cork out of the bottle. Some corkscrews employ levers, springs, gears, or other mechanical devices to lessen the amount of force needed to pull out the cork.

History

Corks were used to seal bottles in the ancient Greek and Roman civilizations. Removing them was not difficult, however, because they extended above the rim of the bottleneck far enough to be grasped firmly. After the fall of Rome in the fifth century, cork bottle stoppers disappeared from use for a thousand years.

During the late sixteenth century, cork bottle stoppers reappeared as in England. Again, no special implement was required to remove them, since they were tapered in shape and

protruded a comfortable distance from the vessel. Blown-glass bottles began to replace barrels and skins as wine storage vessels. The bottles consisted of a squatty chamber topped with a tapered neck. These "shaft and globe" bottles were sealed with tapered corks wrapped with waxed linen, making it easy to grasp and remove the stoppers.

Between the late seventeenth and eighteenth centuries, two developments proceeded simultaneously. It is unclear which was the cause and which was the effect. Manufacturers began mass-producing bottles of uniform size by pouring molten glass into molds rather than hand-blowing bottles one at a time. This manufacturing method allowed production of tall, slender bottles with straight sides and cylindrical necks. This shape of bottle could be laid on its side for storage and shipment, an advantage that boosted the international wine trade. Tighter seals were required so the bottles would not leak, a challenge solved with cylindrical corks that were compressed prior to being forced into the bottle necks. Because of their tighter fit, these corks were harder to remove than the earlier, tapered versions.

The parallel development was the invention and evolution of the corkscrew. It began with the adaptation of the gun worm, a long-handled, helix-tipped tool that could be inserted in the barrel of a musket or pistol to retrieve wadding and unspent bullets. The earliest written reference to a true corkscrew dates back to 1681. Until 1720, when the word "corkscrew" came into use, the tools were called bottlescrews. Bronze and iron were sometimes used to make the worm, but steel became more popular because of its greater strength and its ability to retain a sharp point.

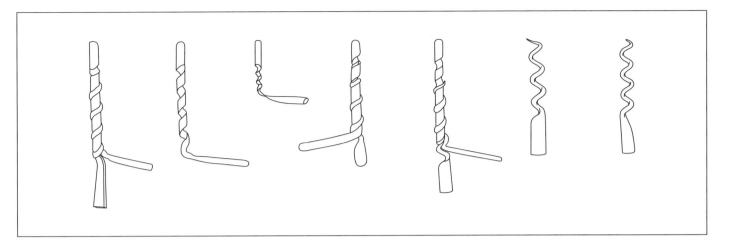

During the early 1700s, pocket corkscrews became popular. A metal or wooden sheath covered the spiral, protecting the worm and the owner's pocket. In some models, the sheath could be inserted in a loop at the end of the worm's shaft to provide an effective T-handle for the corkscrew. During the latter half of the eighteenth century, corkscrews became increasingly elaborate, using materials like silver, gold, exotic woods, ivory, and jewels. Multipurpose tools often combined corkscrews with devices such as pipe tobacco tampers, nutmeg graters, seals, and folding pocket knives. Dainty corkscrews were produced to open small bottles containing perfumes and medicines, for corks were the preferred sealer for all types of bottles until cork-lined metal bottlecaps became popular 1890.

The first corkscrew patent was issued in England to Reverend Samuel Henshall in 1795. It was a simple, T-shaped device with a steel worm protruding perpendicularly from the center of a handle made of bone or wood. Like many corkscrews of that period, brush bristles extended from one end of the handle; the brush was used to clean dust and sealing wax from the cork before opening a bottle. The innovative feature of Henshall's design was a flat disk, or button, mounted on the shaft connecting the worm to the handle. This kept the worm from being screwed too far through the cork; it also established a firmer contact between the corkscrew and the cork, making it easier to pull out the cork.

In 1802, a more complex mechanical corkscrew was patented by British engineer Edward Thomason. A bell-shaped cylinder surrounded the worm; setting the bottom of the cylinder on the top of the bottleneck positioned the worm vertically above the center of the cork. After the spiral had completely penetrated the cork, continued turning of the handle pulled the cork out of the bottle. The user could then hold the corkscrew over a fingerbowl, turn the handle in reverse, and automatically eject the cork without soiling his or her fingers.

A flurry of inventive activity in the late 1800s produced many variations of corkscrews enhanced with levers, gears, springs, and secondary screws to raise the cork out of the bottle. In just over 100 years following the first patent, more than 350 corkscrews patents were granted in England and some 250 were awarded in the United States. One of the more prolific American inventors in the field was W. Rockwell Clough of New Jersey. In 1876, he developed a machine that could bend a single piece of wire into a complete corkscrew; at one end of the helix, the wire was twisted into a fingerloop handle. In a subsequent refinement, he added a wooden sheath so the corkscrew could be carried in a pocket. After the metal bottlecap became popular, he developed a bottlecap remover and attached it to the end of the sheath. Clough's company eventually produced an estimated one billion inexpensive corkscrews, many of which were advertisements with brand names imprinted on the sheath.

Around the end of the nineteenth century, British corkscrew maker Thomas Truelove used a forming machine to forge steel worms. A grooved mandrel (forming rod) was rotated with a hand crank while a red-hot steel rod was inserted through an eyelet.

Two methods of attaching the handles to the worm.

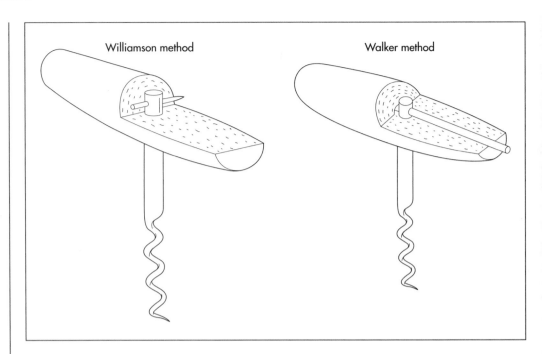

Williamson method   Walker method

The pliable rod was drawn into the grooves of the mandrel, shaping it into a helix.

One of the more significant corkscrew inventions of the twentieth century was the Screwpull, patented by Texas engineer Herbert Allen in 1978. Placing the device on top of the bottle, the user simply pushes a lever down and then pulls it back up to effortlessly insert the worm and extract the cork.

Raw Materials

Steel remains the preferred spiral material, with 440C stainless steel and tempered low-carbon steel being among the most popular choices. Levers and gears may be made from steel or from cast zinc alloy. Handles can be made from many materials, including common or exotic woods, bone, plastic, or various metals.

Design

There are two categories of worms. The auger type is very much like a wood screw, with sharp-edged threads cut into a shank. If the threads are cut so deeply that they extend through the center of the shank, the worm may actually have a hollow center. Auger corkscrew manufacturers claim that their sharp threads help them penetrate corks more easily than round-edge worms. Critics contend that they tend to slice up the center of older corks, ripping out the soft

middle of the cork without removing the entire stopper.

Worms with rounded edges are usually made by wrapping a very hot steel rod around a form to make a helix. The tip of the helix is sharpened to help it penetrate the cork easily. Some manufacturers pull the tip out of the helix's perimeter and position it in the center of its hollow core. This makes it easy to insert it into the center of the cork. However, the rest of the worm cannot exactly follow the path of the tip, so the center of the cork can be damaged by this type of corkscrew.

Some manufacturers of round-edge helixes score one or two shallow grooves into the outer surface of the worm to increase the gripping surface between the worm and the cork.

Round-edge corkscrew worms vary in design. They generally have between three and five turns in a helix that is about 2.5 in (6 cm) long. An open pitch—a comparatively wide spacing between turns—is less likely to cause damage to the cork than a tighter spacing. The outer diameter of the worm is usually 0.3-0.4 in (0.8-1 cm).

The Manufacturing Process

Forming a helix

1 An open worm is formed by forging a steel rod into a helix shape. After sharp-

ening one tip, the rod is heated to soften it for shaping. For example, 440C stainless steel is heated at 1,500°F (650°C) for 30 minutes, then at 2,100°F (1,000°C) for five minutes.

2 The hot rod is wound around a rotating cylindrical mandrel. To help guide the rod into a uniform spiral with the desired angle and spacing between successive coils, a grooved mandrel may be used.

3 Immediately after coiling, the spiral is hardened. This is done by reheating the worm to 1,500°F (650°C) and letting it cool slowly. Mild steel, on the other hand, is quenched by plunging the still- hot spiral into room-temperature water.

Forming an auger

4 An auger-type worm is formed by cutting threads into a tapered steel rod. With the rod secured in a lathe, a cutting head is mounted on the machine's lead screw, which pulls it parallel to the rod at a constant speed. This produces threads of the desired angle and spacing as the head cuts into the rod.

Finishing the worm

5 Whether it is a helix or an auger, the worm has a straight shaft extending beyond the coiled or machined section. This shaft, which will be inserted into a handle or other corkscrew mechanism, is prepared according to the customer's specifications. For example, it may be threaded, flattened, slotted, or drilled with a hole.

6 A surface treatment is usually applied to the worm to prevent corrosion. Typically, the worm is plated with chromium or nickel. To help it slide easily through the cork, the worm may be coated with a non-stick substance like Teflon or Silverstone.

Assembling the corkscrew

7 There are various ways to attach the worm to whatever type of handle or mechanism it will be mounted in. In the simplest case, insertion in a T-handle, two methods are used most commonly. The worm's flat or square shank may be inserted in a hole drilled in the handle and secured with two-part epoxy. Or a flat shank predrilled with a hole may be inserted into the handle and secured by driving a pin through the handle and hole.

The Future

Near the end of the twentieth century, some wineries began using synthetic corks. Because this material is harder than natural cork, popular corkscrew worms did not work well. Elongating the worm by adding an additional turn may have solved this problem. As improvements are made to synthetic cork composition, additional corkscrew modifications may be needed.

Where to Learn More

Books

Giulian, Bertrand B. *Corkscrews of the Eighteenth Century*. Pennsylvania: White Space Publishing, 1997.

O'Leary, Fred. *Corkscrews: 1000 Patented Ways to Open a Bottle*. Atglen, PA: Schiffer, 1997.

Perry, Evan. *Corkscrews and Bottle Openers*. Buckinghamshire, Great Britain: Shire Publications, 2000.

Other

"Corkscrews." http://www.corkscrew.com (October 2000).

ArborFood Wine Web. http://arborfood.com (October 2000).

—*Loretta Hall*

Cotton

Business revenue generated by cotton today is approximately $122.4 billion—the greatest revenue of any United States crop.

Background

Cotton is a shrubby plant that is a member of the Mallow family. Its name refers to the cream-colored fluffy fibers surrounding small cottonseeds called a boll. The small, sticky seeds must be separated from the wool in order to process the cotton for spinning and weaving. De-seeded cotton is cleaned, carded (fibers aligned), spun, and woven into a fabric that is also referred to as cotton. Cotton is easily spun into yarn as the cotton fibers flatten, twist, and naturally interlock for spinning. Cotton fabric alone accounts for fully half of the fiber worn in the world. It is a comfortable choice for warm climates in that it easily absorbs skin moisture. Most of the cotton cultivated in the United States is a short-staple cotton that grows in the American South. Cotton is planted annually by using the seeds found within the downy wool. The states that primarily cultivate cotton are located in the "Cotton Belt," which runs east and west and includes parts of California, Alabama, Arkansas, Georgia, Arizona, Louisiana, Mississippi, Missouri, New Mexico, North Carolina, Oklahoma, South Carolina, Tennessee, and Texas, which alone produces nearly five million bales. Together, these states produce approximately 16 million bales a year, second only to China. Business revenue generated by cotton today is approximately $122.4 billion—the greatest revenue of any United States crop.

The cotton plant is a source for many important products other than fabric. Among the most important is cottonseed, which is pressed for cottonseed oil that is used in commercial products such as salad oils and snack foods, cosmetics, soap, candles, detergents, and paint. The hulls and meal are used for animal feed. Cotton is also a source for cellulose products, fertilizer, fuel, automobile tire cord, pressed paper, and cardboard.

History

Cotton was used for clothing in present-day Peru and Mexico perhaps as long as 5,000 years ago. Also, cotton was grown, spun, and woven in ancient India, China, Egypt, and Pakistan, around 3000 B.C.

Cotton is not native to Western Europe. Around A.D. 800, Arabic traders likely introduced cotton to Spaniards. By the fourteenth century, Mediterranean farmers were cultivating the cotton plant and shipping the fiber to the Netherlands for spinning and weaving. British innovations in the late 1700s include water-powered spinning machinery, a monumental improvement over hand-spinning. An American named Samuel Slater, who worked with British machinery, memorized the plans for a machine spinner and returned to Rhode Island to set up Slater Mill, the first American textile mill to utilize machine spinners. This mill represents the beginning of the U.S. Industrial Revolution, built on the mechanism of the cotton industry.

Two developments spurred the cultivation of American cotton: cotton spinners and the cotton gin. The cotton gin, developed by Eli Whitney in 1793, easily removed tenacious cottonseeds. Southern plantation owners began planting cotton as a result of these innovations, using enslaved labor for harvesting the cotton. Vigorous cotton cultivation in the South using enslaved labor is considered one reason for friction between North and South that led to the Civil War.

Southern cotton was shipped to New England mills in huge quantities. As a result of machine spinning, weaving, and printing, Americans could cheaply purchase calico and it became universally worn. However, labor costs were significant in New England. Mill owners found ways to reduce those costs, first by employing women and immigrants who were often paid poorly, then by employing young children in the factories. After oppressive labor practices were largely halted, many factories moved to the South where labor was cheaper. (Unionizing efforts affected the profits of those mills.) Today, a fair amount of cotton is woven outside the United States where labor is less costly. Polyester, a synthetic, is often used along with cotton, but has little chance of supplanting the natural fiber.

Raw Materials

The materials required to take cotton bolls to spun cotton include cottonseeds for planting; pesticides, such as insecticides, fungicides, and herbicides, to battle disease and harmful insects; and fertilizers to enrich the soil.

There are agricultural requirements for growing cotton in the United States. Cotton has a long growing season (it can be as long as seven months) so it is best to plant cotton early—February in Texas but as late as June in northern cotton-growing states such as Missouri. Cotton should not be planted before the sun has warmed the soil. It performs best in well-drained, crumbly soils that can hold moisture. It can be grown between latitudes of 30° north and 30° south. Good cotton crops require a long, sunny growing season with at least 160 frost-free days and high moisture levels resulting from rainfall or irrigation during the growing season. However, too much rain during harvest or strong winds during picking can damage the open bolls and load the fiber with too much water, which can ruin the cotton in storage. Generally, a cotton farmer must farm about 2,000 acres (20,000 hectares) if the operation is to be economically viable. On average, an acre will produce about 1.5 bales of cotton, or about 750 lb (340 kg).

The Production Process

1 In spring, the acreage is cleared for planting. Mechanical cultivators rip out weeds and grass that may compete with the cotton for soil nutrients, sunlight, and water, and may attract pests that harm cotton. The land is plowed under and soil is broken up and formed into rows.

2 Cottonseed is mechanically planted by machines that plant up to 12 rows at a time. The planter opens a small furrow in each row, drops in seed, covers them, and then packs more dirt on top. Seed may be deposited in either small clumps (referred to as hill-dropped) or singularly (called drilled). The seed is placed 0.75 to 1.25 in (1.9 to 3.2 cm) deep, depending on the climate. The seed must be placed more shallowly in dusty, cool areas of the Cotton Belt, and more deeply in warmer areas.

3 With good soil moisture and warm temperature at planting, seedlings usually emerge five to seven days after planting, with a full stand of cotton appearing after about 11 days. Occasionally disease sets in, delaying the seedlings' appearance. Also, a soil crust may prevent seedlings from surfacing. Thus, the crust must be carefully broken by machines or irrigation to permit the plants to emerge.

4 Approximately six weeks after seedlings appear, "squares," or flower buds, begin to form. The buds mature for three weeks and then blossom into creamy yellow flowers, which turn pink, then red, and then fall off just three days after blossoming. After the flower falls away, a tiny ovary is left on the cotton plant. This ovary ripens and enlarges into a green pod called a cotton boll.

5 The boll matures in a period that ranges from 55 to 80 days. During this time, the football-shaped boll grows and moist fibers push the newly formed seeds outward. As the boll ripens, it remains green. Fibers continue to expand under the warm sun, with each fiber growing to its full length—about 2.5 in (6.4 cm)—during three weeks. For nearly six weeks, the fibers get thicker and layers of cellulose build up the cell walls. Ten weeks after flowers first appeared, fibers split the boll apart, and cream-colored cotton pushes forth. The moist fibers dry in the sun and the fibers collapse and twist together, looking like ribbon. Each boll contains three to five "cells," each having about seven seeds embedded in the fiber.

Most steps involved in the production of cotton have been mechanized, including seeding, picking, ginning, and baling. Samples are taken from the bales to determine the quality of the cotton.

6 At this point the cotton plant is defoliated if it is to be machine harvested. Defoliation (removing the leaves) is often accomplished by spraying the plant with a chemical. It is important that leaves not be harvested with the fiber because they are considered "trash" and must be removed at some point. In addition, removing the leaves minimizes staining the fiber and eliminates a source of excess moisture. Some American crops are naturally defoliated by frost, but at least half of the crops must be defoliated with chemicals. Without defoliation, the cotton must be picked by hand, with laborers clearing out the leaves as they work.

7 Harvesting is done by machine in the United States, with a single machine replacing 50 hand-pickers. Two mechanical systems are used to harvest cotton. The picker system uses wind and guides to pull the cotton from the plant, often leaving behind the leaves and rest of the plant. The stripper system chops the plant and uses air to separate the trash from the cotton. Most American cotton is harvested using pickers. Pickers must be used after the dew dries in the morning and must conclude when dew begins to form again at the end of the day. Moisture detectors are used to ensure that the moisture content is no higher than 12%, or the cotton may not be harvested and stored successfully. Not all cotton reaches maturity at the same time, and harvesting may occur in waves, with a second and third picking.

8 Next, most American cotton is stored in "modules," which hold 13-15 bales in water-resistant containers in the fields until they are ready to be ginned.

9 The cotton module is cleaned, compressed, tagged, and stored at the gin. The cotton is cleaned to separate dirt, seeds, and short lint from the cotton. At the gin, the cotton enters module feeders that fluff up the cotton before cleaning. Some gins use

vacuum pipes to send fibers to cleaning equipment where trash is removed. After cleaning, cotton is sent to gin stands where revolving circular saws pull the fiber through wire ribs, thus separating seeds from the fiber. High-capacity gins can process 60, 500-lb (227-kg) bales of cotton per hour.

10 Cleaned and de-seeded cotton is then compressed into bales, which permits economical storage and transportation of cotton. The compressed bales are banded and wrapped. The wrapping may be either cotton or polypropylene, which maintains the proper moisture content of the cotton and keeps bales clean during storage and transportation.

11 Every bale of cotton produced in the United States must be given a gin ticket and a warehouse ticket. The gin ticket identifies the bale until it is woven. The ticket is a bar-coded tag that is torn off during inspection. A sample of each bale is sent to the United States Department of Agriculture (USDA) for evaluation, where it is assessed for color, leaf content, strength, fineness, reflectance, fiber length, and trash content. The results of the evaluation determine the bale's value. Inspection results are available to potential buyers.

12 After inspection, bales are stored in a carefully controlled warehouse. The bales remain there until they are sold to a mill for further processing.

Quality Control

Cotton growing is a long, involved process and growers must understand the requirements of the plant and keep vigilant lookout for potential problems. Pests must be managed in order to yield high-quality crops; however, growers must use chemicals very carefully in order to prevent damage to the environment. Defoliants are often used to maximize yield and control fiber color. Farmers must carefully monitor moisture levels at harvesting so bales will not be ruined by excess water during storage. Soil tests are imperative, since too much nitrogen in the soil may attract certain pests to the cotton.

Expensive equipment such as cotton planters and harvesters must be carefully maintained. Mechanical planters must be set carefully to deposit seed at the right depth, and gauge wheels and shoes must be corrected to plant rows at the requisite spot. Similarly, improperly adjusted machinery spindles on harvesting machines will leave cotton on the spindle, lowering quality of the cotton and harvesting efficiency. A well-adjusted picker minimizes the amount of trash taken up, rendering cleaner cotton.

Byproducts/Waste

There is much discussion regarding the amount of chemicals used in cotton cultivation. Currently, it is estimated that growers use, on average, 5.3 oz (151 g) of chemicals to produce one pound of processed cotton. Cotton cultivation is responsible for 25% of all chemical pesticides used on American crops. Unfortunately, cotton attracts many pests (most notably the boll weevil) and is prone to a number of rots and spotting, and chemicals are used to keep these under control. There are concerns about wildlife poisoning and poisons that remain in the soil long after cotton is no longer grown (although no heavy metals are used in the chemicals). As a result, some farmers have turned to organic cotton growing. Organic farming utilizes biological control to rid cotton of pests and alters planting patterns in specific ways to reduce fungicide use. While this method of cultivation is possible, an organically grown crop generally yields less usable cotton. This means an organic farmer must purchase, plant, and harvest more acreage to yield enough processed cotton to make the crop lucrative, or reduce costs in other ways to turn a profit. Increasingly, state university extension services are working with cotton farmers to reduce chemical use by employing certain aspects of biological control in order to reduce toxins that remain in the land and flow into water systems.

Where to Learn More

Books

Daniel, Pete. *Breaking the Land*. Champaign, IL: University of Illinois Press, 1987.

Johnson, Guinevere. *Cotton.* Let's Investigate Series. Mankato, MN: The Creative Co., 1999.

Other

The Cotton Pickin' Web. http://ipmwww. ncsu.edu/CottonPickin (January 2, 2001).

Land of Cotton Online Newsmagazine for the Cotton Industry. http://www.landofcotton.com (January 2, 2001).

National Cotton Council of America. Education Materials. http://www.cotton.org/ ncc/education (January 2, 2001).

The Organic Cotton Site. http://www.sustainablecotton.org. (January 2, 2001).

—*Nancy E.V. Bryk*

Cuckoo Clock

Background

The cuckoo clock is a favorite souvenir of travelers in Germany, Austria, and Switzerland, and particularly the Black Forest region of Germany. The clock is prized for a number of its features. The outer worked wood case is usually made of beautiful dark wood that is intricately carved with folk and forest scenes. The clock itself is made in the premier clock- and watch-making area of the world. And, finally, there is the cuckoo and its fellows. On the hour (and often the half- and quarter-hour as well), the charming carved bird pops out of a door to sing the hour in a melodic "Cuckoo! Cuckoo!" call. He is often introduced or followed by a parade of townspeople, forest creatures, or other animals that circle through another door and seem to celebrate the passing of every hour and the timelessness of their carefully crafted clock home.

The cuckoo clock known today is the most popular form of ornamental clock—one that is decorative as well as functional. When the tiny wood cuckoo emerges to call the hour, two small pipes attached to two miniature bellows make his call. The sets of pipe-and-bellows are mounted on either side of the clock with slots cut through the wood frame opposite the bellow vents to allow the sound to be heard. Inside the clock, a finely made set of **brass** clockworks controls the time-telling. Two weights shaped like pine cones that dangle from the ends of chains and a pendulum that is tipped with a leaf add to the traditional appearance, although these are only decorative on modern clocks that are spring-driven.

History

The cuckoo clock has an impressive parent in the Black Forest clock. The provinces of Baden and Württemburg (now the province of Baden-Württemburg) lie deep in the Black Forest region of Germany. Winters there are long, dark, cold, and characterized by deep snowfalls. With forestry and agriculture limited during this season, a cottage industry in the production of clocks grew in the Black Forest. Glass-making was a traditional craft, and clock-making sprang indirectly from this when, in about 1640, a traveler introduced a simple Bohemian clock operated by three wheels on a train (continuous drive), a verge escapement (the device that allows the train to advance a controlled amount by restraining it with weights), and a foliot (a balance bar). The clock was not ornamented.

The local citizens learned how to copy the clock and make the tools to craft it. They also worked together as a group with specialists in frame-making, manufacturing the clockworks, making and painting dials, brass founding, making chains and gongs, finishing metal parts, and performing many supporting tasks. The clockmaker made his own patterns and styles; parts for his clocks were unique and not interchangeable with other makers. By the late 1700s, the clocks were a profitable export for the region and were sold as far away as Russia.

The cuckoo clock may have been invented in about 1730 by Franz Anton Ketterer, a well-known Black Forest clockmaker from Schönwald. The cuckoo's sound was simply incorporated in the contemporary clocks of the day. They had face shields—full front plates that were enameled with the face near the center—rather than the wood frame developed later. Ketterer's clocks were driven by suspended weights shaped like pine

The cuckoo clock may have been invented in about 1730 by Franz Anton Ketterer, a well-known Black Forest clockmaker from Schönwald.

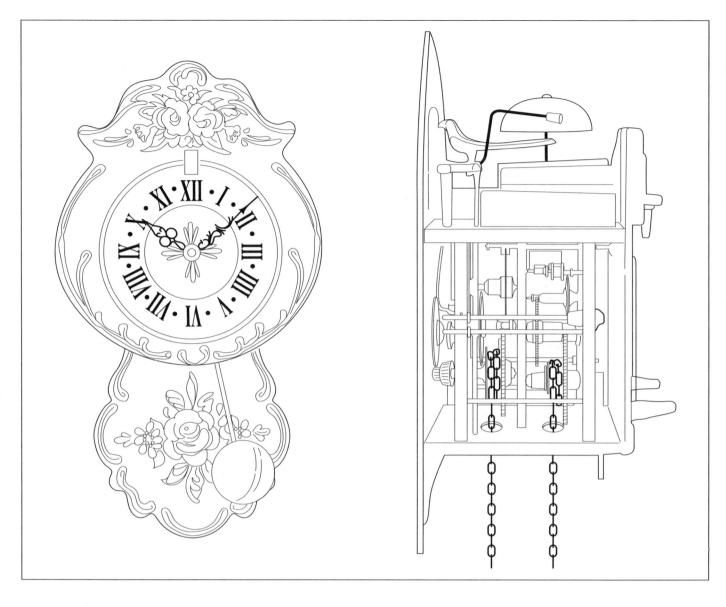

A cuckoo clock and a sideview of
its mechanism.

cones, and these were adapted to the wood-frame style later. Ketterer used the church organ pipe as the basis for the production of the cuckoo's sound, and his clock-making abilities were so skilled that the cuckoo clock became known for its reliability as a timepiece.

The variety of cuckoo clocks reflects clock-making styles of the time. A clock dating from 1770 may have a painted face shield with roses and castles. This was the English style of the day, but was popularized throughout southern Germany and Central Europe; the same design appears on the sides of painted barges. Soon, the decoration was modified to suit the targeted market. For instance, the French liked large bouquets of bright flowers and called the cuck-

oo clocks "Swiss clocks" even though most were made in Germany. Scandinavians preferred hexagonal or octagonal faces, while the Dutch and Belgians liked tin or porcelain dials. In England, the clocks were called "Dutch" clocks (possibly from "Deutsche," meaning German), and they were simple mahogany rims with glasses held in place with brass bezels.

By the mid-eighteenth century, cuckoo clocks moved from the peasant or cottage industry to factories. By 1850, a style called the "hunting lodge" or "chalet" style dominated; the frame shows a lodge at the base of the clock, the clock dial over the roof, and carved trees and animals rising above the dial to the top of the frame. In "The Cricket on the Hearth," author Charles

Dickens describes a clock with the figure of a haymaker with a scythe who moves with the pendulum. One particular style called the "Surrerwerk" or whizzing work strikes with the sound of twelve blows like the sound of small hammers. Usually, the cuckoo clock had two drive trains, one for the clock movement and the other for the so-called striking train, or the sounds and actions produced with the striking of the hour.

The movements became standardized in style, size, and materials. European movements are of brass and steel, and American movements are brass. The numerals on the dial are painted in German gothic style. Modern cuckoo clocks have retained the suspended pine-cone weights. Some large cuckoo clocks made at the end of the nineteenth century also housed barometers. Later clocks from about 1900 have wood frames, brass wheel works, and a wooden carved cuckoo on a sweeping stand that shifted forward to chime the hour. Inlaid wood has also been used to make cuckoo clocks, notably those from Northern Italy in the Ampezzo region. Late in the twentieth century, the cuckoo clock entered the digital age when manufacturers began equipping some models with quartz clocks that play twelve different tunes, one for each hour, and an automatic shutoff to silence the bird for a programmable number of hours during the night.

Raw Materials

Wood is the critical raw material for the manufacture of cuckoo clocks, because the wood casing is the primary feature that distinguishes the cuckoo clock in appearance. Cuckoo clocks are made from the wood of the linden tree, a hardwood that grows in Europe. Some parts of the housing may also be made of walnut. Skilled wood workers purchase the linden and walnut woods well in advance so the wood can be aged for two years. Depending on what the craftsman wants, it may be purchased in logs with the bark removed or in block-like lengths.

The cuckoo clock is also distinguished by the cuckoo and its sounds. The pipes and bellows that make the cuckoo's call are also made of wood. Clocks that play tunes are fitted with music boxes. The music boxes and the mechanical movements for the clocks (as well as small parts like the clock's hands) are produced by specialized subcontractors. The lead pine cone weights and the leaf-shaped weight on the end of the pendulum are made of lead and are produced in metal foundries by pouring a melted lead alloy in tempered metal molds. The foundries that produce these weights are also experts in small, detailed metalwork.

Design

As the history of the cuckoo clock suggests, design of the clock and its highly recognizable parts is based on tradition. Clock manufacturers have developed their own styles of chalets and forest scenes for the wood work as well as particular "casts of characters" for the cuckoo bird itself and the villagers or animals that may share "action scenes" with the cuckoo. New lines or styles of clocks are not likely because customers buy cuckoo clocks for their traditional style. The addition of digital features increases the variety of music and bird songs that the clocks can produce, but clocks with digital enhancements have not yet proven to be more popular than traditional models.

The Manufacturing Process

1 Manufacture of the cuckoo clock begins in the hands of the wood worker. The craftsman selects the pieces of wood to be used for the particular clock and cuts them to the approximate lengths and shapes he will need. Power tools and hand tools are used for this part of the process; hand tools may include measuring tools, saws, rasps, and files for shaping, drilling tools, abrasives including sandpaper, and adhesives and clamps. The box-like case or cabinet for the clock works is cut, fitted, and glued together.

The outer frame—the decorative part of the clock featuring the traditional forest and chalet scene—begins with a stenciled design on paper. The craftsmen make and collect sets of stencils based on their own drawings and those that have been handed down. The sets of stencils are made for specific sizes of clocks. After choosing the stencil for the size and style of clock, the wood worker draws the design on the wood and begins carving and shaping the frame. When the frame and the case are complete, both are stained and left to dry.

2 When the frame and case are dried, the clock is assembled by first mounting the movement in the case. In the old days of village manufacture, the craftsmen who carved the wood and assembled the clockworks probably lived in the same village. The clockmaker poured and handcrafted the internal workings of the clock himself and assembled them. Today, manufacturers buy preassembled clock movements, and the process is reduced to fitting it in the case and properly fixing it in place with wood screws or other fasteners.

3 The sound-making devices are attached to the top of the clock. These include the pipes and bellows for the cuckoo sound and the music box. Attachments that are usually extensions of drive chains are linked to the sets of wire hooks and metal cams and pins that activate the cuckoo and any other moving figures and the doors. The cuckoo is connected to its bellows operation, and the other figures are mobilized by the strike movement. A third movement initiates the playing of the music box. Finally, the pendulum and weight chains are connected to the movement and the lead weights are clipped to the chain ends.

4 The assembled clock is carefully packaged to protect the moving parts and the delicate carved framework. Individually boxed clocks are packed in cartons for shipping and distribution.

Quality Control

Quality control consists of only two steps. Quality is built into every cuckoo clock because each one is hand made. Quality is the mark of the craftsman, and, as with all handcrafted products, gifted and highly trained wood workers will not risk their reputations on poorly made clocks. The final quality step is a complete examination of the finished piece and a trial operation.

Byproducts/Waste

There are no byproducts from cuckoo clock manufacture, and waste is minimal. Some wood scraps and shavings result from crafting the case and carving the frame, but attentive selection of the right pieces of wood for the project and stencils that accentuate the character of the wood limit the volume of scrap. The wood is also too expensive for the wood workers to waste.

The Future

New cuckoo clocks are still among of the most sought-after souvenirs of vacations in the heart of Europe and especially in Germany's Black Forest region. In America, many families can trace their roots to Germany, Austria, Switzerland, and other European localities where cuckoo clocks are traditional ornaments for the home. Consequently, there is a market in America for clocks that represent the best traditions of cuckoo-clock making.

Cuckoo clocks are also highly prized antiques. Hand-crafted clocks with "provenance" (a traceable history) are sought by collectors, but antique hunters also search for factory-made cuckoo clocks. Those dating from the 1850s are highly sought based on the name of the maker; names like Gustav Becker, the United Freiburg Clock Factory (which Becker joined by 1900), Winterhalder & Hofmeier, Kienzle, Junghans, and the Hamburg American Clock Company (which copied American-made clocks for sales in Germany) are among the most collectible. Typically, the antique market also injects life into sales of newly manufactured collectibles because they are more affordable.

Even though some models of cuckoo clocks are now outfitted with quartz movements and electronics, part of the cuckoo clock's charm may be its old-fashioned mechanical movement. When paired with beautifully carved wood and rustic style, the spell of the cuckoo's song on the hour is guaranteed to bring smiles to those who prize childlike delights and exquisite craftsmanship for years to come.

Where to Learn More

Books

Bruton, Eric. *The History of Clocks and Watches.* New York: Rizzoli International Publications, Inc., 1979.

Coggins, Frank W. *Clocks: Construction, Maintenance & Repair.* Blue Ridge Summit, PA: TAB Books, Inc., 1984.

Fleet, Simon. *Clocks*. London: Octopus Books, Inc., 1972.

Hunter, John. *Clocks: An Illustrated History of Timepieces*. New York: Crescent Books, 1991.

Kadar, Wayne Louis. *Clock Making for the Woodworker*. Blue Ridge Summit, PA: TAB Books, Inc., 1984.

Lloyd, H. Allan. *The Complete Book of Old Clocks*. New York: G. P. Putnam's Sons, 1964.

Nicholls, Andrew. *Clocks in Color*. New York: Macmillan Publishing Co., Inc., 1975.

Smith, Alan. *The Antique Collector's Guides: Clocks and Watches*. New York: Crescent Books, 1989.

Tyler, E. J. *European Clocks*. New York: Hawthorne Books, Inc., 1969.

—*Gillian S. Holmes*

Dishwasher

Background

Washing dishes is not the most rewarding task. Cooking can be creative, but cleaning up afterward seems like a waste of time and leaves the person washing complaining about "dishpan hands." The development of the dishwasher has helped relieve some of the monotony, as well as the grease and grime. It operates on a simple principle of washing dishes that have been placed on racks inside the machine with multiple jets of water. The modern dishwasher has features that cater to fine glassware or the toughest pots and pans; multiple cycles that clean, sanitize, and dry; and under-the-counter or stand-alone models for every size, use, and price range. It is far from perfect; tough foods may need personal attention before and after dishes and pans are cleaned in the dishwasher, and few owners of crystal glassware and fine china are willing to trust them to a machine. But the dishwasher, like other kitchen appliances invented and improved in the twentieth century, is a fixture in many kitchens of the twenty-first century.

History

The major obstacle to washing dishes has always been the availability of water. Early civilizations used limited numbers and types of dishes, utensils, and cookware and carried them to streams, ponds, or troughs of water for cleaning. The second choice was to carry the water to the dishes. Women carried water in buckets from communal water sources or from private pumps behind their homes or apartment buildings into the early twentieth century, when indoor plumbing finally brought water indoors, not only for bathing but for kitchen use as well.

The first dishwashers were patented in about 1850, but, like machines for washing clothes, they were large contraptions that used steam power and supplies of heated water to soak many dishes at a time. In some models, the dishes were held on cradles that rocked through the water; others had paddles that sloshed water around the dishes or circular racks that held the dishes and rotated to circulate them through the water. An assortment of propellers, plunging casings bearing the dishes, and plungers that drove water over the dishes were incorporated in other machines. In 1875, C. E. Hope-Vere created a machine that directed sprays of water toward racked dishes; the idea of the water jets was adopted by other inventors including A. W. Bodell, whose model was introduced in 1906. Another, the Blick machine, used a propeller that sprayed jets of water over racks filled with dishes. This basic idea is the one used today.

The first publicly displayed models were introduced in about 1915, but the dishwasher was not widely manufactured and sold to private families until about 1930. The dishwasher was not an immediate hit. The refrigerator was introduced at about the same time and swept America; but this is logical because food preservation is far more important than dishwashing. The machines were also too inefficient to completely eliminate hand work; to be fair, this was not entirely the fault of the dishwashers—soaps of the day were not suited to the task. By the 1950s, special dishwashing soaps that clean without sudsing and rinse away began to be developed especially for dishwashers, and the public began to demonstrate more interest. The automatic dishwasher is still not an absolute in every kitchen, but, by the 1970s when more women

A dishwasher.

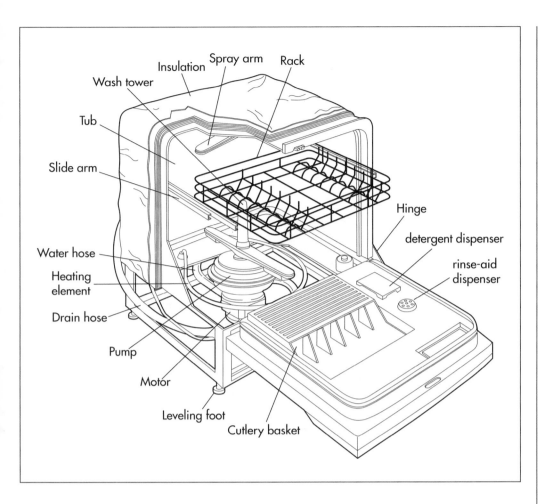

Insulation
Spray arm
Rack
Wash tower
Tub
Slide arm
Hinge
detergent dispenser
rinse-aid dispenser
Water hose
Heating element
Drain hose
Pump
Motor
Leveling foot
Cutlery basket

began working outside the home, the built-in dishwasher was seen as an asset.

Raw Materials

The major components of a dishwasher are made of steel and plastic. The basic structure consists of a steel frame assembly and a steel door panel. Sheets of stainless steel are purchased and fabricated in the required pieces and shapes in the factory; both the door and the wrap-around cabinet for stand-alone models are purchased as coiled sheet steel that has been prefinished in several standard colors. Other small steel parts are designed in house but made by suppliers to the manufacturer's specifications.

The racks that hold the dishes are also made of steel, but it is delivered to the factory as coiled wire. To coat the rack tines to prevent them from scratching dishes, the racks are dipped in plastic in the form of powder polyvinyl chloride (PVC) or nylon.

The inner box that holds the racks and the washer arms is called the tub. It is a single piece (not counting the piece lining the inside of the door) that is injection-molded in the plant. The injection molding is done with pellets of calcium-reinforced polypropylene plastic. This plastic is respected for its strength and for the fact that it is inert; that is, it won't react with chemicals like those in detergents and is resistant to water and heat. Many other parts including the basket for cutlery, containers for detergent, and the wash tower and spray arms are also injection molded.

Motors, pumps, and electrical controls and components are made by subcontractors in accordance with designs by the dishwasher manufacturer.

Design

The engineers who design dishwashers are interested in improving two key features of their products. Efficient cleaning is, of course, the biggest marketing feature, but consumers are also interested in quiet operations. Cleaning systems consist of a wash tower and sprayer arms, but the openings, the power of the water pump, and positions of racks relative to the washers are all design elements. In the late 1990s, consumers became increasingly interested in the dishwasher as a tool for sanitizing dishes, so design efforts have been aimed at adding heating methods for killing germs.

Manufacturers have taken different approaches to keeping their dishwashers as quiet as possible. Maytag, for example, uses a single, powerful motor for all operations and wraps the outside of its machines with heavy insulation. By contrast, Amana Appliances has equipped its dishwashers with two motors (one to operate the water pump and another for the drainage system). Together, the two motors have the same horsepower as single-motor units, but less insulation is needed for quiet operation.

Design of the exterior of a stand-alone model is somewhat more sophisticated because it has to have an attractive outer cabinet. Usually the top of the stand-alone dishwasher is a wooden cabinet top so the machine will function as a spare work surface.

The Manufacturing Process

1 A dishwasher begins to take shape with the injection molding of the tub. Two molds—the cavity relief mold for the outside of the tub and the core relief mold for the inside—have previously been etched into a steel tool, that, when fitted together, contains a void or space that is the shape of the tub. The tool halves are held together in the chamber of the injection molding machine. Pellets of polypropylene are melted in the machine at high temperature and injected by pressure into the void in the tool. The high pressure and liquid state of the plastic forces the plastic into every pocket and crevice in the mold inside the tool. The tool opens to release the tub, which is still hot.

2 The warm tub is conveyed to a cooling area and cooled to a temperature that is easy for assembly workers to handle. Other plastic parts are also made by injection molding, and these smaller pieces are stored in bins (with one kind of part only per bin) that can be moved to the assembly area as needed.

3 In another part of the plant, the steel components of the dishwasher are made. Outer cabinets for stand-alone models and the doors for all models are cut and stamped into shape from stainless steel in the form of coils that are prefinished on one side. Flat steel bars that will be assembled into the dishwasher's frame are sheared to length. The racks are also formed with tools that trim, de-bur, and shape wire into the racks in two welding steps. The perimeter of the rack is called the "mat," and a tool welds all the wire pieces of the mat together at the same time. Similarly, the little pieces or tines that support the dishes are welded into place simultaneously. The completed rack is taken by conveyor to a cleaning station where it is cleaned and prepared to receive its PVC coating. The PVC is in the form of a fine powder that is baked onto the rack. The coated rack is then cured to finish forming the PVC coating and to allow it to cool.

4 Dishwashers are assembled at work stations along an assembly line. The workers are responsible for sets of pieces that are taken from bins alongside the workers. The frame is assembled first, and the motor or motors are attached to special mounts on the frame. The motors are provided to the line workers as completed assemblies. The tub is fitted and fastened into the frame over the motor or motors.

5 With the tub in place, the interior components are installed beginning with the filtering system. The washtower and arms are attached followed by sets of rack rollers to support the racks and allow them to be rolled in and out of the machine so that dishes can be loaded easily. The racks are put in place along with the cutlery basket.

6 The door assembly is completed by installing the detergent dispenser and rinse-agent cups and the controls. The door is attached to the front of the dishwasher. The exterior is completed by finishing the electrical connections and feed lines (for clean and dirty water), and the exterior is insulated to reduce noise and the effects of heat that might warp counter tops and cabinets. Insulation is prefabricated with the insulating fibers wrapped in a foil-like covering. Called "bagged insulation," it is wrapped around the machine and packed inside the toe space. Under-counter models

are now complete. Stand-alone models are finished by attaching the wrap-around cabinet and wood top. Each completed machine is loaded onto a cart to be moved to the packing area.

7 In the packing area, styrofoam bumper sections are placed along the edges of the machine and enclosed by a carton. Packets of instructions and other materials are placed on top of the machine in the carton, and the carton is sealed and moved to a storage area for shipping.

Quality Control

Quality control is assured by three basic processes. First, the assembly line workers are trained in quality issues and can reject parts or partially assembled machines. Second, the assembly process is overseen by line supervisors; when assembly is complete, quality engineers inspect the finished machine and test selected units. The most important part of the quality control process may be a design step that Amana Appliances calls a failure mode effects and analysis (FMEA). As soon as problems are observed during assembly or are reported by customers through the warranty process, corrective steps are taken. The analysis is a highly regimented learning process that continuously cycles improvements, customer feedback, and corrective actions through the marketing and design process so new models and lines benefit from any changes to the old.

Byproducts/Waste

Dishwasher manufacturers produce a range of lines of dishwashers and other appliances but no true byproducts. Waste is virtually eliminated by a thorough recycling program that includes metals, plastics, and paper.

The Future

All industries struggle with the issue of how to attract more customers to their product. For dishwashers, the market is still growing because it is a more open field than for other appliances. Marketers discuss this in terms of market penetration; for example, 99.8% of American households own refrigerators, but only 56.5% have dishwashers. This seems promising for dishwasher manufacturers, but it shows that potential customers who don't have dishwashers may not see that these appliances provide benefits over hand-washing dishes. To attract customers, the latest advance in dishwasher manufacture is the sanitization option with a high heat cycle to kill bacteria. Quiet operation, energy efficiency, and clean dishes without prerinsing are existing features that are continuously being improved.

Where to Learn More

Books

Cohen,Daniel. *The Last Hundred Years: Household Technology.* New York: M. Evans and Company, Inc., 1982.

Weaver, Rebecca, and Rodney Dale. *Machines in the Home.* New York: Oxford University Press, Inc., 1992.

Other

Amana Appliances. http://www.amana.com (August 2000).

—*Gillian S. Holmes*

DNA Synthesis

First developed in the early 1980s, the polymerase chain reaction (PCR) has become a multi-billion dollar industry with the original patent being sold for $300 million dollars.

Background

Deoxyribonucleic acid (DNA) synthesis is a process by which copies of nucleic acid strands are made. In nature, DNA synthesis takes place in cells by a mechanism known as DNA replication. Using genetic engineering and enzyme chemistry, scientists have developed man-made methods for synthesizing DNA. The most important of these is polymerase chain reaction (PCR). First developed in the early 1980s, PCR has become a multi-billion dollar industry with the original patent being sold for $300 million dollars.

History

DNA was discovered in 1951 by Francis Crick, James Watson, and Maurice Wilkins. Using x-ray crystallography data generated by Rosalind Franklin, Watson and Crick determined that the structure of DNA was that of a double helix. For this work, Watson, Crick, and Wilkins received the Nobel Prize in Physiology or Medicine in 1962. Over the years, scientists worked with DNA trying to figure out the "code of life." They found that DNA served as the instruction code for protein sequences. They also found that every organism has a unique DNA sequence and it could be used for screening, diagnostic, and identification purposes. One thing that proved limiting in these studies was the amount of DNA available from a single source.

After the nature of DNA was determined, scientists were able to examine the composition of the cellular genes. A gene is a specific sequence of DNA base pairs that provide the code for the construction of a protein. These proteins determine the traits of an organism, such as eye color or blood type. When a certain gene was isolated, it became desirable to synthesize copies of that molecule. One of the first ways in which a large amount of a specific DNA was synthesized was though genetic engineering.

Genetic engineering begins by combining a gene of interest with a bacterial plasmid. A plasmid is a small stretch of DNA that is found in many bacteria. The resulting hybrid DNA is called recombinant DNA. This new recombinant DNA plasmid is then injected into bacterial cells. The cells are then cloned by allowing it to grow and multiply in a culture. As the cells multiply so do copies of the inserted gene. When the bacteria has multiplied enough, the multiple copies of the inserted gene can then be isolated. This method of DNA synthesis can produce billions of copies of a gene in a couple of weeks.

In 1983, the time required to produce copies of DNA was significantly reduced when Kary Mullis developed a process for synthesizing DNA called polymerase chain reaction (PCR). This method is much faster than previous known methods producing billions of copies of a DNA strand in just a few hours. It begins by putting a small section of double stranded DNA in a solution containing DNA polymerase, nucleotides and primers. The solution is heated to separate the DNA strands. When it is cooled, the polymerase creates a copy of each strand. The process is repeated every five minutes until the desired amount of DNA is produced. In 1993, Mullis's development of PCR earned him the Nobel Prize in Chemistry. Today, PCR has revolutionized the fields of medical diagnostics, forensics, and microbiology. It is said to be one of the most important developments in genetic research.

Background

The key to understanding DNA synthesis is understanding its structure. DNA is a long chain polymer made up of chemical units called nucleotides. Also known as genetic material, DNA is the molecule that carries information that dictates protein synthesis in most living organisms. Typically, DNA exists as two chains of chemically linked nucleotides. These links follow specific patterns dictated by the base pairing rules. Each nucleotide is made up of a deoxyribose sugar molecule, a phosphate group, and one of four nitrogen containing bases. The bases include the pyrimidines thymine (T) and cytosine (C)and the purines adenine (A) and guanine (G). In DNA, adenine generally links with thymine and guanine with cytosine. The molecule is arranged in a structure called a double helix which can be imagined by picturing a twisted ladder or spiral staircase. The bases make up the rungs of the ladder while the sugar and phosphate portions make up the ladder sides. The order in which the nucleotides are linked, called the sequence, is determined by a process known as DNA sequencing.

In a eukaryotic cell, DNA synthesis occurs just prior to cell division through a process called replication. When replication begins the two strands of DNA are separated by a variety of enzymes. Thus opened, each strand serves as a template for producing new strands. This whole process is catalyzed by an enzyme called DNA polymerase. This molecule brings corresponding, or complementary, nucleotides in line with each of the DNA strands. The nucleotides are then chemically linked to form new DNA strands which are exact copies of the original strand. These copies, called the daughter strands, contain half of the parent DNA molecule and half of a whole new molecule. Replication by this method is known as semiconservative replication. The process of replication is important because it provides a method for cells to transfer an exact duplicate of their genetic material from one generation of cell to the next.

Raw Materials

The primary raw materials used for DNA synthesis include DNA starting materials, taq DNA polymerase, primers, nucleotides, and the buffer solution. Each of these play an important role in the production of millions of DNA molecules.

Controlled DNA synthesis begins by identifying a small segment of DNA to copy. This is typically a specific sequence of DNA that contains the code for a desired protein. Called template DNA, this material is needed in concentrations of about 0.1-1 micrograms. It must be highly purified because even trace amounts of the compounds used in DNA purification can inhibit the PCR process. One method for purifying a DNA strand is treating it with 70% ethanol.

While the process of DNA replication was know before 1980, PCR was not possible because there were no known heat stable DNA polymerases. DNA polymerase is the enzyme that catalyzes the reactions involved in DNA synthesis. In the early 1980s, scientists found bacteria living around natural steam vents. It turned out that these organisms, called *thermus aquaticus*, had a DNA polymerase that was stable and functional at extreme levels of heat. This *taq* DNA polymerase became the cornerstone for modern DNA synthesis techniques. During a typical PCR process, 2-3 micrograms of *taq* DNA polymerase is needed. If too much is used however, unwanted, nonspecific DNA sequences can result.

The polymerase builds the DNA strands by combining corresponding nucleotides on each DNA strand. Chemically speaking, nucleotides are made up of three types of molecular groups including a sugar structure, a phosphate group, and a cyclic base. The sugar portion provides the primary structure for all nucleotides. In general, the sugars are composed of five carbon atoms with a number of hydroxy (-OH) groups attached. For DNA, the sugar is 2-deoxy-D-ribose. The defining part of a nucleotide is the heterocyclic base that is covalently bound to the sugar. These bases are either pyrimidine or purine groups, and they form the basis for the nucleic acid code. Two types of purine bases are found including adenine and guanine. In DNA, two types of pyrimidine bases are present, thymine and cytosine. A phosphate group makes up the final portion of a nucleotide. This group is derived from phosphoric acid and is covalently bonded to the sugar structure on the fifth carbon.

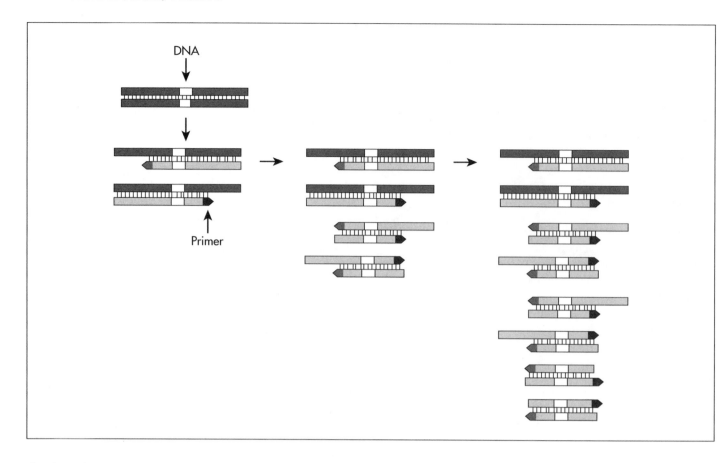

The first phase of polymerase chain reaction (PCR) involves the denaturation of DNA. This "opening up" of the DNA molecule provides the template for the next DNA molecule from which to be produced. With the DNA split into separate strands, the temperature is lowered—the primer annealing step. During the next phase, the DNA polymerase interacts with the strands and adds complementary nucleotides along the entire length. The time required at this phase is about one minute for every 1,000 base pairs.

To initiate DNA synthesis, short primer sections of DNA must be used. These primer sections, called oligo fragments, are about 18-25 nucleotides in length and correspond to a section on the template DNA. They typically have a C and G nucleotide concentration of about 60% with even distribution. This provides the maximum efficiency in the synthesis process.

The buffer solution provides the medium in which DNA synthesis can occur. This is an aqueous solution which contains $MgCl_2$, HCl, EDTA, and KCl. The $MgCl_2$ concentration is important because the Mg_2+ ions interact with the DNA and the primers creating crucial complexes for DNA synthesis. The recommended concentration is one to four micromoles. The pH of this system is critical so it may also be buffered with ammonium sulfate. To energize the reaction, various energy molecules are added such as ATP, GTP, and NTP. These compounds are the same ones that living organisms use to power metabolic reactions.

Other materials that may be used in the process include mineral oil or paraffin wax. After DNA synthesis is complete, the DNA

is typically isolated and purified. Some common reagents used in this process include phenol, EDTA and Proteinase K.

The Manufacturing Process

DNA synthesis is typically done on a small scale in laboratories. It involves three distinct processes including sample preparation, DNA synthesis reaction cycle and DNA isolation. These manufacturing steps are typically done in separate areas to avoid contamination. Following these procedures scientists are able to convert a few strands of DNA into millions and millions of exact copies.

Preparation of the samples

1 To begin DNA synthesis, the various solutions are prepared. This is typically done in a laminar flow cabinet equipped with a UV lamp to minimize contamination. Scientists use fresh gloves during each production step for similar reasons. Typically, all of the starting solutions except the primers, polymerases and the dNTPs are put in an autoclave to kill off any contaminating

organism. Two separate solutions are made. One contains the buffer, primers and the polymerase. The other contains the MgCl$_2$ and the template DNA. These solutions are all put into small tubes to begin the reaction.

DNA synthesis cycle

2 After the reacting solutions are prepared, the PCR cycle is started. The first phase involves the denaturation of DNA. One of the most important initial steps is the complete denaturation of the DNA template. Denaturation of the DNA essentially means breaking apart of the double bonded strand. This "opening up" of the DNA molecule provides the template for the next DNA molecule from which to be produced. An incomplete denaturation will result in an inefficient copy in the first cycle which negatively impacts each subsequent cycle. The initial denaturation is done by heating up the DNA template solution to 203°F (95°C) over one to three minutes. The total time depends on the template composition. In repeat cycles, the denaturation step lasts about two minutes and involves heating the solution to 201°F (94°C). Additional materials may be added to the solution to facilitate DNA denaturation such as glycerol, DMSO, or formamide.

3 With the DNA split into separate strands, the temperature is lowered to 122-149°F (50-65°C). This is known as the primer annealing step and lasts for about two minutes. At this point, the left and right primers match up and chemically link with their complementary bases on the template DNA.

4 The next phase involves the extending step. This part of the reaction is when most of the DNA strand gets copied. The temperature of the system is heated to about 162°F(72°C) and held there depending on the length of DNA to copy. At this stage, the DNA polymerase interacts with the strands and adds complementary nucleotides along the entire length. The time required at this phase is about one minute for every 1,000 base pairs.

5 After this first cycle, the DNA synthesis cycle is repeated. The number of cycles depends of the amount of initial DNA and the amount of DNA desired. If less than 10 copies of the template DNA are available,

Kary Banks Mullis.

Kary Banks Mullis was born in Lenoir, North Carolina, in 1944. Upon graduation from Georgia Tech in 1966 with a B.S. in chemistry, Mullis entered the biochemistry doctoral program at the University of California, Berkeley. Earning his Ph.D. in 1973, he accepted a teaching position at the University of Kansas Medical School in Kansas City. In 1977, he assumed a postdoctoral fellowship at the University of California, San Francisco.

Mullis accepted a position as a research scientist in 1979 with a growing biotech firm—Cetus Corporation, in Emeryville, California—that synthesized chemicals used by other scientists in genetic cloning. While there, he designed polymerase chain reaction (PCR), a fast and effective technique for reproducing specific genes or DNA (deoxyribonucleic acid) fragments that can create billions of copies in a few hours. The most effective way to reproduce DNA was by cloning, but it was problematic. It took time to convince Mullis's colleagues of the importance of this discovery but soon PCR became the focus of intensive research. Scientists at Cetus developed a commercial version of the process and a machine called the Thermal Cycler (with the addition of the chemical building blocks of DNA [nucleotides] and a biochemical catalyst [polymerase], the machine would perform the process automatically on a target piece of DNA).

Cetus awarded Mullis $10,000 for developing the PCR patent, then sold it for $300 million. Leaving Cetus in 1986, Mullis became a private biochemical research consultant and was awarded the Nobel Prize in 1993.

40 cycles are needed. With more initial DNA, 25-30 cycles is sufficient.

6 During the last cycle the sample is held at 162°F (72°C) for about 15 minutes. This allows the filling in (with nucleotides) of any

protruding ends of a new DNA strand. At this stage, the polymerase adds extra A nucleotides on one end of the DNA strands.

DNA isolation

7 When the reactions are complete, the DNA is isolated from the PCR reacting materials such as the DNA polymerase, $MgCl_2$ and the primers. This is done by adding compounds like phenol, EDTA and Proteinase K. Centrifugation is also helpful in this regard.

The Future

While scientists use PCR for DNA synthesis on a regular basis, there is still much that is not understood about DNA replication. In the future, research should elucidate the details of several important steps of the process, such as the components and intermediates involved. Additionally, improved polymerases may be developed, making it possible to create more DNA from smaller starting samples. It is hoped that one day DNA synthesis will help unlock some of the key aspects of living organisms and lead to the development medi-cines that will cure various cancers, viral and bacterial infections.

Where to Learn More

Books

Baker, T. A., and A. Kornberg. *DNA Replication.* San Francisco: Freeman, 1992.

Periodicals

Alberts, B., and L. R. Miake. "Unscrambling the Puzzle of Biological Machines: The Importance of Details." *Cell* 48 (1992): 413-420.

White, T.J. "The Future of PCR Technology Diversification of Technologies and Applications." *Elsevier Trends Journals* (December 14, 1996): 478-483.

Other

DNA Learning Center. Cold Spring Harbor Laboratory. http://vector.cshl.org/resources/biologyanimationlibrary.htm. (December 27, 2000).

—*Perry Romanowski*

Draw Bridge

Background

A bridge over a navigable waterway must allow boats and ships to cross its path, usually by being tall enough to allow them to sail underneath it. Sometimes it is impractical to build a bridge high enough; for example, it may rise too steeply or block the view of an important landmark. In such cases, the bridge can be designed so it can be easily moved out of the way for vessels that are too large to sail under it.

The type of movable bridge that most people think of as a draw bridge is similar to those that spanned medieval castle moats. Technically called "bascule bridges" from the French word for seesaw, they may open at one end and lift to one side (single leaf) or open in the middle and lift to both sides (double leaf). Another common type of movable bridge is the vertical lift span, in which the movable section is supported at both ends and is raised vertically like an elevator. Retractable bridges are made so the movable span slides back underneath an adjacent section of the bridge. Swing bridges are supported on vertical pivots, and the movable span rotates horizontally to open the bridge.

Movable bridges are relatively rare because they are more expensive to operate and maintain than stationary bridges. They also impede traffic—on the water when they are closed and on the roadway or rail line when they are open. Of the 770 bridges for which the New York City Transportation Department is responsible, 25 are movable bridges, including at least one of each of the four types defined above.

History

A few ancient drawbridges were built, including one 4,000 years ago in Egypt and one 2,600 years ago in the Chaldean kingdom of the Middle East. But they were not commonly used until the European Middle Ages. By the end of the fifteenth century, Leonardo da Vinci was not only designing and building bascule bridges but also drawing plans and constructing scale models for a swing bridge and a retractable bridge.

The modern era of movable bridge construction began in the mid-nineteenth century following the development of processes for mass producing steel. Steel beams are light and strong, steel bearings are durable, and steel engines and motors are powerful.

Many of the movable bridges currently in use in the United States were built in the early twentieth century. As they are being refurbished or replaced, two types of improvements can be made. First, more sophisticated design techniques and stronger, lighter materials allow new bridges to be built higher above the water. This means larger vessels can sail under them; consequently, it is not necessary to open them as frequently. Some modern replacements must be opened only one-fourth to one-third as often their predecessors. Second, some new bridges are operated hydraulically rather than being driven with gear mechanisms.

Raw Materials

Draw bridges are made primarily from concrete and steel. Seventy-five hundred short tons (6,804 metric tons) of structural steel and 150,000 short tons (13,6080 metric tons) of concrete were used in the Casco Bay Bridge

Seventy-five hundred short tons (6,804 metric tons) of structural steel and 150,000 short tons (13,6080 metric tons) of concrete were used in the Casco Bay Bridge in Portland, Maine; it has a 360-foot (110-m) tall opening and was completed in 1997.

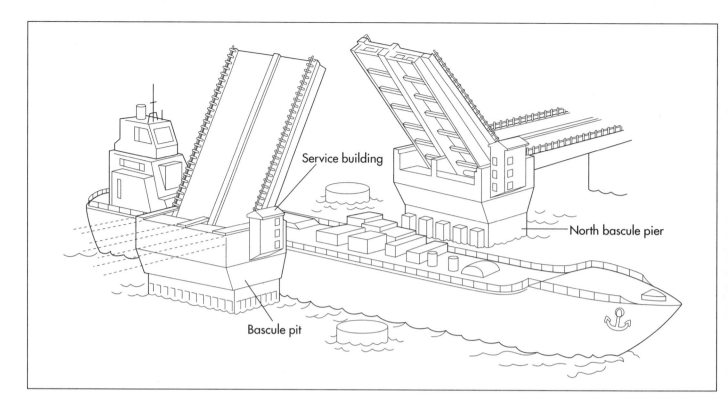

A typical draw bridge.

in Portland, Maine; it has a 360-foot (110-m) tall opening and was completed in 1997.

Design

Each draw bridge is a unique structure designed for its particular location and traffic needs. There are at least half a dozen different design concepts, but the most common is the bascule type. In double-leaf or four-leaf (a double-leaf bridge with separate leaves for each direction of vehicular traffic) bascule bridges, each leaf can be raised and lowered independently.

The energy required to raise and lower the bascule leaves is greatly reduced by counterbalancing each leaf with a compact weight on the opposite side of the pivot axle (trunnion). In various bascule designs, this counterweight might be located above the roadway and allowed to pivot below the roadway as the bridge is raised, or it might be located below the roadway and allowed to descend into a basement level (often well below the waterline) as the bridge opens. The counterweight is a massive concrete box containing chambers into which heavy, metal rods can be inserted to change the weight and its distribution. It might be located adjacent to the trunnion or, for greater

leverage, be set back a few yards (meters). As an example, each pair of 500-ton (450-metric-ton) leaves on the Casco Bay Bridge is balanced with an 800-ton (720-metric-ton) counterweight.

Besides the leaves and the counterweights, the other primary elements of a bascule bridge are the trunnion and the lift mechanism. A single steel trunnion up to 10 ft (3 m) in diameter and 65 ft (20 m) or more in length may be used for one leaf of the movable span; or a separate, short trunnion may be used for each side of each leaf. The lift mechanism is usually a rack-and-pinion gear arrangement driven by electric motors.

The Manufacturing Process

Although each installation is different, the following is a generic description of the construction of a bascule bridge.

Piers

1 If the bascule support piers will be located in the water, a cofferdam is built around the site for each pier. Steel panels are lowered into the water and driven into the riverbed to form a box. A clamshell dig-

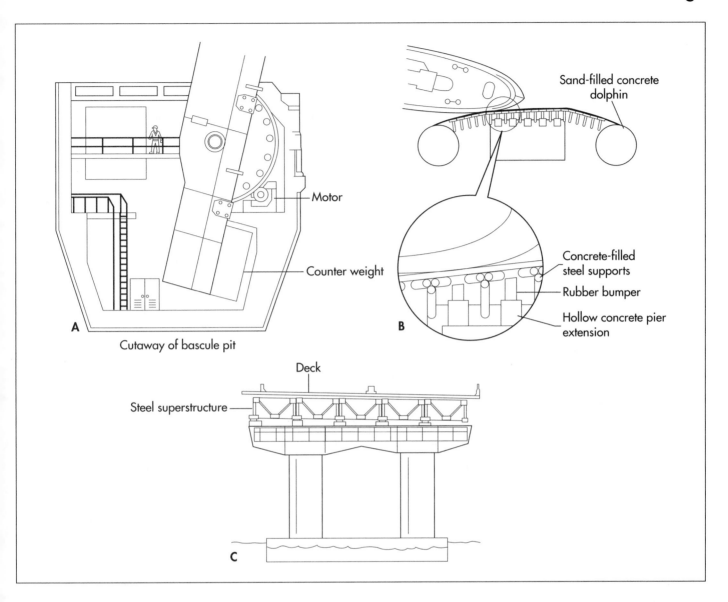

Motor

Counter weight

A

Cutaway of bascule pit

Sand-filled concrete dolphin

Concrete-filled steel supports

Rubber bumper

Hollow concrete pier extension

B

Deck

Steel superstructure

C

A. Bascule pit. B. Fender system.
C. Bridge pier.

ger removes soil inside the cofferdam. Piles are inserted deep into the riverbed to support the great weight of the pier and the bascule leaves. Steel piles may be driven, or reinforced concrete piles may be poured, into drilled holes. The bottom of the cofferdam is sealed with a layer of concrete. The water is pumped out of the cofferdam to provide a dry area for constructing the pier.

2 Forms are built to shape the concrete piers. Steel bars (rebar) are tied together to make a carefully designed reinforcing cage for the interior of the pier. The rebar cage is lowered into position inside the forms. The forms are filled with concrete. When the concrete has hardened, the forms are removed. Around the waterline, a protective layer of an erosion-resistant material,

such as granite, may be attached to the pier. The cofferdam is removed.

3 A fender may be built around the pier to protect it from being hit by errant ships. For example, on the Casco Bridge, large concrete cylinders were erected upstream and downstream from each pier to support the ends of a steel fender. The fender was faced with slippery plastic to deflect minor impacts. Under heavier impacts, the fender can deflect against rubber bumpers and, if necessary, against crushable hollow concrete boxes that would keep the impact from damaging the pier itself.

Bascule leaves

4 One or more trunnions are mounted on supports within the pier.

5 A counterweight is constructed and placed inside the pier.

6 Gear drives and/or hydraulic lift mechanisms are installed in the pier.

7 Two side girders are constructed for the heel section of each leaf of the bridge. A trunnion bearing is mounted in an opening in each girder. The girder may be equipped with gears that will mesh with the lift mechanism, or it may be fitted with paddles that hydraulic rams can push against.

8 The two side girders are lifted into the pier and eased over the ends of the trunnion. The heel section is completed with a crossbeam connecting the two side girders. The counterweight is attached to the heel section.

9 Additional longitudinal girders may be hoisted into position between the side girders and attached to the heel section. Steel braces are attached between the side girders and any other longitudinal girders. As pieces are added to the leaf, an appropriate amount of weight must also be added to the counterweight to maintain stability. This is particularly important if the bridge is being built in the closed position and must be opened during construction to allow marine traffic to pass.

10 The leaf is completed by attaching a tip section that connects the side girders (and any longitudinal girders) at the end opposite the heel. Devices called span locks are mounted on the leaf tips to connect opposite leaves when the bridge is down, so that vehicles driving on the bridge will not make the leaves bounce. Additional locks can secure the leaves in their open position so wind does not force them back down.

Finishing

11 Panels of steel-grate decking are installed atop the leaf. Sometimes a thin concrete surface is added.

12 Final balancing is accomplished by placing heavy iron, steel, or lead rods in the correct counterweight compartments. When properly balanced, the leaf is slightly heavier than the counterweight so gravity gently lowers (closes) the bridge.

Ongoing Adjustments

Throughout the lifetime of the bridge, counterweight adjustments must be made. Short-term adjustments compensate for ice or snow accumulations, for example. Long-term adjustments balance leaf weight changes due to activities such as repaving or painting. When the 250-foot (75-m) long High Street Bridge in Alameda County, California, was refurbished in 1996, 25,000 pounds (11,000 kg) of paint and primer were removed from its two bascule leaves. The counterweights had to be adjusted before and after repainting the span.

A dramatic example of the need to maintain proper counterbalance was shown by an accident on Chicago's Michigan Avenue Bridge on September 20, 1992. The two-level, double-leaf bascule bridge was undergoing repairs, and the concrete paving had been stripped off both the upper and lower decks. A large crane was parked behind the trunnion of one leaf, just above a counterweight that had not been lightened to compensate for the paving removal. Safety locks may also have been improperly engaged or defective. The opposite side of the bridge was opened to allow a boat to pass. When it closed and mated with the side that had remained down, the static half was jarred enough to release its unbalanced energy. The leaf "sprang up without warning, like a gargantuan catapult, hurling equipment and debris hundreds of feet across Wacker Drive into buses, automobiles, and pedestrian traffic," according to an analysis in the *Journal of the American Society of Mechanical Engineers.* The article continued, "The rapid rotation of the bridge ripped it from its trunnion bearings and the entire span slammed to the bottom of the counterweight pit." Six people were injured as they scrambled out of a bus struck by flying debris, and the rear window of an occupied car was smashed by the wrecking ball attached to the crane as it fell from the bridge.

The Future

There are two categories of movable bridge innovations. Refinements of traditional designs include minimizing the construction of large, submerged pits to receive counterweights when the bridge is open. For example, the 17th Street Causeway Bridge in Fort

Lauderdale, Florida, begun in 1998, allows compact counterweights to swing within V-shaped support piers rather than down into basements below bulky piers. The South Eighth Street Bridge in Sheboygan, Wisconsin, completed in 1995, operates without any counterweight despite its comparatively heavy, reinforced concrete deck. Rather than being gear-driven, the 82-ft (25-m) long single-leaf bascule is moved by a powerful hydraulic system.

Other movable bridge innovations introduce entirely new concepts. For instance, the Baltic Millennium Bridge in Gateshead, England (to be opened to the public in 2001), consists of two parabolic arches connected by a series of parallel cables. When the bridge is closed, one arch is horizontal and the other is vertical. The bridge opens by rotating vertically as a complete unit, raising the horizontal arch and lowering the vertical one until both rest approximately 45° and 164 ft (50 m) above the water surface. The steel and aluminum structure is designed to carry pedestrian and bicycle traffic across the 410-ft (125-m) wide River Tyne.

Where to Learn More

Periodicals

"Arched Cable-Stayed Crossing Tilts Sideways to Open." *Civil Engineering* (May 1999): 17+.

Cassity, Patrick A., et al. "Rebound of the Bascule Bridge." *Civil Engineering* (August 1996): 48+.

Studney, Michael J. &lquo;When a Bridge Becomes a Catapult." *Mechanical Engineering* (December 1992): 51+.

Other

17th Street Causeway. http://www.dot.state.fl.us/structures/botm/17thstreet/17thstreet.htm. (May 2, 2000).

Watson, Sara Ruth, and John R. Wolfs. *Bridges of Metropolitan Cleveland.* (1998). http://web.ulib.csuohio.edu/SpecColl/bmc/index.html (May 3, 2000).

—*Loretta Hall*

Duct Tape

By 1999, Manco, the maker of Duck™ brand tape, was selling approximately 5,900 short tons (5,352 metric tons), or 246,217 mi (396,240 km), of tape each year.

Background

Duct tape is a cloth tape coated with a polyethylene resin on one side and very sticky rubber-based adhesive on the other. Unlike other tapes, the fabric backing gives duct tape strength yet allows it to be easily torn. Duct tape is also very malleable and can adhere to a wide variety of surfaces. While it was primarily designed for use in air ducts and similar applications, consumers have found a broad range of uses for this popular product. It can be used for a number of household repair jobs, as a fastener instead of screws or nails, and in car maintenance. Snowmobilers have even been known to apply duct tape to their noses to thwart frostbite and sunburn. The product has generated so much interest, that there are books and web sites dedicated to its unconventional, and often comical, uses. For example, *The Duct Tape Book* describes how to use duct tape to make aprons and trampoline covers.

History

There are conflicting accounts concerning the history of duct tape. According to Manco, Inc. (maker of Duck™ Brand tape), it was created by Permacell—a division of Johnson and Johnson—during World War II in the 1940s. Other experts claim that the tape product was invented in the 1920s by researchers for the 3M Company, led by Richard Drew. Most accounts agree, however, that Permacell perfected duct tape during the war. Using state of the art technology, their research team developed a process to combine multiple layers of adhesive onto a polyethylene coated cloth backing. Some say this early product was nicknamed "duck tape" because it repelled water like the bird's feathers or because the fabric mesh was made from duck cloth.

Regardless of its origin, the military found many uses for duct tape. One of its earliest applications was to hold ammunition boxes together. For this reason, soldiers referred to it as "gun tape." The Air Force found other uses for the product and duct tape was used to cover gun ports on planes to cut down the air friction during take off. Like many other military products, duct tape was originally colored olive green, but after the war it was changed to the more familiar silver color. Manufacturers began marketing it to household consumers who found a variety of new uses. The tape is easier to use and just as effective as screws and bolts when it comes to holding together the kind of ductwork that is found in new homes with forced-air heating.

As the consumer demand grew, marketers began packaging their tapes in a more consumer-friendly fashion. According to Manco, they were the first company to shrink-wrap and label the duct tape so that it could be easily stacked on display shelves. This packaging improvement made it easier for shoppers to distinguish between the different grades. By 1999, Manco was selling approximately 5,900 short tons (5,352 metric tons), or 246,217 mi (396,240 km), of tape each year.

Design

Duct tape is designed for different application based on its grade. The grade is determined by the combination of adhesive type and the strength of the backing material. The strength of the cloth fabric depends on the number of threads it contains; this number determines the rip strength of the cloth. For example, military grade tape has a 40-lb (18-kg) rip strength while the strength of less expensive tapes is in the 20-lb (9-kg) range. Tape designed for use by the federal government must comply with a lengthy 12-page specification guide that dictates the strength and other factors. Commercial grades are less demanding and are typically classified as either utility, general purpose, or premium grade. Other speciality grades include "nuclear tape," which is used in reactors and a "200-mi (322-km) an hour" tape specifically designed for race cars.

Raw Materials

Cotton mesh

Cotton mesh forms the backbone of duct tape. It provides tensile strength and allows the tape to be torn in both directions. Cloth that has a tighter weave and higher thread count is of a higher quality, provides greater strength, and gives a cleaner tear. A premium quality tape can have a thread count of 44 x 28 threads per square inch. Accordingly, tape made with this fabric is more expensive. The cotton fabric is called a "web" when it is spread across the coating machinery.

Polyethylene coating

The cotton fabric is coated with polyethylene, a plastic material that protects it from moisture and abrasion. This plastic coating is flexible and allows the tape to adhere better to irregular surfaces. The polyethylene is melted and applied to the fabric in a preliminary coating operation. The coated fabric is then stored on rolls until the manufacturer is ready to apply the adhesive.

Adhesive compound

The adhesive used in duct tape is unique for two reasons. First, the adhesive itself is formulated with rubber compounds that ensure long-term bonding. Other tapes typically use adhesive polymers that are not as binding. Second, the adhesive is applied to the substrate in a much thicker coating than those used on cellophane or masking tapes. This too serves to increase the adhesive properties of the tape.

The Manufacturing Process

Adhesive compounding

1 The rubber-based adhesive used in duct tape is prepared in a multi-step process. In the first step the adhesive is compounded in a mixer known as a Branbury-type mixer consisting of a stainless steel tank equipped with a steam jacket to heat the compound and a high torque mixer. The rubber compounds are introduced in pellet form, then heated and mixed until they are melted and homogenous. Other ingredients are added to the blend tank as specified in the formulation. These include tackifying agents, viscosity modifiers, antioxidants, and other adjuncts. The final mixture is thick but smooth enough to be pumped to a holding tank connected to the coating equipment.

Adhesive application

2 The adhesive and the fabric are combined using sophisticated coating equipment. First, the adhesive is further softened by heat on a roller mill. A roller mill consists of closely spaced hollow rollers which are made of heavy gauge stainless steel. The rollers are attached to high torque gears and a motor that rotates the them at a set speed. The temperature of the rollers is controlled by continuously pumping water through them. The top roller is held at a temperature of 260°F (127°C) and the center roller is kept at 100°F (38°C). The cylinders are fixed in place so that only a small gap exists between them. As the rollers turn, the rubber compound is fed into this space. The friction generated heats and softens the rubber. This arrangement allows the molten adhesive to form a thin sheet across the surface of the rollers.

3 The cloth is then fed into the coating machine through another set of rollers. It runs against the third roller (held at 199°F [93°C]) that is in contact with the adhesive. During this process, known as coating, the adhesive is transferred to the cloth. By controlling the gaps between the rollers, the machine operator can determine how much adhesive is applied to the cloth. When the adhesive is ready, the cloth backing material—which has been pre-treated with polyethylene—is fed off its storage roll and onto the coating rollers. As the backing material comes in contact with the third roller, it

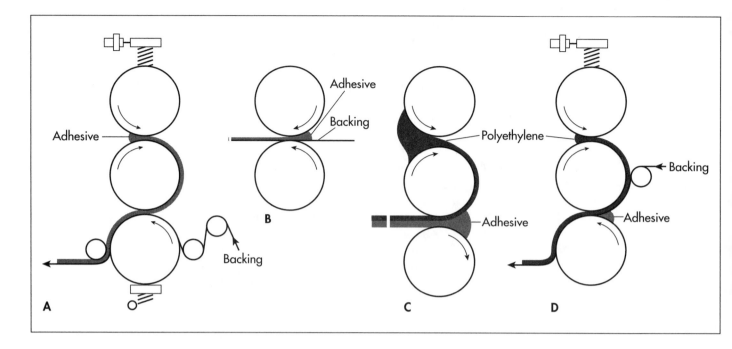

A. Three-roll pressure sensitive tape calender. B. Two-roll adhesive calender. C. Polyethylene and adhesive calender. D. Calender for fabric tape coated with polyethylene.

picks some adhesive up off the transfer roller. This process used to apply such heavy adhesives is known as calendaring. Coatings as thin as 0.002 in (0.05 mm) can be used but they become less uniform below 0.004 in (0.1mm). These fluctuations occur because the rollers experience a small degree of bending as they rotate. This creates uneven gaps between the rollers which cause variations in coating weight across the web.

4 After coating, the tape fabric is wound onto large cardboard cores. When enough tape has been coated and the roll is full, it is removed from its spindle and moved to another area where it can be cut to the proper size. These storage rolls are approximately 5 ft wide and 3 ft in diameter (152 cm x 91 cm).

Respooling and slitting

5 After the coating process is complete, the roles can be cut to their final size. This is done by "unspooling," or unwinding, the large rolls onto a machine equipped with a series of knives. The knives cut the web into more narrow segments which are then rewound on smaller cardboard cores. This process is known as "slitting."

Packaging operations

6 In this final stage the rolls of duct tape are packaged for sale. They are typically shrink wrapped, either singularly or in packages of two or three. These packages are then boxed and marketed for shipping.

Depending on the manufacturer, the steps described above can be combined through automation into fewer steps. For example, Permacell uses a self-contained apparatus which mixes, heats, and fastens the adhesive onto the backing. This method allows the glue to be prepared without pollution-causing solvents.

Quality Control

Duct tape must meet a series of standard tests described by the American Society for Testing and Materials. These methods measure two key properties of the adhesive: its adhesive strength (which determines how well the tape will adhere to another surface) and cohesive strength (which shows how well the adhesive will stick to the cloth backing). One common method of evaluating these properties involves applying the tape to a standardized stainless steel plate, and then measuring the force required to rip it off. The plate is then examined to determine how much, if any, of the adhesive residue is left behind. The adhesive coating itself is monitored to evaluate how well it sticks to its backing. Conditions where the adhesive leaves a residue is known as creeping, crazing, oozing and bleeding. Quality control technicians also watch for fisheyes, the term used to describe an uneven application of adhesive.

In addition to the standard tests, each manufacturer has their own proprietary methods for evaluating their products. For example, companies may measure the duct tape's breaking point. Others evaluate the "scrunch" sound of the tape as it unwinds because consumers believe a noisy rip off the roll is a sign of strength. Other tests are designed to measure quick stick. One way this is done is by shooting ping-pong balls at tape strips with the sticky side up to measure how far they roll before they are stopped by the tape.

Despite duct tape's reputation for superior adhesion, testing done by independent researchers has found that the tape does not work as well as it is intended to. In 1998, researchers Max Sherman and Iain Walker of the Department of Energy's Lawrence Berkeley National Laboratory evaluated 12 different kinds of duct tape. They designed an accelerated aging test that mimicked the temperature conditions of a home or office building from night to day and winter to summer. They used a series of standard finger joints which connected a smaller duct to a larger one. Wrapping each test connection with a different brand of duct tape, they evaluated the seal under temperature and humidity conditions designed to be similar to those found in home heating and air conditioning systems. The researchers also performed a bake test in which the sample joints were baked at higher temperatures. Most of the joints tested were found to be leaking by 50% or more, according to the researchers. Their test results also showed that a large proportion of the tapes failed to function when temperatures dropped below freezing (32°F [0°C]) or rose over 200°F (93°C). However, the tape industry disputed these findings, claiming that for optimal efficiency their tape products should be assisted by collars or clamps.

The Future

Environmental and price considerations drive researches to identify new adhesive formulations at a lower cost—general price for one 2 in (5 cm) wide, 50 yd (46 m) roll of all-purpose duct tape is about three dollars—that maintain high functionality without being damaging to the environment. In addition, marketing considerations drive product improvements. In response to criticism that duct tape does not function well at extreme temperatures, manufacturers are creating more advanced formulations specially designed to withstand severe heat and cold fluctuations. For the first time, formal standards are being created specifically for duct tape to be used on flexible duct work. Tape manufacturers support this plan because they believe that certification by Underwriters Laboratories (a product safety testing organization) will boost their sales.

Where to Learn More

Books

Berg, Jim, and Tim Nyberg. *The Duct Tape Book*. Duluth, MN: Pfeifer-Hamilton Publishers, 1995.

Satas, Dontas. *Handbook of Pressure Sensitive Adhesives*. Second edition. Van Nostrand Reinhold, 1989.

Periodicals

Baird, Christine V. "U.S. Duct Tape Sales at $75 Million a Year and Growing." *Knight-Ridder/Tribune Business News* (October 21, 1996): 1021B0195.

"Duct Tapes Flunk Berkeley Lab Tests." *Air Conditioning, Heating & Refrigeration News* 204, no.18 (August 31, 1998): 1.

Harder, Nick. "Bring On the Duct Tape to Show Creativity." *Knight-Ridder/Tribune News Service* (April 15, 1999): K0467.

Turpin, Joanna R. "Duct Tape: The Ultimate Tool." *Air Conditioning, Heating & Refrigeration News* 201, no.16 (August 18, 1997): 9.

Other

Underwriters Laboratories Inc. http://www.ul.com (January 2001).

—*Randy Schueller*

Electric Blanket

Approximately 4.5 million electric blankets are sold in the United States annually.

Background

An electric blanket is a bed covering with a built-in heating element so that a sleeper can maintain a desired temperature even in a cold room. Many consumers prefer electric blankets because their use can reduce home heating costs. The sophisticated temperature controls of modern electric blankets can sense changes in skin and air temperature and adjust settings accordingly. The most advanced model blankets are programmable, so they can pre-warm the bed at a certain time and shut off at a certain hour. They are also able to adjust temperature for the needs of two people in the same bed. Some blankets even compensate for the different heat needs of different parts of the body; for instance, sending more heat to the sleeper's feet while leaving the head area cooler. Approximately 4.5 million electric blankets are sold in the United States annually.

History

The devices that evolved into today's electric blankets were first intended for invalids. The reputed father of the electric blanket was an American doctor, Sidney Russell, who devised an electrically heated pad in 1912. Russell was trying to find a way to keep his ill patients warm, and he developed a blanket that used electrical wires covered in insulated metal tape to accomplish this. Commercial use of a similar product began in the 1920s, when electrically heated blankets were used on patients in tuberculosis sanatoriums. The sanatoriums' weakened lung patients were advised to get plenty of fresh air, and they slept with windows wide open—sometimes even spending the night outdoors. Electric blankets helped keep the patients warm in these drafty conditions. People, both those sick and well, had long used hot water bottles or stone bed warmers to heat their beds on winter nights. The 1920s and 1930s saw many new, electrified versions of these traditional devices. These included electric warmers shaped like thermos bottles and flattened dome-like bed warmers heated with a light bulb. In the 1930s, electric blankets were produced in the United States and in England, primarily as a luxury item or as the finest accouterment of a sick room. They were generally smaller and much thicker than today's electric blankets, and they were called warming pads or heated quilts. By 1936, one company had introduced a heated quilt with an automatic temperature control. A bedside thermostat responded to temperature changes in the room and cycled the blanket on and off accordingly. These early electric blankets also incorporated several safety thermostats which would switch the blanket off if a portion of it became dangerously warm.

The electric blanket as known today did not develop until after World War II. Research on electrically heated suits for fighter pilots during the war led to safety improvements and allowed manufacturers to make thinner, more easily folded blankets. General Electric was a leading marketer of electric blankets. In 1945, as the war was ending, it began advertising its automatic blanket, emphasizing the connection with its wartime manufacturing of "electrically warm" suits for pilots battling over Japan. The image of the blanket changed, from something needed by the sick or elderly to a modern convenience which made sleep more comfortable for everyone. Though the technology for post-war blankets was better, it was not much different from that used in the warming quilts of the 1930s. A blanket shell encased wires and embedded thermostats, and a bedside control with settings of high or low turned the blanket on and off. Gradually the number of embedded thermometers increased, so that a twin-sized blanket went from having four in the 1950s to possibly 10 in the 1980s. Blankets were developed that had two temperature controls, one for each side of the bed. Beginning in 1984, the technology changed significantly with the application of a thermostatless system. These new blankets used a "positive temperature coefficient wiring system," which enabled the wiring itself to sense temperature changes. So the blanket could sense and respond to body temperature as well as room temperature.

Safety Concerns

Beginning in 1990, electric blanket manufacturers began responding to controversy over the safety of electromagnetic fields. Various studies had raised concerns over the extremely low frequency (ELF) electric and magnetic fields emanating from electronic devices, including electric blankets. One study released in 1990 in the *American Journal of Epidemiology* found an increased risk of brain tumors and childhood leukemia in children whose mothers had used electric blankets while pregnant. Though other studies tracing a cancer link to electric blanket use had negative or conflicting results, the reports were alarming enough to prompt a group of United States congressmen to ask that the blankets be labeled hazardous to women and children. A panel appointed by the Food and Drug Administration concluded that there was not enough evidence to warrant regulation of electric blankets, but research did show some problems. Some brands of electric blankets produced greater electric fields than others. If the blanket's plug was not polarized then it could be plugged in the wrong way producing a significant electric field even when turned off. Though there was no conclusive evidence that the electric fields produced by electric blankets were harmful, the major United States manufacturers altered their products. The major maker of the wiring for electric blankets came up with a new system that used parallel wires holding current flowing in opposite directions. This effectively can-

celed out most of the ELF field. The redesigned blankets produced after 1992 had much weaker electrical fields than their predecessors. After this, the subject of electric blanket safety faded away.

Raw Materials

The materials used in electric blankets are for the most part unique to that industry. An electric blanket consists of three main components: a specially woven cloth called the shell; the heating element, in the form of insulated wire; and a bedside control and power cord. For the shell, manufacturers use a blend of polyester and acrylic which has been formulated for the industry. The wiring is a positive temperature coefficient material that is heat producing and heat sensitive along its entire length, eliminating the need for thermostats. This special wiring is encased in an insulated jacket made of a proprietary blend of plastic. The bedside control and power cord are made of a combination of plastic and metal materials. These are usually manufactured by an outside vendor.

The Manufacturing Process

Making the yarn

1 The blanket manufacturer purchases the raw materials for the blanket shell from a fiber manufacturer. Fibers used for the blanket are then blended and spun.

Weaving the shell

2 The shell is woven on a special high-speed loom which has been designed for this particular use. The blanket shell is structured as a series of 20-30 long hollow tubes or channels, which are lined head to foot. The channels are not sewn together but woven as one piece of cloth. They are formed as a double layer, and the fabric in between the channels is a single layer. The operator starts the loom, and the machine weaves the yarn into the completed shell.

The heating element

3 An electric blanket produces heat when an electric current runs through its heating element. The heating element is a special kind of wire, which acts as its own thermostat. The alloy used in this wire is formed into

Various types of bed warmers patented and popularized during the 1930s and 1940s.

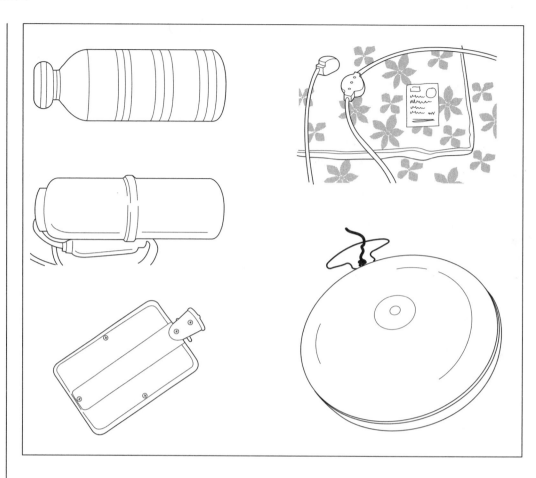

rods, cleaned, and pulled through a wire-drawing machine. The plastic insulating jacketing is formed through an extrusion process. The finished wire is delivered in large spools to the assembly plant, where it is inserted into the blanket shell in the next step.

Inserting the heating element

4 The machinery for threading the heating element into the blanket shell is specialized equipment developed especially for the electric blanket industry. The technology involves a pusher which takes the end of the wire and pushes it through the channels of the shell. Because of the way the shell is woven, the channels are connected and the wire ends up in one continuous zig-zag pattern. In older blankets (those manufactured before the mid-1980s) the heating element was broken up into many sections joined by thermostats. Modern blankets are heated by a single length of continuous wire.

Final assembly

5 Various other small steps are needed to complete the blanket. A manufacturer's label is sewn onto one corner and the edges of the blanket are trimmed. The bedside control and power cord are usually manufactured or partially assembled by outside vendors. Workers take these components and fit them to the blankets. Each blanket is inspected and tested, and then moved to a packaging area, where it is folded, bagged, and finally boxed.

Quality Control

Each element of the blanket undergoes separate testing with the final product as a whole also being tested rigorously. The blanket shell is subjected to various tests, including checks for foreign materials and of the proper thickness throughout. The heating element is tested by its manufacturer as it is being made, so that each inch of the wire has been examined by the time it reaches the blanket assembly factory. It is tested to make sure there are no short circuits, no foreign materials, and that power dissipates through it at the proper rate. The manufacturers of the temperature control and power cord subject these to their own tests. In addition, the electric blanket manufacturer registers its products with Underwriters Laboratory, a private testing agency for electrical devices. The manufacturer agrees to follow certain testing procedures and make its blankets according to specifications outlined by Underwriters Laboratory. Underwriters Laboratory makes sure the manufacturer is sticking to its specifications by making unannounced factory visits. But in general, the manufacturer's own testing is more comprehensive than that required by Underwriters Laboratory. As long as the manufacturer complies with the testing agency's guidelines, it is allowed to display the Underwriter's Laboratory (UL) symbol on its tags, as a guarantee to consumers. Fire safety is an important issue for electric blanket manufacturers. The heating element is completely enclosed in fire and waterproofed insulation.

After each blanket is fully assembled, workers plug it in and turn it on to its full rated power. An inspector observes the blanket for any obvious flaws and constancy of temperature. As far as tests for electrical and magnetic field generation, which caused public concern in the 1990s, these are generally not done. The actual harm of ELF fields remained unproven, and concern over blanket safety ultimately died down with the industry's adaptation of lower energy blankets. Fire safety, ease of use, and general reliability of the blanket are the areas of most concern to quality control inspectors at the assembly plant.

The Future

In the winter of 1999/2000, electric blanket sales increased to 850,000 pieces according to the Winterwarm company. In order to capitalize on this increase in sales for the year 2001, the electric blanket market will see a few changes. Marketing will be geared to a new generation with packages becoming more modern with a younger feel. They will no longer be called electric blankets, but heated blankets with thinner wires. The fabric shells will have brighter patterns and colors. The modernized blankets will also be easily programmed and hopefully appeal to a larger demographic.

Where to Learn More

Periodicals

Chirls, Stuart. "Electric Blanket Market Heats Up." *HFN* (April 19, 1999): 10.

Farley, Dixie. "The ELF in Your Electric Blanket." *FDA Consumer* (December 1992): 22.

Hannam, Sean. "Blanket Coverage." *ER Magazine* 12, 9 (September 2000): 40.

Stix, Gary. "Field Effects." *Scientific American* (December 1990): 122-123.

—*Angela Woodward*

Electronic Ink

The first commercial electronic ink product is the Immedia display. It is an advertising sign that looks and feels just like a paper sign. However, this sign is coated with electronic ink allowing it to be programmed to change its message.

Electronic ink is a special type of ink that can display different colors when exposed to an electric field. It is made through a two step process that involves creating two-toned charged particles and encapsulating them in a transparent polymeric shell. The resulting nanoparticle shells are suspended in a solvent until the ink can be applied to a surface. First developed in the early 1990s, electronic ink promises to revolutionize the printing industry and maybe even change the way we interact with the world.

Background

Ink has been around for centuries and for the purposes of displaying an idea, ink on paper has many advantages over electronic displays. Paper is easy to carry around and can be read almost anytime and anywhere. It does not require a power source and is relatively durable. However, ink on paper has the disadvantage of not being able to be updated. Electronic ink has been designed to maintain the advantages of traditional paper and ink while providing the added advantages of updating and high capacity data storage.

Electronic ink is like traditional ink in that it is a colored liquid that can be coated onto nearly any surface. Suspended in the liquid are millions of microcapsules that contain tiny, two-toned polymeric particles. One side of the particle is a dark color while the other is a contrasting light color. Similar to a magnet, the dark colored side of the particle has an electric charge that is opposite that of the light colored side. When the ink is exposed to an electric field, the particles realign themselves, depending on the charge of the field. When all of the dark colored sides are attracted to the surface, the ink

looks dark. When an opposite electric charge is applied, the light colored sides orient face forward and the surface looks light. This ability to change from white to black or visa versa whenever desired makes electronic ink extremely useful. When a book or other surface is coated with electronic ink, it can be reprogrammed to display different words or pictures.

It has long been known in the printing industry that letters and pictures can be displayed using distinct dots, or pixels. The more pixels that can be placed closer together, the better the image looks. On a standard newspaper about 300 pixels are used in the area of a square inch. When electronic ink is coated on a surface in specific quantities, each of these pixels can be made light or dark depending on how the electronic field is applied. Printing technology is already available to cover surfaces with over 1,200 pixels per square inch of electronic ink. This resolution makes electronic ink suitable for almost any printed work.

The way that an electronic ink display would work is much like a computer screen. Each pixel of ink could be controlled by an attached computer. Groups of adjoining pixels could be turned on or off to creates letters, number, and pictures. While this might be difficult to achieve on a piece of standard paper, a specially designed paper is being developed which will feel and look like paper but actually be a mini-computer complete with a vast array of electric circuits to control each pixel. This special paper would not be essential however because a special scanner could also be developed to have the same effect.

One of the most useful characteristics of electronic ink is that after the electric field is removed, the ink remains in its configuration. This means that only a small amount of power is required as compared to typical electronic displays. The configuration can be changed however, by applying a new electric field whenever desired. This means that if a book was printed with electronic ink, it could contain the words of one book on one day and another book the next. If fitted with memory storage, a single electronic book could contain thousands of different texts.

History

While the printed word has been around for centuries, the idea of electronic ink is a relatively recent invention. In the late 1970s, researches at Xerox PARC developed a prototype of an electronic book. The device used millions of tiny magnets that had oppositely colored sides (black on one, white on the other) embedded on a thin, soft, rubber surface. When an electric charge was introduced the magnets flipped making either a black or white mark similar to pixels on a video screen. The device was never a commercial success because it was large and difficult to use.

Over the next decade various screens were introduced and the idea of an electronic book became a reality. However, these devices still remain more cumbersome than printed paper. In 1993, Joe Jacobson, a researcher at MIT, began investigating the idea of a book that typeset itself. He conceived a variation of the PARC idea using reversible particles. Eventually, he created electronic ink, which utilizes colored polymers encased in a transparent shell. He submitted his idea for patent in 1996 and was eventually rewarded one in 2000.

Jacobson formed E Ink Corporation which was designed to bring electronic ink to the marketplace. The first commercial product is the Immedia display. It is an advertising sign that looks and feels just like a paper sign. However, this sign is coated with electronic ink allowing it to be programmed to change its message. E ink anticipates that electronic ink will eventually be utilized in an area where traditional ink is used such as newspapers, books, magazines and even clothes.

Raw Materials

A variety of raw materials are used in the production of electronic ink. These include polymers, reaction agents, solvents, and colorants.

Polymers are high molecular weight materials which are made up of chemically bonded monomers. To make the charged, colored portions of the electronic ink polyethylene, polyvinylidene fluoride or other suitable polymers are used. These materials are useful because they can be made liquid when heated, solidify when cooled, and will maintain stable dipoles which are long lasting.

Filler materials are added to the polymers to alter their physical characteristics. Since polymers are generally colorless, colorants are added to them to produce the contrast needed for electronic ink. These may be soluble dyes or comminuted pigments. To produce a white color, an inorganic material such as titanium dioxide may be used. Iron oxides can be used to produce other colors like yellow, red, and brown. Organic dyes such as pyrazolone reds, quinacridone violet, and flavanthrone yellow may also be utilized. Other fillers such as plasticizers can be added to modify the electrical characteristics of the polymers. This is particularly important for electronic ink. During production the polymer is heated. For this reason stabilizers are added to prevent it from breaking down. Heat stabilizers include unsaturated oils like soy bean oil. Protective materials that are added include UV protectors such as benzophenones and antioxidants such as aliphatic thiols. These materials help prevent UV degradation and environmental oxidation respectively.

During the electronic ink encapsulation process various compounds are used. Water is used to create an emulsion and provide a vehicle for the encapsulation reaction to take place. Monomers are added to produce the encapsulation shell. Cross-linking agents which cause the monomers to react are utilized. Silicone oil is the hydrophobic material that gets incorporated with the colored particles in the encapsulate. This material provides a liquid medium for the particles to travel through when the electric field is applied. It is clear, colorless and extremely slippery. Other gel or polymeric materials

Made up of tiny, two-toned charged nanoparticles, electronic ink can display different colors or messages when exposed to an electric field. Depending on the type of charge, the particles will either be attracted or repelled from the surface, thus creating different effects.

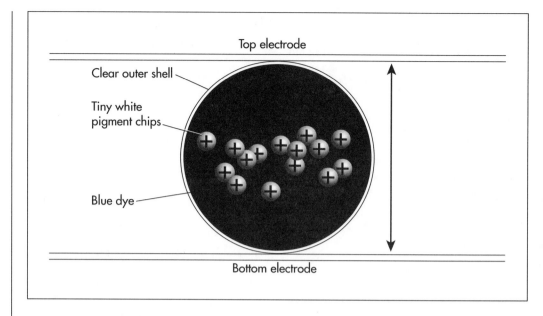

can be added to the encapsulate to improve the stability of the system.

The Manufacturing Process

Electronic ink is made in a step-wise fashion. First, two contrasting inks are given opposite charges. Then the inks are encapsulated in conductive micro spheres and applied to the desired surface.

Producing charged ink

1 Two contrasting liquid polymers are loaded into separate containers which have attached atomizing nozzles. The materials are kept heated so they remain liquid. One of the nozzles has a positively charged potential while the other has a negative potential. Pressure is then used to force the inks through the nozzles causing them to break into tiny particles and also acquire the opposing charges. The containers are situated next to each other so when the inks exit the nozzles they come in contact. Since they have opposite charges, they are attracted to each other and form larger, neutral particles.

2 After the larger particles are formed, the materials are allowed to cool which causes them to solidify. This results in a small two-toned solid particle which has a positive and negative side. The particles are then run past a heating element which reduces surface tension and creates a more perfect sphere.

3 The particles are then run through a set of electrodes to separate out the ones that are imperfectly charged. As they pass the electrodes, the imperfect particles are attracted to the corresponding electrode and then removed. The rest of the particles are transferred to the encapsulating area.

Encapsulating ink

4 The particles are moved into a tank which contains a liquid solution of monomer in a silicone oil. The particles are mixed thoroughly so they are evenly dispersed. This solution is combined with an aqueous phase which creates an emulsion. An emulsion is a semi-stable mixture of oil and water. The electronic ink particles remain in the silicone oil which is surround by water.

5 A cross-linking agent is added to the solution which causes the monomer to react with itself. This produces tiny spheres which contain some silicone oil and the electronic ink particles. The ink particles can then be separated from the aqueous phase for various applications. This can be done by evaporation with subsequent solvent washing.

6 After the reactions are complete, the electronic ink articles are stored in a liquid solvent until they can be applied. Depending on the final product, this application process can involve spreading the liquid

ink on specialized paper, fabric, or other kinds of fibers.

Quality Control

To ensure the quality of the electronic ink, each phase of the production process is monitored. Since it is a relatively new technology, electronic ink is not made in large, rapid quantities. For this reason each step can be thoroughly tested before proceeding to the next. Inspections begin with an evaluation of the incoming raw materials. These materials are tested for things such as pH, viscosity, and specific gravity. Also, color and appearance are evaluated. After the electronic ink is finished, it is tested to ensure that it will properly react to an electric field. The material may be spread on a thin surface and have an electric field applied. The color of the surface should change accordingly. The particle size is also tested using various mesh screens.

The Future

The first products utilizing electronic ink are just being introduced. They are simple two-toned devices that are not more impressive than flat paneled electronic displays. However, future generations promise to have broad applications and may significantly impact the way we interact with the world. The hope that electronic ink manufacturers have is that this material will initially be incorporated into outdoor billboards, handheld computer devices, books, and newspapers. But ultimately, electronic ink will be put onto any surface such as clothes, walls, product labels, and bumper stickers. It will then become ubiquitous to the environment so that any message can be displayed anywhere at anytime.

Since the current electronic ink product is made up of only two colors, the products made with it can not create a full colored display. In the future, more colors of electronic ink will be developed. Scientists still have to work out how to display these differing colors at the proper time, but once accomplished, any surface coated with electronic ink could become as interesting to look at as a television screen.

Where to Learn More

Periodicals

Gregory, P. "Coloring the Jet Set." *Chemistry in Britain* (August 2000): 39-2.

Johnston, M. "Lucent, E Ink Demo Electronic Ink Prototypes." *Digit* (November 21, 2000).

Peterson, I. "Rethinking Ink." *Science News* (June 20, 1998):396- 97.

Wilkinson, S. "E-Books Emerge." *Chemical & Engineering News* (August 21, 2000): 49-4.

Other

E Ink. http://www.eink.com (January 2001).

MIT Media Laboratory. http://www.media.mit.edu/micromedia (January 2001).

—Perry Romanowski

Evaporated and Condensed Milk

Evaporated milk has a shelf life of up to 15 months, while condensed milk is good for two years.

Background

Evaporated and condensed milk are two types of concentrated milk from which the water has been removed. Evaporated milk is milk concentrated to one-half or less its original bulk by evaporation under high pressures and temperatures, without the addition of sugar, and usually contains a specified amount of milk fat and solids. This gives regular evaporated milk—the shelf life differs with the fat content—up to 15 months of shelf life. Condensed milk is essentially evaporated milk with sugar added. The milk is then canned for consumer consumption and commercial use in baking, ice cream processing, and candy manufacture. This product has a shelf life of two years. When concentrated milk was first developed in the mid-1800s before the advent of refrigeration, many used it as a beverage. However, with the exception of some tropic regions, this is rarely the case today.

History

In 1852, a young dairy farmer named Gail Borden was on a ship headed home to the United States from the Great Exhibition in London. When rough seas made the cows on board so seasick that they could not be milked, infant passengers began to go hungry. Borden wondered how milk could be processed and packaged so that it would not go bad. This was a problem not only on long ocean voyages but on land, as well, because at the time, milk was shipped in unsanitary oak barrels and spoiled quickly.

When Borden returned home, he began to experiment with raw milk, determining that it was 87% water. By boiling the water off the top of the milk in an airtight pan, Borden eventually obtained a condensed milk that resisted spoilage. On another trip, this time by train to Washington, DC, to apply for a patent for his new product, Borden met Jeremiah Milbank, a wealthy grocery wholesaler. Milbank was impressed with Borden's ideas and agreed to finance a condensed milk operation. In 1864, the first Eagle Brand Consolidated Milk production plant opened on the east branch of the Croton River in southeastern New York.

Borden's new product was not an unqualified success. In 1856, condensed milk was blamed for an outbreak of rickets in working-class children because it was made with skim milk, and therefore lacked fats and other nutrients. Others complained about its appearance and taste because they were accustomed to milk with a high water content and that had been whitened with the addition of chalk. In spite of this criticism, the idea of condensed milk caught on to the degree that Borden began to license other factories to produce it under his name.

The outbreak of the Civil War proved to be good for business when the Union Army ordered the condensed milk for its field rations. At the height of the war, Borden's Elgin, Illinois plant was annually producing 300,000 gallons of condensed milk.

To differentiate his own product from that of the licensed plants, Borden changed the name of his condensed milk to Eagle Brand. About this time, two American brothers, Charles A. and George H. Page, founded the Anglo-Swiss Condensed Milk Company in Switzerland. One of their employees, John Baptist Meyenberg, suggested that the company use a similar process but eliminate the addition of sugar to produce evaporated

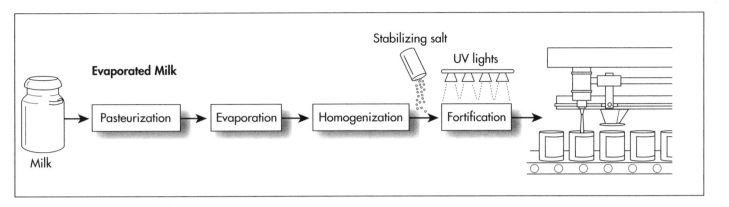

Evaporated Milk

Milk → Pasteurization → Evaporation → Homogenization → Fortification

Stabilizing salt

UV lights

A diagram showing the manufacturing steps involved in making evaporated milk.

milk. Meyenberg's idea was rejected. Convinced that his idea held merit, Meyenberg quit the company and emigrated to the United States. By 1885, Meyenberg was producing the first commercial brand of evaporated milk at his Highland Park, Illinois plant, the Helvetica Milk Condensing Company.

In the late 1880s, Eldridge Amos Stuart, an Indiana grocer in El Paso, Texas, noted that milk was spoiling in the heat and causing illness in children. Stuart developed a method for processing canned, sterilized evaporated milk. In 1899, Stuart partnered with Meyenberg to supply Klondike gold miners with evaporated milk in 16-ounce cans.

An article on homogenization in the April 16, 1904 issue of *Scientific American* had an impact on the concentrated milk industry, which employed the process long before fresh milk plants. Further improvements followed. In 1934, Meyenberg's company, now headquartered in St. Louis, Missouri, and renamed the Pet Milk Company, became the first to fortify its evaporated milk with Vitamin D. This was accomplished by the process of irradiation, developed in 1923 by Harry Steenbock, a chemist at the University of Wisconsin. In this process, the milk is exposed to ultraviolet light, which causes reactions to produce Vitamin D, enriching the milk.

Raw Materials

The primary ingredient is raw cow's milk. Evaporated and condensed milk processors purchase the milk from nearby dairy farms.

A salt, such as potassium phosphate, is used as a stabilizing agent, which keeps the milk from breaking down during processing. Car-

rageenan, a food additive made from red algae (Irish moss) is used as a suspending agent. The milk is also fortified with Vitamin D through exposure to ultraviolet light. Powdered lactose crystals are added to concentrated milk to stimulate the production of lactose, a type of sugar that increases the milk's shelf life.

The Manufacturing Process

Evaporated milk

1 The raw milk is transported from the dairy farm to the plant in refrigerated tank trucks. At the plant, the milk is tested for odor, taste, bacteria, sediment, and the composition of milk protein and milk fat. The composition of protein and fat is measured by passing the milk under highly sensitive infrared lights.

2 The milk is piped through filters and into the pasteurizers. Here, the milk is quickly heated in one of two ways. The High Temperature Short Time method (HTST) subjects the milk to temperatures of 161°F (71.6°C) for 15 seconds. The Ultra High Temperature (UHT) method heats the milk to 280°F (138°C) for two seconds.

Both methods increase the milk's stability, decrease the chance of coagulation during storage, and decrease the bacteria level.

3 The warm milk is piped to an evaporator. Through the process of vacuum evaporation, (exposing a liquid to a pressure lower than atmospheric pressure) the boiling point of the milk is lowered to 104-113°F (40-45°C). As a result, the milk is concentrated to 30-40% solids. Also, the milk has little or no cooked flavor.

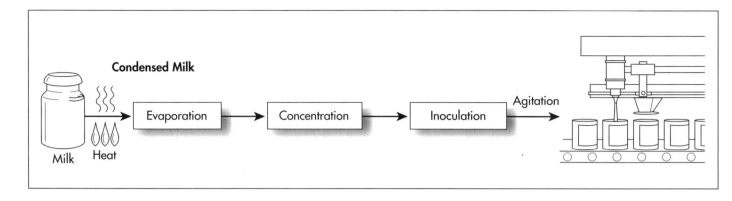

A diagram showing the manufacturing steps involved in making condensed milk.

4 The milk is then homogenized by forcing it under high pressure through tiny holes. This breaks down the fat globules into minute particles, improving its color and stability.

5 Pre-measured amounts of a stabilizing salt, such as potassium phosphate, are added to the milk to make it smooth and creamy. This stabilization causes the milk to turn a pale tan.

6 The milk is passed under a series of ultraviolet lights to fortify it with Vitamin D.

7 The milk is piped into pre-sterilized cans that are vacuum-sealed.

Condensed milk

1 The milk is flash-heated to about 185°F (85°C) for several seconds. It is then piped to the evaporator where the water removed.

2 The milk is then concentrated under vacuum pressure until it measures between 30-40% solid. It now has a syrupy consistency.

3 The milk is cooled and then inoculated with approximately 40% powdered lactose crystals. The milk is then agitated to stimulate crystallization. It is this sugar that preserves the condensed milk.

4 The milk is piped into sterilized cans that are then vacuum-sealed.

Quality Control

The milk industry is subject to stringent regional and federal regulations regarding the prevention of bacteria and the composition of solids and fats. According to the United States Food and Drug Administration (FDA), sweetened condensed milk must contain at least 28% by weight of total milk solids and at least 8% by weight of milk fat. Evaporated milk must contain at least 6.5% by weight of milk fat, at least 16.5% by weight of milk solids that are not fat, and at least 23% by weight of total milk solids. The evaporated milk must also contain 25 International Units (IUs) of vitamin D.

The milk is taste-tested for freshness before it leaves the dairy farm and again when it arrives at the processing plants. Once the milk arrives at the plant, it is not touched by the workers, making its journey from raw milk to evaporated or condensed strictly through pipes, vats, and other machinery. At least one-third of the labor time in the milk industry is devoted to cleaning and sterilizing utensils and machinery. Milk inspectors make frequent inspections.

Where to Learn More

Books

Trager, James. *The Food Chronolgy.* New York: Henry Holts, 1995.

Other

"Borden's Milk." http://www.southeastmuseum.org/html/borden_s_milk.html (March 6, 2000).

"What guidance does FDA have for manufacturers of Sweetened Condensed Milk Products?" http://www.cfsan.fda.gov/~dms/qa-ind5n.html (November 19, 1999).

—*Mary McNulty*

Faucet

Background

A faucet is a device for delivering water from a plumbing system. It can consist of the following components: spout, handle(s), lift rod, cartridge, aerator, mixing chamber, and water inlets. When the handle is turned on, the valve opens and controls the water flow adjustment under any water or temperature condition. The faucet body is usually made of **brass**, though die-cast zinc and chrome-plated plastic are also used.

The majority of residential faucets are single or dual-control cartridge faucets. Some single-control types use a metal or plastic core, which operates vertically. Others use a metal ball, with spring-loaded rubber seals recessed into the faucet body. The less expensive dual-control faucets contain nylon cartridges with rubber seals. Some faucets have a ceramic-disc cartridge that is much more durable.

Faucets must comply with water conservation laws. In the United States, bath basin faucets are now limited to 2 gal (7.6 L) of water per minute, while tub and shower faucets are limited to 2.5 gal (9.5 L).

Faucets run an average of eight minutes per capita per day (pcd), according to a study by the American Water Works Association Research Foundation completed in 1999 that was based on water use data collected from 1,188 residences. In daily pcd use indoor water use was at 69 gal (261 L), with faucet use third highest at 11 gal (41.6 L) pcd. In residences with water-conserving fixtures, faucets moved up to second at 11 gal (41.6 L) pcd. Faucet use was strongly related to household size. The addition of teens and adults increases water use. Faucet use is also negatively related to the number of persons working outside the home and is lower for those who have an automatic **dishwasher**.

History

Plumbing systems have existed since ancient times. Around 1700 B.C., the Minoan Palace of Knossos on the isle of Crete featured terra cotta piping that provided water for fountains and faucets of marble, gold, and silver. Lead pipe systems and personal bath rooms also existed during the Roman period, from about 1000 B.C.–A.D. 476. Rome's public baths also featured silver faucets, with other fixtures of marble and gold. By the fourth century A.D., Rome had 11 public baths, 1,352 public fountains and cisterns, and 856 private baths.

Plumbing systems have changed drastically since then, including faucets. For many years, faucets came with two handles, one for cold water and one for hot water. It wasn't until 1937 that this design changed. That year, a college student named Al Moen living in Seattle, Washington, turned on the faucet to wash his hands and scalded them since the water was too hot. That set an idea in motion in Al Moen's mind for the single-handle faucet.

Moen's first design was for a double-valve faucet with a cam to control the two valves. It was rejected by a major fixture manufacturer since the faucet wouldn't work, and Moen then went to a cylindrical design. From that experience, he resolved to create a faucet that would give the user water at the desired temperature with a piston action. Between 1940 and 1945, he designed several faucets, finally selling the first single-

Around 1700 B.C., the Minoan Palace of Knossos on the isle of Crete featured terra cotta piping that provided water for fountains and faucets of marble, gold, and silver.

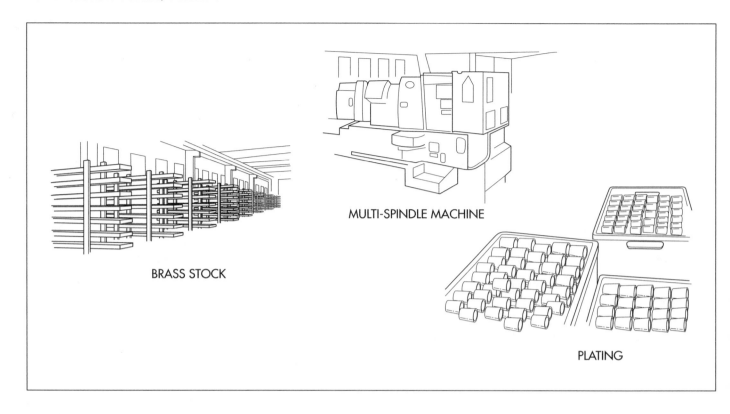

BRASS STOCK

MULTI-SPINDLE MACHINE

PLATING

Using brass stock, the multi-spindle machine automatically forms the faucet parts. Plating increases durability by adding an extra layer of protective coating.

handle mixing faucet in San Francisco in late 1947 to a local plumbing supplier. By 1959, the Moen single-handle faucet was in hundreds of thousands of homes in the United States and sold in approximately 55 countries around the world. Today, single-handle faucets are so popular that they can be found in over 40% of American homes.

Moen came up with a few other inventions during his life, including the replaceable cartridge (eliminating washers in faucets), the screen aerator, push-button shower valve diverter, swivel spray, pressure balancing shower valve, and flow control aerator. But Moen wasn't the only one concerned with faucet improvements. In 1945, Landis H. Perry designed the first ball valve for faucets. Its objective was to provide a combined volume and blending control having a simple and effective means for sealing the valve element. The design also could be easily repaired.

A patent was issued for Perry's ball valve in 1952. Shortly thereafter, Alex Manoogian purchased the rights to the patent and introduced the first Delta faucet in 1954. The Delta single-handle faucet was the first to use a ball-valve design and it proved very successful. By 1958, just four years after the

product was unveiled, Delta's sales topped $1 million.

About 20 years later, a ceramic disc was patented by Wolvering Brass for water control. Unlike cartridges that use rubber in the waterway, ceramicdiscs are lapped and polished to a degree of flatness that can only be measured in lightbands. Such discs last much longer due to their high wear resistance and provide more accurate control. These discs or valves are now in wide use.

Other recent innovations include built-in filter cartridges for reducing chlorine, lead, and cysts; built-in pullout sprays; faucets designed for people with disabilities; and electronic faucets. The latter were introduced in the early 1980s for conservation and hygienic purposes. These faucets are equipped with an infrared beam When a person puts their hands underneath the faucet, the beam is disrupted, which triggers the water to turn on. Battery-operated electronic faucets have also become available in recent years.

Raw Materials

Brass, an alloy of copper and zinc, is the most widely used material for faucets due to its resistance to soft-water corrosion and hard-water calcification. It usually contains some alloying elements—like bismuth—to make it easier to process. Brass is received as bar stock of 0.13-2 in (0.33-5 cm) in diameter, depending on the size of faucet. The majority of the other components that make up a faucet are made of other metals or ceramics and are received as finished parts from other manufacturers.

Design

To meet a variety of consumers' needs, faucets come in a wide range of styles, colors, and finishes. Ergonomic designs may involve a longer spout length and easier to operate handles. The shape of the faucet and its finish will affect the manufacturing process. Some designs will be more difficult to machine or forge than others. A different finishing process may be used to achieve a different look.

For the homeowner, special finishes are available, including brushed nickel, polished nickel, satin black, gold, platinum, and a variety of colors. Consumers also now customize the look of the faucet, combining more than one type of finish. Warranties are longer and more features are available. Prices to the consumer ranged in the spring of 2000 from $40 all the way up to $500.

The Manufacturing Process

The manufacturing process for faucets has become highly automated, with computers controlling most of the machines. Productivity and efficiency have thus improved over the years. The basic process consists of forming the main body of the faucet (sometimes including the spout if no swivel is needed), applying a finish, and then assembling the various components, followed by inspection and packaging. The faucet industry has also been impacted by environmental regulations, which have required special processes to be developed.

Forming

1 There are two methods used to make the faucet bodies. Most manufacturers use a machining process to shape the body into the required size and dimensions. This involves first cutting the bars into short slugs and automatically feeding them into a computerized numerically controlled machining center of multi-spindle and multi-axis design. This machine performs turning, milling, and drilling operations. It typically takes about one minute to make a part.

Larger faucets may require numerous machining operations. For instance, over 32 machining operations are required for some kitchen faucet bodies using a rotary machining center. With the proper machine, it can take as little as 14 seconds to make a part. Some parts, such as cast spouts for kitchen faucets, are also machined in a separate operation before assembly.

2 Some faucet manufacturers use hot forging instead of machining, since this method can produce a near-net shape in about three seconds with little waste. Forging is the process of shaping metals by deforming them in some way. In hot forging, heated metal is forced into a die that is almost the same shape as the faucet body. The pressure is slowly increased over the course of several seconds to make sure the die is completely filled with metal. Only minor machining is required to produce the exact dimensions.

Finishing

3 After machining, the parts are ready for the finishing process. Those components that come into contact with water may first require a special surface treatment to remove any remaining lead. This involves a leaching process that eliminates lead molecules from the brass surface. The conventional finish is chrome since this material is most resistant against corrosion. First a base coating of electroplated nickel is applied, followed by a thin coating of electroplated chromium. The chrome layer is deposited from a plating bath containing certain additives that improve corrosion resistance.

4 If brass plating is used, a clear polymer coating is applied to improve durability. For white and other colored finishes, a similar polymer or epoxy plastic with color added is sprayed onto the faucet in an electrically charged environment. Both coatings then are heat cured.

5 To achieve a polished brass look, physical vapor deposition is used, which ap-

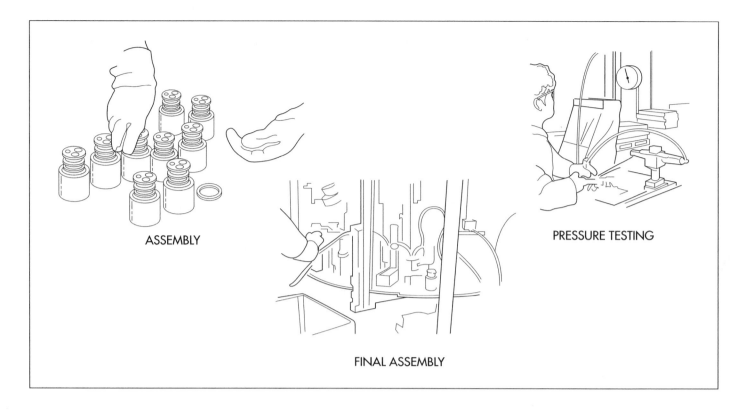

ASSEMBLY

FINAL ASSEMBLY

PRESSURE TESTING

Once the parts are assembled, they are pressure tested.

plies the metal coating in a vacuum chamber. This chamber has four components: a vacuum pump to provide a controlled environment free of contaminants; a tank that emits several types of gases; a target rod acts as the metal source; and racks to hold the faucet parts. The target is made of a corrosion-resistant material such as zirconium.

6 An electric arc heats the target to vaporize the material, then strikes the surface of the faucet at high speed and reacts with the mixture of gases. One gas provides the color and another provides the corrosion resistance. As the target material combines with these gases, it adheres to the faucet part, creating a bond that is virtually indestructible. Some manufacturers use a spiral coil around the target to provide a uniform distribution of the coating.

Assembly

7 After plating, the parts are stored in bins until assembly. Assembly can involve both manual and automated processes. For some faucets, prelubricated rubber seals or o-rings are installed by hand.

8 Finally, the faucets and other components are sent for final assembly. This process takes place on rotary assembly ma-

chines, which are precisely controlled, or by robots. The sprout, if separate, is first installed, followed by the ceramic cartridge. This cartridge is screwed in place with a brass using a pneumatic gun, and then the handle is attached by hand. Sometimes the copper tubes are installed before assembly. After assembly, the faucets are packaged in boxes along with any other components that are needed for final installation.

Quality Control

After the first part is machined, it is checked against the blueprints to ensure it matches all dimensions. A go-no-go gauge is used to make sure the interior and exterior threads fit together. Since machining is automated, random samples are then checked for the more critical dimensions. Before plating, parts are visually checked for surface imperfections, which are removed by sanding. After final assembly, every faucet is pressure tested with air for leaks and tested for durability.

Faucets must also pass several environmental regulations. The National Sanitation Foundation 61 regulation, which limits contaminants in drinking water (lead is 11ppb [parts per billion] in water from endpoint devices), applies to kitchen faucets, lavatory faucets and drinking water dispensers. Other laws are more strict—California's Proposition 65 limits the allowable lead to 5 ppb for a consumer faucet. There are also plumbing codes to deal with, which can vary from city to city. Many now require antiscald tub and shower faucets.

To receive NSF Certification of a faucet, manufacturers first submit a list of all materials including the formulation used in the product. NSF Toxicologists then review the material formulations to determine potential contaminants that may extract from the faucet and into the drinking water. NSF then conducts an inspection of the manufacturing facility to verify material formulations, material suppliers, quality control procedures and operations. Product samples are randomly selected for testing at NSF laboratories.

Faucets undergo a rigorous three-week testing sequence, where they are filled with an extractant water specified in the Standard. Selected water samples are analyzed for contaminants. NSF toxicologists compare the contaminant levels to the maximum allowable levels established in ANSI/NSF Standard 61. If all contaminant levels of the product meet the requirements of the Standard, the product can be certified. Only then is the manufacturer allowed to display the NSF Mark on the product signifying NSF Certification. To become certified, some manufacturers have had to completely modify their manufacturing process, such as switching to a purer brass material or adding a finishing rinse process.

Byproducts/Waste

Scrap metal from the machining or forging process is recycled. The finishing processes may produce waste material that must be disposed of or minimized by recycling. Since the majority of processes are automated, waste is minimized.

The Future

Faucet manufacturers will continue to add value and quality to meet consumers' increasing demands. The number of styles and range in prices will expand, with higher end products becoming more popular as remodeling intensifies. Europe will continue to remain ahead of the United States in the design department, offering more modern styles and colors. Although the overall manufacturing process will remain about the same, more automation will be used.

Novel finishes produced using PVD technology will become more widely available as consumers recognize these finishes can offer both style and durability. This may require manufacturers to improve the economics of the process, since it is more expensive than chrome plating. The popularity of filtration will extend to faucets, as consumers realize the benefits of built- in filters. Eventually these types of faucets will become the norm.

The trend toward battery-operated electronic faucets is expected to continue, and with an overall drop in prices, commercial markets should expand. Technology will continue to improve, making these faucets easier to repair and with improved operation due to fiber optics. With such improvements in technology and price, the electronic faucet will soon enter even the residential market.

Overall, the plumbing industry will continue to consolidate and manufacturers will therefore have to remain flexible. The Internet will continue to play a role in the market and someday faucet manufacturers may even sell their products online directly to the consumer.

Where to Learn More

Periodicals

"Ancient Plumbing." *Plumbing & Mechanical* (1989). http://www.pmmag.com (January 2001).

Ballanco, Julius. "What's All the Fuss About Faucets?" *Plumbing & Mechanical* (June 1998). http://www.pmmag.com (January 2001).

Cummings, James. "Washerless Faucets Work Better, Easy to Install." *Dayton Daily News* (January 27, 2000).

Grochowski, Katie. "What's the Word on Water Use?" *Plumbing & Mechanical* (November 1999). http://www.pmmag.com (January 2001).

Henkenius, Merle. "Running Water: New Faucets Deliver More Value and Better Perofrmance." *Popular Mechanics* (June 1997).

"The History of Plumbing—Roman and English Legacy." *Plumbing & Mechanical* (July 1989). http://www.theplumber.com (January 2001).

Smith, Steve. "Electronic Faucets: Smart Technology Gets Smarter." *Plumbing & Mechanical* (November 1998). http://www.pmmag.com (January 2001).

Smith, Steve. "An Interview with Linda S. Mayer." *Plumbing & Mechanical* (February 2000). http://www.pmmag.com (January 2001).

Smith, Steve. "Issues 2000: Plumbing Execs Talk About a Changing Industry." *Plumbing & Mechanical* http://www.pmmag.com (January 2001).

Other

Delta Faucet Co. PO Box 40980, Indianapolis, IN 46280. (800) 345-3358. http://www.deltafaucet.com.

Interview with Dave Bischof, Faucetcraft Faucet Company. http://www.faucetcraft.com.

Kohler Co. 444 Highland Dr. Kohler, WI 53044. (800) 456-4537. http://www.kohlerco.com.

Moen Inc. 25300 Al Moen Dr., North Olmstead, OH 44070. http://www.moen.com.

—*Laurel M. Sheppard*

Ferris Wheel

Background

A ferris wheel is an amusement park ride consisting of a large vertical wheel with places for people to sit or stand spaced evenly around the outer circumference. In operation, the ferris wheel revolves about a horizontal axis, and the riders are alternately lifted and then lowered as they are carried around the wheel in a circle. When the wheel stops, the people in the seat or platform at ground level exit the ride, and new riders take their place. The wheel then revolves a short distance until the next seat or platform is at ground level, allowing more people to exit and enter. This procedure is repeated until all the seats or platforms are filled with new riders, at which time the wheel is set in motion to undergo several complete revolutions. Although the name "ferris wheel" was not used until the 1890s, the wheel itself has been a part of human festivities for hundreds of years.

History

The earliest designs of wheels used for amusement rides may have been based on the large, circular wheels used to lift water for irrigation. In fact, knowing the human spirit, it is probable that adventuresome children used these water wheels for entertainment from the time they were first developed in about 200 B.C.

English traveler Peter Mundy described what he called a "pleasure wheel" with swings for seats after he visited a street fair in Turkey in 1620. In England, small hand-turned wheels were called "ups-and-downs" as early as 1728.

Whatever they were called, amusement wheels found their way to many parts of the world. One of the first wheels in the United States was built in 1848 by Antonio Maguino, who used it to draw crowds to his rural park and picnic grounds in Walton Spring, Georgia. As the concept of mixing amusement rides with park and picnic facilities caught on, several companies began manufacturing wheels of various designs. In 1870, Charles W.P. Dare of Brooklyn made several wood wheels of 20- and 30-ft (6.1- and 9.1-m) diameters, which he sold as the Dare Aerial Swing. The Conderman Brothers of Indiana made an even larger wheel when they developed a 35-ft (10.7-m) metal wheel in the 1880s.

The race for larger wheels culminated in early 1893 when American bridge builder and engineer, George Washington Gale Ferris, began building a 250-ft (76.2-m) wheel for the 1893 Colombian Exposition in Chicago. Designed like a bicycle wheel, with a stiff steel outer rim hung from the center axle by steel spokes under tension, the wheel could carry as many as 1,440 passengers at a time in 36 enclosed cars. The center axle was 33 in (84 cm) in diameter and 45.5 f (13.9 m) in length. It weighed 46.5 tons (42.2 metric tons) and was the largest steel forging ever produced at the time. The giant wheel opened on June 21, 1893, and drew more than 1.4 million paying customers during the 19 weeks it was in operation. The overwhelming success of Ferris' design ensured that his name would be forever linked with such wheels.

One of the people who rode the ferris wheel at the Colombian Exposition was American inventor and bridge builder William E. Sullivan. Sullivan was fascinated with the wheel and rode it many times. What was es-

In 1893, American bridge builder and engineer, George Washington Gale Ferris, began building a 250-ft (76.2-m) wheel for the 1893 Colombian Exposition in Chicago.

pecially attractive to him was the possibility of making a smaller wheel that could be taken down and moved from one park or fairground to another. Drawing on his experience with bridges, he designed a 45-ft (13.7-m) transportable wheel with twelve three-passenger seats in 1900. In 1906 he formed the Eli Bridge Company and started manufacturing his wheel in Roodhouse, Illinois. Later he moved the company to Jacksonville, Illinois, where it remains in operation today. Most of the ferris wheels found in carnivals and fairs in the United States are made by the Eli Bridge Company.

Raw Materials

Because of the unique design of a ferris wheel, most of the component parts are fabricated by the manufacturer. Steel is the most common raw material and is used to make the trailer chassis, wheel support towers, wheel spokes, and wheel crossmembers. A variety of structural steel shapes are used depending on the application. They include square tubing, round tubing, angles, channels, and wide-flanged beams. Aluminum diamond tread plate is used for the entrance and exit walkways and for the operator's platform.

Aluminum is used to make the seats and the drive rims. The drive rims are rolled out of aluminum angle stock and are attached to the spokes to form a large circle about 10 ft (3 m) smaller in diameter than the outer rim of the wheel itself. Two rubber drive wheels press against the drive rims on each side to rotate the wheel. Aluminum is used in this application because the constant rubbing of the drive wheels quickly removes the paint on the rims, exposing the bare metal. If steel were used, it would rust.

The cushions used on the seats are molded from a self-skinning **polyurethane** foam. This material forms a solid, smooth skin on the outside, while the inside remains a compressible foam. Nylon is used for some of the bushings, and a phenolic plastic is used in some of the electrical components. Support cables within the wheel structure may have a plastic cover for appearance and protection from the elements. The electrical rings that carry electrical power from the hubs to the lights along the rotating spokes are made of copper, and the brushes that bring the power to the rings are made of carbon.

Some ferris wheel components are purchased from other manufacturers and are installed on the ferris wheel when it is built. These include the axles, brakes, tires, and wheels on the trailer. Other purchased components include the electric drive motors, the electrical wires and cables, and the electrical light bulbs and sockets.

Design

Ferris wheels that are designed to be transported on the road from one location to another must conform to the overall width, height, and length restrictions for highway vehicles. Although these restrictions vary from state to state, most states limit the trailer width to 8.5 ft (2.6 m), the height to 13.5 ft (4.1 m), and the length to 55 ft (16.8 m). No matter how big or small the ferris wheel is when it is opened and in operation, it must fold down to meet these restrictions when it is travelling on the highway.

The ferris wheel must also be designed to operate safely. This requires calculations to ensure the horizontal and vertical forces of the fully loaded wheel can be supported when the wheel is in operation. It also requires the design of safety interlocks to prevent the wheel from revolving during loading and unloading operations, and to prevent the operator from inadvertently operating the wheel in an unsafe manner.

The Manufacturing Process

The manufacturing processes used to make ferris wheels varies with the design of the wheel and the manufacturer. Most of the components are built in different parts of the shop before they are brought to the main construction area for final assembly. Here is a typical sequence of operations used to build a transportable ferris wheel used in carnivals and county fairs. In operation, the wheel described is about 60 ft (18.3 m) in diameter with a capacity to carry up to 48 riders in 16 seats.

Building the chassis

1 The trailer chassis forms the base for the ferris wheel, both when it is being transported on the highway and when it is in operation. The component parts of the chassis

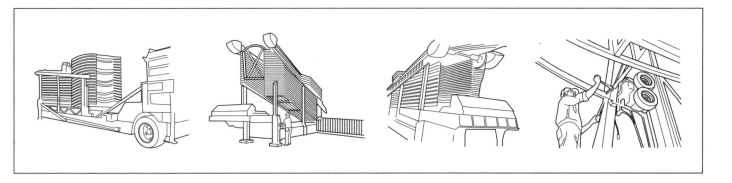

Raising a ferris wheel.

are cut to length, either with a metal-cutting saw or with a torch, and are welded together. Two vertical support posts are welded to the forward section of the chassis. These posts hold the upper end of the two wheel support towers when they are in their lowered position for travelling.

2 The completed chassis is then sandblasted to remove any scale and spatter formed during the welding operation. This ensures a smooth surface appearance and prevents the scale from chipping off later and leaving patches of bare steel.

3 The chassis is then coated with a rust-inhibiting primer. After the primer has dried, one or more coats of finish paint are applied in the desired color.

Installing the towers

4 The two wheel support towers are fabricated and painted elsewhere and are lifted into position on the chassis. The lower ends are attached to hinges on each side of the chassis, and the upper ends rest on the two support posts. The towers include ladders welded along one edge to provide access to the electrical rings and brushes at the wheel hubs and to the electrical drive motors and wheels that turn the drive rim on each side. The center axle is then installed between the wheel hubs at the tops of the two towers.

5 A long hydraulic cylinder is attached between the chassis and the wheel support tower on each side, about halfway along the length of the tower. These hydraulic cylinders are used to raise the towers into their upright position when the ferris wheel is being set up for operation. The cylinders are secured in place with a pivot pin at each end.

6 A separate lateral support arm is attached near the top of each wheel support tower. These arms each consist of two pieces of square tubing, with one piece slightly smaller in cross section so it slides inside the other. When the wheel support towers are raised for operation, the lateral support arms are pulled out to the side and the inner section of each is extended and locked in place with a pin. Two other pieces of square tubing are hinged to the chassis frame on each side and swing out to attach to the bases of the lateral supports. This gives the ferris wheel the required side-to-side stability it needs.

7 Hydraulic and electrical lines are routed inside the chassis frame pieces where they will be protected. The operator's control station is installed and connected. The chassis axles, brakes, tires, wheels, and stabilizer jacks may be installed at this time or they may be installed after all other work is complete.

Installing the spokes

8 Sixteen pairs of spokes run from the center hubs at the tops of the towers out to the seats. To install the spokes in the factory, the first pair of spokes is laid flat on the factory floor, and two crossmembers are installed between the spokes. One crossmember is located at the point where the drive rims will be attached, which is about 5 ft (1.5 m) in from the outer end of the spokes. A pair of curved sections of the drive rims are also bolted in place on each side at the same point. Only one end of the drive rim sections are bolted, leaving the other end free. This procedure is repeated for the remaining spokes, crossmembers, and drive rim sections until they form a stack. The inner ends of each pair of spokes are pinned

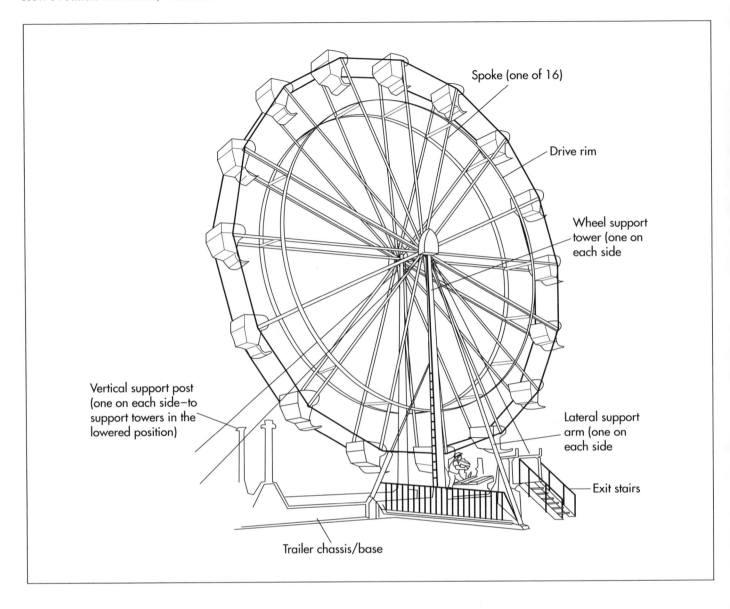

Spoke (one of 16)

Drive rim

Wheel support tower (one on each side

Vertical support post (one on each side—to support towers in the lowered position)

Lateral support arm (one on each side

Exit stairs

Trailer chassis/base

A ferris wheel.

to the pair below it. V-shaped lighting booms are installed between the center of every other outer crossmember as the stack is assembled. This overlapping pattern of lights produces a double-star effect.

9 The stack is then lifted onto the trailer with an overhead crane, and the top pair of spokes is pinned to the hubs. In operation, the spokes are all pulled into the vertical position when the towers are raised. The spokes are then pinned to the hubs, one pair at a time, and the free ends of the drive rim sections are swung down and bolted to the adjacent spokes to form the wheel — like a paper fan being unfolded.

10 Electrical cables are connected from the electrical rings at the wheel hubs

to each lighting boom. Mechanical support cables are installed between the ends of the spokes around the outer circumference of the wheel. Other mechanical cables are installed in an x-pattern between each pair of spokes to give additional stability.

Finishing the wheel

11 The entrance and exit stairs and walkways, safety fences, and trim pieces are fabricated, painted, and installed. The seats are fabricated and painted. In operation, four of the seats are carried attached to the wheel. The remaining seats are carried separately on the trailer and are manually lifted and pinned into place after the wheel is erected.

Safety Considerations

As with any amusement park ride, safety is the primary concern of both the manufacturer and the operator. Current safety regulations governing ferris wheels vary from city to city and state to state. The American Society for Testing and Materials (ASTM) is in the process of developing a comprehensive standard for the design, testing, manufacturing, and operation of all amusement park rides. Ferris wheel manufacturers and amusement park operators are actively participating in this process.

The Future

Having provided entertainment for several hundred years, if not several thousand years, the ferris wheel will probably continue to be a pleasurable experience for many years to come. Although roller coasters and other thrill rides may dominate amusement parks, the ferris wheel will still give riders the gentle thrill of being carried up in the air in an open seat to hang high above the crowds on a warm summer evening.

Where to Learn More

Books

Anderson, Norman D., and Walter R. Brown. *Ferris Wheels.* New York: Pantheon Books, 1983.

Periodicals

Marks, D., and J. Barfield. "Riding High." *People Weekly* (November 15, 1999): 62-63.

Other

Eli Bridge Company. http://www.elibridge.com (October 13, 2000).

—*Chris Cavette*

Flashlight

The modern battery powered flashlight was created in 1898 by Joshua Lionel Cowen, the original owner of the American Eveready Battery Company.

Background

A flashlight is a portable, battery-operated device used for illumination. A typical unit consists of one or more dry cell batteries arranged in a line inside a battery compartment that forms the handle of the light. The flow of electricity from the batteries to the bulb at the front end of the light is controlled through a switch mechanism placed between the batteries and the lamp.

History

Practical, portable light sources have been sought throughout history. Torches and candles were early sources of light but these were largely replaced with lanterns as people learned to burn various animal and mineral oils. However, it was not until the nineteenth century that electricity was harnessed to create light. The modern battery powered flashlight was created in 1898 by Joshua Lionel Cowen, the original owner of the American Eveready Battery Company. Cowen originally developed an idea for a decorative lighting fixture for potted plants. His fixture was composed of a metal tube with a lightbulb and a dry cell battery. Cowen passed his idea to one of his Eveready salespersons, Conrad Hubert, who turned the metal tube, lightbulb, and battery into the world's first flashlight and started selling the batteries and the flashlight. In the last hundred years, advances in technology have resulted in flashlights with hundreds of different styles and features. For example, flashlights are now made with rechargeable batteries that can be used multiple times. Other lights are designed for special operations, such as working underwater or in high-temperature conditions.

Design

The most common flashlight design is the simple household light that consists of a tube-like handle that contains the batteries. This handle is affixed to a threaded head assembly that houses the bulb mechanism. These units operate on standard batteries and provide a typical light output. Special designs are required for industrial or professional use. These lights are made from heavier gauge materials, and are more durable. They are also intended to produce a brighter beam of light. Flashlights with brighter beams are used by the police, firefighters, and the military. Camping lanterns are larger units, usually powered by heavy duty batteries. These frequently use fluorescent tubes as their light source because of their greater energy efficiency, however, this type of bulb does not cast as strong or directed a beam of light. Furthermore, the increased size and weight of this type of lantern limits its portability. Novelty flashlights are designed for use by children. These tend to be made of lightweight plastic and are notable for their visual design. The body of the light may be adorned with decorative plastic overlays that increase the child-appeal of the unit. Their designs are often based on favorite characters from popular cartoons or children's books. Finally, there are a variety of specialty lights designed for particular uses. For example, snake lights, flexible tubes that can be bent or twisted to provide light in hard to reach places. Others are designed to be small enough to fit on a key chain to illuminate keyholes.

Factors to consider when designing flashlights include light output, durability, and the ability to operate in special environments. Battery life is also an important factor, and some lights are designed to be plugged into an electrical outlet to be recharged or to maintain a charge until it is needed. Other lights use special bulbs, like the new generation flashlights built with light-emitting diodes. These are not as bright as conventional incandescent bulbs, but their power consumption is so low that they can last hundreds of hours on a set of conventional batteries, as compared to a few dozen hours for incandescent bulbs.

The Manufacturing Process

Plastic housing

1 The plastic components used in flashlight construction are typically injection molded using polystyrene and other durable polymers. In this process, plastic pellets are mixed with plasticizing agents and colorants. This mixture is liquefied by heating and then injected into appropriately shaped molds via an injection plunger. The mold is then subjected to high pressure to assure that the molds are completely filled, and to hold the molds together against force of injected liquid plastics. The end closures are also molded, where usually both internal and external threads are molded. Pressures as high as 2,500 tons may be used for high-speed or multiple-cavity production molders.

2 After the injection process, the molten plastic is cooled by forcing water through channels in the mold. The plastic hardens as it cools and the pressure is released. At this point, the two halves of the mold are separated and the plastic part can be removed for finishing. The plastic polymers used in this process are thermoplastic, meaning they can be repeatedly melted so the scrap pieces can be reworked to make additional parts. Therefore, there is very little wasted plastic in this process. Subsequent operations may be required to polish, cut, and finish the plastic parts.

Light source

3 Incandescent bulbs are the most common light source used in flashlights. These consist of a metal filament sealed in a glass bulb. When the filament is exposed to an electric current the resistance of the wire causes it to heat up and emit light in the visible wavelengths. The filament is welded to two wires that pass though holes in a cylindrical glass bead that forms the base of the bulb. This structure is placed in a fixture and a cylindrical glass envelope that is closed at one end is placed over the filament. The open end of the glass envelope rests against the glass bead.

4 The structure is placed inside a vacuum chamber and heat is applied to seal the glass envelope to the glass bead. The heat causes the glass to soften, and may cause the filament to be displaced to one side. Therefore, care must be taken to ensure the filament is properly aligned or the bulb will not project a beam of light in the right direction.

Other possible light sources include fluorescent bulbs, which are often used in camping lanterns. These bulbs emit light due to the excitation of gas molecules inside the bulb. LEDs, or light emitting diodes, are used in some specialty lights; these emit light when exposed to extremely low levels of electric current. The bulb is often fitted in front of a polished aluminum reflector that helps to focus the light during operation.

Switch and controls

5 The electronic circuitry of a flashlight varies depending on its design. Simple lights rely on an off/on switch to make the connection between the wires connecting the battery terminals to the wires extending from the base of the bulb. This type of switch is most commonly a slide type that moves up or down to make the proper connection. The switch assembly is more complicated in the more sophisticated lights. One United States patent describes a flexible metal strip that is depressed to create a contact between the wires.

Assembly

6 Depending on the design and the manufacturers capability, units may be assembled on an automated conveyor line or by hand. Some models, particularly those using small watch batteries, have the battery inserted during assembly. Otherwise, the unit may be assembled without the batteries that are inserted later by the consumer. This operation involves screwing the lamp assembly onto the threads on the casing.

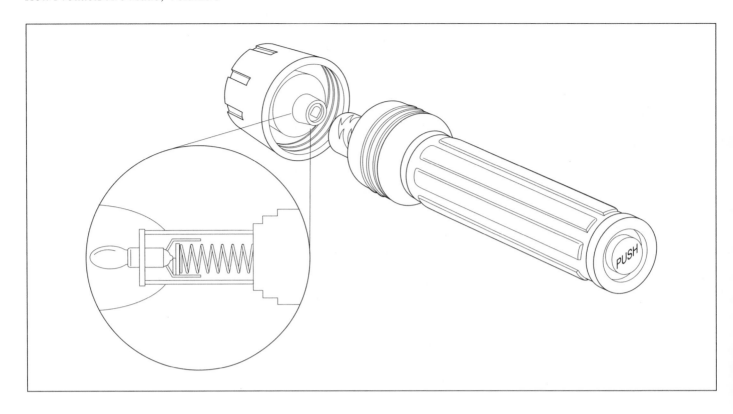

A flashlight.

Packaging

7 Assembled units may be placed in some form of outer packaging, such as a clear plastic blister pack or clam shell. The plastic shell may then be attached to a cardboard display card or packed in a box prior to shipping.

Quality Control

Completed flashlights undergo a series of quality control tests to ensure they function properly. First, the bulb must be checked to ensure it is properly aligned with the reflector; if it is misaligned performance may suffer. Second, the switch assembly is evaluated to determine if it makes proper contact with the electrical leads. Third, the seal on the battery compartment must be checked to determine if moisture will not inadvertently enter the battery compartment. This seal must allow venting of gasses that may be formed during battery operation.

The bulb itself must meet separate quality standards. Generally, Division 2-approved flashlights are temperature-rated as T1 to T6, where T1 is a temperatures less than, or equal to, 842°F (450°C) and T6 is less than, or equal to, 185°F (85°C). Testing labs used

by flashlight manufacturers include Factory Mutual Research Corporation, Underwriters Laboratories, and Demko.

Hazardous environment

Any flashlight that will be used in a hazardous environment or confined space must be properly tested to ascertain that it meets or exceeds all applicable safety standards for those locations. Hazardous Locations are defined by the National Electric Code and include the following classifications. Class I locations are areas where flammable gases may be present in sufficient quantities to produce explosive or flammable mixtures. Class II locations can be described as hazardous because of the presence of combustible dust. Class III locations contain easily ignitable fibers and filings. Hazardous atmospheres are further defined by "groups." These include atmospheres containing acetylene, hydrogen, or gases or vapors of equivalent hazard, such as ethylether vapors, ethylene, cyclo-propane, gasoline, hexane, naptha, benzene, butane, propane, alcohol, acetone, benzol, lacquer solvent vapors, or **natural gas**. Metal dust, including aluminum, magnesium, and their commercial alloys, may also create hazardous atmospheres. Environments containing carbon black, coal or coke dust, flour, starch, or grain dusts, are classified by the Code. Flashlights designed for use in these environments are individually tested before leaving the factory.

The Future

Manufacturers continue to improve upon the design of their flashlights. New models with improved power sources are becoming increasingly popular. For example, the power plant of a new self-powered flashlight is the revolutionary Freeplay Generator, which is a unique and patented mechanism that stores kinetic energy in a carbon steel **spring** as the user turns a winding handle. This energy is released as electrical energy when the light is turned on, thus powering the light without any other external power source.

Other improvements in flashlights include tougher polymers for improved durability and smarter computer technology that will allow automatic shut off mechanisms to conserve battery life. Finally, increasingly sophisticated molding techniques will allow the creation of novelty flashlights in a wider variety of shapes and colors.

Where to Learn More

Books

Ahmstead, B.H. *Manufacturing Processes.* New York: John Wiley & Sons, 1977.

Other

"Inventors." http://inventors.about.com (January 2001).

—*Randy Schueller*

Footbag

By 1995, several million footbags were sold every year.

Background

Footbags are small, soft pliable bags filled with pellets or other small solid objects. Also known as Hacky Sacks—the brand name for certain footbags—they are a little bigger than a golf ball, a few inches in diameter and an ounce or so in weight. Footbags come in two varieties: a crocheted version and paneled kind made of one of several kinds of leather or artificial materials.

The modern footbag was developed as a rehabilitation tool in the 1970s. They were soon marketed as sporting toys and became popular on a large scale. To play with the footbag, it is bounced or kicked with the foot or other parts of the leg with the goal of keeping it off the ground. Leather court shoes specially laced are the preferred shoes for using footbags. Shoes are an important part of using footbags so that optimal control is maintained.

Footbags are now the center of several games (including net sack), and organized international competitions are held regularly. Soccer players also use footbags to train for their sport.

History

The footbag originated in Asia during the dynastic era in ancient China. Imperial guards stayed alert during their overnight assignments by kicking about a small round object stuffed with hair. A similar object was used to train Chinese soldiers in 2600 B.C. About 2,000 years ago in Asia, a game called shuttlecock began to be played. At its center was a disc with feathers on it that was kicked between players. Shuttlecock is still played in parts of that continent.

Other footbag-like games are still part of Asian culture. In Malaysia, the national sport is *sepak takraw*. It is played with plastic or bamboo ball that is light and hollow. Similar sports are played in Philippines, Myanmar, and Singapore.

The history of the footbag in the United States began in the early 1970s. Mike Marshall had taken a trip to Asia and seen one of the footbag-like games there. In 1972, Marshall met John Stalberger Jr., a former football player recovering from a knee injury who was looking for a rehabilitation exercise. They came up with the footbag. This early version was a small sock stuffed with dried corn and tied. Stalberger and Marshall tested out several versions of the bag, a few of which were beanbag-like, and experimented with various ways to use it.

Stalberger dubbed the small bag a Hacky Sack because when he and Marshall would play with the bag, they said they were going to "hack the sack." Despite Marshall's death in the mid-1970s, Stalberger continued to market the footbag to schools and sports stores in his local (Portland, Oregon) area. Receiving an enthusiastic response, Stalberger patented the Hacky Sack in 1979. (He sold the rights to the name to Wham-O! in 1983.)

Early footbags were made of a heavy cordura-like fabric, though cowhide leather soon dominated the market. The leather was used to make panels that were sewn together, not unlike a soccer ball. In 1981, crocheted footbags (called granny sacks, because they were handsewn, allegedly by grandmothers) were introduced. Crocheted footbags were seen as an improvement on the leather footbags, which were hard to break in. Crocheted footbags were already soft.

Over the course of the 1980s and 1990s, artificial materials began to be used more often than leather. Ultrasuede does not require the same kind of break-in period that leather footbags do. Paneled footbags became made of vinyl, pigskin, water buffalo skin, snake skin, kangaroo skin, and various artificial materials.

While the crocheted type of footbag are popular among some footbaggers, others believe that paneled footbags are better and last longer. Different footbags have different characteristics, especially concerning bounce. Both types of footbags are usually filled with plastic pellets, but have been filled with other materials like cherry pits. In 1995, sand-filled footbags were introduced, and soon became popular among freestyle enthusiasts because they are conducive to certain tricks. By 1995, several million footbags were sold every year.

Footbag games

Footbags became a popular fad in North America by the early 1980s. A million footbags were sold by 1983. Though interest in footbags was concentrated in the United States, it soon became a worldwide phenomenon. Though faddish interest in footbags faded, a core constituency has remained. Those who use footbags value how it increases agility, endurance, coordination, balance, and concentration. Others enjoy footbags because they are not competitive in the same way most sports are: users have to cooperate with others to keep the footbag off the ground.

One way that footbags are used are in hack circles. The footbag is passed around a circle via foot, using one of the five basic kicks (the inside kick, knee kick, toe kick, outside kick, or back kick). Circles can be comprised of as few as two and as many as 25 or more players, and is often played on college campuses. Hack circles evolved into footbag freestyle. Footbaggers show off tricks like toe- stalls and clippers, linked together in routines, while keeping the footbag off the ground. Competitors are judged on difficulty of routines as well as artistic merit.

Freestyle is just one event in footbag competitions. Another is Footbag Consecutive, played in singles or pairs. The object is to keep the bag going as long as possible. A world record in pairs set in 1995 had two people completing 123,456 kicks in 19 hours, 19 minutes, and 20 seconds.

Some footbag games are individual while others involve teams. Net-sack was invented by Stalberger in the late 1970s. It is essentially volleyball with a footbag. Played on a badminton-sized court with a 5 ft (15.2 m) high net, net-sack features singles and doubles brackets and is scored like volleyball. Footbag golf is played and scored like golf on a course with holes and obstacles.

Design

Crocheted footbags have designs woven into them, including stripes, names and symbols. The patterns are set before production.

While some paneled footbags have brand names or logos silk-screened to them, the primary design process concerns how the panels are shaped, colored, and placed, and the number of panels. The panels can be shaped like squares, triangles, octagons, circles, and pears. Colors are limitless. There is even a footbag made with a glow-in-the-dark exterior. Early paneled footbags were sometimes made of only two panels, but more panels have become more common. Eight to 14-paneled footbags are very popular on the retail level. Footbags made of 32 panels also have a large following among accomplished footbaggers. The more panels, the rounder the footbag.

Raw Materials

Crocheted footbags are often made of double stranded heavy duty rayon. Most paneled footbags are made of ultrasuede, though they can also be made of split-grain cowhide, vinyl, pigskin, water buffalo skin, snake skin, kangaroo skin, facile, multifuzz polymer suede, or other artificial materials. Paneled footbags are sewn with a tough, durable synthetic thread, not unlike dental floss. Faceted styrene plastic beads the size of the BBs are most often used to stuff footbags. Sometimes footbags are filled with plastic regrind pellets, plastic polyresin filler pellets, small rocks, corn, or sand.

Examples of crocheted and paneled footbags.

The Manufacturing Process

Crocheted footbags

1 Rayon is hand crocheted according to a pattern with a certain number of stitches. A hole is left on top.

2 The faceted styrene plastic beads are measured by volume, sometimes in a plastic jig. They are put in the footbag through the hole by hand.

3 After the bag is filled, the whole on top is crocheted closed. It takes about 20 minutes to one hour to finish one bag.

Paneled Footbags

1 To create the panels, bolts of ultrasuede (or other fabric or leather) are loaded on to a machine with an automated, repetitive punch and die mechanism. The punch is lowered through the fabric to create the shaped panels. This process is often guided and supervised by a worker.

2 Some of the panels are set aside and a logo or other stamp is silk-screened to them. This silk-screen process is done by hand.

3 The panels are assembled for sewing by hand. They are sewed inside out, leaving the final four or five stitches loose. (While most panels are sewn by hand, a few companies sew paneled footbags on machine, especially those that will be filled with sand.)

4 The footbag is turned inside out. About 1.1-1.8 oz (32-50 g) of plastic pellets (or other filling material) is measured and inserted by hand through the space created by the loose stitches.

5 The final stitches are completed with an interior knot.

6 Depending on the manufacturer, the finished product is packaged for retail sale, then boxed for shipping. Smaller manufacturers pack the footbags as is for shipping.

Quality Control

Crocheted footbags are dimension tested for size and tight crocheting. The beads must be the same size and right fill weight. Any rejected crocheted footbags are unraveled and redone.

Panel footbags are often sample checked at the end of the manufacturing process for the correct stitching, weight, and size.

Byproducts/Waste

There are no byproducts in the crocheted footbag process. Extraneous materials are reused.

Scraps from the panel creation process are thrown away, but the waste is minimal.

The Future

Paneled footbags will be improved as more new fabric materials are put on the market. More durable, washable ultrasuedes and other fabrics will mean better, longer-lasting footbags. New designs will also be continually introduced. Different fillers for both crocheted and paneled footbags are regularly developed.

Where to Learn More

Books

Cassidy, John. *The Hacky-Sack Book.* Palo Alto: Klutz Press, 1982.

Periodicals

Berg, Scott. "Footbag: Kickin' Up a Storm." *Washington Post* (August 6, 1999): N63.

Colton, Michael. "The Goodwill Game." *Los Angeles Times* (September 18, 1995): E1.

Najarian, Ara. "Kicking It: For Simple Fun With Your Feet, Footbag is a Sport with Sole." *Los Angeles Times (Orange County Edition)* (May 28, 1999): 28.

Other

World Footbag Association (1998). http://www.worldfootbag.com (December 27, 2000).

—Annette Petruso

Geodesic Dome

One of the most recognizable geodesic domes is the 165-ft (48-m) diameter sphere at Walt Disney World's Epcot Center.

Background

A geodesic sphere is an arrangement of polygons that approximates a true sphere. A geodesic dome is a portion of a geodesic sphere. Buildings or roofs have been constructed out of geodesic domes that range from 5-100% of a sphere. Domes used for houses are usually arrays of triangles that form three- or five-eighths of a geodesic sphere.

Geodesic domes are efficient structures in several ways. The triangle is a very stable shape; for example, a force applied to the corner of a rectangle can deform it into a parallelogram, but the same force will not deform a triangle. This makes geodesic dome buildings highly resistant to such forces as snow coverings, earthquakes, wind, and even tornadoes. The surface area of a geodesic dome is only 38% of the surface area of a box-shaped building enclosing the same floor space. There is less surface exposed to outdoor temperature fluctuations, making the building cheaper to heat and cool than a rectilinear structure. Geodesic domes can be constructed quickly without heavy equipment. Using prefabricated components, it takes just a few people to erect the dome for a 2,000-sq ft (185-sq m) home in 10 hours or less.

A geometric dome supports itself without needing internal columns or interior load-bearing walls. This property makes such structures appealing for use as churches, sports arenas, and exhibition halls. The aesthetic appeal of lofty ceilings makes them attractive as homes, and full or partial second-story floors are easily suspended halfway up the enclosure without any support other than attachment to the dome itself.

History

In 1919, seeking a way to build a larger planetarium, German engineer Walter Bauersfeld decided to mount movable projectors within a stationary dome. Until that time, planetarium domes rotated while external light entered through holes on the dome shell to simulate stars and planets. This limited the practical size of the dome and the number of people it could hold. Bauersfeld's concept of interior projection would work in a much larger dome. The first model constructed was more than half of a sphere; 52 ft (16 m) in diameter. Bauersfeld solved the problem of how to construct such a large sphere by approximating it with an icosahedron (20-sided solid with equal triangular faces) and subdividing each face into smaller triangles. He framed the triangles from nearly 3,500 thin iron rods. To construct a spherical shell over this framework, he erected a spherical wooden form inside the frame and sprayed on a pasty concrete mixture. The shell was designed to be the same proportional thickness as that of an eggshell compared to its diameter, a ratio later considered appropriate for geodesic domes.

Thirty years later, R. Buckminster Fuller, an American architect, engineer, poet, and philosopher, independently invented a similar structural system. Following World War II, Fuller wanted to design affordable, efficient housing that could be built quickly from mass-produced components. Willing to look outside of conventional approaches, Fuller began to work with spherical shapes because they enclose a given space with a minimum of surface area. He first framed spheres with a network of strips approximating great circles (circles on a sphere with centers that coincide with the sphere's center); the strips formed triangles as they crossed one another. He called the product a geodesic dome because great circles are known as geodesics (from a Greek word meaning earth dividing). Eventually, Fuller began forming spheres from hexagons and pentagons (like the panels on a soccer ball) and dividing them into triangles for strength and ease of construction.

In 1953, Fuller used his new system to cover the 93-ft (28-m) diameter courtyard surrounded by Ford Motor Company's headquarters building. The building was not designed to support the great weight of a traditional dome, but Fuller's creation weighed 95% less. He completed the design and construction in only three months. A temporary mast erected in the center of the courtyard supported the dome during construction, and the structure was incrementally raised and rotated following completion of each new section. The frame consisted of 12,000 aluminum struts weighing a total of 3,750 lb (1,700 kg) that were connected to form triangles and then lifted into position and riveted to the growing frame. When the dome was completed, it was gently lowered onto mounts that had been installed on the existing building. A clear fiberglass panel was installed in each triangle to complete the dome.

In 1954, Fuller received a patent on geodesic domes. During the 1960s and 1970s, an era in which unconventionality was prized, geometric domes became popular as an inexpensive way for environmentally conscious people to build their own homes. Instructions were widely available, but the quality of materials (including such strange choices as paper mâché and discarded tin cans) and the skill of do-it-yourself builders were inconsistent. Amateur-built domes tended to leak when it rained, insufficient use of insulation limited their energy efficiency, and inadequate numbers of skylights left interiors dreary.

Fuller predicted that a million geodesic domes would be built by the mid-1980s, but by the early 1990s, estimates placed the worldwide number somewhere between 50,000 and 300,000. A small but persistent contingent of unconventional homebuilders continue to build geodesic dome homes, pri-

marily from kits. However, *Newsday* reported in 1992 that the majority of geodesic dome structures have been built for greenhouses, storage sheds, defense shelters, and tourist attractions. One of the most recognizable of these is the 165-ft (48-m) diameter sphere at Walt Disney World's Epcot Center. Built of composite panels of ethylene plastic and aluminum in 1982, the structure houses a ride called Spaceship Earth, a termed coined by Fuller himself.

Raw Materials

Geodesic domes range in size from the 460-ft (143-m) Poliedro de Caracas sports arena in Venezuela to temporary shelters that are 15 ft (5 m) or less in diameter. Consequently, construction materials vary widely. Simple, movable structures may be built of polyvinyl chloride (PVC) pipe or galvanized steel conduit frames covered with plastic sheeting or parachute canopies. Large, permanent structures like arenas and factories have been built from materials like aluminum and steel frame struts covered with aluminum, copper, structural gypsum, acrylic, or Plexiglas panels.

Most residential dome kit manufacturers use wood components, primarily kiln-dried Douglas fir struts covered with 0.5-in (1.3-cm) exterior- or structural-grade plywood. Such kits include various designs of connectors to securely fasten the wood struts together in the proper configuration; high-strength aluminum, or steel coated with zinc, epoxy, or industrial primer are commonly used for connectors. Zinc-plated steel bolts secure the connectors and paneling is nailed on.

A few kit manufacturers use alternative materials to make prefabricated panels that combine the frame and exterior covering. One, for example, makes molded fiberglass panels. Another supplies reinforced concrete panels; steel mesh extending from the panel edges is overlapped with mesh from the adjoining panel, and the joint is sealed with concrete.

Most dome kits are built atop concrete foundation slabs. Often, these slabs are recessed into the ground to provide a basement level. Foundation walls and riser walls (vertical walls below the dome that may be used to raise its overall height) are usually made of concrete or wood. Interior insulation general-

Building a geodesic dome.

ly consists of fiberglass batting or sprayed-on urethane, cellulose, or Icynene plastic foam.

Design

Although dome homes are built from manufactured kits, designs are flexible. As many as half of the triangles in the dome's lowest row can be removed without weakening the structure, so door and window openings can be plentiful. Vertical-walled extensions can be built out from such openings to increase the floor space. The dome can sit directly on ground-level footings (short walls recessed into the ground to bear the building's weight), or it can be erected atop a riser wall up to 8 ft (2.5 m) tall.

Space must be provided between the interior and exterior walls to accommodate insulation. Some manufacturers create this space by making the struts from wood that is 4-8 in (10-20 cm) thick. Others make this space 14.5-21 in (37-53 cm) thick by using compound struts consisting of two strips of lumber joined with plywood gussets.

The Manufacturing Process

The following is a composite of techniques used by several individuals using kits from various manufacturers.

The substructure

1 After clearing and leveling the home site, a trench is dug for the foundation footing, following detailed drawings supplied by the kit manufacturer. The base of the dome is not circular; rather, it is outlined by five short walls alternating with five long walls (twice the length of the short walls). Forms are placed for the footings; many builders like to use permanent Styrofoam forms that need not be removed. Concrete is then poured in the footing forms.

2 A layer of sand may be used to further level the surface and provide a base for the foundation slab. Reinforcing steel bars are tied together in a grid, and concrete is poured to form the foundation.

3 Foundation walls are built atop the footings, up to approximately ground level. If desired, riser walls (which are provided as part of the kit) are installed atop the foundation walls and bolted to one another.

4 Floor joists are installed by standing wooden 2x12 (1.5x11.5 in [3.8x29.2]) boards 16 in (40 cm) apart above the foundation. The joists are nailed to a perimeter wooden frame and a wooden crossbeam. Three-quarter-inch (1.9-cm) thick plywood sheets are laid across the joists and nailed in place.

The superstructure

The superstructure typically consists of 60 triangular panels. Depending on the desired size of the dome, the panels are usually 6-10 ft (1.8-3 m) on a side. They may be prefabricated with the exterior panels installed, or they may constructed on site from precut lumber and metal connectors.

If dome panels were supplied with the kit, they are set atop the foundation or riser walls and connected to one another in a sequence prescribed by the manufacturer. Until enough panels are connected to support themselves, they must be braced with poles radiating out from a block in the center of the floor. The following steps describe the more common case of frame erection followed by exterior panel installation:

5 Base plates are installed atop the foundation or riser walls. These precisely beveled 4x6-in (10x15-cm) wood strips provide a transition between the horizontal top edge of the walls and the slightly tilted triangles of the dome's bottom strip of panels.

6 Matching the color coding on the kit's wooden struts and metal connectors, a triangle is formed and secured with bolts. The triangle is lifted into position and bolted to the wall and/or to the adjacent triangles. Successive rows of triangular elements are placed until the dome is completely formed. Because of their light weight, the triangles do not need supplementary bracing to hold them in place during construction.

7 Wooden studs are nailed inside each triangle. Running perpendicular to one side of the triangle, they are placed about 16 in (40 cm) apart. If an odd number of studs is used, the center one is secured against a perpendicular block near the triangle's vertex, rather than extending to the vertex.

Richard Buckminster Fuller.

Richard Buckminster Fuller was born July 12, 1895, in Milton, Massachusetts. He entered Harvard University in 1913 but was expelled two years later. In 1917, he married Anne Hewlett and formed a construction company. In 1923, Fuller invented the stockade brick-laying method—bricks with vertical holes reinforced with concrete. In 1927 he designed a factory-assembled Dymaxion house, a self-contained unit suspended from a central mast with a complete recycling system. Dymaxion was a term he used for anything deriving maximum output from minimum input. Fuller also designed a Dymaxion car—an omnidirectional vehicle that gave minimum wind resistance—in 1928. This vehicle could seat 12, make 180° turns, run easily at 120 mph (193 kmph), and average 28 mi/gal (12 km/L) but was unprofitable because automobile manufacturers wouldn't mass produce it.

Fuller's financial break came in 1940 with his Dymaxion Deployment Unit (DDU), a circular self-cooled living unit with pie-shaped, corrugated steel rooms. The British used DDUs in World War II. In 1949, Fuller started work on geodesic domes. He applied for a patent in 1951 and received a contract in 1953 from Ford Motor Company to build a dome over their headquarters' courtyard in Detroit. The U.S. Defense Department became Fuller's largest customer, using domes as temporary housing units and to protect radar equipment from harsh environments.

At the time of his death in 1983, Fuller's domes were used worldwide. In 1985, fullerene was discovered. Fullerenes are carbon atoms arranged in sphere-like shapes with pentagonal and hexagonal faces, similar to geodesic domes. These "bucky balls" have up to 980 carbon atoms.

8 Matching color-coded edges, the plywood panels are lifted into position on the exterior of each triangle and nailed into place. By working downward from the top of the dome, the worker can stand on the open framework below while attaching each panel.

9 Vertical walls and roofs are framed for any desired extensions that will project outward from the dome. Plywood panels are nailed to the exterior faces of the extensions. Dormer extensions can also be erected for second-story windows.

Finishing

10 Windows, skylights, and exterior doors are installed.

11 The roof is covered with rubber sheeting, and conventional roofing material (such as shingles or tiles) is applied.

12 Conventional siding material (such as stucco or vinyl siding) is applied to the exterior of the riser walls.

13 Insulation is placed between the struts and studs inside the dome and extension walls.

14 Walls are framed to divide the interior into rooms. Conventional drywall sheets are cut according to patterns included in the kit, and they are nailed to the interior walls and the inside surfaces of the dome and riser walls. Because of the many angles between triangular sections of the dome, amateur builders often hire a professional to tape the drywall joints.

Quality Control

A quality geodesic dome structure is airtight and structurally sound. These are the factors that lower energy costs, the main consideration when building a geodesic home. Because the structure is basically airtight, condensation can sometimes be a problem. Normally it is controlled by the heating and cooling system but when the house has been closed up for a few days, moisture can build up. This is easily solved by turning the air system on or opening a door or window.

The Future

Future refinements in geodesic dome construction may come from improved building materials. For example, in 1997 a concrete block manufacturer developed a hollow, beveled, triangular block with scored edges that could interlock with adjacent blocks. Properly shaped, such blocks could be used to construct domes.

Another innovation involves designing domes based on a different mathematical premise. In a true geodesic dome, the edges of the triangular elements align to form great circles. Although not geodesic, a new design patented in 1989 uses hexagons and pentagons to form domes with an elliptical cross section. Because of its mathematical derivation, this design is called geotangent.

Although geodesic domes maximize strength while minimizing construction materials, elliptical-profile domes offer two different advantages. They can cover a circular area without rising as high as a spherical dome. And they can cover elongated or irregularly shaped areas that vary in elevation. Located in northern Mexico, the world's largest industrial domes are a pair of manufacturing buildings covered with elliptical roofs 735 ft (224 m) long and 260 ft (80 m) wide.

Where to Learn More

Periodicals

DiChristina, Mariette. "Elliptical Dome." *Popular Science* (January 1990): 74.

Horton, Ted. "The Dome." *Mother Earth News* (June/July 1999): 64.

Sieden, Lloyd Steven. "The Birth of the Geodesic Dome: How Bucky Did It." *The Futurist* (November/December 1989): 14+. http://www.lsi.usp.br/usp/rod/bucky/geodesic_domes.txt. (January 6, 2000).

Other

"An Introduction to Geodesic Domes." http://0www.dnaco.net/~michael/domes/intro.html (December 2, 1999).

"Design and Construction of Alpine Dome Homes." Alpine Domes. http://www.freeyellow.com/alpinedomes (January 6, 2000).

"Home Sweet Dome." http://future.newsday.com (January 6, 2000).

Timberline Geodesics. http://www.dome-home.com (April 4, 2000).

—*Loretta Hall*

Gyroscope

The gyroscope is not just a toy, but a part of many scientific and transportation-related instruments.

Background

The gyroscope is a familiar toy that is deceptively simple in appearance and introduces children to several mechanical principles, although they may not realize it. Something like a complex top made of precisely machined metal, the gyroscope is a spinning wheel that may be set within two or more circular frames, each oriented along a different line or axis. The framework can be tilted at any angle, and the wheel—as long as it is spinning—will maintain its position, or attitude.

But the gyroscope is not just a toy. It is a part of many scientific and transportation-related instruments. These include compasses, the mechanisms that steer torpedoes toward their targets, the equipment that keeps large ships such as aircraft carriers from rolling on the waves, automatic pilots on airplanes and ships, and the systems that guide missiles and spacecraft relative to Earth (that is, inertial guidance systems).

The gyroscope consists of a central wheel or rotor that is mounted in a framework of rings. The rings are properly called gimbals, or gimbal rings. Gimbals are devices that support a wheel or other structure but allow it to move freely. The rings themselves are supported on a spindle or axis at one end that, in turn, can be mounted on a base or inside an instrument. The property of the rotor axle to point toward its original orientation in space is called gyroscopic inertia; inertia is simply the property of a moving object to keep moving until it is stopped. Friction against the air eventually slows the gyroscope's wheel, so its momentum erodes away. The axle then begins to wobble. To maintain its inertia, a gyroscope must spin at a high speed, and its mass must be concentrated toward the rim of the wheel.

History

The gyroscope is a popular children's toy, so it is no surprise that its ancestor is the spinning top, one of the world's oldest toys. A single-frame gyroscope is sometimes called a gyrotop; conversely, a top is a frameless gyroscope. In the sixteenth through eighteenth centuries, scientists including Galileo (1564–1642), Christiaan Huygens (1629–1695), and Sir Isaac Newton (1642–1727) used toy tops to understand rotation and the laws of physics that explain it. In France during the 1800s, the scientist Jean-Bernard-Léon Foucault (1819–1868) studied experimental physics and proved Earth's rotation and explained its effect on the behavior of objects traveling on Earth's surface. In the 1850s, Foucault studied the motions of a rotor mounted in a gimbal frame and proved that the spinning wheel holds its original position, or orientation, in space despite Earth's rotation. Foucault named the rotor and gimbals the gyroscope from the Greek words *gyros* and *skopien* meaning "rotation" and "to view."

It was not until the early 1900s that inventors found a use for the gyroscope. Hermann Anschütz-Kaempfe, a German engineer and inventor, recognized that the stable orientation of the gyroscope could be used in a gyrocompass. He developed the gyrocompass for use in a submersible for undersea exploration where normal navigation and orientation systems are impractical. In 1906, Otto Schlick tested a gyroscope equipped with a rapidly spinning rotor in the German torpedo boat *See-bar*. The sea caused the torpedo boat to roll 15° to each side, or 30° total; when his gyroscope was operated at full speed, the boat rolled less than 1° total.

In the United States, Elmer Ambrose Sperry (1860–1930)—an inventor noted for his achievements in developing electrical locomotives and machinery transmissions—introduced a gyrocompass that was installed on the U.S. battleship *Delaware* in 1911. In 1909, he had developed the first automatic pilot, which uses the gyroscope's sense of direction to maintain the course of an airplane. The Anschütz Company installed the first automatic pilot—based on a three-frame gyroscope—in a Danish passenger ship in 1916. In that year, the artificial horizon for aircraft was designed as well. The artificial horizon tells the pilot how the airplane is rolling (moving side to side) or pitching (moving front to rear) when the visible horizon vanishes in the clouds or other conditions.

Roll-reduction was needed for ships, too. The Sperry Company had introduced a gyrostabilizer that used a two-frame gyroscope in 1915. The roll of a ship on the ocean makes passengers seasick, causes cargo to shift and suffer damage, and induces stresses in the ship's hull. Sperry's gyrostabilizer was heavy, expensive, and occupied a lot of space on a ship. It was made obsolete in 1925 when the Japanese devised an underwater fin for stabilizing ships.

During the intense development of missile systems and flying bombs before and during World War II, two-frame gyroscopes were paired with three-frame instruments to correct roll and pitch motions and to provide automatic steering, respectively. The Germans used this combination on the V-1 flying bomb, the V-2 rocket, and a pilotless airplane. The V-2 is considered an early ballistic missile. Orbiting spacecraft use a small, gyroscope-stabilized platform for their navigation systems. This characteristic of gyroscopes to remain stable and define direction to a very high degree of accuracy has been applied to gunsights, bombsights, and the shipboard platforms that support guns and radar. Many of these mechanisms were greatly improved during World War II, and the inertial navigation systems that use gyroscopes for spacecraft were invented and perfected in the 1950s as space exploration became increasingly important.

Raw Materials

The materials used to manufacture a gyroscope can range from relatively simple to highly complex depending on the design and purpose of the gyroscope. Some are made more precisely than the finest watch. They may spin on tiny ball bearings, polished flecks of precious gemstones, or thin films of air or gas. Some operate entirely in a vacuum suspended by an electrical current so they touch nothing and no friction develops.

A gyroscope with an electrically powered motor and metal gimbals has four basic sets of components. These are the motor, the electrical components, electronic circuit cards for programmed operation, and the axle and gimbal rings. Most manufacturers purchase motors and electrical and electronic components from subcontractors. These may be stock items, or they may be manufactured to a set of specifications provided to the supplier by the gyroscope maker. Typically, gyroscope manufacturers machine their own gimbals and axles. Aluminum is a preferred metal because of its expansion and strength characteristics, but more sophisticated gyroscopes are made of titanium. Metal is purchased in bulk as bar stock and machined.

Design

Using the electrical and mechanical aspects of gyroscopic theory as their guides, engineers choose a wheel design for the gimbals and select metal stock appropriate for the design. The designs for many uses of gyroscopes are fairly standard; that is, redesign or design of a new line is a matter of adapting an existing design to a new use rather than creating a new product from the most basic beginning. Design does, however, involve observing the most fundamental engineering practices. Tolerances, clearances, and electronic applications are very precise. For example, design of the gimbal wheels and design of the machining for them has a very small tolerance for error; the cross section of a gimbal must be uniform throughout or the gyroscope will be out of balance.

The Manufacturing Process

1 The gimbals and gimbal frames are machined from aluminum bar stock using

An example of a gyroscope.

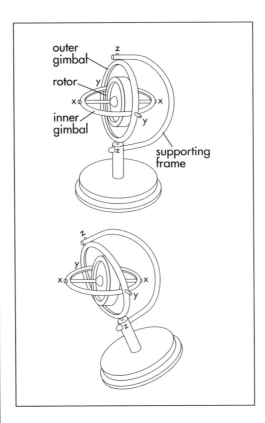

tools developed as part of the design process. They are polished and cleaned and stored in bins until assembly. For assembly, the bins are moved to appropriate locations along the assembly line.

2 Gyroscopes are manufactured in a straightforward assembly line process that emphasizes the importance of "touch labor" over automation. Gyroscopes are assembled from the inside out. The motor is the heart of the gyroscope and is installed first. A "typical" gyroscope motor is synchronized to spin at 24,000 revolutions per minute (rpm). It must be perfectly synchronized, and the motor is typically bench-tested before assembly. Electrical connections are added to the motor.

3 The gimbals and frames are assembled next, beginning with the inner gimbal and ending with the outer gimbal frame. Bearings are put into place. The "end play" of the bearings (the looseness of fit) typically has a very small tolerance of 0.0002-0.0008 in (0.006-0.024 mm).

4 The outermost electrical connections are attached on the assembly line, and circuit cards are added. Finally, the gyroscope is calibrated at the end of the assembly process. The suspension of the bearings and

calibration are hand checked; manufacturers have found that, for even calibration, human observation, testing, and correction are more trustworthy than automated methods.

The gyroscope is an elegant example of an application of simple principles of physics. Because it is simple, manufacturers closely guard any proprietary techniques. Because the gyroscope is a simple device with wide-ranging uses, some require more manufacturing processes. The manufacturing steps described above take about 10 hours and result in a free gyroscope for an application such as missile guidance. A more exotic gyroscope may require 40 hours of assembly time.

Quality Control

Quality control is essential throughout the design and assembly processes in manufacturing gyroscopes because the instruments are part of manned aircraft, unmanned missiles, and other transportation and weapons devices that could cause catastrophes if they fail. Engineers, scientists, and designers are highly educated and trained before they are hired and while on the job. Assembly-line workers must pass initial training to be hired, and they have regularly scheduled, ongoing training sessions. Many of the quality standards that must be met in gyroscope manufacture can be measured, so in-process inspection is performed throughout manufacture. Quality control at the highest level is performed by inspectors from outside the company and includes government inspectors. Customers also perform their own inspections and acceptance testing; if the manufacturer's product fails the customers' tests, the failed gyroscopes are returned.

Byproducts/Waste

Gyroscope manufacturers do not produce byproducts, but they tend to make full lines of gyroscopes for a wide variety of applications. They also do not produce much waste. Machining the gimbals and rings produces some aluminum chips, but these are collected and returned to the aluminum supplier for recycling.

Safety Concerns

Manufacturers observe the mandates of the Occupational Safety and Health Administration (OSHA) for light, ventilation, and ergonomics (comfortable seating and work benches that reduce the likelihood of repetitive stress injuries). Humidity must be maintained in the plant to prevent electrostatic discharge. Minor quantities of cleaning solvents are required, but citrus-based cleaners that are benign (harmless) are used.

The Future

Uses for gyroscopes are increasing with the number of devices that require guidance and control. Although the basics of the gyroscope are grounded in the laws of physics and can never change, the technology is evolving. Mechanical and electrical methods for providing the spinning mass that makes the gyroscope work are gradually being replaced by ring lasers and microtechnology. Coils of thin optical fibers hold the key to compact, lightweight gyroscopes that might have applications in navigation systems for automobiles. The gyroscope is such a simple but sophisticated instrument for keeping so many tools in transportation, exploration, and industry in balance that, seen or unseen, it certainly has a place in the future.

Where to Learn More

Books

Campbell, R. W. *Tops and Gyroscopes.* New York: Thomas Y. Crowell Company, 1959.

Langone, John. *National Geographic's How Things Work: Everyday Technology Explained.* Washington, DC: National Geographic Society, 1999.

Sparks, James C., Jr. *Gyroscopes: What They Are and How They Work.* New York: E. P. Dutton & Co., Inc., 1963.

Walton, Harry. *The How and Why of Mechanical Movements.* New York: Popular Science Publishing Company, E. P. Dutton & Co., Inc., 1968.

Periodicals

"A Gyroscope's Gravity-defying Feat." *Science News* 137, no. 1 (January 6, 1990): 15.

Scott, David. "Optical-fiber Gyro." *Popular Science* 230 (June 1987): 25.

Other

Gyroscopes as Propulsion Devices. http://www.gyro-scope.co.uk (July 2000).

Gyroscope Study Guide. http://clubknowledge.com/study/gyro.html (July 2000).

How a gyroscope works. http://www.accs.net/users/cefpearson/gyro.htm (July 2000).

—*Gillian S. Holmes*

Halogen Lamp

One hundred and eighty novel halogen lamps were used in the Times Square Ball for New Year's Eve in 1999.

Background

A halogen lamp is a type of incandescent lamp. The conventional incandescent lamp contains a tungsten filament sealed within a glass envelope that is either evacuated or filled with an inert gas or a mixture of these gases (typically nitrogen, argon and krypton). When electrical power is applied to the filament, it becomes hot enough (generally over 3,600°F [2,000°C]) to become incandescent; in other words, the filament glows and emits light. During operation, the tungsten evaporating from the hot filament condenses on the cooler inside bulb wall, causing the bulb to blacken. This blackening process continuously reduces the light output over the life of the lamp.

A halogen lamp comes with a few modifications to eliminate this blackening problem. The bulb, made of fused quartz instead of soda lime glass, is filled with the same inert gases as incandescent lamps mixed with small amounts of a halogen gas (usually less than 1% bromine). The halogen chemically reacts with the tungsten deposit to produce tungsten halides. When the tungsten halide reaches the filament, the intense heat of the filament causes the halide to break down, releasing tungsten back to the filament. This process—known as the tungsten-halogen cycle—maintains a constant light output over the life of the lamp.

In order for the halogen cycle to work, the bulb surface must be very hot, generally over 482°F (250°C). The halogen may not adequately vaporize or fail to adequately react with condensed tungsten if the bulb is too cool. This means that the bulb needs to be smaller and made of either quartz or a high-strength, heat-resistant grade of glass known as aluminosilicate. Since the bulb is small and usually fairly strong due to its thicker walls, it can be filled with gas to a higher than usual pressure. This slows down the evaporation of the tungsten from the filament, increasing the life of the lamp.

In addition, the small size of the bulb sometimes makes it economical to use heavier premium fill gases such as krypton or xenon—which help retard the rate of tungsten evaporation—instead of the cheaper argon. The higher pressure and better fill gases can extend the life of the bulb and/or permit a higher filament temperature that results in better efficiency. Any use of premium fill gases also results in less heat being conducted from the filament by the fill gas. This results in more energy leaving the filament by radiation, slightly improving the efficiency.

Halogen bulbs thus produce light that is whiter and brighter, use less energy, and last longer than standard incandescent bulbs of the same wattage. They can last from 2,000-4,000 hours (about two to four years) compared to conventional incandescent bulbs, which only operate for 750-1,500 hours or three hours a day for about a year. However, halogen bulbs cost more.

Most halogen lamps range in power from 20-2,000 watts. Low voltage types range from 4-150 watts. Some halogen lamps are also designed with a special infrared reflective coating on the outside of the bulb to ensure that the radiated heat, which otherwise is wasted, is reflected back to the lamp filament. The filament burns hotter so less wattage is required. These lamps can last up to 4,000 hours.

Although more efficient than other large incandescent lamps, tungsten halogen lamps are inefficient relative to fluorescent and high intensity discharge (HID) lamp types. Halogen lamps can also pose a safety threat, as the heat generated can range from 250-900°F (121-482°C).

History

Oil lamps with glass chimneys were the predecessors to electric lamps. Gas lamps were also common but had obvious disadvantages. In the early nineteenth century, a lamp using an electrically heated wire (platinum) was developed. More efficient lamps became possible as different filament materials were used. In 1860, an English inventor by the name of Swan demonstrated a carbon filament lamp. Both he and Thomas Edison finally improved this lamp for practical use around 1878. Edison installed the first successful electric lighting system in 1880.

Later, these carbon filaments were replaced with tantalum and then tungsten filaments, which evaporate slower than carbon. After a process for drawing tungsten wire was perfected, the first tungsten-filament lamps were introduced in 1911. These were vacuum lamps. In 1913, General Electric Corporation introduced tungsten filament lamps using inert gas and coiled filaments. Six years later, the annual production of light bulbs in the United States exceeded 200 million. Today, almost all electric incandescent lamps are made with tungsten filaments.

The tungsten-halogen cycle used in halogen lamps was first devised and tested 40 years ago. Some of the first commercial halogen bulbs were introduced in 1959. Applications since then have included studio lighting, projection lamps and vehicular headlamps. The latter led to a different type of glass, called aluminosilicate, which was first introduced in lamps during the early 1970s. The lower softening or working temperature of these glasses allowed high speed automated production of halogen lamps.

A bulb industry emerged early in the twentieth century, as electric power became available to the general public. By the early 1980s, around 70 United States companies were selling over $2 billion worth of bulbs and tubes each year. Over the next decade, due to a decline in the early 1990s, the total bulb market only grew to around $2.9 billion. The market reached nearly $4 billion in 1994 but remained relatively flat for the next several years.

In 1992, the United States passed the National Energy Security Act, mandating the use of advanced bulbs that were more efficient. The act sought to prevent the sale of inefficient fluorescent light bulbs beginning in 1994 and other energy-inefficient bulbs by 1995. It also banned several types of fluorescent light tubes, some incandescent reflector lamps, and various flood lamps. The passing of this act also increased the price of bulbs by 4-6%.

This act, as well as declining profits, inspired lamp makers during the mid-1990s to offer lamps that could reduce energy consumption, improve lighting, boost longevity, and minimize environmental impacts. Compact fluorescent and halogen bulbs were two types that offered growth. Thus, during the 1993–1998 period halogen shipments increased nearly 15% per year. The overall United States market for lighting equipment was over $10 billion in 1998.

In mid-1997, the Consumer Products Safety Council coordinated a recall of halogen torchieres for in-home repair due to the fire hazards caused by poor fixture design and hot bulbs. The purpose of this recall was to retrofit existing torchiere lamps with a protective wire bulb guard (lamps manufactured after the recall already included these guards).

Other types of light bulbs, including halogen, have continued to improve over the years and are being designed for special applications. The latest advancement in halogen lamp technology is the halogen infrared-reflecting (IR) lamp. These lamps can provide the same light output (lumens) for much less power (watts) or conversely, substantially increased lumens for the same watts as standard halogen lamps. Only 10-15% of the power used in incandescent and halogen lamps produce visible light. The majority of the power is radiated as heat (infrared energy).

These new lamps have an infrared reflecting coating applied to the outside surface of the lamp capsule which reflects much of the wasted infrared energy back into the capsule and onto the tungsten filament. This redi-

A graph showing the difference in power between halogen and incandescent light.

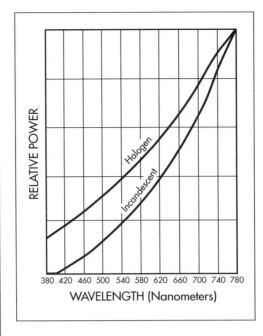

RELATIVE POWER

Halogen

Incandescent

WAVELENGTH (Nanometers)

380 420 460 500 540 580 620 660 700 740 780

rected energy increases the filament temperature thus producing more light without any additional wattage. Today, these lamps are primarily used in large retail applications for general illumination and accent or display lighting. Recently, 180 novel halogen lamps were used in the Times Square Ball for New Year's Eve in 1999. A double-envelope design makes heat distribution of these lamps similar to that of incandescent lamps.

Raw Materials

Depending on the type of halogen lamp, the bulb material is either quartz (fused silica) or aluminosilicate glass. Quartz glass has the appropriate temperature resistance for the tungsten-halogen cycle, which produces bulb temperatures of up to 1,652°F (900°C). For lamps of low wattage up to about 120 watts, aluminosilicate glass can be used. Either glass comes in the form of cylindrical tubes that are precut to the desired length or cut to length by the lamp manufacturer.

Tungsten is used for the incandescent filament. The tungsten is received in the shape of wire that is fabricated using a doping (adding tiny amounts of other materials) and heat treatment process. The dopants produce the ductility needed for processing the tungsten into coils and help prevent distortion during operation. Molybdenum—used for sealing—is received in the form of foil and wire on spools. Bases made of ceramic, glass, or metal are prefabricated.

Gases used during manufacture include argon, nitrogen, krypton, xenon, bromine, hydrogen, oxygen, and natural or propane gas. Most of these gases are supplied in tanks or cylinders, some in liquid form. **Natural gas** is piped in from the gas company.

Design

The electrical properties of the lamp are determined by the filament wire dimensions and shape or geometry. The higher the operating voltage, the longer the wire must be. For higher wattages, a thicker wire is required. The filament is wound into the shape of a coil of different configurations, depending on the lamp application.

The most common configurations are known as round core, flat core and double filament. In special cases, other configurations are used, either modulated (for maximum efficacy of light generation) and segmented (for uniform distribution of light). Filaments are also oriented in two ways, axial or transverse. The orientation is always axial in double-ended cylindrical lamps. In single-ended lamps the orientation is determined by the application.

The Manufacturing Process

Some lamp components are made at different locations and shipped to the factory where the final assembly takes place. The degree of manufacturing automation depends on the lamp application, sales volume and selling price. The process for single-end quartz halogen lamps will be discussed.

Making the coil

1 Since a thin, straight wire has poor emission characteristics and it is difficult to fit into the lamp bulb, the wire is wound into the shape of a coil using automated machines that resemble high speed bobbins. To make a round-core filament, each turn is spirally laid adjacent to the next one on a cylindrical rod. A rectangular rod is used for a flat-core filament. For a double filament, the wire is first wound into a very fine primary coil, and this is then wound once more around a second, thicker core. A large amount of wire can thus fit into a very small space.

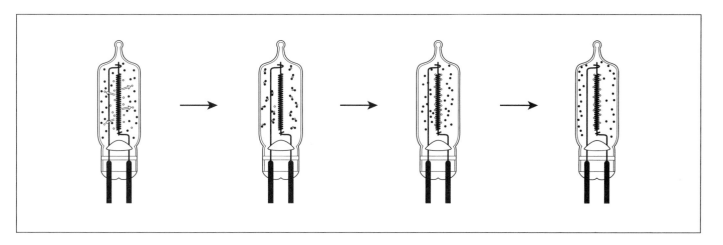

Forming the bulb

2 After the glass tube is cut to length, an exhaust tube must be attached to the top. First the top of the tube is heated using gas/oxygen fires. A tungsten carbide wheel folds the softened glass over to form a dome shape containing a small hole.

3 A smaller glass tube called the exhaust tube is placed in the hole and joined to the larger tube by melting. This small diameter tube is used as a means to flush the air out of the lamp during the sealing operation and evacuate the air and introduce the fill gas during the exhaust process. This process is performed on special rotary machines.

Making the mount

4 Next, the mount is fabricated. First, the bridge is made by embedding preformed tungsten wires in a small cylindrical quartz rod. The filament is welded to these support wires and welded to the outerlead assembly consisting of the molybdenum sealing foils and outerleads.

5 The completed mount is sent through a hydrogen furnace at 1,925°F (1,050°C) for cleaning. This process removes any oxides that can damage the tungsten filament during lamp operation.

Sealing

6 A machine called a press seal is used to hermetically seal the mount inside the bulb. The mount is inserted into the bulb and both parts held securely. The bottom portion of the bulb is then heated to around 3,272°F (1,800°C) using gas/oxygen burn-ers to soften the quartz. Stainless steel press pads, operating at pressures of 20-60 psi, press the quartz to the molybdenum foils forming the hermetic seal. During this operation, the bulb is being flushed with an inert gas (nitrogen or argon) to remove the air and prevent the mount from oxidizing. The outerleads protrude from the end of the press and provide a means to electrically connect the lamp to the lamp base.

Evacuating and filling the pressed bulb

7 The pressed bulb is filled with the halogen gas on the exhaust machine. This machine employs vacuum pumps to evacuate the air from the bulb and a filling system to introduce the halogen gas mixture into the bulb through the exhaust tube. The high internal lamp pressure is achieved by first filling the lamp above atmospheric pressure and then spraying or dipping the bulb into liquid nitrogen which cools and condenses the fill gas below atmospheric pressure. Gas/oxygen fires then melt the exhaust tube at the top of the bulb forming the tip and trapping the gas in the bulb. The gas expands as it warms to the ambient temperature and thus results in a pressurized lamp.

Attaching the base

8 The base of a lamp provides the electrical connection and mounting. The geometry is defined in national and international standards. There are several different types of bases. For single-ended lamps, glass, ceramic or metal bases are used. These are usually bonded to the glass bulb with special cement that has good resistance to high tem-

A halogen bulb is made of fused quartz and filled with the same inert gases as incandescent lamps mixed with small amounts of a halogen gas. The halogen reacts with the tungsten deposit to produce tungsten halides, which breaks down when it reaches the hot filament. The breakdown releases tungsten back to the filament—known as the tungsten-halogen cycle—and maintains a constant light output over the life of the lamp.

peratures, moisture and thermal stress or attached mechanically. A cement-free connection is used for special applications.

Packaging

9 After final testing, the lamps are manually or automatically packaged into boxes, depending on application. Lamps sold to retail stores are individually packaged.

Quality Control

A pressure test (at 40-100 atmospheres depending on fill pressure) is conducted after the press/seal process to ensure that the lamp does not burst during operation. A random sample is usually taken though some lamps are 100% tested. After the filling process, the lamps are tested for leaks by placing them onto a rotary machine and lighting them for a few minutes. If there is a major leak, the lamp will turn a white yellow color. If there are any major mechanical defects, the lamp will usually arc out. A random sample from each lot is also tested to make sure all specifications (watts, temperature, light output, and life) are met.

Byproducts/Waste

Defective quartz is disposed of or recycled. Sometimes the exhaust tubes are reused. Waste tungsten is salvaged and sold as scrap. Completed lamps that fail testing are discarded. However, lamp manufacturers continue to use more environmentally friendly materials to reduce non-recyclable wastes.

Some halogen lamps are made with lead solders in the base of the lamp. Since lead is a highly toxic material, products containing lead must pass the Environmental Protective Agency's TCLP (toxicity characteristic leaching procedure). If they do not, they must be classified as hazardous waste and follow special disposal regulations in some states. Some lamp manufacturers avoid this problem by using lead-free solder.

The Future

Shipments of tungsten halogen lamps are forecast to increase 7.7% per annum to 58 million units in 2003, outpacing shipments of incandescent lamps. This reflects the growing acceptance of halogens in residen-

tial and commercial applications, such as track and recessed lighting, table and floor lamps, and other general and task lighting.

Despite the increasing use of halogen lamps in a number of applications, unit shipments have slowed significantly from the mid 1990s pace, due to the increase in imports from countries such as China, South Korea, Taiwan, Japan, the Philippines, Mexico, Germany, and Hungary. In addition to competition from imports, other factors will contribute to falling unit prices, which will limit value gains for shipments to 5.3% per year to $180 million in 2003. In an effort to capture market share, some manufacturers will limit price increases. Additionally, improved economies of scale and production techniques will help to lower unit prices.

Halogen lamp manufacturers will also continue to develop bulbs with superior lighting characteristics, greater efficiency and improved longevity that cost less. New and improved designs will be offered to meet the needs of special applications. Lamps will continue to be made that are more environmentally safe and manufacturing processes made more efficient to reduce waste.

The global market for lighting products was expected to reach around $28 billion by the turn of the century. The United States is expected to increase its share of this market beyond the current 30%. United States bulb and lamp companies are also expanding overseas, by forming joint ventures or acquiring facilities. The market for North American lighting equipment is expected to reach over $15 billion by 2005.

Incandescent lamps will remain dominant in the United States market, with over 80% of unit sales and over 50% of market value, based on their substantial use in the large residential and transportation equipment markets. Due to the maturity of the incandescent market, competition from other lamp types, and the slowing of the housing and motor vehicle sectors, growth in incandescent lamp demand will trail industry averages.

Where to Learn More

Books

"Electric Lighting and Wiring Equipment." In *U.S. Industry Profiles.* The Gale Group, 1998.

Klipstein, Donald. *The Great Internet Light Bulb Book.* 1996.

Waymouth, John, and Robert Levin. *Designers Handbook: Light Source Applications* Danvers, MA: GTE Products Corporation, 1980.

Periodicals

Cable, Michael. "Mechanization of Glass Manufacture." *Journal of the American Ceramic Society* 82, no. 5 (May 1999): 1107-1108.

Other

The Freedonia Group, Inc. 767 Beta Drive, Cleveland, OH 44143-2326. (440) 684-9600. http://www.freedoniagroup.com.

Frost & Sullivan. http://www.frost.com.

Osram Sylvania Products Inc. 100 Endicott Street, Danvers, MA 01923. (800) 544-4828. http://www.sylvania.com.

—*Laurel M. Sheppard*

Hard Hat

Although 20 million Americans wear hard hats while working, approximately 120,000 on-the-job head injuries occur each year, and nearly 1,500 of them are fatal.

An industrial hard hat is a helmet worn to protect the head of a worker from falls or from impacts by sharp or blunt objects. Typical users include construction laborers, repairmen, and warehouse workers.

Background

Although 20 million Americans wear hard hats while working, approximately 120,000 on-the-job head injuries occur each year, and nearly 1,500 of them are fatal. Worn properly, a hard hat provides two types of protection. Its hard shell resists penetration by sharp objects. And its suspension system lessens the consequences of a localized blow by distributing the force over a broader area. The most common type of suspension system, a network of straps connected to a headband attached to the helmet, holds the shell at least 1.25 in (3 cm) away from the wearer's head.

In 1997, the American National Standards Institute (ANSI) revised its performance standards for hard hats. Although conformance to the standards is voluntary, most manufacturers choose to comply so they can label their products as providing a certain class of protection. Under the 1997 standards, Type I hard hats provide specified levels of protection from impact and penetration to the top of the head; Type II hard hats also provide specified levels of protection for impact and penetration to the side of the head. Three class designations indicate the degree to which a hard hat protects the wearer from electrical current. ANSI-compliant hard hats must also meet flammability criteria.

In addition to meeting manufacturing specifications, hard hats must be properly cared for to ensure their continuing effectiveness. One requirement for ANSI approval is that an instruction booklet be supplied with each hard hat, explaining how to care for the helmet, how to inspect it for signs of damage, and how to make certain it fits correctly. Even with proper care and no damaging impacts, a hard hat should be replaced after five years of use.

History

The steel helmet Edward Bullard brought home from World War I was more than a souvenir. His doughboy headgear was the inspiration for a revolution in industrial safety. For 20 years, Bullard's father had been selling equipment to gold and copper miners. The miners, who wore hats similar to modern baseball caps with shellacked hard-leather brims, needed more protection from falling objects. In 1919, Bullard patented a "hard-boiled hat" that was made by using steam to impregnate canvas with resin, gluing several layers together, and varnishing the molded shape. That same year, the United States Navy asked the Bullard Company to develop some sort of head protection for shipyard workers, and hard hat usage began to spread. Bullard soon developed an internal suspension system to make the hard-boiled hat more effective.

In 1933, construction began on San Francisco's Golden Gate Bridge. The project's chief engineer, Joseph Strauss, was committed to making the workplace as safe as possible. He installed safety nets, an innovation that saved 19 lives on the project. And he became the first supervisor to require workers to wear hard hats. Cooperating with Strauss to meet workplace needs, Bullard designed a special hard hat for sandblasters to use; it covered their faces, provided a window for vision, and used a pumping system to bring fresh air into the enclosure. Aluminum hard hats were introduced in 1938, providing improved durability with lighter weight, although they could not be used where electrical insulation was important. During the 1940s, fiberglass became popular for hard hats, only to be largely replaced by thermoplastics (plastics that become soft and easy to shape when heated) a decade later.

A change in ANSI testing procedures in 1997 opened the door to development of a hard hat with ventilation holes to keep the wearer's head cooler. The first ventilated hats were produced in the United States the following year. During the late 1990s, manufacturers sought to make hard hats more attractive by decorating them with sports team logos. One company even produced an ANSI-approved model shaped like a cowboy hat.

Accessories for hard hats are becoming more sophisticated. Common accessories include transparent face shields, sun visors, sound-muffling ear covers, and perspiration-absorbing cloth liners. Recent innovations have taken a high-tech turn, introducing such attachments as pagers, AM-FM radios, and walkie-talkies. A digital package introduced in 1997 links a hat-top camcorder camera to a handheld computer, and provides a visor-mounted viewing screen.

Raw Materials

Depending on the intended use and the manufacturer, modern hard hat shells may be made of a thermoplastic such as polyethylene or polycarbonate resin, or of other materials like fiberglass, resin-impregnated textiles, or aluminum. Because it is strong, lightweight, easy to mold, and nonconductive to electricity, high-density polyethylene (HDPE) is used in most industrial hard hats. The suspension system for industrial hard hats consists of strips of woven nylon webbing and bands of molded HDPE, nylon, or vinyl. Together with the strap suspension system, most Type II hard hats use a foam liner made of expanded polystyrene (EPS).

Brow pads attached to the front of the helmet's headband increase comfort for the wearer. Various materials are used for brow pads, including foam-backed vinyl, foam-backed **cotton** terry cloth, and specialty fibers (e.g., CoolMax or Sportek) designed for sweat absorption in athletic clothing and accessories.

The Manufacturing Process

The following description of the production of Type I industrial hard hats is based largely on the manufacturing techniques of one major manufacturer. However, some details have been expanded to include variations used by other manufacturers.

The shell

1 The appropriate shell mold for the model being produced is selected. After adjusting a date-of-manufacture dial inside the mold, the form is positioned in an injection molding press. Electric lines are connected to the mold, as are lines carrying chilled water that will cool the mold.

2 High-density polyethylene pellets are pulled from a supply hopper by a vacuum system. Pellets of colorant are drawn from another supply hopper and mixed with the HDPE pellets in a ratio of 4% to 96%. The vacuum system then transfers the pellet mixture into the injection molding press.

3 Within the press, the pellets are heated to melt them. The molten plastic is injected into the mold to form the hard hat shell. The press opens the mold and ejects the shell onto a conveyor belt.

4 A worker picks the shell up and cuts off the sprue (a lump formed where the molten plastic entered the mold). The worker glues a label inside the shell; the label identifies the manufacturer and the appropriate ANSI type and class designations. The suspension system.

5 Component parts of the suspension system are produced. Injection molding machines form headbands, plastic "keys" that will be used to attach the suspension system to the shell, and nylon strips and gears for the ratchet mechanism that will allow headband size adjustment to fit the hard hat user. Nylon webbing (0.75-1 in [1.9-2.5 cm]

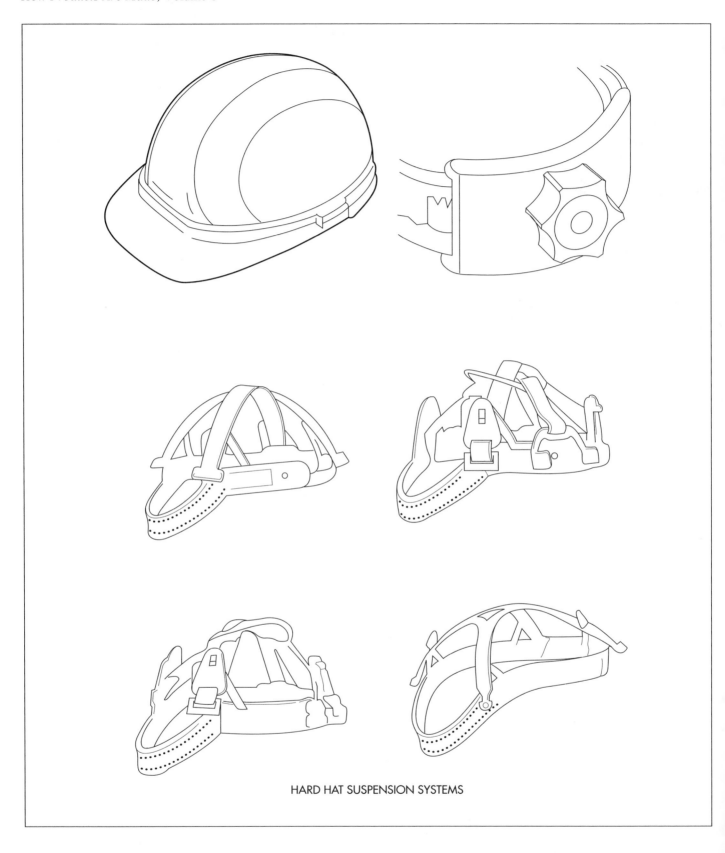

HARD HAT SUSPENSION SYSTEMS

Different types of hard hat suspension systems help to lessen the consequences of a blow to the head by distributing the force of it over a broader area.

wide) from large spools is fed into a cutting machine that produces strips of the appropriate length (approximately 15 in [38 cm]). A die-cutting machine produces browpads.

6 A worker threads one end of a webbing strap through a slot in the end of a key. The worker folds the strap end back and sews it to the strap with a buttonhole machine, securing the key in a loop of the strap. The same process is repeated on the other end of the strap.

7 Depending on the model being produced, the appropriate number (4, 6, or 8) of nylon straps is arranged in a star pattern on a holding fixture, and they are secured to each other by a line of stitching at the crossover point. Alternatively, they may be threaded through a slotted, circular pad (called a crown cushion) that will rest on top of the user's head.

8 A worker inserts both ends of the headband strip into the ratchet mechanism.

9 A worker attaches a browpad to the front of the headband by folding its tabs over the headband and hooking slots in the browpad over nodules protruding from the headband.

10 A worker attaches webbing strips to the headband by mating slots on the keys with nodules on the headband. On a six-point suspension system, only four keys are attached to the headband; the other two keys will attach only to the hard hat shell.

11 An instruction booklet and the suspension assembly are placed inside the hat, and these components are placed into a plastic bag and a box for shipment. After purchase, the user will attach the suspension to the shell by sliding the keys into slots.

Quality Control

A sample of hard hats from each batch or production shift is set aside for testing according to the ANSI criteria. Some samples are cooled to 0°F (-18°C) for a two-hour period prior to testing, and others are heated to 120°F (49°C) for two hours before testing. The Type I impact test involves dropping an 8-lb (3.6 kg) steel ball from a height of 5 ft (1.5 m) on the top of the hat as it sits on a head form; no more than 1,000 lb (4,400 N) of peak force can be transmitted to the head form, and no more than 850 lb (4,000 N) of average force can be transmitted. The Type I penetration test involves dropping a 2.2-lb (1-kg) pointed steel penetrator with a 60 angle on the top of the hat from a distance of 8 ft (2.4 m); it must not make contact with the head form. In addition, Type II impact and penetration tests involve dropping helmeted head forms onto steel anvils and pointed steel penetrators.

Under earlier ANSI standards, electrical conductivity was tested by measuring current in bodies of water inside and outside the hat. Since 1997, the test has been performed using metal foil on opposite surfaces. The conductivity test is performed on a sample that has already withstood the impact test. The most rigorous criterion (for the highest class designation) requires the hat to withstand 20,000 volts for three minutes with no more than nine milliamps of current leakage, followed by exposure to 30,000 volts with no burn-through permitted. For the flammability test, a hard hat is positioned on a head form and exposed to a 1,550°F (843°C) flame for five seconds. There must be no visible flame on the helmet five seconds after removal of the test flame.

The Future

Manufacturing techniques will be refined, perhaps incorporating a greater degree of automation if efficient machines can be designed. Helmet design changes, such as elimination of the need for sewing the suspension straps, may contribute to this effort.

Manufacturers hope to find new materials that have better qualities. In addition to strength and low weight, they look for resistance to heat, chemicals, and ultraviolet radiation.

Where to Learn More

Periodicals

Cravens, Catherine P. "Let Protection Go to Your Head." *Occupational Health & Safety* (March 1998): 40 ff.

"High-Tech Hard Hats." *The Futurist* (May/June 1997): 11.

Meade, Vicki. "Heads Up: What's New in Protective Gear." *Occupational Health & Safety* (July 1995): 33 ff.

Other

"ANSI-Z89.1-1997 Standard Highlights." Custom Hard Hats. http://www.customhardhats.com/ansi.html (March 8, 2000).

"History of the Hardhat." Bullard Company. 1998. http://www.bullard.com/company Info/hardHatHistory.html (March 27, 2000).

—Loretta Hall

Harpsichord

Background

The harpsichord is the distinguished, classical ancestor of the piano. Its shape, described as a large wing shape, was developed hundreds of years before the similar shape of the grand piano. But the operation of the harpsichord and its history are far different from those of its descendant.

The piano player makes music by fingering keys that strike tightly stretched strings within the piano, and by pushing pedals with the feet that change the dynamics (loudness, softness, and length of tone) of the struck strings. Within the harpsichord, the back of the key is attached not to a hammer but to a vertical jack that has a vertical slot containing a swinging tongue. The tongue grips a plectrum, or pick. As the player's finger strikes the key, the jack rises, and the plectrum lifts up and plucks the string. As it falls back past the string, the swinging tongue moves to pass the string without touching it and producing a sound. A lightweight **spring** pushes the tongue back to its original position so the plectrum is ready to pluck the string with the next stroke of that key. In the first 500 years or so of the harpsichord's history, the plectrum was a quill from the wing of a turkey, eagle, raven, or crow; later plectrums were made of leather or plastic. After plucking the string (which is not as tightly bound as a piano string), the jack has a release device that returns it to the rest position. The harpsichord's tone depends on where the string is plucked along its length, and the material composing the plectrum. The harpsichord does not have pedals to modify its dynamics; after the string is plucked, its sound dies quickly. Large harpsichords were better able to produce changes in dynamics, but did not come close to the range of dynamics possible with a piano.

The apparent limitations of the dynamics of the harpsichord caused composers who wrote for the harpsichord to be creative, and skilled players can also enhance the dynamics to a certain degree. Composers used music filled with trills and other ornamentation to make a more continuous sound. Players learn to make joined and detached sounds called *legato* and *staccato*. While the lack of dynamics seems to limit the harpsichord, the instrument also has a uniquely beautiful tone that is prized by professional musicians and other admirers who want the elegant instrument in their homes and even purchase kits for constructing their own harpsichords.

The cases of harpsichords are beautifully shaped and, historically, have been elegantly ornamented and painted. But the case is also critical to the sound. The case has five parts: the long straight side to the player's left is the spine; the short end is also straight and is called the tail; the bentside to the player's right forms a long, gentle curve (like the underside of the wing shape); another short, straight piece called the cheek is immediately to the player's right; and the bottom, which closes the instrument, forms both a structural and acoustic base for the keyboard. The wrest plank is another wooden component that holds the keyboard in place so it is seated on the bottom. The case must provide the strength to resist the tension of the strings, so, internally, the case contains a bracing system to balance the tensions.

History

The history of the harpsichord is distinguished by type of instrument, the century

About 25 professional harpsichord builders are active in the United States, and about 100 instruments per year are made in the United States for universities, orchestras, other music organizations, and private players.

in which it was made and played, and national school. The national schools of the greatest importance are French, Italian, German, Flemish, and English. The harpsichord's close relatives include the clavicymbalum, the virginal, the lautenwerk, the clavichord, the spinette, and, of course, the later instrument, the piano. The first of these—the clavicymbalm—is mentioned in documents dating from 1397 in Padua, Italy. The oldest clavicymbalm that still exists was built in Bologna, Italy, in 1521. The earliest instrument called a harpsichord was mentioned in 1514. It was short in length, had a thick case, and was a so-called single manual, meaning it had one set of keys. Instruments called double manuals that had two sets of keys like a modern organ and stops, also like an organ, were known at about the same time. For example, a harpsichord listed among the expenses for the court of England's King Henry VIII in 1530 was called "a pair of virginals in one coffer with four stops." The sets of manuals were matched internally with sets of strings called choirs; that is, a single manual had a single choir, and a double manual was connected to a double choir of strings.

Venice, Italy, and Antwerp, Belgium, were two centers of production of harpsichords in the sixteenth and seventeenth centuries. The Venetian style had a long, thin body that was made of cypress wood and had an ornately decorated outer case. The Venetian instrument had either a single 8-ft-long (2.4-m-long) choir or two choirs measuring 8 ft (2.5 m) in length (the length of the choir plus the depth of the keys was the approximate total length of the harpsichord). The Flemish school based in Antwerp was led by the Ruckers family. They built both harpsichords and virginals that had thicker bodies, painted cases, and double choirs and manuals. The Ruckers's harpsichords were valued for their beautiful resonance and tone and were exported all over Europe. In England and France, the Ruckers creations were popular and copied. The French Blanchet family made its own versions of the Ruckers harpsichord that were even more elaborately painted and lacquered; by 1750, they were the official harpsichord makers for the royal court of France.

The Germans also made prized instruments. Hamburg was their center of manufacture, and they favored large, heavy instruments with extra registers, pedal-type keyboards, and as many of five choirs of strings plucked by three sets of manuals. These instruments were the ones favored by Johann Sebastian Bach (1685–1750), perhaps the greatest composer for, and friend of, the harpsichord. English-style instruments lacked the painting and ornamentation of the Continental styles and had cases faced with walnut or oak veneer.

By the early 1800s, the harpsichord had fallen out of favor and the piano was becoming increasingly popular. From 1809 until well into the twentieth century, harpsichords were not played in Europe or America. As an industry, harpsichord making simply vanished until the present, modern revival in which instrument builders pride themselves on reproducing the great historic harpsichords of the national schools and, primarily, of the eighteenth century.

Raw Materials

Wood is the chief material composing a harpsichord. Wood from the American trees basswood and yellow poplar, Northern European linden, and the European tulip poplar are used to make harpsichord cases for most types except German harpsichords. The cases of German-style instruments are made entirely of pine; American makers use Ponderosa pine. Harder woods including oak, maple, walnut, beech, and spruce are used for structural supports inside the cases.

Traditionally, soundboards are made of Norway spruce, which grows over much of the European continent as far south as the Apennine mountains in Northern Italy. An American species, the Ingleman spruce, is similar to the Norway spruce and is sometimes used in the United States. Fir trees are also used occasionally. Many American makers import the Norway spruce, but supplies are becoming more limited as pollution threatens the spruce forests.

Other materials include ebony, basswood, and ivory for the key tops. Animal glues are used (modern synthetic glues do not work as well), and metal is used for the strings. **Brass** wire is drawn in a wrought process; harpsichord makers work with local brass founders to make sure the manufacturing process is correct. Other hardware includes wood screws, turning pins that are made by a European supplier, parts of the harpsichord action like jacks and jack slides that are purchased in large quantities, hooks that are also common to piano-making, and "roses" (ornamental pieces that cover the opening in the soundboard).

To finish harpsichords, the natural wood may be varnished and polished, but the outside of most harpsichords is painted, and painting begins with laying down a gesso finish. Gesso is a mixture of finely ground chalk and glue. Colored paints are usually so-called "Japanned colors" made of pigment and oil that is applied over the gesso and that produces a high gloss. Many coats may be used. The leather rose is gilded, and gilt work may also be a part of the painted trim or other ornamentation.

Design

Design of a harpsichord is based exclusively on tradition and existing, historical instruments; that is, there is no such thing as a new harpsichord pattern, style, or sound. Harpsichord builders do make adjustments to existing designs, but most of these are out of necessity because historic materials are not available or are not desirable. For example, lead-based paints were used to decorate harpsichords in the past, and these are no longer desirable for health and environmental reasons.

To reproduce an existing harpsichord, for example, an instrument made in Paris in 1707, the harpsichord maker obtains drawings and measurements from the museum or institution where the 1707 model is presently housed. Museum experts have often restored historic instruments and used modern techniques to analyze the harpsichord's construction. X rays are useful in identifying types of internal fasteners, and fiber optics can be employed to look through the rose and into the guts of the instrument.

If data does not exist, the builder may request permission to do a detailed examination of the instrument. The builder makes drawings of every visible part of the instrument, starting by measuring the width and then proportioning the other parts. The builder has to keep in mind that the 1707 harpsichord was built on the old inch system, so any information that may be available from original construction must be converted to measurement systems in use today. The original builders did not have paper readily available to document each construction and probably relied on memory or a master book of guidelines. By simply proportioning, the modern maker can measure the parts of the case (including width, length, string length, and internal geometry) to within ±0.03 in (±1 mm); this is well within tolerances for modern handiwork. There is no need to be more precise than the original; sometimes a builder can overwork a design or restoration and spoil it by being too exact.

The Manufacturing Process

1 Construction of a harpsichord begins with selecting the wood based on the type of harpsichord ordered. As noted above, different kinds of wood are used depending on the national style or model, and they certainly affect the sound quality of the finished instrument. The wood must be aged for at least two years to reduce the natural moisture content that would cause it to warp or curl. The bentside—the side of the case that curves—is steam bent to fit a jig, which is a preconstructed form with the correct curve for a particular instrument.

The harpsichord maker's shop resembles that of a cabinet maker, and harpsichord construction has a lot in common with ship building and cabinet making. The process of steaming wood and forming it to specific curves is part of the shipbuilder's art of constructing the curved prow of a ship. The process of carving and fitting the smaller pieces of the harpsichord requires the skills of a cabinet maker or master woodworker. The shop is also filled with cabinet-making tools like planes, chisels, rasps, and files. The major difference between the shops of a modern harpsichord builder and one from the eighteenth century is the presence of electrically powered tools, particularly band and table saws, in the present-day shop.

Other smaller pieces are also cut and curved by steam bending in sets of smaller jigs while the bentside is being made. The other sections of the case are straight; these in-

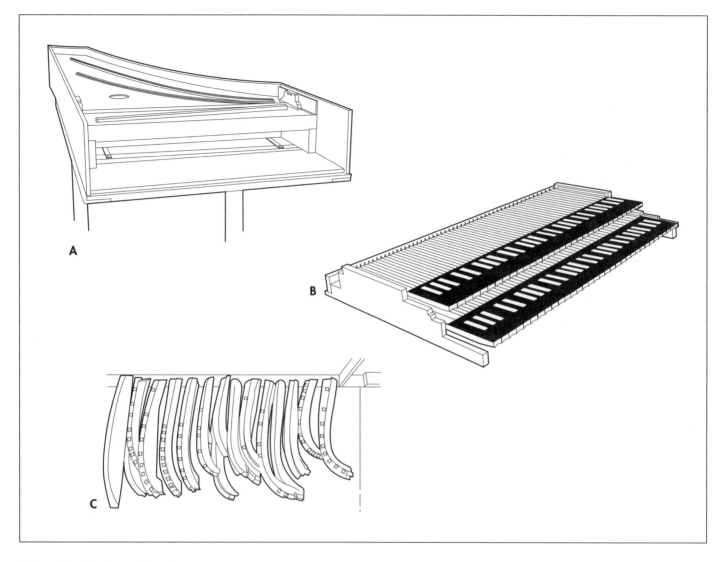

A. Harpsichord with soundboard.
B. Double manual keyboard. C.
Jigs.

clude the tail, the cheek, the front end, the wrest plank (the heavy slab of wood crossing the top of the instrument and connecting the spine to the cheek), and the corner keys (strengthening blocks that are used to reinforce the case corners) that are cut, flattened, and fitted together by hand. When the case sides have all been assembled, the outline of the harpsichord is complete.

2 The soundboard is an elaborate construction made of a large piece of very thin wood. The piece is made up of a series of slats that are 0.08-0.09 in (2.5-3 mm) thick. The slats are glued together, and a pattern or template is marked on the soundboard to cut it to shape. The shape is slightly larger than the case of the harpsichord because the soundboard has to fit snugly and curve, or crown, inside to prevent it from buzzing when the instrument is played. The sound-

board is fitted to the instrument first; later, it is glued to the liner inside of the case. An equally elaborate scheme of ribs supports the soundboard and holds its curvature. A large rib called the cutoff bar is also fitted into place; it stiffens the soundboard, and also cuts the acoustical or sound-producing hollow of the harpsichord into two areas for high and low sounds. The ribs are cut with a band saw and shaped and tapered by hand. The bridges are glued to the soundboard and are held in place until the glue dries with a system of go-bars that work like large clamps along each bridge and have a number of contact points to fit the curvature of the bridge. The rose hole is also cut through the soundboard after the bridges are secured.

3 The lid and front flap are fitted to the instrument so they can be hand painted while construction of the harpsichord case

continues. The soundboard may also be hand painted, depending on the style of instrument. When painting is complete, the brass rose is installed in the rose hole and supported with fabric strips. It is gilded with great care to prevent damage to the adjacent, painted ornamentation. The rose and its opening are mistakenly thought to have some acoustical purpose, but the harpsichord is actually a closed instrument. Some makers allow the soundboard to dry in sunlight after painting and before installation to shrink it slightly as a method of preventing cracks from appearing during later aging and drying.

4 The harpsichord case still does not have a bottom or a top (the soundboard) installed at this point. The system of framing and bracing is cut and hand-fitted in place to keep the shape of the case and to support the soundboard. Lower frame pieces cross the bottom of the instrument from the bentside to the spine to strengthen the case. The case also has a liner—a collection of pieces similar to the pieces that make up the case—and is inset from both the top and bottom. The liner provides a ledge to support the soundboard—except at the front end of the harpsichord, where the soundboard is supported by the upper belly rail. Rails are additional interior wood pieces that anchor the strings. After the liner sections have been glued to the case sides, upper frame members are glued into position; they cross from the bentside to the spine to hold the structural integrity of the upper part of the soundbox.

5 The bottom of the harpsichord is put into place with a combination of wood screws and glue. Although the lid is made of one piece of wood, primarily for appearance, the bottom is still an acoustical part of the instrument. It is made of two pieces of wood for ease of installation and for sound properties. At the back of the instrument (from the lower belly rail to the tail), the grain of the wood runs lengthwise. From the front of the belly rail to the front of the harpsichord, the wood grain runs across the instrument. The two pieces meet at a joint along the belly rail.

6 The soundboard is glued into place. Go-bars clamp it at a number of places along the liner, while pony clamps secure the front edge of the soundboard to the top of the

upper belly rail. After the glue is dry, hitch pins are installed to ready the instrument for stringing. But first, the exterior of the case (which is now complete except for installation of the lid and flap) must be prepared for the final surface finish. The painted soundboard is protected with plastic sheeting, and the other surfaces of the instrument are coated with gesso, a mixture of chalk and glue. The gesso seals the wood and provides a base for the final finish.

7 In another distinct operation, the key blanks and key frames are cut. The frames are constructed with mortise and tenon joints. Most of the harpsichords reproduced today are "double manuals," meaning they have two keyboards. A detailed keyboard pattern is marked on the single piece of wood used for each keyboard. The pattern shows the positions of the holes in the keys as well as the shape of the keys. The holes are drilled, and the keys are finished along their fronts before they are cut. The sharp cuts are marked, and the platings (surface coverings) for the natural keys are glued onto the blanks before the keys are cut using a band saw. Concentration and remarkable skill are needed for key-making because 1,008 holes are required for a set of two keyboards and the harpsichord keys have narrower spacing than those on a piano. The cross cuts along the fronts of the sharps are made and trimmed, the plating for the sharps is trimmed, and the fronts of the naturals are undercut so they have lighter weight and better balance in the frames.

The naturals are mounted on the key frames and leveled. The keys for the sharps are then fitted between the naturals and leveled. The sharp rise (the raised part of the sharp keys) is glued onto each sharp last. The platings on harpsichord keys also differ from those on pianos. French, German, and Flemish instruments have ebony or boxwood platings on the naturals and ivory on the sharps. Only Italian instruments seem to favor white or ivory naturals and black sharps. This difference, along with the narrower key spacings, makes it impossible to use factory-made piano keys for harpsichords.

8 To finish the keys, felt is glued onto the back of the keys, cut apart between each key, and trimmed. The upper keys have weights in the ends and are guided by pins

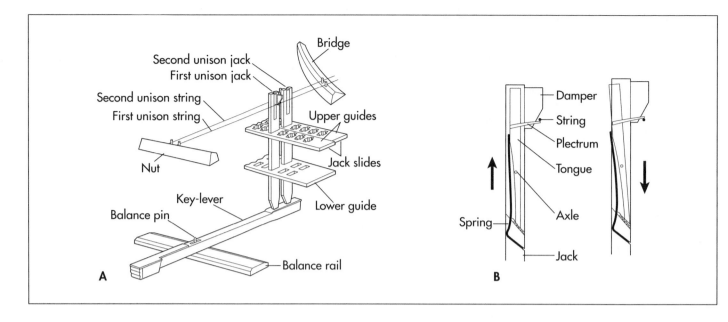

A. Harpsichord mechanism. B. When a key is pressed, the jack at the back of the key raises. When released, the jack moves downward, causing the plectrum to hit the string.

through slots in the keys. The guide system for the lower keys is a rack in the back with a pin in the back of each key that fits through a slot in the rack. Piano bridge pins can be used in harpsichords as hitch pins, piano center pins work as bridge pins, and zither pins are used as harpsichord tuning pins.

The harpsichord is strung and the jacks that pluck the strings are installed before the keys are mounted in the instrument and connected to the jacks to complete the instrument action. The instrument is voiced (tuned) with minor trimming of the tip of the plectra and adjustment of the action.

9 Decoration of the case is completed by adding case moldings, completing exterior painting, and gilding the moldings. Other gold bands may also be applied to the sides of the instrument. The lid and flap are hinged and attached to the instrument after the exterior finish is completed. A harpsichord also differs from piano in that it has no attached legs. In another woodworking and finishing operation, a stand or L-shaped supports are made to match the exterior of the instrument.

10 Each harpsichord is shipped upon completion. A quilted fabric cover is used to wrap the instrument completely, and padded fabric sheets protect the stand. Piano-movers or other experienced movers handle the instrument, which may be valued at over $30,000.

Quality Control

The harpsichord is a creation by artists for artists. Harpsichord makers are highly skilled wood workers, painters, and artists in many supporting crafts; they also play the instrument themselves and hold great reverence for fellow musicians who lovingly reproduce the music of great composers who wrote works unique to the harpsichord's sound. Their ability to channel this respect back into every aspect of construction of a single harpsichord is highly effective quality control.

Byproducts/Waste

Harpsichord makers typically produce different models in the styles of the national schools and various eras in harpsichord history. The manufacture of harpsichord kits for the home craftsman is a separate industry, and most harpsichord makers do not also make kits, although they may well have learned the craft from experimenting with kits. They also do not supply their handmade harpsichord parts to others; parts (like jacks, key felt, or wire) can be either purchased from specialized suppliers or the harpsichord maker produces his or her own parts for private use.

Waste is very limited. Some wood scrap is generated, but the value of these rare woods prompts wood workers to use them efficiently. There is also little waste of paint, gesso, and other finishing supplies because they are hand-mixed in the quantity needed.

The key to the production of a beautiful musical instrument is to be especially careful early in the process so that tiny mistakes do not multiply and create difficulties in completing construction or in voicing and playing the instrument.

Safety is an issue in the operation of electrically powered tools and in the surface finishing of harpsichords. Caution with the electrical supply and tools like power saws and sanders is essential. When dust is generated during wood working, craftspeople wear masks and sometimes respirators. All painting is done in a ventilated paint room.

The Future

The harpsichord's future seems secure for the moment. The development of the harpsichord kit has fostered a new group of enthusiasts who love the voice of the instrument, the opportunity to use their own skills in handcrafting such a project, and the chance to own a grand piece of musical history. A wide range of persons buy harpsichords from professional builders out of similar appreciation. About 25 professional harpsichord builders are active in the United States, and about 100 instruments per year are made in the United States for universities, orchestras, other music organizations, and private players.

According to harpsichord maker John Phillips (a self-taught builder who began with harpsichord kits), the greatest potential threat to the future of the harpsichord is the level of musical culture, especially in the United States. Music education in schools is being cut because of cost, and there is no doubt that private music training is expensive for most families. But it is through music education beginning at an early age that children come to appreciate fine music and explore less familiar instruments like the harpsichord. Hope rests in the fact that, once heard, the evocative sound of the harpsichord is seldom forgotten.

Where to Learn More

Books

Bragard, Roger, and Ferdinand J. De Hen. *Musical Instruments in Art and History*. New York: The Viking Press, 1967.

Clemencic, Réné. *Old Musical Instruments*. Translated by David Hermges. London: Weidenfeld & Nicholson; New York: G. P. Putnam's Sons, 1968.

Crombie, David. *Piano*. San Francisco: Miller-Freeman, 1995.

Dearling, Robert, ed. *The Illustrated Encyclopedia of Musical Instruments*. New York: Schirmer Books, 1996.

Hubbard, Frank. *Three Centuries of Harpsichord Making*. Cambridge: Harvard University Press, 1965.

Kottick, Edward L. *The Harpsichord Owner's Guide: A Manual for Buyers and Owners*. Chapel Hill, NC: The University of North Carolina Press, 1987.

Ripin, Edwin M., Denzil Wraight, and G. Grant O'Brien, et al. *Early Keyboard Instruments*. New York: W. W. Norton & Co., Inc., 1989.

Sachs, Curt. *The History of Musical Instruments*. New York: W. W. Norton & Co., Inc., 1940.

Unger-Hamilton, Clive. *Keyboard Instruments*. Minneapolis: Control Data Publishing, 1981.

Zuckermann, Wolfgang Joachim. *The Modern Harpsichord: Twentieth-Century Instruments and Their Makers*. New York: October House, Inc., 1969.

Other

A Harpsichord Primer. http://www.bigduck.com (January 2001).

John Phillips Harpsichords. http://www.home.earthlink.net/~jplectra (July 2000).

Peter Tkach, Harpsichord Maker. http://www.tkach-harpsi.com (June 2000).

Zuckermann Harpsichords International. http://zhi.net/kits/tour.html (June 2000).

—*Gillian S. Holmes*

Hockey Puck

Background

Hockey pucks are flat, solid, black disk-shaped objects made of vulcanized rubber. Regulation National Hockey League (NHL) pucks are black, 3 in (7.6 cm) in diameter, 1 in (2.54 cm) thick, and weighing 5.5-6 oz (154- 168 g). The edge has a series of "diamonds," slightly raised bumps or grooves. The diamonds give a taped hockey stick something to grip when the puck is shot. The blue pucks used in junior hockey are sometimes only 4 oz (143 g).

During a game, each team keeps a supply of pucks in a freezer at all times. When a professional hockey team receives their supply of pucks for a season, they are rotated so that the older pucks are used first. During games, pucks are kept frozen in an ice-packed cooler, which usually sits on the officials' bench. All pucks are frozen to reduce the amount of bounce.

Though no one knows exactly how the hockey puck got its name, many believe that it was named for the character in William Shakespeare's *A Midsummer's Night Dream*. Like the impish flighty Puck, the hockey disk moves very quickly, sometimes in unexpected directions.

History

Played in Europe for several hundred years, field hockey is a predecessor of ice hockey, which sprang up in Great Britain during the 1820s. The game blossomed in the British protectorate of Canada in the second half of the nineteenth century. In Canada where long, cold winters are a certainty, ice hockey soon became the national game. Hockey also became popular in the

northern parts of the United States during the same time period.

At first, amateurs dominated hockey and the rules were ever changing. The first professional league was organized in 1904 and called the International Hockey League. It only lasted three years. In 1917, the National Hockey League (NHL) was created, and is still the top level of professional hockey played in North America today. With the establishment of the NHL came codified rules and regularization of the game. Today, hockey is played by all ages, both men and women, throughout North America and many parts of the world.

In the early years, c. 1860–1870s, a rubber ball was the object used in hockey. Because the ball bounced too much, a block of wood was sometimes used instead. The modern hockey puck was invented around 1875. There are two different versions of its origination. One story claims that in 1875, students at Boston University sliced a rubber ball in half to make a puck. Another version places the evolution in Montreal, Quebec, Canada. The owner of one of the first indoor ice rinks, Victoria Rink, also allegedly sliced a rubber ball in half. In any case, the first recorded use of a flat disk was in Montreal in March 1875.

Early pucks were made by gluing two pieces of rubber together (sometimes from recycled tires). Because of this construction, the pucks could split when they hit the goal post. During the 1931–1932 season, a puck with beveled (sloped) edges was used. By midseason, complaints by players and teams led to the return of the original puck. Though there was no official NHL puck until the

1990–1991 season, the basic construction from the early 1900s remained the same.

The FoxTrax puck

During the 1995–1996 NHL season, a slightly different puck was introduced. While the outside of the puck remained the same, the inside and effect was totally different. That year, the Fox television network obtained the rights to air the NHL All-Star Game and the Stanley Cup playoffs. Fox believed that to attract new viewers to the game, the network had to make the small-looking puck easier to follow on television. To that end, they developed an enhanced puck called the FoxTrax puck. It contained a computer board and battery at its center and 20-pin holes all over the puck (12 on the edges, four on top, and four on the bottom) that guided infrared emitters, each beeping approximately 30 pulses per minute. These emitters communicated with 16 sensoring devices placed around the rink to follow the puck's movement. The sensoring devices were linked by fiber optics to computers outside in the "Puck Truck."

When processed by computer, the FoxTrax puck had a completely different look to the television audience. It had a translucent blue halo, which was supposed to make the puck more visible on a small screen. When a player shot the puck at speeds exceeding 50 mph (80 kph), a red tail appeared on television. If the puck reached speeds over 75 mph (120 kph), the tail was green. When put into play, each FoxTrax was remotely activated by a wireless controller. Unlike standard pucks, which were used until they went into the stands or otherwise damaged, FoxTrax pucks could only be used for about 10 minutes before the battery ran out. While Fox-Trax pucks weigh about the same as NHL regulation pucks, they cost much more to make. Each puck had a value of about $400.

From its first use, players complained that the FoxTrax puck did not move the same way a normal puck did. The FoxTrax puck also did not hold the cold as well. FoxTrax pucks became bouncy much more quickly than their regulation counterparts. When the Fox network declined to renew its contract to air the NHL All-Star Game and play-offs after the 1998–1999 season, the FoxTrax puck was no longer used or manufactured.

Raw Materials

A hockey puck is made of vulcanized rubber. The top and bottom of some pucks are decorated with team and/or league logos. These logos are silk-screened on to the rubber. The silkscreen process uses a rubber-based ink and four-color processing.

In addition to the rubber and silk-screened ink, the FoxTrax puck included several computer components—a lithium battery; 20 infrared emitters; a ceramic oscillator; an accelerometer; CMOS logic and switching; a four-layer, silver-dollar sized circuit board; surface mount parts; and a flexible epoxy to pot the board.

Design

The design of NHL regulation pucks was regularized in 1940 by Art Ross. Though pucks remained basically the same, Ross's innovation was a puck that was easy to manufacture and acted with some consistency when used in play.

Logos that are silk-screened on the puck are designed by the various professional hockey leagues (including the NHL) and individual teams.

The Manufacturing Process

Currently, hockey pucks are only made in four countries: Canada, Russia, China, and the Czech Republic. There are two kinds of manufacturing processes for pucks. One is for practice and souvenir pucks. The other is for regulation NHL and other professional league pucks that are used in games.

Practice/souvenir pucks

1 Rubber arrives at the factory in the form of cord packed into 40 ft (361.9 m) long tubes. The cord is fed by hand through a pultrusion machine.

2 A worker monitors as an automated, timed machine pulls the rubber and slices long pieces of rubber 4- 5 in (10-13 cm) thick.

3 The machine grabs 4 in (10 cm) of the rubber cord and drops it into a two-part (male-female) mold. The mold is heated. The two parts of the mold are compressed

A FoxTrax hockey puck.

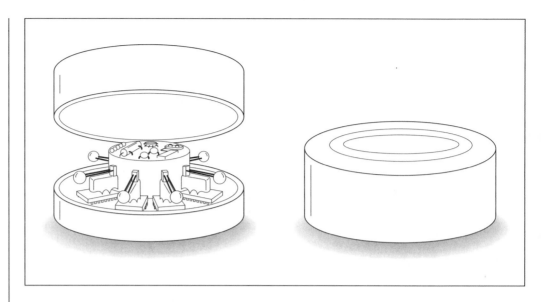

together. Approximately 10,000 pucks can be made in one day.

4 After the puck is made, they are silk-screened with a rubber-based ink. The pucks are fed into one of four kinds of silkscreen machines, depending on the number of colors included in the team or league's logo. (There is a hand silkscreen machine, as well as three-color, six-color, and eight-color silk-screening machines.) The logo is placed on the puck.

5 The pucks are packed for shipping in cases of 100. Wax paper is placed between rows to separate the pucks so that the logos are not marred.

Regulation NHL pucks

1 Granular rubber is mixed with special bonding material by hand.

2 The mixture is put in a two part (male-female) mold. A molding pallet of 200 mold cavities is filled by hand. The mold is cold compressed. (This procedure actually takes place at room temperature.) About 5,000 pucks can made per week.

3 In a separate procedure, the pucks are silk-screened with a rubber-based ink. The pucks are fed into one of four kinds of silkscreen machines, depending on the number of colors included in a team or league's logo. (There is a hand silk-screen machine, as well as three-color, six-color, and eight-color silk-screening machines.) The logo is placed on the puck.

4 The pucks are packed for shipping in cases of 100. Wax paper is placed between rows to separate the pucks so that the logos are not marred.

FoxTrax pucks

1 The pucks are manufactured in the same way as regulation pucks.

2 The pucks are cut in half by hand.

3 The center of the puck is carved out by hand. Sometimes this was done with a routing bit and a routing machine.

4 A drill with a special drill bit carves the 20-pin holes that are the paths for the infrared emitters. This is done by hand.

5 The computer board, driver circuits, battery, and other high tech parts are placed inside the puck by hand. The board is potted with an epoxy that is flexible by hand. The epoxy is made from materials similar to the puck. The puck is glued back together with a mixture of flexible epoxy and filler.

6 The puck is silk-screened. The pucks are fed into one of four kinds of silkscreen machines, depending on the number of colors included in the team or league's logo. (There is a machine designed for silk-screening by hand, as well as three-color, six-color, and eight-color silk-screening machines.) The logo is placed on the puck.

7 The pucks are packaged for shipment. Wax paper is placed between rows to separate the pucks so that the logos are not marred.

Blue (junior hockey) pucks

1 Rubber and blue-colored plastics, both in a granular state, are mixed with special bonding material. This is done by hand.

2 The mixture is put in a two part (male-female) mold. A molding pallet of 200 mold cavities is filled by hand. The mold is cold compressed. (This procedure actually takes place at room temperature.) About 5,000 pucks can made per week.

3 In a separate procedure, the pucks are silk-screened with a rubber-based ink. The pucks are fed into one of four kinds of silkscreen machines, depending on the number of colors included in a team or league's logo. The pucks are fed into one of four kinds of silkscreen machines, depending on the number of colors included in the team or league's logo. (There is a hand silkscreen machine, as well as three-color, six-color, and eight-color silk- screening machines.) The logo is placed on the puck.

4 The pucks are packed for shipping in cases of 100. Wax paper is placed between rows to separate the pucks so that the logos are not marred.

Quality Control

Pucks are checked for the regulation size and weight. If regulation pucks do not meet prescribed standards, they are recycled and the rubber is reused to make pucks. After regulation pucks are made, certain specimens are frozen for 10 days, then bounced. The tester ensures that the pucks bounce the same ways as those in previous batches. A consistent product is important in the production of pucks. Every puck must act the same way on the ice.

During the silk-screen process, the ink can be affected by the moisture in the air, dust particles, and hair. The pucks are checked for the effect of any of these qualities. Any effected pucks are washed with paint thinner and go through the silk-screening process again.

In the past, Russian-made pucks sometimes had metal fragments in them. These pucks

Wayne Gretzky.

Wayne Douglas Gretzky was born on January 26, 1961, in Brantford, Ontario. Able to skate at two, he signed with World Hockey Association's Indianapolis Racers at 17. After eight games, Gretzky was sold to the Edmonton Oilers, who were admitted to the National Hockey League (NHL) in 1979. In his nine seasons with the Oilers, from 1980 to 1988, Gretzky scored 583 goals and handed out 1,086 assists. For six of those years he averaged 73 goals and 130 assists a season and led the Oilers to four Stanley Cup championships. Gretzky led Team Canada to win Canada's Cup in 1987. He was traded to the Los Angeles Kings in August of 1988 and in his first year scored 54 goals and passed for 114 assists. The next season, Gretzky broke Gordie Howe's all-time scoring record of 1,850 points. In February 1996, he was traded to the St. Louis Blues, and the next season signed with the New York Rangers as a free agent.

Gretzky retired on April 16, 1999, having established records of 2,857 points; 1,963 assists; 894 goals; and played in 1,486 games all over his 20 seasons in the NHL. Gretzky was awarded nine league MVPs (Most Valuable Player), three All-Star Game MVPs, and 10 Ross awards. The three-year waiting period was waved, and Gretzky was inducted to the Hockey Hall of Fame on November 22, 1999.

were rejected for use by North American markets. Pucks with air bubbles or softer rubber in the middle were rejected for similar reasons.

Byproducts/Waste

Any excess rubber from the manufacturing process is collected, re-shred, and used again to make pucks.

The Future

The future does not involve much change to the actual puck, its composition, or manufacture. While a blue-colored puck is currently made for junior hockey, pucks of different colors serve no purpose in the game and are not likely to be manufactured on a large scale. Any improvements to the silkscreen process would result in changes in the decoration of pucks. Plated souvenir pucks might be available in the future.

Where to Learn More

Books

Duplacey, James. Puck. In *The Annotated Rules of Hockey.* New York: Lyons & Burford, 1996, pp. 52-54.

Periodicals

Modoono, Bill. "Puck's History is Hardly the Stuff of Legends or Lore." *Star- Tribune Newspaper of the Twin Cities Minneapolis-St. Paul* (December 13, 1992): 3C.

Vizard. "Hockey's Chip Shot." *Popular Mechanics* (May 1996): 40.

—Annette Petruso

Horseshoe

Background

Horseshoes protect a horse's hooves from wear on hard or rough surfaces. They are also designed to improve a horse's gait, to help its conformation (how the horse stands), and to control interference problems (when hooves or legs collide with each other). For instance, for a condition called winging in—where hoof flight is to the inside of the stride—can be corrected by a square-toe shoe.

A horseshoe is a U-shaped metal plate, usually made of steel though aluminum is also used. Aluminum shoes are more expensive than steel. There are also plastic-coated metal shoes, and even all-plastic horseshoes. Two-piece shoes are made of steel with a removable plastic insert. The all-purpose steel shoe used for pleasure riding is known as a keg shoe, since they used to come in a wooden keg. The shoes can be attached with nails since the hoof is very dense and contains no nerves. The cost to shoe a horse ranges from $80-$280, depending on the type of shoe and material.

There are about 15 other types of horseshoes, depending on the breed of the horse and its function. Some breeds use different shoes on the front hooves than on the back. Some shoes come with a traction device so the horse can grip the ground better. Thoroughbred racehorses use what is known as a racing plate made from aluminum on the front. For the hind feet, an aluminum shoe known as a caulk or cauk is used that is embedded with steel to help improve abrasion and wear resistance. Some aluminum shoes are more squared off to resemble a hoof from a healthy wild horse.

Standardbred racehorses use a lightweight steel shoe in front that is designed with a half round section or a grooved section called a swage. Sometimes a combination of these two is used. Steel is used in the back. Different shoes are used whether the horse is a trotter or pacer. Horses used for hunting or jumping use a shoe similar to the keg but a bit wider. Sometimes a traction device called a stud (like cleats) is used that is screwed into the shoe before the show. Draft horses use a thicker shoe since they are much heavier and larger than most other breeds. Saddlebreds, which have special gaits, use a toe weight shoe, which is 1 in (2.54 cm) wide in the front and 0.5 in (1.3 cm) wide elsewhere.

History

The horse was a major means of transportation in the United States until the automobile was invented. The horse population declined from 1910–1960, as they were replaced with cars. However, in the early 1960s, the population increased as horse racing and riding became popular as a means of recreation. Over the last decade or so, the horse population has remained relatively stable. The decline in horse racing due to the replacement of other forms of gambling has been offset by the increased popularity of pleasure riding and horse shows.

The process of forging and attaching horseshoes became an important craft in medieval times, and played a major role in the development of metallurgy. Blacksmiths (iron was called black metal) made most of the iron objects used in everyday life though farriery, (farrier, which comes from the Latin word for iron, *ferrum*) or horseshoeing, was the most frequent occupation. A farrier's equipment consists of a furnace or forge, an anvil (a heavy block of steel or

The cost to shoe a horse ranges from $80-$280, depending on the type of shoe and material.

iron), tongs, and hammers. First, the sole and rim of the horse's hoof is cleaned and shaped with rasps and knives. The horseshoe is heated in a forge until it is soft enough to shape with the hammer to fit the hoof, cooled by quenching it in water, and attached to the hoof with nails.

As machines took over the blacksmith's job, and horses disappeared from use in agriculture and transport, the need for farriery declined. Improvements in horseshoe design also have reduced the amount of forging work required by farriers.

Raw Materials

A low carbon mild steel designated A-36 is used for the most popular type of horseshoe. The steel comes in the form of round bars ranging from 0.5-0.7 in (1.27-1.8 cm) in diameter, depending on the type of shoe to be made. These bars are then cut to various lengths, again depending on the shoe type.

Design

Horseshoes are designed according to the breed of the horse, and will vary in size, shape and thickness. Since most are mass-produced, these designs are standard and are made using a two-part die that matches the dimensions and shape desired. A new die is only required if it wears out. In a few instances, custom-designed shoes are still made by hand at several companies. Some farriers also custom make horseshoes for lame or ill horses, which are typically shaped like a heart instead of a U.

The Manufacturing Process

Forging, one of the oldest metal forming methods, is the primary process used for horseshoes. This process shapes metals by deforming them with a hammer, a press, or rollers. Smith forging consists of making a part by banging on the heated metal with a hammer. This is the familiar forging process performed by blacksmiths (farriers) on horseshoes. Hammer forging is simply a larger and more machine-produced version of the same thing.

Cutting & bending

1 The bar is cut to length according to the type of shoe using shear blades. The cut-

ting equipment is operated either mechanically or manually by a foot pedal. After cutting, the bar is heated in an induction furnace or gas-fired forge to a temperature of around 2,300°F (1,260°C). The softened metal is then wrapped around a block of steel in the shape of the shoe using custom designed equipment. Another type of bending equipment uses a plunger in the shape of the shoe, which forces the bar into the desired shape.

Forging

2 Drop forging, the most common of the forging processes, is used for most horseshoes. After the bar is bent, it is then forced into a die with the required dimensions by a powered hammer. One half of the die is attached to the hammer and the other half to the anvil. A cam mechanism determines the length of the stroke of the ram or how close the dies come together.

Finishing operations

3 Once the shoe is bent, it is transferred to a punch press for making the nail holes. Usually eight holes are made per shoe. The punch tooling is custom designed for each type of shoe. Then a trim press is used to remove excess material, or flash. At this point, the shoe is still hot, around 1,900-2,000°F (1038-1093°C). After trimming, it is air cooled for 45-60 minutes.

Packaging

4 After cooling, horseshoes are packaged by hand in 25 or 50 pound boxes. A 50-lb (23-kg) box typically holds 80 shoes. Some manufacturers package 15 or 30 pairs per box.

Quality Control

The raw material must meet the chemical specifications as designated, which is verified by testing by the raw material supplier before shipping. To prevent rust, the round bars are stored under tarps. During forging, the operators constantly check the die visually for wear and other defects. Dimensional tolerances of the shoe (width and thickness) must be within 1/16 inch. The nail holes must be checked for correct position. The hole is monitored for any cracks, which

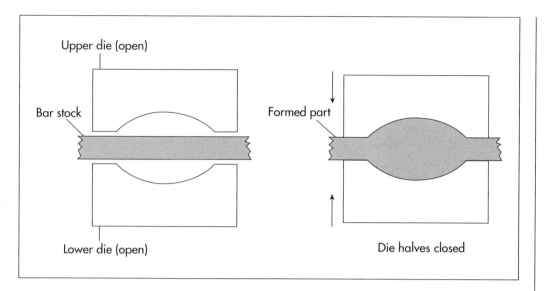

causes missing holes. During the trimming operation, the press is also monitored for cracks, which can cause burrs.

Byproducts/Waste

The flash, or excess material, that is trimmed from the shoe is collected and sent back to a scrap yard for remelting and rerolling. Defective horseshoes are also reprocessed and reused.

The Future

The steel shoe is expected to remain the most popular shoe over the long term. The basic manufacturing process therefore will not change much, though forging presses are becoming more automated, with programmable control of how much force is used and when. Horseshoe designs will continue to be modified to meet specific markets.

The horse population is expected to remain relatively stable over the next few years, though certain parts of the United States will see declines. For instance, in states like New Jersey the price of real estate is becoming so high that one can no longer afford to buy enough land to keep horses. In Michigan,

the horse population has declined by almost 20% from 1984 to 1997, as farm land is developed and casinos replace race tracks as the gambler's preference.

Where to Learn More

Periodicals

Cain, Charlie. "Urban Sprawl and Demise of Racetracks Cuts Horse Population." *Detroit News* (September 30, 1997).

"A Cobbler's Touch: Custom Horseshoes Help Lame, Ill Horses." *The Dallas Morning News* (February 11, 1999).

Van Wyk, Anika. "Making the Shoe Fit/ Modern Touches to an Old Trade." *The Calgary Sun* (2000).

Other

Equestrian Records and Curiosity. http://www.mrhorse.com/Recordsen.htm (21 February 2000).

International Equine Resource Center. http://www.horseshoes.com (January 2001).

—*Laurel M. Sheppard*

Hula Hoop

Over 100 million Hula Hoops were sold worldwide in 1958 alone.

The late 1950s saw one of biggest fads documented by sociologists, the Hula Hoop. Like many fads, the hoop is deceptively simple. It is made of hollow plastic, usually very bright in color, and sometimes with the hollow carrying a couple of ball bearings, bells, or other noise makers. With a variety of sizes the diameter of a hoop toy ranges from 20 in (51 cm) to about 3 ft (91 cm).

Background

Fads interest sociologists because they are adopted by a broad range of persons. To be called a fad, the idea must be a "key invention" that has the possibility of generating many offshoots. The Hula Hoop certainly fits this description because enough variations in size, color, and ornamentation of the hoop enabled each child to own a special version, and children can create their own styles of spinning the hoops. Hula Hoop-spinning contests were held at local fairs, and records were established for the most hoops kept in motion for the longest time periods. These contests are reminders of yet another use of the hoop; jugglers have spun small-diameter hoops on their arms, legs, necks, and on sticks to the delight of circus crowds for generations.

History

Varieties of hoops have always been toys. Along with the ball, the hoop may be among the most popular toys. The ancient Greeks were the first to popularize the hoop, and many of their documents—including illustrations on pottery—show the hoop in action. The hoop was a toy for Greek children, but it was also an exercise device. Hoop-rolling was thought to be a light and benefi-

cial exercise for people not strong enough for more intense exercise or sport. Roman children also played with hoops, and both Greek and Roman versions were made of metal fashioned from scrap strips.

Native Americans used hoops for more than just toys. Eskimos played a game in which a hoop is rolled and poles are thrown through it as it rolls. This game, for children and adults, taught practical skills needed in harpooning and other hunting. North American Indians used the hoop in many ways. Like the Eskimos, the Indians used it as a target for teaching accuracy in shooting arrows and in throwing. Among the Lakota Indians, hoop dancing became a sophisticated art form that is still practiced today. To the Lakota, the hoop represents the circle of life, the vast circle of the horizon as the viewer turns to look all around, and the many repeating patterns in nature like the cycle of the moon. In the hoop dance, the dancer may use 12-28 hoops to forms symbols and figures.

Like the Hula Hoop, the hoops made by the hoop dancer must be large enough to move over the shoulders and around the body; hoops that are about 28 in (71 cm) in diameter are made of natural materials like willow, rattan (a flexible but strong vine), or plastic tubing. Rattan or willow is soaked in water until it softens and can be shaped into a circle. The ends are wrapped with binding. The tubing easily takes the round shape, and a short length of wooden dowel is inserted into the matching ends to even the alignment and form a strong joint. This is also wrapped with binding. Colored binding is wrapped around the entire tube so patterns can be used in the dance. White, yellow, red, and black are the colors of the four directions (north, south, etc.) and the four races of human-kind, according to the Lakota.

Children's hoop toys in Western Europe were made of wood. Hoop-rolling also achieved fad status in England in the 1800s, and those hoops were wood fitted with metal strips or tires on the outer edge. Hoop rolling was called bowling a hoop. The hoop was propelled along the sidewalk, street, or ground with the hand or with a stick called a skimmer. This same fad traveled to the United States, and antique hoops are now favorite toys of collectors. Push hoops were used to help teach babies to walk. Usually, hoops for the very young contained bells or made other sounds to hold the child's interest. Another popular design had pieces of wood shaped much like the spools that hold sewing thread on the spokes of the push hoop. As the hoop turned, the spools slid back and forth on the spokes to make a jingling sound. The rolling hoop was patented in 1871 by Albert Hill. Hill's rolling hoops were about 12-20 in (20-51 cm) in diameter and were pushed with handles that were 20-27 in (51-69 cm) long. The handles and hoops were made of wood with a natural finish, but the noise-making spools were brightly painted.

Other hoops uses and games have long histories but are still known today. Hoops can be thrown, as in the game called *quoits*, or spun. They are used as targets in games like basketball, and, in football, suspended hoops or tires are targets for improving the aim of quarterbacks. Also in football, hoops or tires laid on the ground are used to improve foot mobility, coordination, and speed among players.

The toy known as the Hula Hoop was born out of the brainstorm of two American toy inventors who learned about an Australian practice. Arthur "Spud" Melin and Richard Knerr heard that Australian children used rings made of bamboo for exercise. They produced a plastic hoop in 1958 and promoted it around the Los Angeles, California, area by going to playgrounds, demonstrating the hoop to the kids, and giving away Hula Hoops. Their playground-to-playground salesmanship produced the biggest toy fad the United States has ever witnessed. In four months, over 25 million Hula Hoops were sold in the United States for $1.98 each;

worldwide, over 100 million were sold in 1958 alone. In Japan, the hoop was banned, and the Soviet Union described it as evidence of the decadence of American culture. At the peak of its popularity, Wham-O, Inc. produced 20,000 hoops per day; it is estimated that the plastic tubing for all the Hula Hoops sold would stretch around the world more than five times.

Numerous records in the *Guinness Book of World Records* have involved hoop spinning; in 1999, Lori Lynn Lomeli spun 82 Hula Hoops at the same time for three complete revolutions, a feat that garnered her a place in the book. The Hula Hoop phenomenon never completely disappeared off toy shelves, but it has ebbed and flowed in popularity like most fads. In the late 1990s, the Hula Hoop again experienced a renaissance and appears to be going strong long after its 40th birthday.

Raw Materials

The only materials in most hula hoops is plastic, pigments for coloring the plastic, any inserts like ball bearings, staples to close the circles, and paper labels with adhesive backing. Plastic is used to make both the hoop and the dowel-like insert forming the joint. Some hoops have ball bearings, beads, stars, glitter, bells, or other noise-makers inside the hollow tube. These add extra visual interest and motion or sound as the twirler spins the hoop. Metal staples and the paper labels are provided by outside, specialty suppliers.

Design

Like all toys, the Hula Hoop, even in its simplicity, adapts to changing trends. Color trends change the color combinations in the hoops every few years (Wham-O changes the colors of its hoops every year), and toy designers look for other ways of varying and remarketing the toy to keep it among the top sellers. One recent design features fruit-scented hoops that give off a slight, pleasantly fruity smell as the plastic warms in play or in the sun. The scent matches the color of the hoop (grape scent for purple hoops, orange for tangerine-colored plastic, and so forth). Others are wrapped with glittery paper.

Children are not the only target market for design and sales of spinning hoops. Their

Hoops have been used by many different peoples as toys, in ceremonies or dances, in games, etc.

benefits as an exercise toy are celebrated, and one manufacturer has added a calorie counter to the hoop. Fitted in the joint where the two ends of the plastic tube meet, the calorie counter uses a microprocessor to count the number of revolution, duration of spinning, and calories expended. The microprocessor also offers spoken encouragement as the spinner reaches "personal bests" in number of revolutions and time spent twirling, and it plays five choices of music for added incentive. An AA battery powers the counter. The health-conscious hoop is 36 in (91 cm) in diameter, and the tube has a larger diameter than the child's toy at 1-1.5 in (3.8 cm). It also weighs 2.25 lb (0.84 kg).

The Manufacturing Process

1 The process for making Hula Hoops is as straightforward as the toy itself. The factory receives high-density polyethylene (HDPE) plastic in pellets that are already colored or that can be tinted with pigments in the factory. The plastic beads or pellets are fed into a hopper. From the hopper, they flow into the barrel of an extrusion machine. The pellets are heated by mechanical means using heat of friction in the barrel, and pig-

ments are added as needed. Dies inside the extruder shape the thickness of the tube that is extruded as a continuous piece of plastic. In a proprietary process, Wham-O uses a twist machine so that a stripe in the plastic twists around the plastic hoop.

2 The tube cools quickly and is cut into lengths equal to the circumference of the particular hoop by a high-speed cutoff, which is much like a circular saw. Wham-O makes three different lengths of tube resulting in hoops of three different diameters for a range of sizes and ages of hoopsters.

3 The lengths of hoop are loaded on to large carts much like those used to carry lengths of steel. They are fed manually through a benching machine that curves the tube lengths into hoop shapes.

4 Ball bearings or other noise and motion pieces are inserted in the hoops. The hoops are then fitted with a plastic dowel or insert that is another extruded piece that is also hollow but slightly smaller in outer diameter than the hoop. The insert is stapled in place to attach the ends. The staples are held in place by friction, and a label is fastened over the insert and staples for added safety.

5 The finished hoops are conveyed to the and packing department; the three sizes of hoops are all nested together in the same packing box to reduce the amount of packing materials that are needed and to save retailers storage space.

Byproducts/Waste

Plastics manufacture is a toy-making speciality. Makers of Hula Hoops usually make a number of colors, sizes, and other varieties. They also produce other plastic toys that use similar extrusion and molding techniques. Waste is minimal. When colors are changed in the extruder, the old color is wasted, but this amounts to only 0.1% of the volume of plastic used in hoop manufacture. Hoops that are defective are pulled from the manufacturing line and collected in bins for recycling.

Employee safety is carefully controlled by government regulation, employee training, and distance. Regulations by the Federal Occupational Safety and Health Administration (OSHA) limit employee exposure to the high heat and pressure of the extrusion machine. Safety guards keep employees a safe distance from the cutoff, twist machine, and other heavy machinery. The employees are also well educated in their own protection. In the 50-year history of the Wham-O Hula Hoop, there have been no factory injuries or injuries caused by the toy itself.

Quality Control

Quality enters the process during design when extrusion dies and other tools are made to low tolerances for error. This helps reduce irregularities during manufacture and waste of plastic. Inspectors are stationed at each machine to observe the product at every step. All employees have the responsibility of taking faulty hoops out of any stage of manufacture. A final quality audit is performed before the hoops are packed.

The Future

The Hula Hoop seems to have established a firm place in the American way of life and childhood. It now has a steady sales pattern and seems destined to remain a part of our play. The hoop's popularity is helped by modern emphasis on health and exercise. In the future, manufacturers expect to emphasize play patterns so the hoops can be used more like a game and to introduce new products to help make the hoop a purely individual toy. Just as older civilizations celebrated the symbolism of the circle, American children have found their own hoop dance to add to play and exercise.

Where to Learn More

Books

Barenholtz, Bernard, and Inez McClintock. *American Antique Toys.* New York: Harry N. Abrams, Inc., 1980.

Left Hand Bull, Jacqueline, and Suzanne Haldane. *Lakota Hoop Dancer.* New York: Dutton Children's Books, 1999.

Other

Hula Hoops. http://www.hula-hoops.com (January 2001).

Wham-O, Inc. http://www.wham-o.com (January 2001).

—*Gillian S. Holmes*

Ice Cream Cone

Background

Today, the ice cream cone is a standard in any ice cream store or stand. This tasty treat is known as a way to cool down in the summer and makes an edible container for a cold snack. The frosty smoothness of the ice cream complements the crispy crunch of the cone for an interesting taste combination. There are almost as many stories of how the ice cream cone was invented as there are flavors that it holds.

History

The ice cream cone would seem to be a simple and unpolitical a treat, yet it's origin is hotly contested. The most favored folk tale regarding the invention of the ice cream cone takes place at the 1904 World's Fair held in St. Louis, Missouri. Two food vendors had stalls next to each other. Arnold Fornachou made and sold ice cream. His neighbor, Ernest A. Hamwi, had come to the United States from Damascus, Syria. Hamwi made sweet wafers (much like today's wafer-like cookies) that Syrians call "zalabias." Hamwi cooked the wafers on a waffle iron heated over a coal fire, coated them with sugar, and rolled the wafers while they were still hot so they were easy to eat and carry. When Fornachou ran out of dishes to hold his ice cream, Hamwi rolled his wafers into a cone shape instead of a tube, and the gentlemen topped the wafer with scoops of Fornachou's ice cream. Zalabias became "World's Fair Cornucopias," and the cone concept was born.

With over 50 ice cream vendors at the Fair, Hamwi was soon doing a land-office business. He started his own cone company after the Fair called the Cornucopia Waffle Oven Company, but tired of business and went to work for the competition, Heckle's Cornucopia Waffle Oven Company in St. Louis. The cornucopia or waffle name was replaced with the word cone in 1906. Meanwhile, Hamwi promoted cones at fairs all across the United States. Returning to his own business in 1910, Hamwi started the Missouri Cone Company of St. Louis. He died in 1943 after amassing a fortune founded on ice cream cones.

A second contender, David Avayou also claims to be the cone's creator. Avayou owned an ice cream parlor in New Jersey where he made both ice cream and cones. He took his wares to the St. Louis World's Fair and claims to have been selling them there when Fornachou and Hamwi stumbled on their joint product.

Still a third contestant is Abe Doumar, another immigrant who had moved with his family of 12 brothers and sisters from Lebanon to St. Louis. Doumar's favorite treat from his homeland was a pita bread rolled into a cone shape and filled with fruity jam. He approached another of the Fair's zalabia-makers and suggested applying the same concept by rolling a waffle and filling it with ice cream. Doumar later developed a variety of waffle machines, moved to New York, and sold ice cream cones at Coney Island. By the 1930s, Doumar owned a number of restaurants along the East Coast; the new trend for "fast food" that grew with the popularity of the automobile almost drove him out of business until he got the idea to make waffle cones in the front windows of his restaurants. The baking process and the girls in the windows rolling cooked waffles into cones became attractions that saved the restaurants.

Opposing these charming stories is a solid fact. In 1903 (the year before the World's Fair), Italo Marchiony was awarded a patent for the "pastry cornet," which he developed to hold his frosty wares. Marchiony was an Italian immigrant who lived in New York City. His product was lemon ice that he scooped onto small glasses and sold to customers along Wall Street. After consuming the ice, the customer returned the glass, and it was washed and used again. Breakage and the continual task of washing dishes frustrated Marchiony; he substituted paper cones, but these (and littering consumers) made a messy problem. As early as 1896, Marchiony invented a fully consumable alternative. By 1903, he had made a machine that created cones like the sugar cone known today. The machine resembled a long waffle iron with spaces to cook 10 cones. Later, Marchiony opened a cone factory in Hoboken, New Jersey. He is also credited with building the first ice cream sandwich with two waffle squares.

Apart from his patent from the United States government as proof, Marchiony has history and sentiment on his side. His business of selling lemon ice in glass scoops is part of a tradition in Italy dating back to the early 1800s. The Penny-Ice Men became common across Europe from about 1820 to 1860, as revolution and economic hard times drove immigration. Part of this wave consisted of Italians who left their homeland for Europe's major cities. They pushed carts through the streets beginning as early as 7 A.M. during the summers and sold flavored ice seated on tiny glass goblets. A goblet cost a penny, the people consumed the ice, and the goblet was returned to the vendor. In Italy, the Penny-Ice Men cried, "Ecco un poco, che un poco" (Here's a little for so little [money]), and this cry became distorted by non-Italians into the word hokeypokey. In New York and other American cities—where the custom had migrated by the mid-1800s—the Penny-Ice Men were known as Hokeypokey Men. Their trade and their use of the tiny glass goblets are a direct link to the development of the ice cream cone.

After the World's Fair, cone-making machines were regularly sold in catalogs for $8.50. Individual vendors could afford these, so the street vending of ice cream now accompanied by cones grew enormous-

ly. In 1912, Frederick Bruckman devised a machine that rolled the cones hot from the waffle iron automatically; 245 million ice cream cones were sold in 1924 alone.

Raw Materials

Three main dry ingredients compose all types of cones. Wheat flour, tapioca flour, and sugar are chosen for baking quality, strength, and relative sweetness, respectively. Tapioca is made from the cassava plant, which has a starch-like root. The root is processed into the tapioca "pearls" familiar in pudding and also into finely ground flour. The cassava grows only in tropical climates so cone manufacturers import it from South America and Southeast Asia. Manufacturers purchase both tapioca flour and sugar in large bags, but wheat flour is bought by the tanker-truck load and is unloaded by air pressure that blows it from the tanker into storage silos. During World War II, wheat flour was needed for priority items like bread; as a substitute, ice cream cone makers used popcorn that was ground to a flour-like consistency.

The quantity of sugar is a major distinguishing feature between cone types. Sugar and waffle cones are made of one-third sugar. Not only does this influence the sweet flavor, but it affects the brown finished color and the crispy texture. Cake cones have less than 5% sugar.

Wet ingredients (and others added with the wet materials) include water, shortening (edible fat or grease), **baking powder** (a dry ingredient but one that begins to react as soon as it is mixed with water so it is added last to avoid contact with any moisture in the air), coloring, flavoring, and salt. Both the coloring and flavoring are natural products made by outside specialists.

Before any liquid is added, air compressors are again used to mixed these dry ingredients in large coolers. The compressors are computer-controlled to regulate the quantities, and different combinations of ingredients are used to make waffle/sugar cones and cake cones, so separate coolers are used to mix each type. The combined dry ingredients are termed cone filler or cone batter. Some specialty suppliers premix cone filler and sell it to cone bakers.

Design

There are three principal types of ice cream cone; the cake cone (also called a molded or flat-bottomed cone), the waffle cone, and the sugar cone. The waffle cone is characterized by a rough or unfinished top edge. The sugar cone is made with the same ingredients and process as a waffle cone but has a finished top edge and sometimes a chocolate lining.

The waffle pattern on all types of cones, the finished edge of the sugar cone, and the shape of the flat-bottomed cone (as well as cornet varieties of the cake cone) greatly influence the ease with which the finished cones pop out of their molds. Cone designers refine the waffle pattern and other shape characteristics and make trial batches to find the best design that releases from the mold without burning, breaking, or creating weak spots that won't hold ice cream or will break when the scoop is applied. The molded cone has a lip around the top that keeps drips contained inside the cone. The row of teeth helps firmly seat the scoop of ice cream and provides added strength where the upper lip of the cone meets the cylindrical base.

The flat bottom of the cake cone is now an accepted industry standard, but it was not invented until the late 1940s. Before this, cake cones were also cone shaped, but Joseph Shapiro of the Maryland Cup Corporation (later the Ace Baking Company) made the flat base especially for the Diary Queen chain. Filling cone-shaped cones and handing them to customers is a two-handed business, but the flat-bottomed cone stands on its own and can be filled more easily.

Shapes and patterns also affect baking characteristics. The finished cone should be uniform in color as well as shiny on the outside. It should bake uniformly so that all sides (including the flat bottom) are thoroughly cooked. The size is important because cones are expected to hold single, double, and triple scoops. The first scoop has to fill the cone and weight the bottom without vanishing completely into the cone, and the third scoop should not overpower the cone and cause it to break or tip too easily. The filled cone should look equally appetizing whether it has one, two, or three flavors atop it.

Strength is an important characteristic, not only to the consumer who holds it. Cones must withstand prefilling in the factory if they are used for frozen treats like the Drumstick. Unfilled cones have to be packed together by mechanical devices. The cones must "nest" (fit one inside another) neatly to allow efficient packing. Minimal packing materials are used to cushion the cones, mainly because of cost.

Taste is the key design factor. Cake cones should be crisp instead of spongy and tasty like a mild cereal. Waffle cones should be crunchy and sweet but not too hard or overpowering in sugar content. The ice cream is the featured food, and the cone must complement its quality.

The Manufacturing Process

1 The batter for all cones is mixed in large vat-like mixers and stored in coolers. Air compressors blow the dry ingredients into the mixers. Separate mixers and coolers are used to combine and store the batter for cake cones and for waffle and sugar cones together. The air compressors that pump in all ingredients are computer-controlled so the recipe for each cone is correct. Computers also control all the other machines in the factory; in the mixing area, they tell the compressors when the coolers are running low on batter, so the next batch is mixed automatically.

2 As the dry ingredients are blown in, water is added, and the mixers begin to stir the batter. The dry ingredients and water are mixed for nine minutes before the other ingredients are added. The computer signals to a worker when the nine-minute mix is done, and the worker inspects the partial batter then adds the remaining ingredients by hand. This is one of the few hands-on parts of the process; it is essential to the character and quality of the finished cones. The worker resets the mixer when the ingredients have been added, and the mixer beats the batter for a few minutes at high speed, not only for perfect blending but to add just the right amount of air to the batter. The mixer for cake cones yields about 300 lb (112 kg) of batter, and the waffle/sugar cone batter is mixed in 150-lb (56-kg) batches. The mixed batter is then pumped to its cooler; the mixer shuts itself off automatically and resets itself for the next batch.

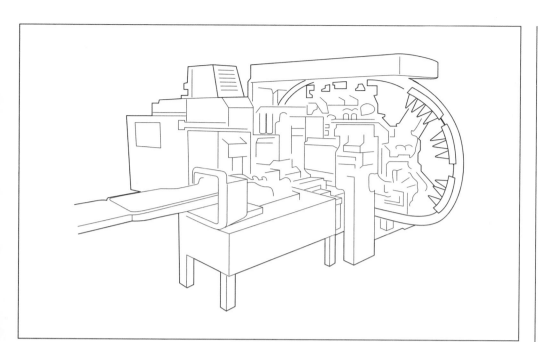

An example of a fully automated ice cream cone maker.

3 From the coolers, batter is pumped to storage tanks next to the baking ovens. It is then pumped through a pipe. Cake-cone batter is pumped into the cake-cone molds, and waffle/sugar-cone batter flows onto plates much like the bottom plates of waffle irons. The pumping system applies a pre-measured amount of batter to either the mold or the plate. The cake cones bake for about 90 seconds and emerge in their finished shape and ready to be packed. Waffle and sugar cones bake for about 82 seconds, but they take longer to finish because they have to be shaped. The flat, hot, baked circles are rolled into cone shapes by specialized cone-rolling machines in a process that takes about 20 seconds. These cones cannot be handled for packing until they are completely cooled, and they are air-cooled for 2 minutes. Cooling makes the cones firm to hold their shape.

A large cone-making plant will be equipped with as many as 40 ovens that will produce 5 million cones per day. The plants usually operate 24 hours per day and every day of the year except significant holidays. Total production from a major manufacturer can be 5 billion cones per year.

4 Finished cones travel along conveyors to the packing area. Cake cones are relatively strong and are nested inside each other, wrapped with clear paper that is sealed to be air tight, and placed in boxes. Waffle and sugar cones are crisp and delicate, so each one is individually packed in a Styrofoam container with a bottom bowl and a lid (a "clamshell" container). These packages are also boxed. All boxes have been preprinted by an outside printer and box manufacturer. The outer design is not only decorative but carries the nutritional information required by the United States government for a single-cone serving. The boxes are bulk-packed into larger cartons called master packs for shipping and distribution.

Some boxed cones are sent to the dairy-pack industry which fills the cones with chocolate liners and ice cream, freezes their products, and repackages them for individual sale and bulk sale in boxes in grocery stores. The best known of the dairy-pack products are probably Drumstick and Nutty Buddy. Boxed cones are also distributed to food service businesses like Dairy Queen, Baskin-Robbins, and McDonalds. These businesses (like the dairy-pack trade) fill cones individually with their own ice cream and soft-serve products. Amusement parks are also part of the food service business that fills cones with frozen treats on the spot.

Finally, packaged cones are sold in bulk to retail businesses like grocery, chain, convenience, and drug stores. These retailers usually do not fill or modify the cones; they sell the boxed cones directly to the consumer who can make custom desserts and snacks with the cones at home.

Quality Control

Although cone-making is computer-controlled, workers are essential to quality control. The correct addition of ingredients is the most obvious quality control step, but throughout the process specially trained quality control inspectors watch cone making and baking, taste-test cones occasionally, and reject any that are misshapen, broken, or over/under-cooked. Whole cones are also removed from the process and cut and broken apart to check that cones are truly perfect inside and out.

Byproducts/Waste

Cone makers usually do not produce byproducts although they commonly make the three types of cones. There is some breakage, and some are rejected during the quality control process. During the period from 1920 to about 1950, cone makers bagged the broken cones and sold them as a snack byproduct. Families could buy the broken pieces and eat them like chips or crackers or crush them more finely and use them as toppings for ice cream, pudding, and fruit. During the Depression in the 1930s, crushed cones were a tasty substitute for expensive nut toppings.

As manufacturers' volumes have increased and crushed cones have become less desired by the public, cone makers have found another use for discarded cones. The cones are ground up and sold to farmers for animal feed. Paper goods from the packaging process and wooden pallets for storage are recycled, so the industry produces virtually no waste.

Safety in the cone factory is also a lesser concern because most processes are fully automated. Workers are trained about safety issues related to electrical and mechanical equipment and the heat of the baking ovens. They are also required to protect the safety of the product and wear clothing, hair covers, and gloves to keep the cones sanitary.

The Future

The ice cream cone is such a fixture of worldwide desserts, entertainment, amuse-ment, and relaxation that it is unlikely to fall out of favor. Ice cream and its cone are consumed year-round—with Americans eating about 23 qt (22 L) per person a year—although summer is certainly the prime season. A mark of the acceptance of the ice cream cone may be its stature as a highly recognizable icon or symbol. In 1945, the Macy's Thanksgiving Day Parade featured a helium-filled balloon shaped like an triple-scoop ice cream cone. It stood 40 ft (12.2 m) high and 16 ft (4.9 m) wide. Hot-air balloon races and festivals have also been treated with scoop-shaped hot-air balloons and cone-shaped baskets. In 1962, the Swedish-born sculptor Claes Oldenburg displayed a Pop-art version of a "Store" filled with everyday objects that were greatly oversized and made of foam rubber covered with canvas. Oldenburg chose an enormous ice cream cone to represent American life.

The tried-and-true types of cones are not likely to change. Of course, manufacturers are constantly improving their products, but they stick to the varieties that are popular with the public. The sugar, waffle, and cake cones perfectly complement the changing flavors within the ice cream world while adding their own support, taste, crunch, and sense of tradition.

Where to Learn More

Books

Dickson, Paul. *The Great American Ice Cream Book*. New York: Atheneum, 1972.

Liddell, Caroline, and Robin Weir. *Frozen Desserts*. New York: St. Martin's Press, 1995.

Wardlaw, Lee. *We All Scream for Ice Cream*. New York: Harper Trophy, Harper-Collins Publishers, Inc., 2000.

Periodicals

Belleranti, Shirley W. "A treat from Marco Polo." *Hopscotch* 8, no. 2 (August/September 1996): 9.

Gustaitis, Joseph. "Who Invented the Ice Cream Cone?" *American History Illustrated* 23, no. 4, (Summer 1988): 42-45.

Other

Dairy Queen Corporation. http://www.dairy
queen.com (January 2001).

The Joy Cone Company, Hermitage, PA.
http://www.joycone.com (January 2001).

—*Gillian S. Holmes*

Indigo

Egyptian artifacts suggest that indigo was employed as early as 1600 B.C. and it has been found in Africa, India, Indonesia, and China.

Background

Indigo, or indigotin, is a dyestuff originally extracted from the varieties of the indigo and woad plants. Indigo was known throughout the ancient world for its ability to color fabrics a deep blue. Egyptian artifacts suggest that indigo was employed as early as 1600 B.C. and it has been found in Africa, India, Indonesia, and China.

The dye imparts a brilliant blue hue to fabric. In the dying process, **cotton** and linen threads are usually soaked and dried 15-20 times. By comparison, silk threads must be died over 40 times. After dying, the yarn may be sun dried to deepen the color. Indigo is unique in its ability to impart surface color while only partially penetrating fibers. When yarn died with indigo is untwisted, it can be seen that the inner layers remain uncolored. The dye also fades to give a characteristic worn look and for this reason it is commonly used to color denim. Originally extracted from plants, today indigo is synthetically produced on an industrial scale. It is most commonly sold as either a 100% powder or as a 20% solution. Through the early 1990s, indigo prices ranged near $44/lb ($20/kg).

History

The name indigo comes from the Roman term indicum, which means a product of India. This is somewhat of a misnomer since the plant is grown in many areas of the world, including Asia, Java, Japan, and Central America. Another ancient term for the dye is *nil* from which the Arabic term for blue, *al-nil*, is derived. The English word aniline comes from the same source.

The dye can be extracted from several plants, but historically the indigo plant was the most commonly used because it is was more widely available. It belongs to the legume family and over three hundred species have been identified. *Indigo tinctoria* and *I. suffruticosa* are the most common. In ancient times, indigo was a precious commodity because plant leaves contain only about small amount of the dye (about 2-4%). Therefore, a large number of plants are required to produce a significant quantity of dye. Indigo plantations were founded in many parts of the world to ensure a controlled supply.

Demand for indigo dramatically increased during the industrial revolution, in part due to the popularity of Levi Strauss's blue denim jeans. The natural extraction process was expensive and could not produce the mass quantities required for the burgeoning garment industry. So chemists began searching for synthetic methods of producing the dye. In 1883 Adolf von Baeyer (of Baeyer aspirin fame) researched indigo's chemical structure. He found that he could treat omega-bromoacetanilide with an alkali (a substance that is high in pH) to produce oxindole. Later, based on this observation, K. Heumann identified a synthesis pathway to produce indigo. Within 14 years their work resulted in the first commercial production of the synthetic dye. In 1905 Baeyer was awarded the Nobel Prize for his discovery.

At the end of the 1990s, the German based company BASF AG was the world's leading producer, accounting for nearly 50% of all indigo dyestuffs sold. In recent years, the synthetic process used to produce indigo has come under scrutiny because of the harsh chemicals involved. New, more environmentally responsible methods are being sought by manufacturers.

Raw Materials

The raw materials used in the natural production of indigo are leaves from a variety of plant species including indigo, woad, and polygonum. Only the leaves are used since they contain the greatest concentration of dye molecules. In the synthetic process, a number of chemicals are employed as described below.

The Manufacturing Process

Natural extraction

1 Plant extraction of indigo requires several steps because the dye itself does not actually exist in nature. The chemical found in plant leaves is really indican, a precursor to indigo. The ancient process to extract indican from plant leaves and convert it to indigo has remained unchanged for thousands of years. In this process, a series of tanks are arranged in a step wise fashion. The uppermost tank is a fermentation vessel into which the freshly cut plants are placed. An enzyme known as indimulsin is added to hydrolyze, or break down, the indican into indoxyl and glucose. During this process carbon dioxide is given off and the broth in the tank turns a murky yellow.

2 After about 14 hours, the resulting liquid is drained into a second tank. Here, the indoxyl-rich mixture is stirred with paddles to mix it with air. This allows the air to oxidize the indoxyl to indigotin, which settles to the bottom of the tank. The upper layer of liquid is siphoned away and the settled pigment is transferred to a third tank where it is heated to stop the fermentation process. The resultant mixture is filtered to remove impurities and dried to form a thick paste.

Historically, the Japanese have used another method which involves extracting indigo from the polygonum plant. In this process the plant is mixed with wheat husk powder, limestone powder, lye ash, and sake. The mixture is allowed to ferment for about one week to form the dye pigment which is called sukumo.

Synthetic production

3 A variety of synthetic chemical processes have been used to produce indigo. All these processes involve combining a series of chemical reactants under controlled conditions. The reactants undergo a series of reactions which result in the formation of the indigo molecule. A number of other chemical byproducts are also produced in this reaction.

4 These synthesis reactions are conducted in large stainless steel or glass reaction vessels. These vessels are equipped with jackets to allow steam or cold water to flow around the batch as the reactions progress. Because of the complexity of these chemical processes, the dye is usually made in batch quantities. There are, however, a few methods invented by the Germans for continuous process manufacturing.

Types of reactions

5 The first commercial method of producing indigo was based on Heumann's work. In this method, N-phenylglycine is treated with alkali to produce indoxyl, which can be converted to indigotin by contact with air. However, the amount of dye yielded by this process is very low. Another, more efficient, synthesis route utilizes anthranilic acid. This process was popular with major manufacturers, such as BASF and Hoechst, for over 30 years. A variation of this method (which has become widely used) involves the reaction of aniline, formaldehyde, and hydrogen cyanide to form phenylglycinonitrile. This material is then hydrolyzed to yield phenylglycine which is then converted to indigotin. Currently, a method which uses sodamide with alkali to convert phenylglycine to indoxyl. Sodamide reacts with excess water, thus lowering the overall reaction temperature from almost 570°F (300°C) to 392°F (200°C). This results in a much more efficient reaction process.

Finishing operations

6 After the chemical reaction process is complete, the finished dye must be washed to remove impurities and then dried. The dried powder can be packed in drums or reconstituted with water to form a 20% solution and filled in pails.

The chemical symbol for indican, the compound found in the leaves of the indigo plant that is used to make indigo dye.

$$O\!-\!C_6H_{11}O_5$$

Quality Control

During indigo manufacture, the reaction process is continuously monitored to ensure the chemicals are combined in the proper ratios. Key elements that must be controlled include the pH (or acid/base quality of the batch), the temperature (which controls the speed of the reaction), and the reaction time (which determines the degree of completion). If any of these variables deviate from specifications, the resulting reaction product can be affected. Typically, poor quality control results in lower yield of the dye, which increases costs for the manufacturer.

To ensure that manufacturers can consistently purchase the same shade of dye, indigo is assigned a Color Index number that defines its shade. It is designated as "CI Natural Blue CI 75780."

Byproducts/Waste

Indigo production produces a variety of waste products which must be handled carefully. In addition to the reactants described above, there are other reaction side products that are produced along with the indigo. Some of these materials are considered to be hazardous and must be disposed of in accordance with local and federal chemical waste disposal guidelines. These waste chemicals can enter the environment in at least three different ways. The first is during the actual manufacture of the molecule. The second is when the dye is applied to the yarn, and the third is when the dye is eluted into the wash water during the initial stonewashing or wet processing of the fabric. This last route typically occurs during the production of denim fabric.

The Future

Much of the need for indigo is being met with other types of blue dyes and today most of the indigo used by the world is made outside the United States. Researchers are concentrating on new methods of indigo manufacture that are more environmentally friendly. One promising future method involves using biocatalysts in the dye reaction process. Indigo dye may be one of the first high-volume chemicals made through a biological route. Genencor International, of Rochester New York, is evaluating a process to produce indigo using biotechnology. According to Charles T. Goodhue, Genencor's Program Director/Biocatalysis Research and Development, indigo produced by this method is chemically the same as the regular synthetic dye and behaves identically in dyeing tests. However, at this time the technology is expensive and production costs could be prohibitive. Genencor is seeking a major market partner to work with them in the development of this new technology.

Manufacturers who use indigo in dying operations are also seeking to improve their use of the dye. For example, Burlington's Denim Division introduced a technology in 1994 they call "Stone Free," which allows indigo dye in the fabric to break down 50% faster in the stonewash cycle. Compared to traditional methods of stonewashing fabric dyed with indigo, their new process uses few, if any, pumice stones which help give the fabric its faded look. Therefore, pumice stone handling and storage costs are reduced, along with time required to separate pumice from garments after stonewashing. It also uses much less bleach. Therefore, this new process not only reduces garment damage, but also reduces waste produced by the stones and bleach.

Where to Learn More

Books

Kirk, R. E., and D. F. Othmer (ed.) *Encyclopedia of Chemical Technology: Alkoxides, Metal to Antibiotics (Peptides).* Wiley-Interscience, John Wiley and Sons, 1978.

Periodicals

Guilbaut, G. B., and D. W. Kramer. "Resorufin Butyrate and Indoxyl Acetate as Fluorogenic Substrates for Cholinesterases." *Analytical Chemistry* 37 (1965):120-23.

McCurry, John. "Burlington Debuts Stone Free Denim." *Textile World* 144, no.3 (March 1994): 120- 123.

Rotman, David, and Emma Chynoweth. "The Quest for Reduced Emissions, Greener Processes." *Chemical Week* 153, no.1 (July 7, 1993): 117.

—*Randy Schueller*

Industrial Hemp

Currently 32 countries, including Canada, Great Britain, France, and China, allow farmers to grow industrial hemp.

Background

Hemp is a distinct variety of the plant species *Cannabis sativa L.* that grows to a height anywhere from 4-15 ft (1.2-4.5 m) and up to 0.75 in (2 cm) in diameter. The plant consists of an inner layer called the pith surrounded by woody core fiber, which is often referred as hurds. Bast fibers form the outer layer. The primary bast fiber is attached to the core fiber by pectin—a glue-like substance. The primary fibers are used for textiles, cordage, and fine paper products. The wood-like core fiber is used for animal bedding, garden mulch, fuel, and an assortment of building materials.

Due to the similar leaf shape, hemp is frequently confused with marijuana, another cannabis plant. The major difference is their tetrahydrocannabinol (THC) content, the ingredient that produces the high when smoked. Marijuana can contain as much as 20% THC, compared to less than 1% for industrial hemp. Despite this difference, some countries are reluctant to legalize growing of hemp (especially the United States), since there is a fear this will make it more difficult to control the use of the drug. Most hemp varieties also have a hollow stalk that have a very high fiber content (35%), in contrast to marijuana varieties that usually have a solid stalk having low fiber content (15%).

Canada is one country that has legalized hemp, though with certain restrictions. The maximum allowable THC concentration is 0.3% and all hemp farmers are required to undergo a criminal-records check, as well as obtain a license from Health Canada. Despite these restrictions, hemp production has increased threefold in just a year, from 6,175 acres (61.75 hectares) harvested in 1998 to nearly 20,000 acres (200 hectares) in 1999. Over 95% of the acres grown in 1999 in Canada were for hemp grain.

Farmers who grow hemp claim it is a great rotation crop and can be substituted for almost any harvest. It grows without requiring pesticides and is good at aerating the soil. On a per-acre basis, one estimate claims hemp nets farmers more income ($250-$300) than either corn or soybeans ($100-$200). A full crop of hemp only takes 90 days to grow, yielding four times more paper per acre, when compared over a similar 20 year period with redwood trees in the northwest United States. However, there are other varieties of trees that yield two to three times more than hemp.

Advocates of hemp claim that it can be used in 25,000 different products, from clothing to food to toiletries. Until the nineteenth century, hemp was used in 90% of ships' canvas sails, rigging, and nets (and thus it was a required crop in the American colonies). Today, hemp fiber is being used as a replacement for fiberglass in automotive components and made into cloth for window dressings, shower curtains, and upholstery. China is the world's largest producer of hemp fabric, whereas India produces the most hemp overall.

Other products made from hemp fiber include: insulation, particleboard, fiberboard, rope, twine, yarn, newsprint, cardboard, paper, horse stable bedding, and compost. Hemp bedding has been found superior to straw and other materials for horse stalls in reducing the smell of ammonia. Hemp seed is used to make methanol and heating oil, salad oil, pharmaceuticals, soaps, paint, and ink.

Currently 32 countries, including Canada, Great Britain, France, and China, allow farmers to grow industrial hemp. The current hemp market for sales and exports in North America is estimated at between $50-$100 million per year. Unites States imports of industrial woven fabrics made from hemp totaled $2.9 million in 1997. Import volume jumps to around $40 million when other products—such as paper, shampoo, and oil—are included. Textile uses of hemp represent 5% of hemp products produced in Canada.

History

Hemp was the first plant to be domestically cultivated around 8000 B.C. in Mesopotamia (present-day Turkey). Hemp was grown for fiber and food. It was recorded as being harvested in central Asia around 6500 B.C. Several centuries later, China started growing hemp as a crop and later used it in medicine. By 2700 B.C., the Middle East, Africa, and most of Asia used hemp for fabric, rope, medicine, and food. Hemp was introduced to Europe 400 years later. The oldest surviving piece of paper, a 100% Chinese hemp parchment, was dated to A.D. 770.

From 1000 B.C. to the nineteenth century, hemp was the world's largest agricultural crop, where it was also used for paper and lamp oil. During this period, several well-known books, including the Bible and *Alice in Wonderland,* were printed on hemp paper, and several famous artists painted on hemp canvas. The first crop in North America was planted by a French botanist in Nova Scotia in 1606. Thomas Jefferson drafted the United States Declaration of Independence on hemp paper and grew hemp himself. Two centuries later, the United States and Canada put a stop cannabis cultivation in 1937 with the Marijuana Tax Act (this put a one dollar per ounce tax on any hemp manufacturers), which was later lifted during the World War II effort.

Global production of hemp has been declining since the 1960s, from over 300,000 short tons (272,160 t) of hemp fiber and tow in 1961 to 69,000 short tons (62,597 t) in 1997. China accounts for 36% of this production and 73% of grain production. This has dropped from 80,000 to 37, 000 short tons (72,576 to 33,566 t) over the same period. Around 1994, there were 23 paper mills using hemp fiber, at an estimated world production of 12,000 short tons (10,886 t) per year. Most of these mills were located in China and India for producing printing and writing paper. Others produced specialty papers, including cigarette paper. The average hemp pulp and paper mill produces around 5,000 short tons (4,536 t) per year, compared to wood pulp mills at 250,000 short tons (226,800 t) per year.

However, in the last decade, the number of companies trading in and manufacturing hemp products has increased dramatically. The North America market is still in its infancy since Canada just legalized hemp production and sale in 1998. Hemp cultivation tests in the United States began a year later though it is still illegal to grow it commercially.

Raw Materials

Fiber processing uses few chemicals, if any at all. However, the fiber may be blended with other materials, such as synthetic fibers or resins as binders, depending on the final product being made. For paper making, water and chemicals (sodium hydroxide or sulfur compounds) are mixed with the fibers to remove the natural glue components.

The Manufacturing Process

Cultivation and harvesting

Hemp is an annual plant that grows from seed. It grows in a range of soils, but tends to grow best on land that produces high yields of corn. The soil must be well drained, rich in nitrogen, and non-acidic. Hemp prefers a mild climate, humid atmosphere, and a rainfall of at least 25-30 in (64-76 cm) per year. Soil temperatures must reach a minimum of 42-46°F (5.5-7.7°C) before seeds can be planted.

1 The crop is ready for harvesting high quality fiber when the plants begin to shed pollen, in mid-August for North America. Harvesting for seed occurs four to six weeks later. Fiber hemp is normally ready to harvest in 70-90 days after seeding. A special machine with rows of independent teeth and a chopper is used. To harvest hemp for textiles, specialized cutting equipment is required. Combines are used for harvesting

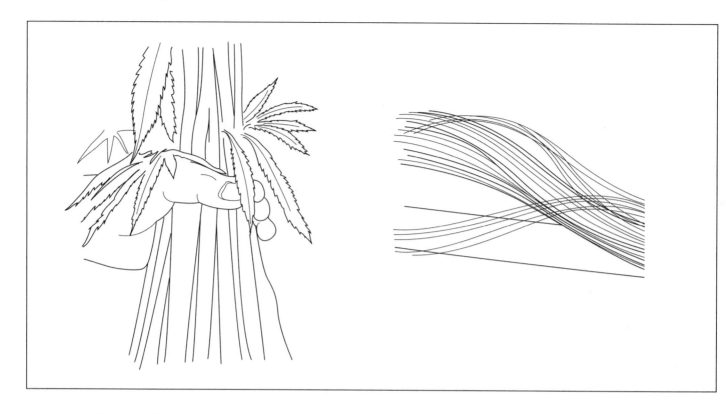

An example of hemp and hemp fibers.

grain, which are modified to avoid machine parts being tangled up with bast fiber.

2 Once the crop is cut, the stalks are allowed to rett (removal of the pectin [binder] by natural exposure to the environment) in the field for four to six weeks—depending on the weather—to loosen the fibers. While the stalks lay in the field, most of the nutrients extracted by the plant are returned to the soil as the leaves decompose. The stalks are turned several times using a special machine for even retting and then baled with existing hay harvesting equipment. Bales are stored in dry places, including sheds, barns, or other covered storage. The moisture content of hemp stalks should not exceed 15%. When planted for fiber, yields range from 2-6 short tons (1.8-5.4 t) of dry stalks per acre, or from 3-5 short tons (2.7-4.5 t) of baled hemp stalks per acre in Canada.

Grain processing

3 Hemp seeds must be properly cleaned and dried before storing. Extraction of oil usually takes place using a mechanical expeller press under a nitrogen atmosphere, otherwise known as mechanical cold pressing. Protection from oxygen, light, and heat is critical for producing a tasty oil with an acceptable shelf-life. Solvent extraction methods are also emerging for removing oil since they achieve higher yields. Such methods use hexan, liquid carbon dioxide, or ethanol as the solvent. Refining and deodorizing steps may be required for cosmetics manufacturers.

4 A dehulling step, which removes the crunchy skin from the seed using a crushing machine, may be required. Modifications to existing equipment may be required to adequately clean the seeds of hull residues.

Fiber processing

5 To separate the woody core from the bast fiber, a sequence of rollers (breakers) or a hammermill are used. The bast fiber is then cleaned and carded to the desired core content and fineness, sometimes followed by cutting to size and baling. After cleaning and carding, secondary steps are often required. These include matting for the production of non-woven mats and fleeces, pulping (the breakdown of fiber bundles by chemical and physical methods to produce fibers for paper making), and steam explosion, a chemical removal of the natural binders to produce a weavable fiber. Complete processing lines for fiber hemp

have outputs ranging from 2-8 short tons/hour (1.8-7.2 t/hr).

Packaging

6 The primary fiber is pressed into a highly compressed bale, similar to other fibers like **cotton**, wool, and polyester. Other products, such as horse bedding, are packaged in a compressed bale.

Paper making

7 Bast fibers are usually used in paper, which are put into a spherical tank called a digester with water and chemicals. This mixture is heated for up to eight hours at elevated temperature and pressure until all fibers are separated from each other. Washing with excess water removes the chemicals and the extracted binding components (pectin). The clean fibers are then fed into a machine called a Hollander beater, which consists of a large tub equipped with a wheel revolving around a horizontal axis. This beating step, which lasts for up to 12 hours, cuts the fibers to the desired length and produces the required surface roughness for proper bonding. Bleaching chemicals are sometimes added during this step or to separate tanks with the fibers. The bleached pulp is then pumped to the paper machine or pressed to a dryness suitable for transportation to a paper mill at another location.

Quality Control

Hemp fibers are tested for tensile strength, fineness (fiber diameter), and the color is recorded. Moisture content is recorded during every stage of the growing and production process. The THC content of the plant is also contiguously tested to make sure that the level does not exceed the 0.3% mark. Research is still being conducted on the effects that hemp would have on the industry. Set standards are constantly being altered and changed.

Byproducts/Waste

The harvested hemp not used is burned. During fiber processing, the core fiber is saved and usually used to make paper, horse bedding, or construction materials. Most hemp producers recycle the core fiber by removing dust, then baling and packaging. The dust can be pressed into pellets used for fuel. The dirt and small chips of core are also used as a high nutrient soil additive.

The Future

Where it is legal, the hemp industry has been growing at an annual growth rate of 20%. Other potential uses are being developed. For instance, hemp meal has demonstrated it can be used as a food ingredient for aquiculture farms, specifically freshwater fish and shrimp. Even hemp beer has entered the Canadian market, though it is expected to remain a small part of beer sales. Composite materials for the building industry are also being investigated.

Using hemp as a source of food may become the largest application, since hemp seeds have much nutritional value. The seed contains essential fatty acids, protein, calcium, iron, zinc, and vitamins B, C, and E. Hemp seed can be made into oil or flour and can also be eaten whole, since it tastes similar to pine nuts or sunflower seeds.

The outlook for hemp in the United States is uncertain since it is still illegal to grow it. There are 10 states that passed legislation in 1998 to allow growing hemp for research purposes—Arkansas, California, Hawaii, Illinois, Minnesota, Montana, New Mexico, North Dakota, and Virginia—and a number of other states are considering it. However, federal law still prohibits growing industrial hemp. The Drug Enforcement Agency will have to change its mind before any market can be developed in the United States. Once that happens, hemp could become a billion dollar crop if there is enough investment and interest, prices are competitive, and high quality products can be made. Processing technology also needs to be upgraded for higher value-added products.

Where to Learn More

Books

Schreiber, Gisela.*The Hemp Handbook*. Munich, Germany: Wilhelm Heyne Verlag GMBH & Co. KG, 1997.

Periodicals

Adams, John. "Dope Idea: U. Minnesota Could Research Uses of Industrial Hemp." *Minnesota Daily* (March 30, 1999).

Anonymous. "Ag Study: Market for Hemp is Thin." *Dese Moines Register* (January 30, 2000).

Kane, Mari. "Hemp Industry Prepares to Grow." *In Business* (November/December 1999).

Katz, Helena. "Smoking Out New Hemp Markets." *Marketing* (November 22, 1999).

Nickson, Carole. "All-purpose Hemp a Retail Find." *Home Textiles Today* (November 15, 1999).

Sturgeon, Jeff. "Hemp-Gooods Shop Capitalizes on Plant's Versatility." *The Roanoke Times* (August 8, 1999).

von Roekel, Jr., Gertjan. "Hemp Pulp and Paper Production." *ATO-DLO Agrotechnology* (1994).

von Sternberg, Bob. "In Canada, hemp hasn't lived up to the hype." *Star Tribune* (October 16, 1999).

Ward, Joe. "Hemp Advocates Assail U.S. Report."*Courier-Journal* (January 26, 2000).

Other

Geofrey G. Kime, President, Hempline Inc. 11157 Longwoods Rd., Delaware, Ontario, Canada, N0L 1E0. (519) 652-0440. http://www.hempline.com. info@hempline.com.

North American Industrial Hemp Council, P.O. Box 259329, Madison, WI 53725-9329. (608) 258-0243. http://www.naihc.org. info@naihc.org.

Peter Dragla. *A Maritime Industrial Hemp Product Marketing Study.* Canadian Department of Agriculture and Marketing, 1999. http://agri.gov.ns.ca/pt/agron/hemp/hemp-masaf. htm (January 2001).

—*Laurel M. Sheppard*

Jawbreaker

Background

The jawbreaker is a type of hard, round candy that is ideally so difficult to bite down on that it must be sucked. Jawbreakers range from the size of a hazel nut to the size of a golf ball, and come in many flavors and colors. They are popular with children, and often sold in vending machines. Though originally a trade name, the term jawbreaker became so widespread that it is considered a generic name for any brand candy of this type.

History

Both written and pictorial records indicate Egyptians prepared sweets with honey, sweet fruits, spices, and nuts. Sugar was not known in Egypt, and the first written evidence of its appearance dates to A.D. 500 in India. The method of making sugar from the boiled syrup of the sugarcane plant spread from India through the Arab world, and sugar was introduced to Europe sometime around A.D. 1100 It was first thought of as a spice, and even up through the fifteenth century, sugar was so rare that it was used, for the most part, only medicinally, prescribed in minute doses by physicians. By the sixteenth century, widespread sugarcane cultivation and the technology for refining sugar developed sufficiently that sugar was not such a precious commodity. Small manufacturers produced crude candies in Europe at that time. The methods used were all simple, and produced the kinds of candies that could still be made at home today. By the late eighteenth century, entrepreneurs had developed candy-making machinery, and more complex candies were made and on a greater scale.

Candies are distinguished in broad categories by their hardness, and this corresponds to the temperature to which the sugar is heated. Sugar cooked at a low temperature results in chewy candy; medium heating results in a soft candy; and sugar cooked at a high temperature becomes hard candy, where the sugar is fully crystallized. The jawbreaker, being a type of hard candy, is similar to many candies popular in the United States in the mid-nineteenth century. These hard candies were generally sold singly. A storekeeper pulled out the desired number of pieces from a loose bunch in a glass case or jar. By the mid-1800s, there were close to 400 candy factories operating in the United States, turning out penny candy and other types.

The jawbreaker was made famous by the Ferrara Pan Candy Company of Forest Park, Illinois. The origin of the name, however, is obscure. The word jawbreaker first showed up in the English language in 1839, used to mean a "hard-to-pronounce word." Later, it was used as a slang or derogatory term for a dentist. Ferrara Pan was founded by an Italian immigrant to the United States, Salvatore Ferrara, in 1919. Ferrara came to the United States in 1900. Though he was a skilled confectioner, for years he worked various odd jobs, including as dishwasher and as a railroad foreman. Eventually, he saved up enough money to open his own pastry shop in Chicago in 1908. Among his products was a kind of sugar-coated almond known in Italy as *confetti*. These became so popular that Ferrara started a separate company to make them. In 1919, Ferrara teamed up with his two brothers-in-law, and founded the Ferrara Pan Candy Company. The new corporation focused on making candies in the hot pan and cold pan process. Ferrara Pan produced many well known confections, including Boston Baked Beans and Red Hots, as well as its

Though originally a trade name, the term jawbreaker *became so widespread that it is considered a generic name for any brand candy of this type.*

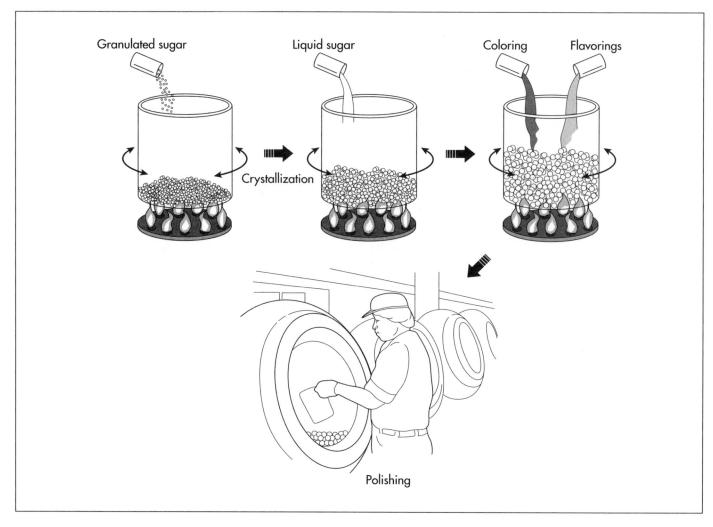

Granulated sugar   Liquid sugar   Coloring   Flavorings

Crystallization

Polishing

During the hot panning process, sugar is slowly crystallized into balls that grow into jawbreakers.

original Jaw Breakers. These candies became so popular that the earlier meanings of the term jawbreaker disappeared, and it began to be applied to all candies of this type. There are many manufacturers of jawbreakers today, though Ferrara Pan remains the leading maker of hot pan candies in the world.

Raw Materials

The crucial ingredient in the jawbreaker is sugar. All other ingredients form only a tiny percentage of the finished candy. Jawbreakers use natural and artificial flavors and a variety of artificial colors. Manufacturers may also add calcium stearate, a binding agent, and a wax such as carnauba wax, to provide a shiny, polished surface.

The Manufacturing Process

Jawbreakers are made by the hot pan process, and the type of pan used is very im-

portant. Candy-making pans are little like pans found in an ordinary kitchen. They are huge spherical copper kettles with a wide mouth. The pans rotate constantly over a gas flame so the sugar inside is kept tumbling. The worker who makes candy in using these pans is known as a panner.

Pouring the sugar

1 A worker puts granulated sugar into the pan while the pan heats over its gas flame. Each grain of sugar in the pan will eventually become a jawbreaker as it crystallizes, and other grains crystallize around it in a spherical pattern. The panner begins this process by filling a beaker with hot liquid sugar. Using a ladle, the panner carefully pours the liquid sugar into the pan along its edges. The liquid sugar adheres to the sugar grains, and the jawbreakers begin to grow. But this is a lengthy process. With the pans continually rotating, the panner keeps

adding liquid sugar at intervals over a period of 14-19 days. In total, the panner may add liquid sugar more than 100 times. The panner or another worker inspects the jawbreakers visually, to make sure the candies are growing perfectly round, and not lopsided.

Adding other ingredients

2 Most jawbreakers are colored only in the outer layers. The panner adds the color and flavor ingredients to the pan when the jawbreakers are almost their finished size. The coloring and flavoring are pre-measured into a small bottle or beaker, and the panner pours them in carefully along the edge of the pan. As the pan rotates, all the jawbreakers in the pan receive the coloring and flavoring equally.

Polishing

3 After approximately two weeks, the jawbreakers have reached their desired diameter, and they are removed from the hot pan to a polishing pan. This pan looks essentially the same as the hot pan. A worker pours the jawbreakers into the polisher and sets it to rotate. Food-grade wax is added, and coats each individual candy as the polisher revolves. After polishing, the jawbreakers are finished, and are now ready for packaging.

Measuring

4 The first step of packaging is to measure the jawbreakers into small batches. This is done by a measuring machine. A worker loads the finished jawbreakers onto a tilted ramp. All the different colors can be mixed together at this point, so that the small batches hold an assortment. The jawbreakers roll down and fall into the central chute of the measuring machine. From the chute, the candies fall into trays that are arranged on spiral arms around the central chute. Each tray will only hold a specific weight, for example one pound. As soon as the weight is reached, the tray swings out of the way and the next tray loads. As the top trays fill, the bottom trays dump into the bagging machine.

Bagging

5 Bagging is done automatically on a large machine that holds a wide spool of thin plastic on a revolving drum. The plastic is in a single layer at this point. The bagging machine forms the bags out of this material, fills them, and then seals them. The plastic may be imprinted with the logo of the candy manufacturer and any other necessary information. The machine unwinds a section of plastic from the roll and pulls it across a form that causes the plastic to fold lengthwise in two. Heated jaws press along the fold and melt the two sides together, forming the side seam. The folded plastic is then drawn upwards again, and another pair of heated jaws clamp the bottom, forming another seam. Now the machine automatically cuts the top of the bag and holds it open. The pre-measured amount of jawbreakers from the measuring machine drops in, and more heated jaws then clamp the bag shut along the top. The filled and sealed bags then drop onto a conveyor belt. Workers take them off the belt and toss them into packing boxes. At this point the jawbreakers are ready for distribution or storage.

Quality Control

Quality control is generally simple for jawbreakers. They are a relatively pure product, since they are close to 100% sugar. Workers rely on visual inspection to make sure a batch of jawbreakers is forming correctly. Since the process of making these candies takes about two weeks, and the pans are open, workers have many opportunities to observe the jawbreakers and see that they are shaped right. Each day, a worker may remove several jawbreakers from the batch in process and break them open. The crystalline structure inside should look like concentric rings. Workers also do a taste test. Making jawbreakers is a process that requires little technology, and quality control does not demand any elaborate chemical or physical analysis.

Byproducts/Waste

If quality control reveals any defective jawbreakers, they cannot be melted down and reused. Since the sugar is crystallized throughout the product, it would have to be ground down. So there may be a small amount of waste in the process, if a portion of the product has to be thrown out. Otherwise, the manufacturing process creates no byproducts.

Where to Learn More

Books

Broekel, Ray. *The Great American Candy Bar Book.* Boston: Houghton Mifflin Company, Inc., 1982.

Mintz, Sydney W. *Sweetness and Power.* New York: Penguin Books, 1985.

—*Angela Woodward*

Kaleidoscope

The kaleidoscope makes magic with light and mirrors. It may be considered a child's toy (or a toy for all ages), but it is also a simple optical device with technical applications for designers and pattern-makers. Greek words are the source of the name; it comes from *kalos*, *eidos*, and *skopios* meaning beautiful, form, and view, respectively.

Background

The body of the kaleidoscope has two main parts, the viewing tube (with an eyepiece at one end) and the object box or case at the opposite end of the tube. The object box is a thin, flat box made of two glass disks and a band circling the edges and holding the disks and the objects enclosed. Those objects are fragments of colored glass, beads, tinsel, or other reflective materials.

The outer disk of the object box is ground so it diffuses the incoming light; that is, it acts like a screen. The viewing tube has a glass eyepiece at one end; it may be ordinary glass or an optical lense with magnifying properties. Inside the tube, three strips of mirrors are joined to form a triangle; the angles of the mirrors also affects the view through the kaleidoscope. Typically, they are angled at either 45° or 60°. When the object box is turned or tapped, the glass or objects inside move and tumble freely. As the viewer looks through the eyepiece toward a light source, the mirrors produce symmetrical order out of the tumbling objects and multiply them six, eight, or more times depending on the angles of the mirrors.

A variation of the kaleidoscope—the teleidoscope—replaces the object box with another lens that allows the viewer to look at a distant object and view it in multiples. Still other variations use more mirrors. Two mirrors have the advantage of producing a centered pattern; multiple mirrors split and duplicate the image many times over. The kaleidoscope is infinitely entertaining because the patterns and combinations are endless and are not permanent unless photographed.

History

Although the ancient Greeks, including the mathematician Ptolemy, had contemplated the effects of abutting multiple mirrors, the kaleidoscope is the creation of one man. David Brewster (1781–1868) was born in Scotland and educated to become a minister at the University of Edinburgh. University studies exposed him to the wonders of science, however, and he abandoned the church in favor of studying the properties of light. He became an expert in polarization of light (the linear and planar properties of light), reflection of light using metal, and light absorption. For his scientific discoveries, Brewster was elected a fellow of the Royal Society (Britain's leading scientific organization) in 1815 and knighted in 1831.

Brewster invented and patented the kaleidoscope in 1816. He described its structure and operations in a 174-page scientific paper titled *Treatise on the Kaleidoscope*. In his treatise, Brewster calculated that 24 fragments of glass in the object box of a kaleidoscope could create more than 1.4×10^{33} fleeting views. He also described the most effective combinations of colors for kaleidoscopes based on light properties. In the 1840s, he used two lenses to produce a three-dimensional effect in creating the stereoscope. He was also a leading advocate of the flat Fresnel lens adopted by the British for lighthous-

David Brewster calculated that 24 fragments of glass in the object box of a kaleidoscope could create more than 1.4×10^{33} fleeting views.

es and was credited with saving thousands of lives by protecting vessels against shipwrecks. Brewster taught at the University of Edinburgh and the University of St. Andrews in Scotland, was one of the first editors of the *Encyclopedia Britannica,* and published many books and scientific papers.

Following its invention in 1816, the kaleidoscope grew in popularity around western Europe, and the first one to appear in the United States was reported in 1870. It became a favorite toy for children but also an entertainment for adults in parlor games like viewing stereoscopic photographs and playing charades. The most famous kaleidoscopes, other than Brewster's originals, were made by Charles G. Bush of Boston. The Bush kaleidoscope was constructed of a viewing tube of banded black cardboard, a **brass** wheel to turn the object box, and a wooden stand. It was the objects that made Bush's version unique (and a valuable collectible today). He used 35 objects of various colors and shapes, but some were filled with liquid containing air bubbles. The air bubbles moved through the liquid even when the observer held the object case still. Bush secured the patents for the liquid-filled objects (ampules), for his method of adding and subtracting objects without taking the box completely apart, and for stands and other kaleidoscope accessories.

As a tool for designers, the kaleidoscope produces ranges of colors and patterns used to create rugs, stained glass, jewelry, architectural patterns, wallpaper, woven tapestries, and ideas for painters. The kaleidoscope fell out of popular interest in the early twentieth century, but it revived in the late 1970s when new styles and the collectible character of antique kaleidoscopes fired the curiosity of new generations. Bush's kaleidoscope with the liquid-filled objects sold in 1873 for $2.00; collectors in the early twenty-first century willingly pay over $1,000.

Raw Materials

The types of materials that can be used to make kaleidoscopes are almost as endless as the images its mirrors produce. The viewing tube can be made of paper, cardboard, plastic, acrylic, wood, plexiglass, brass, copper, sterling silver, and other metals and materials. The endcap containing the eyehole is made of material that is compatible with the viewing tube. The eyehole or peephole and the two faces of the object case are usually made of plastic or glass. The objects in the box can be fragments of rock or minerals, gemstones, beads, glass or plastic ampules (filled or unfilled), shells, bits of glass, bits of metal, tiny trinkets, or any combination of these. The objects can be chosen based on similarity or variety in color. Inside the tube, mirrors are essential for forming the images. Tape holds them together, and some kind of padding or stuffing like newspaper, **cotton,** or styrofoam keeps the mirrors from rattling against the inside of the tube. Tape can also be used to hold the endcaps and object box in place. Other connectors or fastening materials, attachments, and trims can be made to match the look of the kaleidoscope.

Design

The designer of the kaleidoscope chooses the size, materials for the case, orientation of the mirrors, type of object case or rotating wheel, and objects to shape the views. All of these choices affect the price of the kaleidoscope (to produce as well as purchase) as much as the kind of images the designer is trying to create. Miniature kaleidoscopes that can be attached to a key chain to those as tall as a person have been made. Cases can be manufactured of simple materials like paper, plastic, and wood; many varieties of metals are chosen from brass to sterling silver and gold-plated models. If the kaleidoscope end is a rotating wheel, that wheel may be made of gemstones, stained glass, thin slices of agate and other minerals and rocks, and more. Object boxes may be filled with crystals, glass containing embedded threads (laticcino glass), baubles, gemstones, seashells, chips of glass, or liquid-filled capsules (ampules). Kaleidoscopes can also use light sources or light filters other than natural light. Electronic scopes, oil-suspension scopes, polarized light scopes, and projector scopes are examples.

Designers may have an individual style, a material they prefer to work with, a particular type of image or view they want to create, or objectives related to pleasing their customers. Some are made to be ideal gifts, and some are unique creations for collectors. Again, many issues and ideas may motivate the kaleidoscope designer. Internationally known designer Carolyn Bennett makes kaleidoscopes from acrylics. Her viewing tubes are often square or rectangular and look like interesting sculptures from the outside as well as producing gorgeous images within. To keep costs down, she uses stock sizes of plastics and other materials, but the design always reflects the environment or the character of the collector or designer over the ease of manufacture. About half of her designs are made to suit the colors and budgets of customers that include museums and corporations; her designs made for stores suit her artist's eye with some consideration for engineering aspects and constructability.

Kaleidoscopes have generated a vocabulary describing their light-adjusting capabilities, materials, and construction. Dichromatic glass makes different colors depending on the angle light strikes it. Flashed glass is a mirror made of two colors, one overlying the other. The oil-suspension scope contains oil in the object case along with bits of glass or other materials that float in the oil. Slumped glass is heated until it bends; pieces of slumped glass are held in the object box of some designs. Hot glass is scrap glass that is heated until it fuses and then painted. The teleidoscope (combination of a telescope and kaleidoscope) replaces the object case with another lens so the tree or bird viewed through the teleidoscope is reflected in multiples. These terms and many others characterize the variety available in kaleidoscopes.

The Manufacturing Process

The kaleidoscope described in this section is a simple version, and many of the materials are not itemized in detail. As the sections above suggest, the choice available to budding kaleidoscope designers is almost without limits; and, although some kaleidoscopes (especially toys) are mass-produced, the "manufacturing" of a kaleidoscope is creative and artistic.

1 The inner diameter of the tube determines the size of the mirrors that will be inserted. The tube is selected or made, and the inner circle is drawn on a piece of paper. A compass is used to bisect the circle and subdivide it into six equal parts to measure and plan a system of three mirrors forming an equilateral triangle (a triangle with sides of equal length and internal angles that are all 60°). By connecting every other dot out of the six subdivisions of the circle, the outline of the three mirrors can be drawn and the width of the mirrors is measured. About 0.13-0.25in (0.32-0.64cm) is subtracted from the width of each mirror to allow for the thickness of the mirror. Three pieces are used to create a clear image, although the maker can choose many other configurations. The lengths of the mirror are equal to the length of the tube minus the space at the end of the tube for the object chamber. On the order of 0.5-1in (1.27-2.54 cm) should be allowed for a basic, hand-held kaleidoscope.

2 The kaleidoscope maker can cut his or her own mirrors or have them cut at a glass shop. First-surface mirror with the silvering on the surface is usually chosen. Next, the mirrors are taped together to fit inside the viewing tube. This is done by fitting the mirrors together with one hand imitating the tube and with the other free to tape. The mirrors are aligned so the edge of one mirror is seated on the surface or face of the adjacent one, which in turn has its edge seated on the face of the third. If the mirrors are connected edge to edge rather than edge to face, the image will be skewed. With the three mirrors held together, tape is wound around the outside in a spiral.

3 The endcap with the eyehole is made by cutting out a circle as large as the outer diameter of the viewing tube from a material that matches or is compatible with the material forming the tube. A concentric, smaller circle is cut out of the end cap, and a piece of plastic is taped or glued to hold it over the opening on the inside of the cap. This hole can be left uncovered, but a clear covering is a good safety precaution to keep any objects from the box or the inside of the tube from falling into the viewer's eye. Acrylic, mylar, acetate, or an optical lense can be used for the peephole. The endcap can then be carefully glued or taped to the end of the tube; in more elaborate designs, these are formally made into fittings that can be screwed on or specially mounted to the tube.

4 With the viewing end now closed, the maker slides the assembled set of mirrors into the tube. If the triangle of mirrors rattles

Differing angles of the mirrors create different views through the kaleidoscope.

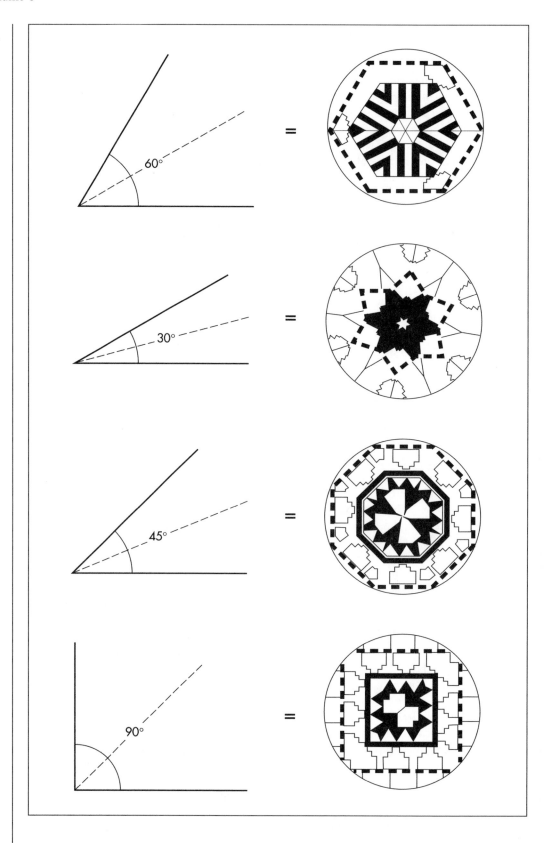

in the tube, it is taken out, and paper, fabric, or other protective material is wrapped around the outside of the mirrors until they fit securely. Plastic styrofoam "popcorn" or pieces of paper can be pushed down alongside the mirrors to secure them, but they are likely to be off center, which, again will skew the image. The mirrors are pushed in to abut the endcap and to leave space at the other end of the tube for the object box.

5 The drawing of the inner diameter of the tube that was made previously van now be used to make plastic pieces for the two ends of the object box. A circle of clear plastic is cut out and inserted in the tube to fit against the ends of the mirrors; it must fit so that it is perpendicular to the tube walls. The distance between this piece of plastic and the open end of the tube is the thickness of the chamber. After measuring this thickness, the maker cuts a piece of cardboard that is as wide as the chamber thickness and as long as the inner circumference of the tube. The maker softens the cardboard by working it between the fingers so it can be wound around the inside of the tube. It should be tested to fit smoothly, then taken out, glued, and affixed in the tube. This cardboard piece holds the inner plastic end of the object box in place, and it also smooths out any gaps in that plastic piece so they can't be seen. The outer plastic circle is then cut. This may also be made of clear plastic or it can be sandpapered to make the surface translucent. If it is left clear or transparent, part of the outside world will show through the object case. A translucent finish will blur the incoming light (and its image) so the inner beads are the objects in focus.

6 The kaleidoscope maker can now begin adding beads or other materials to the object case. As the descriptions above suggest, almost any colorful items can be chosen as the objects. The maker adds beads or objects to the case and peeks through the viewing end to check the color balance. When the right combination of objects is selected, the frosted end circle is taped or glued in place. For this simple kaleidoscope, the object case is not twisted relative to the viewing tube. Instead, the whole tube is rotated to change the images.

Byproducts/Waste

Kaleidoscope-making generates little waste. The manufactured versions are made to strict specifications so tubes and mirrors are cut to those standards. Some materials (particularly plastics) can be reground if they are faulty and recycled. Some wastage occurs in the beads and items used for the object cases. A company like C. Bennett Scopes, Inc., takes spare beads to schools for use in art projects. The majority of waste at a kaleidoscope factory comes from packing boxes that are also recycled.

Quality Control

Safety for employees is a critical issue. If plastics or acrylics are used, acrylic solvents are necessary, and adequate ventilation must be provided around each worker in accordance with regulations by the Occupational Health and Safety Administration (OSHA) and the Environmental Protection Agency (EPA), both Federal agencies. Other safety issues involve equipment mechanics; bandsaws may be used particularly to cut parts for prototypes, and a sonic welder that is noisy and is used to seal liquid chambers is operated inside a box insulator. Plastic and lengths of mirror have sharp edges, and care needs to be taken in handling them.

Regulations concerning the quality of the mirrors and lenses are also checked. The lens must be firmly attached to guarantee that no objects are able to fall into the eye. Also chips or fractures in the mirrors must be replaced with new materials. The object box is also checked to make sure that any ampules are not leaking.

The Future

In the early 1970s, kaleidoscopes experienced a rebirth in interest that then had smaller peaks and valleys in the following 30 years. In the twenty-first century, kaleidoscopes seem to have a well-established following as an art form, potential gifts, and objects of curiosity for children of all ages. Conventions held by the Brewster Society are an excellent measure of the kaleidoscope's well-being; over 30 shops, countries including Japan and Switzerland, and hundreds of manufacturers and aficionados meet annually to support Sir David Brewster's invention and exchange ideas. Society members, like other hobbyists and enthusiasts, firmly believe in the infinite variations that are possible in the kaleidoscope which they call "candy for the eye."

Where to Learn More

Books

Baker, Cozy. *Through the Kaleidoscope... And Beyond.* Annapolis, MD: Beechcliff Books, 1987.

Baker, Cozy, and Sara Macfarland. *Kaleidoscopes: Wonders of Wonder.* Concord, CA: C & T Publishing, 1999.

Bennett, Carolyn, and Jack Romig. *Kaleidoscopes.* New York: Workman Publishing, 1994.

Boswell, Thom, ed. *The Kaleidoscope Book: A Spectrum of Spectacular Scopes to Make.* New York: Sterling Publishing Company, Inc., 1992.

Newlin, Gary. *Simple Kaleidoscopes: 24 Spectacular Scopes to Make.* New York: Sterling Publishing Company, 1996.

Periodicals

Andrews, Jeanmarie. "The Kaleidoscope." *Early American Homes* 28, no. 6 (December 1997): 35.

Kripalani, Manjeet. "A Rather Strange Object." *Forbes* 150, no. 13, (December 7, 1992): 232.

Novak, William. "Confessions of a Kaleidoscope Collector." *Forbes* (May 17, 1999): 348.

Other

The Brewster Society. P.O. Box 1073, Bethesda, MD 20817. http://www.kaleido.com/brewster.htm.

C. Bennett Scopes, Inc. 609 West State Street Media, PA 19063. (800) 272-6737. info@cbennettscopes.com. http://www.cbennett scopes.com.

—*Gillian S. Holmes*

Litmus Paper

Litmus paper is the most recognized member of chemical indicators. Like most pH paper, litmus changes color when exposed to an acidic or basic solution. The simple pH scale ranges from 0-14 with 0 being the most acidic, 7 being neutral, and 14 being the most basic or alkaline. Litmus paper is commonly used in educational science classes. Because it has such wide recognition, it has become a cultural reference in our society as well. It is common to use the term litmus test when referring to a test in which a single factor determines the outcome.

Background

Litmus paper allows an observer the opportunity to assess a sample's pH. pH is a way to characterize the relative acidic or basic nature of a substance based on its hydrogen ion concentration. An ion is an atom that carries an electrical charge and is therefore reactive with its environment. An acidic substance releases hydrogen ions (H+) in water. Acids are known as proton donors because the H+ ion has one extra positively charged proton trying to stabilize itself by combining with a negatively charged ion. A basic substance releases a hydroxide ion (OH-) in water. Bases are called proton acceptors because the hydroxide ion will accept a proton to stabilize itself. Interestingly enough, when acids and bases are combined, the result is a neutral salt. For example, a strong acid like hydrochloric acid combined with sodium hydroxide (a strong base) results in a neutralization reaction with the byproducts sodium chloride (table salt) and water.

pH is an important biological indicator because most life forms have a very small range of pH in which they can survive. For example, the acid-base ratio in the human body is a delicate balance. Even a slight change in the blood's pH in either direction can result in death. Plants are also susceptible to minute pH changes in the soil. That is why soil that is too acidic for a plant is neutralized with calcium carbonate fertilizer, a base.

The simple pH scale ranges from 0-14 with 7 being neutral. Numbers less than 7 are considered to be acidic and numbers greater than 7 are considered basic. The smaller the number the more acidic the solution. This means that a substance with a pH of 1 would have a greater ability to donate a proton to another molecule or ion than a substance with a pH of 4. For instance, sulfuric acid is very effective at transferring a hydroxide ion, while acetic acid (vinegar) is not. Therefore, sulfuric acid is considered to be a strong acid and acetic acid is considered a weak acid. Similarly, there are also strong and weak bases. A strong base like potassium hydroxide, with its more abundant hydroxide ions, will more readily accept protons than a weak base like ammonia. The greater the number, the stronger the base.

While litmus paper is effective at indicating whether a substance is acidic or basic, it cannot report an exact numerical pH value. Universal indicators or pH meters are used for this purpose. Universal indicators are composed of a variety of materials, each changing different colors at different pH values which allows the observer to determine more precisely where the solution in question falls on the pH scale. Universal indicators can be impregnated onto paper and made into pH paper or they can be used in the liquid form. A reference color card is provided with each universal indicator that

The simple pH scale ranges from 0-14 with 0 being the most acidic, 7 being neutral, and 14 being the most basic or alkaline.

correlates a particular color with a pH range. Generally speaking, most universal indicators are accurate to within two values on the pH scale. For example, a green result could indicate a pH from 8-9. This means universal indicators can determine the pH of a sample quantitatively within a certain range.

pH meters allow for even more precise quantification by using electricity to determine a numerical pH value. A probe is put in the test sample and a current of electricity flows through the probe. Since electricity is composed of electrons, which have a negative charge, the force of current flowing through the meter is directly proportional to the hydrogen ion concentration. The more H+ ions in the solution, the more current will flow through the meter. This number is then converted into a numerical pH value that can be read by the observer.

History

The term litmus comes from an Old Norse word meaning "to dye or color." This is fitting since the lichens used to make litmus have also been used to dye cloth for hundreds of years. Very little information is available about the beginnings of litmus. There is some data that suggest that litmus paper was developed by J.L. Gay-Lussac, a French chemist during the early 1800s. Gay-Lussac is best known for his Law of Combining Volumes, which states that whenever gases are formed or react with one another at a constant temperature and pressure, their volumes are in small whole number ratios. In other words, when gases combine, they always do so in the same way provided that the temperature and pressure stays the same.

Raw Materials

The primary raw materials used for making litmus paper are wood cellulose, lichens, and adjunct compounds. Litmus paper, as its name implies, is primarily composed of paper. The paper used to make litmus paper must be free of contaminants that could change the pH of the system it is measuring. Like most paper, litmus paper is made from wood cellulose. The wood is treated with solvents prior to paper manufacturing in order to remove resinous material and lignin from the wood. One of the most common

solvents in the United States is a sulfate—either sodium sulfate or magnesium sulfate.

The ability of litmus paper to change color when exposed to an acid or base is a result of litmus paper being infused with lichens. In the plant world, lichens are unique in that they are actually two distinct organisms, a fungus and an alga, living as one. Botanists classify lichens as fungi because it is the fungi that are considered to be responsible for sexual reproduction. However, each lichen has its own distinct name. Approximately 15,000 different types of lichens have been identified. Lichens can be found growing on rocks, trees, and walls, in the soil and even under water in virtually all types of climates. Lichens are commonly used as gauge for environmental quality because they are sensitive to various pollutants. Several varieties of lichen are used to produce litmus including *rocella tinctoria*, native to the Mediterranean, and *lecanora tartarea*, a common lichen in the Netherlands. In fact, the Netherlands is one of the largest producers of litmus paper products.

Design

Most litmus paper and other types of pH indicators are sold through scientific supply houses. Litmus paper is available in both red and blue varieties. The natural color for litmus paper is blue. When put in an acidic solution the blue paper turns red. Red litmus paper is first mixed with an acid when it is made. This causes the paper to appear red. When put in the presence of a base, the paper returns to its natural blue color.

The Manufacturing Process

The production of litmus paper has many features in common with paper manufacturing. In this process, the wood pulp is converted to paper, the paper is infused with the lichen solution, and the paper is dried and packaged.

Converting wood pulp

1 In this first step, wood is shredded and mixed with a solvent and water under steam pressure. The resulting mass is called wood pulp. The pulp is spread on a belt of wire mesh and passed over rollers. This cre-

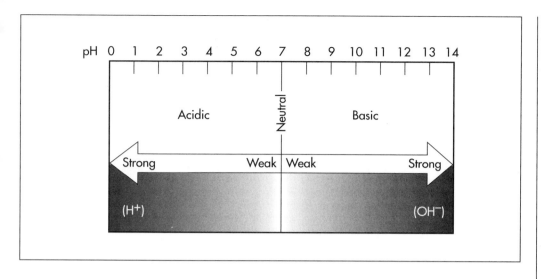

ates a thin layer that can be dried and made into usable paper. A wooden box beneath the belt catches water runoff. Since there is salvageable fiber in this water, it is re-mixed with the pulp.

2 To speed up the drying process, air suction pumps beneath the belt are used. As the paper moves along the belt it passes under a mesh or wire cylinder called a dandy roll. The purpose of the dandy roll is to give the paper a weave or watermark that identifies the paper grade and the manufacturer. The paper continues on its journey down the conveyor belt where it is pressed between two felt-covered rolls (couching rolls) that force the fibers to bind together by expressing out additional water.

3 From these couching rolls the paper is pushed through two sets of smooth metal press rolls leaving the paper with a smooth finish. The paper is completely dried by subjecting it to heated rollers, cut by revolving cutters and wound onto reels.

Infusion of lichens

4 To make the paper pH active, it is then infused with an aqueous solution consisting mostly of lichens. This is done by running the paper through a bath of the solution. It absorbs the solution and is then passed to allow it to ferment and dry.

5 The lichens are allowed to ferment in the presence of potassium carbonate and ammonia. After fermentation, the mass has a blue color and is then mixed with chalk. Blue litmus paper is prepared by impregnat-

ing white paper in an infusion of the litmus mixture mentioned above. The paper is then carefully dried in open air. Red litmus is similarly prepared but a small percentage of sulfuric or hydrochloric acid is added to cause it to turn red.

Packaging

6 After the paper is prepared, it is sent to a final packaging station. Litmus paper is typically sold in pre-cut strips. Manufacturers place the strips in re-sealable plastic vials. It is important that the packaging prevents the strips from becoming exposed to moisture since any liquid that comes into contact with the litmus paper could cause the indicator to change color. It is possible, although not as common, to purchase litmus paper in rolls that can be cut by the user. Manufacturers also supply written directions with every package of litmus paper so that the user will know how to use the product correctly. In the case of universal indicator papers or solutions, written instructions are not enough. A color reference card is also supplied so the user can match the test result to the reference card to determine the pH.

The Future

Litmus paper will most certainly continue to be used extensively in education due to its reasonable cost and ease of use. However, some varieties of lichens are becoming extinct. As a result, it is possible that manufacturers of litmus paper may switch to synthetic materials in the future. This is already being done by manufacturers of other types of pH papers. Additionally, because litmus cannot give quantitative results, it cannot replace other pH papers and pH meters. In fact, the trend is to make pH indicators that are even more accurate and less subjective. One such trend is to utilize fiber optic probes in pH meters in order to make them even more sensitive.

Where to Learn More

Books

Brady, George S. *Materials Handbook.* 14th Edition. New York: McGraw Hill, 1997.

Daub, William G., and William S. Seese. *Basic Chemistry.* 7th Edition. Upper Saddle, NJ: Prentice Hall, 1996.

LaRoe, Edward T. *Our Living Resources.* Washington, DC: U.S. Department of Interior-National Biological Service, 1995.

"Lichen." In *Van Nostrand's Scientific Encyclopedia.* 8th Edition. New York: Douglas M. Considine, 1995.

Other

Botanical.com. http://www.botanical.com (January 2001).

Hanna Instruments Online. http://www.hannainst.com (January 2001).

Kiwi Web Chemistry and New Zealand. http://www.chemistry.co.nz (January 2001).

Precision Labs. 9889 Crescent Park Drive Westchester, OH. (513) 777-3034.

—*Sandy Delisle and Perry Romanowski*

Lollipop

Background

Lollipops, or suckers as some call them, are essentially hard candies with a short stick of some sort. The tightly wrapped white paper stick serves as a handle, and the hard candy lollipop is either sucked or bitten apart until consumed. Lollipops take an astonishing array of forms. There are the very small and popular "Dum Dum" lollipops with fruit and other flavors; the Tootsie Pop—slightly larger and filled with a chocolate chewy center; the Blow Pop with its gum center; and very large suckers that take all day to eat—such as those often found at circuses and carnivals. No matter what size, the lollipop is made primarily of sugar, water, corn syrup, and flavorings.

Lollipops are not complicated to make and do not really require special equipment for home production. Sugar-corn syrup solutions are cooked until the concentration of the solution reaches a high level, and this supersaturation of sugar remains upon cooling. A gas or electric stove may be used, with the temperature monitored using a hand-held candy thermometer until it reaches 310°F (154°C), or what is referred to as the *hard crack stage*. When the concoction is hot (and it is very hot—hot enough to severely burn the skin) it is plastic or malleable, and may be poured into molds that can be purchased in a variety of shapes. As the solution cools, it takes the shape of the mold, becoming "glass-like," as it may be broken or cracked like a piece of glass. The home lollipop-maker may add any desired colors or flavorings just before the lollipop is poured into the mold. Colorings and flavorings are commercially available. Recently, molds for domestic lollipop-makers have been developed and are easy to obtain.

While lollipops may be made at home, most people purchase inexpensive suckers at a local store. They are beloved by both adults and children. Children love them for their sweetness and novelty. Adults have increasingly turned to them to kick addictions to nicotine, because the motion of taking the sucker in and out of the mouth mimics the motion of the hand when smoking. Manufactured lollipops are consumed in huge quantities. Spangler Candy Company produces over one billion Dum Dum suckers a year, and the world's largest lollipop maker, Tootsie Roll Industries, turns out 16 million lollipops per day.

History

It is difficult to know when lollipops were first made by home chefs. Charles Dickens refers to candies on a stick in his novels of the mid-nineteenth century. These sweet hard candies were sometimes put on the end of pencils and sucked on and were popular around the time of the American Civil War. Older cookbooks make it clear that these *lollypops* were frequently made at home as hard candies that were simply dropped onto wax paper in globular form, with a wooden stick inserted into the hot syrup until set. No molds were necessary and thus the lollipop forms were rather haphazard. We cannot be certain which company first began to mass produce these confections. However, it is known that George Smith, a candymaker who liked to eat a competitor's chocolate caramels on a stick, attached a hard candy to a stick and referred to this creation as a *lollypop* (named after a favored racehorse of his named Lolly Pop).

Dum Dum Lollipops were first manufactured by the Akron Candy Company in

Spangler Candy Company produces over one billion Dum Dum suckers a year, and the world's largest lollipop maker, Tootsie Roll Industries, turns out 16 million lollipops per day.

1924. Apparently, even at that early date marketers were wise to the fact that the name meant everything—Dum Dum was believed to be a name that any kid could say, and ask for by name. The Spangler Candy Company purchased that company in l953 and continues to expand the line. By l931, Tootsie Roll Industries had inserted their chewy Tootsie Roll into the center of the traditional lollipop, which is also still going strong. Refinements and variations on the traditional lollipop are myriad. Some have jawbreakers embedded in them, gum in the center, sour centers, sizzling candies inserted within, and a new twist within the last year is the lollipop inserted into a radio that turns on only when the lollipop is sucked on. Sugar-free suckers are now produced, too, in order to help limit tooth decay. An interesting innovation to the traditional sucker is the manufacture of flexible cellophane strips in place of the stiff paper stick in order to prevent puncture of the child's mouth. Other developments have included lollipops impressed with Halloween or other holiday motifs, so that some kinds of lollipops are manufactured only in season.

Raw Materials

The ingredients used in the production of lollipops varies by manufacturer. The ingredients in a plain, hard-candy lollipop with no special center include: water, sugar, corn syrup, flavorings (both natural and artificial), and malic or citric acid. The paper sticks are generally constructed using tightly-wrapped bright white paper that has been bleached and coated with a fine layer of wax. Wrappers vary in style. Some are clear cellophane, while others are made of printed and waxed paper.

The Manufacturing Process

1 Lollipops are made primarily from sugar and corn syrup. Most manufacturers produce lollipops in enormous quantities and the raw materials are brought into the factory in bulk. First, dry sugar is brought into the plant in huge rail cars called gondola cars. These cars are attached to a liquidizer. The sugar drops down into 180°F (82°C) water and is dissolved. This is a lengthy process—it may take about nine hours to melt 180,000 lb (81,650 kg) of sugar (a typical amount brought in on the rail car). Once the sugar is dissolved into the hot water, the sugary water is pumped into the pre-cooker.

2 Corn syrup, a key ingredient in the manufacture of lollipops, is also delivered by huge trucks and pumped, in liquid form, into the pre-cooker. Eventually, there is 55% liquid sugar and 45% corn syrup in the pre-cooker waiting to be heated.

3 Once the corn syrup-liquid sugar mixture is in the pre-cooker, the solution is heated to around 228°F (109°C). The pre-cooker is essentially a set of coils through which the sweet slurry is sent. The coils are heated by steam.

4 This heated syrup is then pumped from the pre-cooker and sent to the final cooker where it is cooked under a vacuum for about four minutes to a temperature of about 290°F (143°F). The vacuum is essential to this phase as it removes the moisture and heat from the candy. Large final cookers can cook up to 150 lb (68 kg) of candy in a batch.

5 After the candy is cooked and just before it is mixed, color, flavor, and citric acid or malic acid are added. (One flavor of lollipop is made at a time.) These flavors and colors are in liquid form and have been carefully pre-measured in a vial before being added, by hand, to the candy batch. Citrus acid and malic acid are extremely important to the flavor of these pops. Citrus acid promotes the flavor of the citrus-based flavored lollipops and cuts the excessive sweetness as well. Malic acid is used to enhance the flavor of non-citrus flavors. The candy batch, now with flavor and color added, is thoroughly mixed using two huge arms that push the candy around and lift it up, mimicking human kneading. Mixing not only thoroughly distributes flavor and color but reduces temperature and removes air bubbles produced from cooking and mixing. Human touch (using clean gloves) is essential in order to feel the batch to ensure that the candy is at the right consistency and the right temperature to undergo the next step, extrusion.

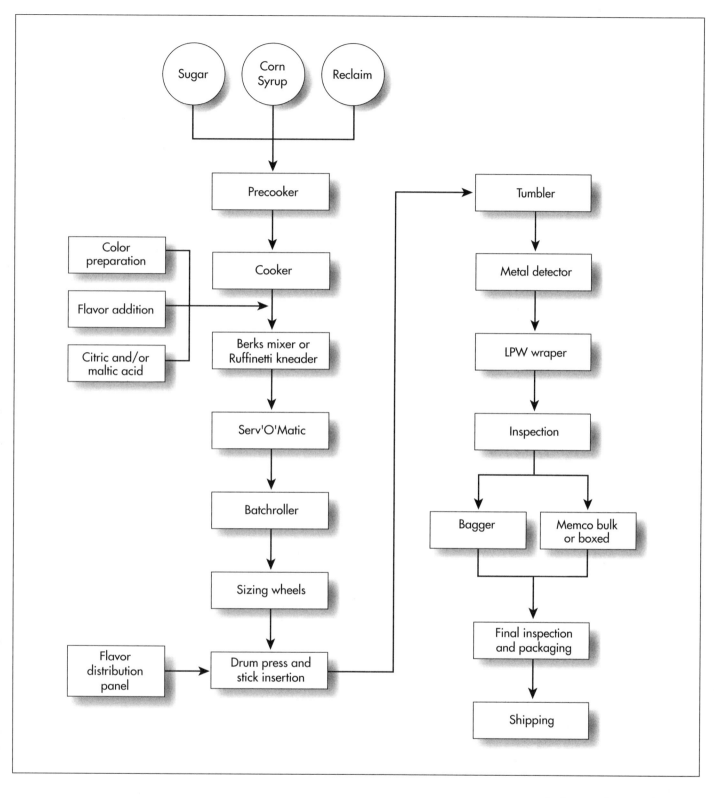

A diagram depicting the manufacturing steps involved in making lollipops.

6 Now the candy is made ready to be formed into a lollipop. The cooling candy is sent to the batch roller, which extrudes the candy through cones and rolls it into a fat rope. Then, the batch is sent to the sizing wheels, which reduces the rope to a smaller size.

7 The candy rope is then sent to the forming machine, which presses the heads into a spherical shape and inserts a paper stick at the same time. The machine is capable of forming 1,000 pops per minute. The lollipops are then cooled in a cooling drum that rotates slowly while the pops are ex-

posed to cool air. After four minutes, they are at room temperature and come out one end of the cooling drum.

8 After the lollipops cool, they are moved by conveyor belts and dropped into the wrapping machine. The pops are automatically wrapped and sent to either a boxing operation or a bagging operation, depending on how they are to be retailed.

9 Some companies purchase their sticks for the pops. Others manufacture their own. Manufacturing of the tightly wrapped sticks often occurs at the same plant as the candy manufacture. First, a roll of paper goes into a cutting machine that cuts 2.5-in (6.4-cm) strips that are 15 in (38 cm) long. Then, the strips move into a large drum in which water is applied. The paper is then rolled like a pencil, and the roll becomes tighter and tighter. It is finally cut, dried, and waxed with a fine coat so that it will not dissolve when exposed to human saliva.

Quality Control

There are two places for quality control—in the laboratory and on the floor of the plant. The labs check the quality of all raw ingredients sent to the manufacturing floor. They check the sugar quality and make sure they have what was ordered. The chemists perform heat tests on the corn syrup, since poorly processed syrup turns brown and can ruin the color of the lollipop. The laboratory also pulls samples of the candy batch from the cooker and analyzes the moisture content, because too much corn syrup will make the candy too malleable and it will melt in warmer weather. Flavors and colors are carefully checked and tested, and measured precisely for inclusion in the batch.

In the factory, operators ensure that the machinery is clean and running properly. Some machines turn off the processes at certain temperatures or when the batch reaches a certain weight, so these machine-tripping devices must be carefully maintained as well. Operators use their eyes to discard heads that are not properly filled out (a problem in pressing the head) or whose sticks are not properly inserted. Random checks on suckers are performed regularly as well.

Byproducts/Waste

Any candy that falls onto the floor cannot be consumed by humans. It is usually sent to a landfill. Candy that is determined to be inferior in shape or color is generally ground up, melted down, filtered, and pumped back into the pre-cooker so that it may be re-used in a lollipop.

The Future

While each company produces new flavors in different ways, one company revealed how it actively develops new flavors that sell well immediately upon introduction. New flavors are developed using a *hedonic taste panel,* which tries a variety of potential new flavors. Each member of the panel ranks these possible products in a variety of categories based on whether they like or dislike it extremely. Once the potential flavors are narrowed down to a few, trained panels of tasters are asked to test them again as a group. Then, decisions are made based on the results of the two panels.

Where to Learn More

Other

Spangler Candy Company. http://www.spangler candy.com (December 19, 2000).

Tootsie Roll Industries, Inc. http://www. tootsie-roll.com (December 19, 2000).

—*Nancy E. V. Bryk*

Matryoshka Doll

Background

The matryoshka doll is a symbol of Russia and its culture. It is truly a doll—a child's plaything—but it began its history just over 100 years ago as a highly collectible art form. The matryoshka doll (or, simply, the matryoshka) is a nested doll with two halves that can be pulled apart. The outer figure contains increasingly smaller versions of itself. The largest figure is usually on the order of 2-12 in (5-30 cm) tall, although larger ones up to several feet tall have been made. And the smallest may be very tiny—less than 0.25-in (0.6-cm) tall.

The painted image on the dolls is most often a woman wearing traditional Russian costume. The woman is a mother; the names Matryona and Matryoshka were common Russian country names for generations. Both come from the Latin root *mater* for mother. So matryoshka has come to mean "little mother" based on the idea that the outer or largest doll holds her babies inside like an expectant mother and that each daughter in turn becomes a mother. They are symbols of fertility and motherhood and have a modified egg shape.

From the largest doll to the smallest in a set, each resembles the others, but they are not necessarily identical. The outer doll may wear a costume that is red, the next one green, the third blue, and so forth. Or the costumes may be the same, but each doll may carry something different in her hands. For example, the outer doll may hold a loaf of bread (a symbol of welcome in Russia), the next may carry a bowl of salt (representing welcome and the family's offering of its wealth to guests—salt was once very rare), the third doll may hold several large beets (a traditional Russian vegetable symbolizing the richness of the earth), and a fourth may carry a basket of strawberries (for the sweetness of the garden).

Flowers are one of the most traditional themes with particular flowers representing the cities where the dolls are crafted; usually, the flowers are painted as designs on the shawls and aprons of the matryoshka. The most highly prized artistic collectibles may not have faces; instead, they tell a story, perhaps of a Russian fairytale, all around the exterior. A different scene from the tale appears on each nest; stories are also told in the apron panels of traditional doll styles. The sets of nested dolls may include as few as three or as many as 25 nests or dolls; historically, sets containing up to 1,800 dolls are known. A typical set contains three to twelve dolls.

Souvenir and toy matryoshka also depict many other kinds of images other than the traditional Russian mother. Sets have been made showing great Russian leaders (from Vladimir Putin, the Russian president elected in 2000, back to the czar Peter the Great), household pets (with the dog usually the largest and a cat, bird, fish, and mouse inside), a traditional Santa Claus (called Saint Nicholas or Father Snow in Russia) with his wife and elves as inner dolls, many scenes from Russian folk tales, or images of historical landmarks like Saint Basil's Cathedral in Moscow or the Hermitage Art Museum in St. Petersburg. And the figures and scenes shown are not always Russian. Some matryoshka are sets of American baseball or football players or images of paintings from the Italian Renaissance. Russian artists are, for after all, eager to appeal to the buying public and eager to show the quality of their

The sets of matryoshka dolls may include as few as three or as many as 25 nests or dolls; historically, sets containing up to 1,800 dolls are known.

artwork. Although the majority of matryoshka show figures that are both Russian and traditional, the origin of the nested doll is neither Russian nor particularly old.

History

The "Russian" matryoshka doll came to Russia from Japan at the end of the nineteenth century. Little more than 100 years ago, Russia was experiencing an economic boom and a rising sense of culture and national identity. New artistic trends were developing, and a "Russian style" was growing and focusing on the revival of traditions that were in danger of being lost. In St. Petersburg, Russia, in December 1896, an exhibition of Japanese art opened. Among the exhibits was a doll depicting a Buddhist wise man named Fukuruma. The sage was shown as a bald-headed old man with a wooden body that could be split at the waistline into two halves; nested inside were the images of the man when he was younger and bearded and still with hair on his head. The doll came from the island of Honshu; the Japanese claim that they are the inventors of nested dolls or matryoshka, but they also generously admit that the first nested dolls made on Honshu were carved and painted by a Russian monk. That first set of dolls showing Fukuruma is in the Artistic Pedagogical Museum of Toys (APMT) in Sergiyev Posad, a city in Russia that is a cultural center for the making of matryoshka dolls.

Meanwhile, the matryoshka began developing its Russian identity thanks to an industrialist named Savva I. Mamontov (1841–1918). Mamontov was also a patron of the arts and a believer in traditional and nationalistic artistic expression. He established an art studio at his Abramtsevo estate near Moscow. This studio was also an innovation and was the first of a number of "artistic units" around the country where folk craftsmen and professional artists worked together to preserve the skills, techniques, and traditions of Russian folk art including peasant toys. Mamontov's brother, Anatoly Ivanovich Mamontov (1839–1905) created the Children's Education Workshop to make and sell children's toys. The first Russian matryoshka set worked by Vassily Zviozdochkin and painted by Sergei Maliutin (an illustrator of children's books) was

made at the Children's Education Workshop and shows a mother carrying a red-combed rooster—inside are her seven children, the smallest being a sleeping, bundled baby.

Whether the first matryoshka was Japanese or Russian, Russian artists have clearly made nested dolls a symbol and souvenir of Russia. Woodworking and turning is an ancient Russian craft, and the first paintings by Maliutin all came from archaeological and ethnographic (ethnic tradition specific to different regions) sources. Embroidery, clothes, historic dyes and colors, and peasant culture were sources of inspiration for him. Clothing for the dolls that are traditional motherly figures includes an apron, a brightly colored scarf, an embroidered shirt, and sarafan (the national dress of Russia). Lace, flowers, fruit and vegetables, traditional embroidery patterns, and bright colors and complicated designs are copied in detail by matryoshka painters.

The Children's Education Workshop was closed in the late 1890s, but the tradition of the matryoshka simply relocated to Sergiyev Posad, the Russia city known as a toy-making center since the fourteenth century. Sergiyev Posad is located about 45 mi (73 km) from Moscow and is the site of a famous monastery, the Trinity-St. Sergius Monastery. The founding monk, St. Sergius Radonezhsky, carved wooden toys himself, using the rich woodlands surrounding the monastery for materials. His so-called "Trinity" toys became famous among pilgrims who came to the monastery and were even collected by generations of children of the czar. In the 1930s under the Soviet political system, Sergiyev Posad was renamed Zagorsk, and with the fall of the Soviet Union, the city reverted to its traditional name in 1991. With this long tradition of wooden toy-making, the artists of Sergiyev Posad quickly adopted matryoshka with the closing of the Children's Education Workshop. Dolls from this center are called Sergiyev Posad or Zagorsk matryoshka.

In 1900, Russia participated in the World Exhibition in Paris and entered various styles of matryoshka dolls. The nation's exhibit won a medal and many admirers for the nested dolls. The Russian Craftsmen Partnership opened a shop in Paris, and, by 1911, matryoshka—or dolls *la Russe*—were being sold to customers in 14 countries. Until about 1930, matryoshka dolls continued to be very individual. Under the Soviet regime, emphasis shifted to the mass production of nested dolls. In the 1980s, the opening of Russia and the other Soviet countries to the West introduced more freedom, and the "author's matryoshka," with the highly individual style of the particular artist, began to dominate again. Today, matryoshka dolls are collected much like paintings or icons on the reputation of the specific artist over the school or style.

Other major centers for the turning and painting of matryoshka dolls are the city of Semyonov, the Russian region of Nizhegorod (especially the villages of Polkhovsky Maidan and Krutets), and the Mordvinia, Vyatka, and Tver' areas. The popularity of matryoshka painting has spread from Russia to some of the other former republics of the Soviet Union, particularly the Ukraine (known for its delicately painted Easter eggs), Mari El, and Belarus.

Raw Materials

Matryoshka dolls are made of wood from lime, balsa, alder, aspen, and birch trees; lime is probably the most common wood type. These woods share softness, light weight, and fine grain texture. In early spring, the trees for matryoshka-making are marked for cutting. They are felled in April when they are full of sap. After cutting, the trees are stripped of most of their bark, although a few inner rings of bark are left to bind the wood and keep it from splitting. The top and butt ends of the trunks are smeared with sap to keep them from cracking. The logs are stacked in piles in such a way as to leave clearance between the logs so air can circulate.

The logs are aerated in the open for at least two years. A master woodworker decides when they are seasoned enough to be worked. The tree trunks are cut into lengths appropriate for the heights of the matryoshkas to be made and transported to the woodworker's shop.

Raw materials for treating the worked dolls before painting include oil to retain the moisture and a starch-based glue primer. The artist uses tempera paints, oil paints,

gold leaf, and less often, watercolors. Lacquer and sometimes wax are used to provide protective layers on the painted artworks.

Design

The source pieces of wood dictate design somewhat in that they may limit the height, diameter, thinness of the shells of the dolls, and other factors. The master woodworkers are extraordinarily skilled in choosing the right wood for the work. Although matryoshka dolls usually take one of several basic shapes, the turner is free to choose all aspects of shape and size. In painting, the author's style dominates; that is, the individual artist is able to select the theme, story, or character of the doll and to decorate it as he or she wishes. Design limitations vanished with the collapse of the Soviet Union and the opening of Russia to the world marketplace.

The Manufacturing Process

Turning

1 It is essential that the full set of matryoshka be made from one piece of wood because the expansion-contraction characteristics and moisture content of the wood are unique; making a set of dolls from different pieces of wood would result in a set that almost certainly would not fit together properly. Matryoshka-making begins with the smallest doll—the one is that is a solid piece and cannot be taken apart. This smallest figurine is shaped on a turning lathe first, and her shape and size determine those of all the larger dolls that follow. The bottom half of the next doll (the smallest one that can be taken apart) is turned first. The last portion of this lower half that is made is the ring fitting the bottom to the top. When the ring on the lower half is finished, the upper part of the matryoshka is made and the inset for the ring is carved. Each doll is turned at least 15 times.

The craftsman uses few tools, including the turning lathe and a variety of woodcarving knives and chisels of different lengths and shapes. The woodworker completes his job by putting the upper part of the matryoshka doll on its lower half and allowing the wood to dry. This tightens the ring to its upper fitting so the halves of the doll will close se-

Matryoshka dolls.

curely. Turning the dolls on a lathe and sizing them to fit each other takes skill, intuition, and a master woodworker's experience. No measurements are made during the manufacture of a set of dolls.

Treating

2 The worked doll is almost pure white because of the color of the source wood. She is oiled to retain moisture and prevent cracking and left to cure over time. After curing, she is cleaned thoroughly, and one or more coats of starchy glue are painted over the outer surface as a primer for painting. The primer is very carefully applied to create a smooth surface and to prevent smudging.

Painting

3 In the history of the matryoshka doll, the early dolls were prized for the skills of the turner and his ability to make a thin shell for the matryoshka. Woodworking was prized above painting. By the 1980s, this balance had shifted and the painting was considered to add more value than the wood turning. There were also two schools of emphasis in painting; one puts more importance on the doll's face, and the other features the costume and its details. Matryoshka artists are often also painters of religious icons (images of Jesus Christ, the Virgin Mary, and other religious figures) that are revered in churches and private homes. Thus, the detail they can achieve in their chosen style is amazing.

The painter is the next craftsman to work on the matryoshka. Early matryoshkas were painted with gouache, an opaque form of watercolor; today, high-quality tempera (colloid-based paint like poster paint), oil, and other paints (the same as those used by artists on canvas) are used to color the dolls. Watercolors are also used, but watercolor dolls are more rare and expensive because watercoloring wood is a difficult technique. The painters are true artists who know the character of the wood, the tradition of the matryoshka and other wooden toys, and national costume and folk tales, as well as their own individual artistic strengths. The themes used to paint the matryoshka are usually typical of the studio of the artist and the region and are suited to the size and shape of the dolls. The artistic style may be very coarse or extremely fine—sometimes, only a single hair from a brush is used to add eyelashes and threads of lace. Gold leaf is also added to enhance the detailing.

Some styles of matryoshka are colored with aniline (synthetic organic) dyes instead of paint. The dye has a lighter texture, more like watercolor, and dolls that are dyed usually have a more childlike style. The colors tend to be basic green (from a vegetable dye), fuchsine (a brilliant bluish red), blue, and yellow. Early examples of dolls colored with dye were also coated with glue that dried the colors to dark hues.

Although the majority of matryoshkas are painted all over, some are not primed so the native wood is exposed. The wood becomes the background or thematic color of the doll, and paint is added to give her a face and costume. A heated poker is also used in some designs to burn in details of the doll including facial features and costume details. The doll may be left with only the poker work designs as her character, or the poker outlines may be filled with paint. All painted dolls are covered with lacquer to finish them and protect the paint. Some dolls with unpainted wood and poker detailing are not lacquered.

Finishing

4 The painter completes his or her set of matryoshka by adding his signature to the bottom of the largest doll as well as a number showing the number of nests in the set. After the paint has dried, the dolls are finished with a protective coating. Wax and varnish are used rarely, and lacquer is the most common finish. For the artistic sets, at least five coats of lacquer are applied.

Byproducts/Waste

Matryoshka making does not produce any byproducts although the artistic centers where they are crafted usually make other wood products. The seasoning of the wood is time consuming, and, when the wood is ready for turning, woodworkers avoid waste whenever possible. Similarly, the painters are highly skilled craftsmen and little paint or lacquer waste is generated.

The Future

The production of matryoshka dolls experienced a huge upsurge with the collapse of the Soviet Union and the greater availability of Russian products to a worldwide audience. Matryoshka dolls are highly prized by collectors; essentially, they have become artistic works auctioned through Sotheby's and other leading auction houses, with far less expensive versions sold to tourists and as toys. Unfortunately, the price gap between the artistic and toy versions is large, and there is no middle ground.

The same open market that encourages artists to make matryoshka dolls also discourages them, however; the painters particularly are very gifted artists and often have experience painting icons and other products that command a still higher price. The rebirth of religion in the countries of the former Soviet Union has pulled many artists back into the painting of icons. The competition among manufacturers, then, is in keeping the artists interested in nested dolls as a form of artistic and cultural expression. Only 15 to 20 artists produce the top-quality matryoshka dolls, and, like a painting bound for a museum, each set is a unique masterpiece that may command $2,000. Clearly, interest in matryoshka dolls is well-established and has come to represent Russian culture. In the future, it remains to be seen whether the availability of skilled artists can meet the demand.

Where to Learn More

Books

Soloviova, L. N. *Matryoshka*. Moscow, Russia: Interbook Business, 1997.

Soloviova, Larissa, and Marina Marder. *Russian Matryoshka*. Moscow, Russia: Interbook, 1993.

Periodicals

Dunn, Jessica, and Danielle Dunn. "Matryoshka Dolls: Russian Folk Art." *Skipping Stones* 9, no. 5 (Novemeber/December 1997): 26.

Other

Exclusive Art of Matryoshka. http://matryoshka.itgo.com (December 20, 2000).

Russian Treasure. http://www.russiantreasure.com (December 20, 2000).

Russian World. http://www.russianworld-online.com (December 20, 2000).

—*Gillian S. Holmes*

Mayonnaise

Mayonnaise is regulated by the U.S. Food and Drug Administration's Standard of Identity. It must contain at least 65% oil by weight, vinegar, and egg or egg yolks.

Background

Mayonnaise is a cold, emulsification used as sauce or as a condiment. It is made by blending egg yolks and oil, then flavored with varying combinations of vinegar, mustard, herbs and spices. Mayonnaise is often used as base for creamy-type salad dressings.

History

Food historians offer four possible theories for the origin of mayonnaise. The most popular story dates to June 28, 1756, when the French Duke Richelieu captured Port Mayon on the Spanish island of Minorca. When preparing the victory feast, the duke's chef was forced to substitute olive oil for cream in a sauce. Unexpectedly pleased with the result, the chef christened the result "mahonnaise" in honor of the place of victory.

Caràme, a French food writer and author of *Cuisinier parisien: Trarté des entrées froids* believed the word was derived from the French verb *manier*, meaning to stir. Another food expert, Prosper Montagné maintained that the origin lay in the Old French word *moyeu*, meaning egg yolk.

Still others insist that the creamy sauce was a specialty of the town of Bayonne in southwest France. Thus, what was originally called bayonnaise was later modified to mayonnaise.

Regardless of its origins, mayonnaise quickly because a popular sauce and spread in European cuisine. In the early 1900s, a German immigrant named Richard Hellmann opened a delicatessen in New York City. The salads that his wife made with her homemade may-onnaise were particularly popular items. When customers began to ask if they could purchase the mayonnaise itself, the Hellmans produced it in bulk and sold it by weight in small wooden butter-measuring vessels.

Eventually the Hellmans were packing their mayonnaise in glass jars. In 1913, they built their first factory in Astoria. A company in California, Best Foods Inc., was also enjoying success with their version of mayonnaise. In 1932, Best Foods acquired the Hellman's brand.

A variation of mayonnaise, made with a cooked based and labeled **salad dressing**, was developed by National Dairy Products in 1933 at the Century of Progress World's Fair in Chicago. It would eventually become known as Kraft Miracle Whip Salad Dressing.

Raw Materials

Mayonnaise is regulated by the U.S. Food and Drug Administration's Standard of Identity. It must contain at least 65% oil by weight, vinegar, and egg or egg yolks. Spices and other natural seasonings may added with the exception of turmeric and saffron. These would give the mayonnaise a yellow hue and thus appear to contain added egg yolks.

The FDA's Standard of Identity requirements for salad dressing are 30% vegetable oil, 4% egg yolk, vinegar or lemon juice, and spices.

Soybean oil is the most common type of oil used in the production of mayonnaise. Vinegar is distilled from distilled alcohol. Lemon or lime juice is diluted with water.

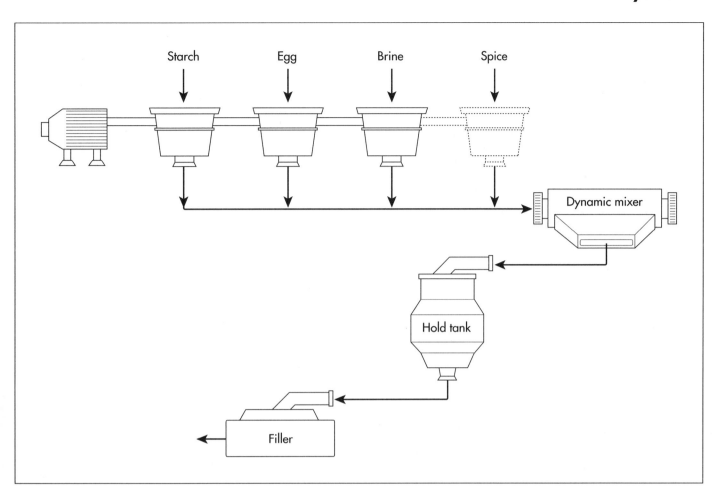

The eggs are subjected to pasteurization by heating them without actually cooking them. Low-fat versions are made with egg whites. To replace the fat from the missing egg yolk, modified food starches are added so that the low-fat mayonnaise retains the creamy texture and thickness of real mayonnaise. These food starches can be made from corn, gums, or agar agar (a seaweed extraction).

Salt is added to enhance flavor. The amount equals approximately 1/16 teaspoon of salt per tablespoon serving of mayonnaise. Preservatives such as calcium disodium EDTA are added to increase shelf-life.

The Manufacturing Process

Creating the emulsion

1 A continuous blending system is employed to sustain the correct degree of emulsification. An emulsion (known technically as a colloid) occurs when the blending of two liquids, in this case vinegar and oil, causes one of the liquids to form small droplets that are dispersed throughout the other liquid.

The blend of vinegar and oil moves continuously through a series of positive replacement pumps. These pumps feature a cavity or set of cavities fitted with rotary impellers. A regulated pumping action causes the cavities to fill and empty. The impellers move the blended fluid from one cavity to another

Adding ingredients

2 Pre-measured ingredients are piped in through openings in the sides of the pumps or from spigots overhead.

Bottling the mayonnaise

3 The mayonnaise is moved through the pumping system to the bottling station. Pre-sterilized jars move along conveyer belts as premeasured amounts of mayonnaise are poured into the jars. The jars are

sealed with metal screwcaps. They are not vacuum-sealed.

Quality Control

All raw materials are check for freshness when they arrive at the processing plant. Stored materials are tested periodically as well. Samples of the mayonnaise are drawn off and taste-tested during the manufacturing process.

Where to Learn More

Books

Coyle, L. Patrick Jr. *The World Encyclopedia of Food* New York: Facts on File, 1982.

Trager, James. *The Food Chronology.* New York: Henry Holt, 1995.

Other

Association for Dressings and Sauces. http://www.dressings-sauces.org/ (October 2000).

Kraft Unit Operations. http://www.kraftunitops.com/ (October 2000).

—*Mary McNulty*

Mouthwash

Mouthwash is a liquid oral product designed to freshen breath. Certain varieties may also kill bacteria and/or whiten teeth. Mouthwashes are made by combining the appropriate raw materials in large, stainless steel tanks and then filling the product into individual packages. First used by ancient societies, technological advances in chemistry have resulted in steadily improving formulas. In 1998, Americans spent over $652 million on mouthwash

Background

The need for mouthwash is a result of a condition called halitosis, or bad breath. It is estimated that over half the population occasionally has foul-smelling breath. This typically occurs upon first awakening or after a meal with garlic or onions. It has been found that bad breath is mostly due to bacterial activity in an unclean mouth. Specifically, anaerobic bacteria that grow on the protein-rich food debris stuck between the teeth or on the tongue. As the bacteria breaks down the proteins, those containing sulphur give off foul odor molecules such as methylmercaptan and hydrogen sulphide which result in bad breath.

Mouthwashes are designed to eliminate bad breath in two ways. First, they relieve it by killing the bacteria responsible for producing the foul odor. The best of these products prevent bad breath for as long as eight hours. The second way that mouthwashes help reduce bad breath is by masking the odor. This is a much less effective method which lasts no more than 30 minutes.

History

Products used for freshening breath or cleaning teeth have been in existence for centuries. Many of the ancient societies—including the Egyptians, Chinese, Greeks, and Romans—had recipes for such preparations. They used a variety of ingredients; from edible materials like fruit, honey, or dried flowers to less appealing compounds such as ground lizard, minced mice, or urine. These products were generally ineffective and in some cases were harmful to the sensitive enamel which coats each tooth.

While tooth cleaning preparations steadily improved over the years, it was not until the early 1800s—when the modern toothpaste was developed—that truly effective oral products became available. The first mouthwashes were basically solutions of grain alcohol and were likely developed accidentally during this era. One of the most famous brands, Listerine, was developed during the 1880s and is still sold today.

The antibacterial effect of fluoride was an important discovery for the development of modern mouthwashes. In the early 1900s a dentist named Frederick McKay found that some of his patients had a condition called mottled enamel. He found that this condition was linked to a reduction in tooth decay. In 1931, he tested the drinking water that these patients consumed and found a high level of natural fluoride. By the early 1940s, other workers had determined that fluoride in drinking water at one part per million would reduce tooth decay without causing mottling. Various testing went on during the rest of this decade and by the 1950s it was recommended by the United States federal government that all public water sources be fortified with fluoride. This discovery led to the development of toothpastes and mouthwashes that contained fluoride compounds. During the years that

In 1998, Americans spent over $652 million on mouthwash.

followed, various raw materials have been developed that have an antibacterial effect but do not contain alcohol. Additionally, materials that prevent tartar, whiten teeth, and reduce cavities have also been discovered and added to mouthwash formulas.

Raw Materials

Mouthwashes are generally composed of diluents, antibacterial agents, soaps, flavorings, and colorants. The primary ingredient in most mouthwashes is water, a diluent, making up over 50% of the entire formula. The water is specially treated to remove various particles and ions which might impact flavor. Water treated as such is called deionized water. The sources of water vary, coming from reserves such as underground wells, lakes, and rivers. Alcohol is another diluent typically used in up to 20% of the formula. While early mouthwashes used alcohol extensively; today its use is limited because of governmental regulations and consumer desires.

Numerous antibacterial agents have been employed in mouthwash formulations. These include ingredients like phenols, thymol, salol, tannic acid, hexachloraphene, chlorinated thymols, and quaternary ammonium compounds. Chlorinated phenols like parachlormetacresol have both an antibacterial effect and a desired flavor. Thymol, which is obtained from volatile oils, is used at low concentrations and in conjunction with other ingredients. Hexachloraphene is substantive to the mucous membrane which makes it ideal for longer lasting formulas. Quaternary compounds are often used because of their non-toxic and non-irritant nature. They are effective against plaque.

Color and flavor are added to the formulas to improve the consumer acceptability of the mouthwash ingredients. Flavor is an essential feature of a mouthwash because it has the most consumer perceptible impact. In the United States, flavorants such as peppermint, menthol, methyl salicylate, and eugenol are commonly used. The most common colors, blue and green, are the result of adding governmentally approved and certified FD&C dyes. Some mouthwash formulas also include a synthetic detergent to give extra foaming and cleansing action.

Design

In general there are three types of mouthwashes. There are antibacterial products that reduce the bacterial population of the mouth. These products have a fresh taste and improve breath odor. The second type are fluoride mouthwashes, which help to improve the fluoride layer on tooth enamel. Finally, there are remineralizing mouthwashes that help repair various lesions in the mouth.

Mouthwashes are sold in a variety of flavors and colors. The most popular is the golden colored, medicinal-tasting Listerine. Blue or green mint varieties are also common. There are mouthwash products that are geared specifically toward smokers. These products are designed to remove tobacco tar stains as well as freshen breath and whiten teeth. There are also tartar control mouthwashes and those that kill the germs responsible for gingivitis and plaque.

The packaging for mouthwashes is typically a clear, plastic bottle. Since the products generally contain alcohol, child resistance and tamper-evidence closures are typically used. Additionally, some mouthwashes have a dosing feature which lets the consumer squeeze out an exact amount each time it is used.

The Manufacturing Process

After a mouthwash formula is designed, it is tested to ensure that minimal changes will occur over time regardless of the storage conditions. This testing, called stability testing, helps detect physical changes in such things as color, odor, and flavor. It can also provide information about product performance over time. In the United States, the Food and Drug Administration (FDA) requires that specific stability testing be done to ensure product performance during long term storage.

In general, the process for creating a mouthwash occurs in two distinct steps. First, a large batch of mouthwash is made then it is filled in the appropriate packaging.

Compounding

1 Mouthwash is made via a batch process in an area of the manufacturing plant called compounding. Here operators, called com-

Top Mouthwash Products Leaders

Mouthwash sales rose 3.8% to over $670 million last year for the year ended December 5, 1999 according to Information Resources Inc., Chicago. Sales and unit figures are measured in millions.

Product	$ Sales	$ Change	$ Unit Sales	$ Sales
Listerine	$293	7.3%	74	5.8%
Private label	$133	1.3%	64	-1.7%
Scope	$111	2.3%	41	2.5%
Plax	$38	-5.2%	11	-7.1%
Act	$19	11%	3	9.4%
Targon	$17	23.4%	4	21.7%
Colgate Peroxyl	$10	12%	1.5	6.7%
Cepacol	$8	-5.5%	3	-4.3%
Act for Kids	$8	11.8%	2	10.0%
Biotene	$5	22.7%	0.8	22.3%
Category total	**$677**	**3.8%**	**218**	**1.3%**

pounders, make batches of 2,000-3,000 gal (7,571-11,356 L) of mouthwash following specific formula instructions. The raw materials are delivered to the compounding area by fork lift trucks. Compounders add them to the main batch tank where they are thoroughly mixed. Depending on the formula instructions, the batch is heated and cooled to get the raw materials to rapidly combine.

2 Materials which are used in large quantities—such as alcohol or water—are then pumped directly into the tank. This is done by simply setting computer controls to the appropriate amount and pushing a button. Computers also control the mixing speed and temperature of the batch. Depending on the size of the batch and the number of raw materials, a mouthwash can take anywhere from one to three hours to make.

Batch analysis

3 When the batch is completed, a sample is taken to the Quality Control (QC) lab. The appearance and flavor of the batch is examined to ensure that it meets the specifications laid out in the formula. QC chemists may also run pH determinations and viscosity checks. If some characteristic of the batch is found to be out of a specified range, adjustments may be made at this point. For example, colors can be modified by adding more dye.

4 After the batch is approved, it is pumped from the main tank to a holding tank. This holding tank may be directly hooked up to the filling lines where the product is put into individual packaging.

Filling

5 At the beginning of the filling line there is a large bin called a hopper which contains the empty bottles that will be filled. In this bin, the bottles are physically manipulated so that they come out standing upright on a conveyor belt. They are then moved to the filling carousel which contains the bulk mouthwash product.

6 The filling carousel has a series of piston filling heads that are designed to deliver an exact amount of mouthwash. As the bottles move around the carousel, the piston moves down and the mouthwash product is dispensed into the bottle.

7 After the bottles are filled, they are sent on a conveyor belt to a capping machine.

The caps are also held in a large bin and correctly aligned. As the bottles pass the capping hopper, the caps are put on and either twisted or pushed in place.

8 From the capping station, the bottles are moved to a labeling machine. The labels are held on large spools and threaded through the machine. As the bottles pass by, the label is either stuck on using an adhesive or heat pressed.

9 Beyond labeling, the bottles are next moved to a boxing station. They are typically gathered in a group of 12 or 24 and dropped into a box. The boxes then move to a palleting machine and stacked. The pallets are moved via fork lifts to large trucks and shipped to distributors. High speed production lines like these can produce over 20,000 bottles per hour.

Quality Control

While quality control is a critical step in the batching process, it is also done at other points during manufacture. Workers are stationed at various points on the filling lines to inspect the production process. They examine things such as bottle quality, fill levels, and label placement. They also make sure that all the caps are put on correctly. Microbial contamination is also routinely checked during the filling process. Additionally, the packaging is checked for things such as bottle thickness, appearance, and weight to make sure the final product has the desired characteristics.

The Future

With advances in chemical technology, mouthwashes of the future will be designed with a larger array of and more improved functions. In the past, mouthwashes were primarily powerful breath fresheners. They eventually evolved into tooth protectors. Today, products are available to not only fight bad breath but whiten teeth and help battle cavity formation and gum recession.

Some new technologies that will undoubtedly be adapted to mouthwash products have recently been discovered. For example, researchers have found a peptide known as p1025, which can bond to the teeth and prevent the growth of naturally occurring bacteria. This prevents the cavity-causing bacteria to adhere to the tooth and thus inhibits cavity formation. Using this technology, they have created a mouthwash that may prevent tooth decay for up to three months.

Another new mouthwash may actually contain a good bacteria to kill the odor- and cavity-causing germ *Streptococcus mutans*. Using genetic engineering, scientists at the University of Florida College of Dentistry developed this bacterium and are now testing it in humans to determine whether it can be used. Ultimately, this new bacteria may be added to mouthwash products and thereby revolutionizing oral care.

Where to Learn More

Periodicals

Hickey, James. "Oral Care Market." *Happi* (February 2000). http://www.happi.com/special/feb002.htm (January 2001).

Hickey, James. "Total Domination." *Happi* (February 1999). http://www.happi.com (January 2001).

Shaw, Anita. "Oral Care Market is All Smiles." *Soap & Cosmetics* 75, no.6 (June 1999): 28.

Travis, John. "Wash That Mouth Out With Bacteria!" *Science News* 157, no.12 (March 18, 2000): 190.

Wilson, Jim. "Mouthwash Cancels Cavities." *Popular Mechanics* 177, no.2 (February 2000): 15.

—*Perry Romanowski*

Natural Gas

Background

Natural gas is a mixture of combustible gases formed underground by the decomposition of organic materials in plant and animal. It is usually found in areas where oil is present, although there are several large underground reservoirs of natural gas where there is little or no oil. Natural gas is widely used for heating and cooking, as well as for a variety of industrial applications.

History

Natural gas was known to early man in the form of seepages from rocks and springs. Sometimes, lightning or other sources of ignition would cause these gas seepages to burn, giving rise to stories of fire issuing from the ground. In about 900 B.C. natural gas was drawn from wells in China. The gas was burned, and the heat was used to evaporate seawater in order to produce salt. By the first century, the Chinese had developed more advanced techniques for tapping underground reservoirs of natural gas, which allowed them to drill wells as deep as 4,800 ft (1,460 m) in soft soil. They used metal drilling bits inserted through sections of hollowed-out bamboo pipes to reach the gas and bring it to the surface.

The Romans also knew about natural gas, and Julius Caesar was supposed to have witnessed a "burning spring" near Grenoble, France. Religious temples in early Russia were built around places where burning natural gas seepages formed "eternal flames."

In the United States, the first intentional use of natural gas occurred in 1821 when William Hart drilled a well to tap a shallow gas pocket along the bank of Canadaway Creek near Fredonia, New York. He piped the gas through hollowed logs to a nearby building where he burned it for illumination. In 1865, the Fredonia Gas, Light, and Waterworks Company became the first natural gas company in the United States. The first long-distance gas pipeline ran 25 mi (40 km) from a gas field to Rochester, New York, in 1872. It too used hollowed logs for pipes. The development of the Bunsen burner by Robert Bunsen in 1885 led to an interest in using natural gas as a source of heating and cooking, in addition to its use for lighting. In 1891, a high-pressure gas deposit was tapped in central Indiana, and a 120 mi (192 km) pipeline was built to bring the gas to Chicago, Illinois.

Despite these early efforts, the lack of a good distribution system for natural gas limited its use to local areas where the gas was found. Most of the gas that came to the surface as part of oil drilling in more remote areas was simply vented to the atmosphere or burned off in giant flares that illuminated the oil fields day and night. By the 1910s, oil companies realized that this practice was costing them potential profits and they began an aggressive program to install gas pipelines to large metropolitan areas across the United States. It wasn't until after World War II that this pipeline program had reached enough cities and towns to make natural gas an attractive alternative to electricity and coal.

By 2000, there were over 600 natural gas processing plants in the United States connected to more than 300,000 mi (480,000 km) of main transportation pipelines. Worldwide, there are also significant deposits of natural gas in the former Soviet

In 2000, there were over 600 natural gas processing plants in the United States connected to more than 300,000 mi (480,000 km) of main transportation pipelines.

269

Union, Canada, China, and the Arabian Gulf countries of the Middle East.

Raw Materials

Raw natural gas is composed of several gases. The main component is methane. Other components include ethane, propane, butane, and many other combustible hydrocarbons. Raw natural gas may also contain water vapor, hydrogen sulfide, carbon dioxide, nitrogen, and helium.

During processing, many of these components may be removed. Some—such as ethane, propane, butane, hydrogen sulfide, and helium—may be partially or completely removed to be processed and sold as separate commodities. Other components—such as water vapor, carbon dioxide, and nitrogen—may be removed to improve the quality of the natural gas or to make it easier to move the gas over great distances through pipelines.

The resulting processed natural gas contains mostly methane and ethane, although there is no such thing as a "typical" natural gas. Certain other components may be added to the processed gas to give it special qualities. For example, a chemical known as mercaptan is added to give the gas a distinctive odor that warns people of a leak.

The Manufacturing Process

The methods used to extract, process, transport, store, and distribute natural gas depend on the location and composition of the raw gas and the location and application of the gas by the end users. Here is a typical sequence of operations used to produce natural gas for home heating and cooking use.

Extracting

1 Some underground natural gas reservoirs are under enough internal pressure that the gas can flow up the well and reach Earth's surface without additional help. However, most wells require a pump to bring the gas (and oil, if it is present) to the surface. The most common pump has a long rod attached to a piston deep in the well. The rod is alternately pulled upward and plunged back into the well by a beam that slowly rocks up and down on top of a verti-

cal support. This configuration is often called a horse head pump because the shape of the pulling mechanism on the end of the rocking beam resembles a horse's head.

2 When the raw natural gas reaches the surface, it is separated from any oil that might be present and is piped to a central gas processing plant nearby. Several hundred wells may all feed into the same plant.

Processing

3 About 75% of the raw natural gas in the United States comes from underground reservoirs where little or no oil is present. This gas is easier to process than gas from oil wells. Regardless of the source, most raw natural gas contains dirt, sand, and water vapor, which must be removed before further processing to prevent contamination and corrosion of the equipment and pipelines. The dirt and sand are removed with filters or traps near the well. The water vapor is usually removed by passing the gas through a tower filled with granules of a solid desiccant, such as alumina or silica gel, or through a liquid desiccant, such as a glycol. After it has been cleaned and dried, the raw gas may be processed further or it may be sent directly to a compressor station and pumped into a main transportation pipeline.

4 If the raw natural gas contains a large amount of heavier hydrocarbon gases, such as propane and butane, these materials are removed to be sold separately. The most common method is to bubble the raw gas up through a tall, closed tower containing a cold absorption oil, similar to kerosene. As the gas comes in contact with the cold oil, the heavier hydrocarbon gases condense into liquids and are trapped in the oil. The lighter hydrocarbon gases, such as methane and ethane, do not condense into liquid and flow out the top of the tower. About 85% of the propane and almost all of the butane and heavier hydrocarbons are trapped this way. The absorption oil is then distilled to remove the trapped hydrocarbons, which are separated into individual components in a fractionation tower.

5 At this point, the natural gas contains methane, ethane, and a small amount of propane that wasn't trapped. It may also contain varying amounts of carbon dioxide,

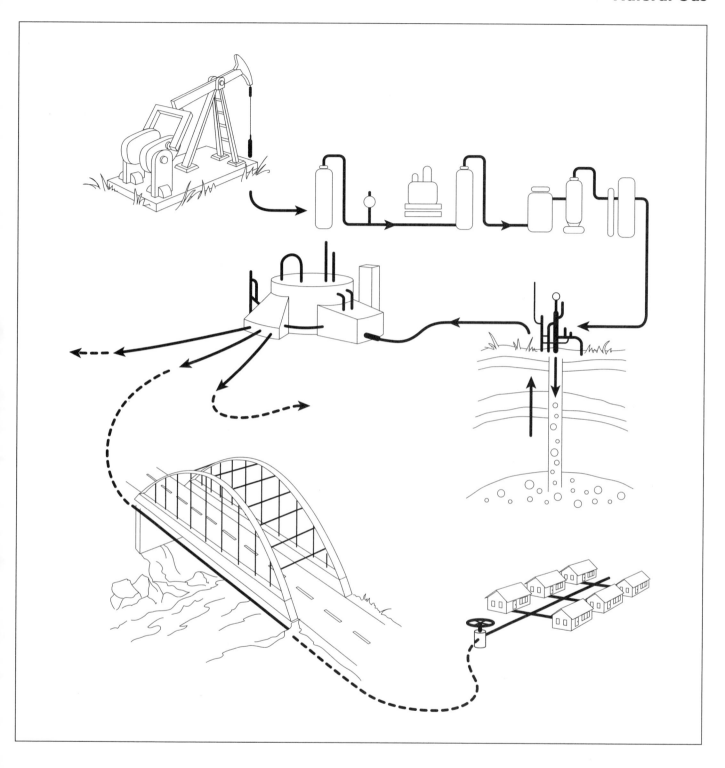

hydrogen sulfide, nitrogen, and other materials. A portion of the ethane is sometimes removed to be used as a raw material in various chemical processes. To accomplish this, the water vapor in the gas is further reduced using one of several methods, and the gas is then subjected to repeated compression and expansion cycles to cool the ethane and capture it as a liquid.

6 Some natural gas contains a high percentage of carbon dioxide and hydrogen sulfide. These chemicals can react with the remaining water vapor in the gas to form an acid, which can cause corrosion. They are removed by flowing the gas up through a tower while a spray of water mixed with a solvent, such as monoethanolamine, is injected at the top. The solvent reacts with the

A diagram depicting the production of natural gas from underground source to household usage.

chemicals, and the solution is drained off the bottom of the tower for further processing.

7 Some natural gas also contains a high percentage of nitrogen. Because nitrogen does not burn, it reduces the heating value of the natural gas. After the carbon dioxide and hydrogen sulfide have been removed, the gas goes through a low-temperature distillation process to liquefy and separate the nitrogen. Together, the processes in steps 6 and 7 are sometimes called "upgrading" the gas because the natural gas is now cleaner and will burn hotter.

8 If helium gas is to be captured, it is done after the nitrogen is removed. This involves a complex distillation and purification process to isolate the helium from other gases. Natural gas is the primary source of industrial helium in the United States.

Transporting

9 Mercaptan is injected into the processed natural gas to give it a distinctive warning odor, and the gas is piped to a compressor station where the pressure is increased to about 200-1,500 psi (1,380-10,350 kPa). The gas is then transported across country through one of several major pipelines installed underground. These pipelines range from 20 to 42 in (51 to 107 cm) in diameter. About every 100 mi (160 km), another compressor boosts the gas pressure to make up for small pressure losses caused by friction between the gas and the pipe walls. This keeps the gas flowing.

10 When the pressurized natural gas reaches the vicinity of its final destination, it is sometimes injected back into the ground for storage. Depleted underground gas and oil reservoirs, porous rock layers known as aquifers, or subterranean salt caverns may be used to store the gas. This ensures a ready supply during the colder winter months.

Distributing

11 When gas is needed, it is drawn out of underground storage and is transported through pipelines at pressures up to 1,000 psi (6,900 kPa). These pipelines bring the gas into the city or area where it is to be used.

12 The pressure is reduced to below 60 psi (410 kPa), and the gas is distributed in underground pipes that run throughout the area. Before the gas is piped into each house or business, the pressure is further reduced to about 0.25 psi (1.7 kPa).

Quality Control

Natural gas burns readily in air and can explode violently if a large quantity is suddenly ignited. Entire buildings have been leveled by powerful blasts resulting from natural gas leaks. In other cases, people have suffocated in closed rooms that slowly filled with natural gas. Because natural gas is odorless, foul-smelling mercaptan is added to the gas so that even a small leak will be immediately noticeable. To protect high-pressure underground gas pipelines, a bright yellow plastic tape is buried in the ground a few feet above the pipeline to warn people who might be digging in the area. That way, they will uncover the tape before they actually strike the pipeline below. Warning signs are also placed at ground level along the entire length of the pipeline as an additional precaution.

The Future

Because natural gas is clean burning, it is being considered as an alternative fuel for motor vehicles. Compressed natural gas (CNG) cars and trucks are already on the road in many areas. Companies using industrial processes that require high temperatures are also turning to natural gas instead of other fuels in order to reduce the air pollution emitted by their plants. This includes companies involved in manufacturing steel, glass, ceramics, cement, paper, chemicals, aluminum, and processed foods.

Where to Learn More

Books

Kroschwitz, Jacqueline I., and Mary Howe-Grant (eds.). "Gas, Natural." In *Encyclopedia of Chemical Technology*. 4th ed., vol. 12. New York: John Wiley and Sons, Inc., 1993.

Tussing, Arlon R., and Bob Tippee. *The Natural Gas Industry: Evolution, Structure, and Economics*. 2nd ed. Tulsa, OK: PennWell Publishing, 1995.

Other

Natural Gas Information and Educational Resources. http://www.naturalgas.org (November 1, 2000).

Pacific Gas and Electric Company. "How Our Gas System Works." http://www.pge.com/006_news/006c2_gassys.shtml (November 12, 2000).

—Chris Cavette

Needle-free Injection System

The first air-powered needle-free injection systems were developed during the 1940s and 1950s. These devices were gun-shaped and used propellant gases to force fluid medicines through the skin.

Needle-free injection systems are novel ways to introduce various medicines into patients without piercing the skin with a conventional needle. They can take the form of power sprays, edible products, inhalers, and skin patches. While hypodermic needles were first introduced during the 1800s, needle-free systems are relatively recent inventions. Today, they are a steadily developing technology that promises to make the administration of medicine more efficient and less painful.

Background

People are given injections to protect them from influenza, tetanus, cholera, typhoid, and other diseases. When a needle is inserted through the skin, the vaccine (or drug) it carries provides systemic immunity. This is because the vaccine gets into the bloodstream and provokes the body to create antibodies that are carried throughout the entire body.

In the United States, children may get over 13 vaccine injections by the age of 16. Unfortunately, there are a variety of problems associated with the hypodermic needles used for these injections. One of the most significant drawbacks is the relatively high cost of the needles. The cost results in a lower vaccination rate, especially for children in developing countries. Another problem with traditional needles is the lack of reusability. If a needle syringe is not sterilized, reusing it can lead to the spread of disease. Additionally, many people have a fear of needles which causes them to avoid treatment. These drawbacks have led to the development of alternative delivery systems to needle injections.

Needle-free systems are designed to solve these problems making them safer, less ex-

pensive, and more convenient. It is anticipated that these systems will increase the incidence of vaccination and reduce the amount of prescribed antibiotics. Moreover, they should reduce the number of needle stick accidents that have resulted in some health care workers contracting diseases.

More than a dozen companies have developed alternatives to needle injections. Some of the different designs include nasal sprays, nose drops, flavored liquids, skin patches, air forced and edible vaccine-packed vegetables.

The needle-free systems that are most like traditional injections involve the direct transfer of the medicine through the skin. One company offers an injection system where the drug is dispersed through the skin as a fine mist or powder. In this system, a tube-shaped device is held against the skin and a burst of air forces the molecules of medicine into the body. The device is designed to force the medicine far enough through the skin so it enters the bloodstream. An application for which this system is particularly useful is for patients who need daily doses of growth hormone.

Patches have been introduced as needle-free delivery systems. These devices, which look like bandages, slowly transfer medicine through the skin. In one type of patch, thousands of tiny blades are imbedded on its surface. The patch is covered with medicine and then placed on the skin. The blades make microscopic cuts in the skin that opens a path for drugs to enter through. When an electric current is applied, the medicine is forced into the body. This process, called iontophoresis, does not hurt.

Nasal sprays, suppositories, and eye and nose drops are forms of needle free systems that deliver medications through the mucous membrane, where 90% of all infections occur. The mucous membrane is found throughout the body and includes the lining of the respiratory tract, digestive tract, and urinary and genital passages. These needle free systems prompt the body to produce both antibodies at the mucosa surfaces and system-wide.

The nasal shot may be the first needle-free flu shot. It is a syringe-like device that has an aerosol sprayer substituted for the needle. It delivers a weak flu virus directly to the nasal passages and creates immunity to the flu with minimal side effects.

Inhalers are another type of needle-free delivery system. In these systems, liquids or powders are inhaled and delivered into the lungs. These devices are good for delivering protein drugs because the lungs provide a rapid absorption into the bloodstream. In one system there is a pump unit that atomizes a powdered medication. This allows the patient to inhale the proper amount of medicine without it getting trapped in the back of the throat. For diabetics who require daily injections of insulin, an aerosol inhaler has also been introduced.

Oral vaccines are needle-free systems that may replace vaccine injections. This technology has been difficult to perfect for many reasons. The primary problem with this type of delivery system is that the environment of the digestive system is harsh and typically destroys vaccines and other drugs. Also, vaccines do not work as well in provoking antibody production in the digestive lining. One of the latest oral vaccines involves freeze drying the medicine and mixing it with a salt buffer to protect it when it is in the stomach. Other edible forms include a sugar solution of a vaccine against the bacterium that causes ulcers. For travelers, a typhoid-vaccine capsule ·has been developed as an alternative to the two painful shots typically required.

Genetic engineering has enabled the production of oral vaccines in food. In 1998, potatoes were produced that contained genes from the virus that causes cholera. These potatoes showed efficacy in protecting people from this disease. This is particularly useful for developing countries where potatoes are a dietary staple and the refrigeration that is typically required for transporting vaccines is not readily available.

History

As long as drugs have been known to cure diseases, people have searched for better methods of delivering them. During the early nineteenth century researchers made a series of discoveries that eventually led to the development of the hypodermic needle by Alexander Wood in 1853. This device was used to give morphine to patients suffering from sleeping disorders. In subsequent years, the hypodermic needle underwent significant changes which made them more efficient to use, safer, and more reliable. However, needles still have significant drawbacks which prompted researchers to find needle-free alternatives.

The first air-powered needle-free injection systems were developed during the 1940s and 1950s. These devices were gun-shaped and used propellant gases to force fluid medicines through the skin. Over the years, the devices have been modified to improve the amount and types of medicines delivered, and the efficiency and the ease of use.

Raw Materials

Since these devices directly contact the body, they must be made from materials that are pharmacologically inert. The materials also must be able to withstand high temperatures because they are heat-sterilized. Air-forced injection systems are available in different shapes as sizes. The outer shell of the device is made from a high strength, lightweight thermoplastic such as polycarbonate. Polycarbonates are polymers produced synthetically through various chemical reactions. To make the polymer easier to mold, fillers are added. These fillers make plastics more durable, lightweight, and rigid. Colorants are also incorporated into the plastic to modify the appearance. Prior to manufacture, the plastics are typically supplied in pellet form with the colorants and fillers already incorporated. Air-forced systems typically use carbon dioxide or helium gas to propel the medicine into the body.

The three different types of injections.

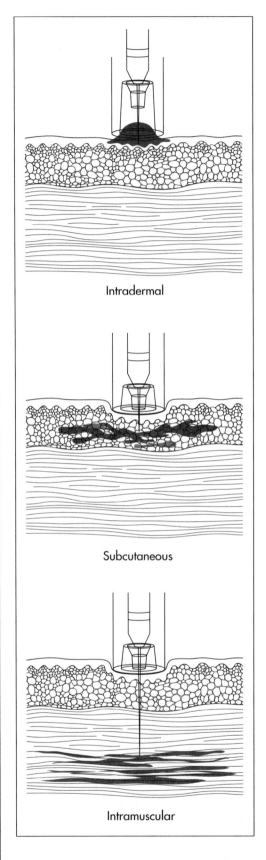

Intradermal

Subcutaneous

Intramuscular

Certain types of medicines work better with needle-free injection systems than other. Insulin, which must be administered daily to diabetics, can be incorporated into an inhaler system. Lidocaine hydrochloride, a local anesthetic is suitable to be delivered needle free. Other medicines suitable for needle free systems include Fentanyl (an opioid analgesic), Heparin (an anticoagulant) and a variety of vaccines. Various adjunct ingredients included in these medicines include cyclodextrins, lactose, liposomes, amino acids and water.

Design

The air-forced needle-free injection systems are typically made up of three components including an injection device, a disposable needle free syringe and an air cartridge. The injection device is made of a durable plastic. It is designed to be easy to hold for self-administration of medicine. The needle-free syringe is also plastic. It is sterilized and is the only piece of the device that must touch the skin. The syringe is made to be disposed after every use. For portable units, pressurized metal air cartridges are included. Less mobile devices have air hook-ups that attach to larger containers of compressed air. Some air-forced systems use a re-usable **spring** to generate the pushing force instead of pressurized air cartridges.

The Manufacturing Process

There are numerous methods of producing each needle-free injection system. The following process focuses on the production of an air-forced system. These systems are made through a step by step procedure which involves molding the pieces, assembling them, and decorating and labeling the final product. The individual pieces are typically produced off-site and assembled by the needle free injection system manufacturer. All of the manufacturing is done under sterile conditions to prevent the spread of disease.

Making the pieces

1 The first step requires the production of the component plastic pieces from plastic pellets. This is done by a process called injection molding. Pellets of plastic are put into a large holding bin on an injection molding machine. They are heated to make them flowable.

2 The material is then passed through a hydraulically controlled screw. As the screw rotates, the plastic is directed through a nozzle which then injects it into a mold. The mold is made up of two metal halves that form the shape of the part when brought together. When the plastic is in the mold, it is held under pressure for a specified amount of time and then allowed to cool. As it cools, the plastic inside hardens.

3 The mold pieces are separated and the plastic part falls out onto a conveyor. The mold then closes again and the process is repeated. After the plastic parts are ejected from the mold, they are manually inspected to ensure that no significantly damaged parts are used.

Assembling and labeling

4 The parts are next transported to an assembly line. In this production phase various events occur. Machines apply markings that show dose levels and force measurements. These machines are specially calibrated so each printing is made precisely. Depending on the complexity of the device, human workers or machines may assemble the devices. This involves inserting the various pieces into the main housing and attaching any buttons.

Packaging

5 After the assembly step, the injection devices are put into packaging. They are first wrapped in sterile films and then put into cardboard or plastic boxes. Each part is packaged so movement is minimal to prevent damage. For consumer products, an instruction manual is included along with safety information. These boxes are then stacked on pallets and shipped via truck to distributors.

Quality Control

Quality control checks are done throughout the manufacturing process. Line inspectors check the plastic components to assure they conform to predetermined specifications. Visual inspections are the first test method, but measuring equipment is also used to check the dimensions including size and thickness. Instruments that can be used include laser micrometers, calipers and microscopes. Inspectors also check to make sure the printing

and labeling is correct and that all the parts are included in the final packages.

Since these devices can have various safety issues, their production is strictly controlled by the Food and Drug Administration (FDA). Each manufacturer must conform to various production standards and specifications. Announced and unannounced inspections may occur to ensure that these companies are following good manufacturing practices. For this reason detailed records must be kept related to production and design.

The Future

Many of these needle-free alternative technologies are in the development stage. Companies are still working on producing devices that are safer and easier to use. They are also working on alternatives which can deliver even more types of medicines. Inhalers are being improved as are nasal sprays, forced air injectors and patches. In the future, other foods may be genetically enhanced to deliver vaccines and other drugs. These include foods like bananas and tomatoes. In fact, bananas are being looked at as carriers for a vaccine to protect against the Norwalk virus. Tomatoes that protect against hepatitis B are also being developed. In addition to new delivery systems, scientists are also investigating methods for producing longer lasting drugs that will reduce the number of needle injections.

Where to Learn More

Periodicals

Henry, C. "Special Delivery." *Chemical & Engineering News* (September 18, 2000): 49-65.

Potera, C. "Making Needles Needless." *Technology Review* (September/October 1998): 67-70.

Potera, C. "No-Needle Vaccine Techniques." *Genetic Engineering News* (August 1998): 19.

Seppa, N. "Edible Vaccine Spawns Antibodies to Virus." *Science News* (July 22, 2000): 54.

Other

Bioject (December 2000). http://www.bioject.com (January 2001).

—Perry Romanowski

Paintball

The paintball industry brought in an estimated $700 million in 1999, and as the number of teenagers is expected to increase over the next decade, overall sales are expected to hit the $1 billion mark within a few years.

Background

Paintball is a game developed in the 1980s that soon became popular worldwide. Players shoot pellets of paint from airguns at opposing players in a strategic game similar to the children's classic Capture the Flag. A trademarked version of paintball, called the Survival Game, is the standard version, though the game is played with many variations. Most games are played outdoors in special paintball fields. The owners of the field typically rent all the equipment necessary, and charge players a fee for use of the area. The basic equipment for the game includes a specially designed airgun, paintballs, carbon dioxide cartridges to expel the paintball, and safety goggles. Most people play paintball in teams. In officially sanctioned events, team size is 15 players, but nonofficial games often attract much larger teams, of 40-50 players, or even more. Paintball has been used as a way to build team spirit, and so it is popular with business groups hoping to increase corporate communication. Though people of all ages and genders can play, paintball's main enthusiasts are adolescent boys. The paintball industry brought in an estimated $700 million in 1999, and as the number of teenagers is expected to increase over the next decade, overall sales are expected to hit the $1 billion mark within a few years. As paintball went from a faddish extreme sport to a more mainstream pastime in the 1990s, paintball equipment moved from specialty stores into large retailers such as Kmart. In the United States, there are a number of magazines devoted to paintball, and many players gather information about the sport from prominent websites.

History

The game of paintball was first played in 1981. It was invented by Charles Gaines, the author of the bodybuilding classic *Pumping Iron*; Hayes Noel, a New York stockbroker; and a ski shop owner named Robert Gurnsey. Gurnsey, Noel, and Gaines were old friends who had often discussed ways of testing survival in a combat or outdoor situation. They got the idea for the game after seeing an advertisement for a paint gun used to blast paint pellets at steers for marking purposes. This paint gun was developed much earlier by Charles Nelson, of the Michigan-based Nelson Paint Company.

Charles Nelson founded his paint company with his brother Evan in the 1930s. He was an eager entrepreneur, always looking for new ways to use or market paint. In the 1950s, Nelson developed a paint marker for the Forestry Service. The Forestry Service was a major consumer of paint that was used to mark trees for cutting or clearing. The Forestry worker's lot was often hard, as he had to lug a five-gal (19 L) bucket of paint through dense woods, sometimes wading through streams or scrambling up steep banks. Nelson devised a simple paint squirter that allowed forestry workers to mark their trees from a more comfortable distance. Apparently, sales were not what Nelson hoped. But he thought of another market for the device, cattle herders. Cattle needed to be marked often, to distinguish which animals were to be sold, for example, or which to be separated for inoculation or artificial insemination. Traditionally, cowboys rode up close to the animals and marked them with chalk. Nelson modified his first paint gun for the cattle industry. Instead of a squirter, which produced a wide splat of paint, he developed paint-filled pellets that could be shot out of an air gun. The pellets would break on impact, leaving a paint mark. Nelson made wax prototypes of the pellets, and eventually had them manufactured by a Michigan pharmaceutical company, R. P. Scherer. He advertised his "Nel-Spot Pellet Pistol" in farming and ranching magazines, boasting that the gun was fast, safe, and economical. It could hit the animal accurately from about 75 ft (23 m) away, and was useful not only for cattle ranchers but for wildlife game managers and animal census takers.

At some point, Gaines, Noel, and Gurnsey saw an advertisement for Nelson's paint markers, and decided to organize a survival game using them. They rounded up nine friends and played a capture-the-flag-type game on 100 acres (40 hectares) of New Hampshire woods in June of 1981. The three originators soon formed a corporation, the National Survival Game, Inc., and popularized the sport. It received tremendous media attention early in the 1980s, and grew in epidemic proportions through the decade. By 1989, an estimated 75,000 people were playing paintball every weekend in the United States, with many more enthusiasts playing in Canada, Europe, Australia, and beyond. Specialized playing fields and stores for the equipment sprang up across the country, with Southern California alone boasting more than 50 playing fields. Different versions of the game developed, including "Civil War," where players faced each other across a field and loaded their pellets one at a time, in the style of weapons used during the Civil War. The companies that arranged therapeutic paintball sessions for their executives included many bastions of corporate America, such as Rockwell International and Sears. Though paintball used guns, backers emphasized that it was played for fun, and was not a war game or combat training. Even church groups went on paintball excursions by the early 1990s. By the end of the 1990s, paintball had grown to a multimillion dollar international industry. The use of paintballs spread beyond the game, and by the late 1990s, media reports surfaced of paintball big game hunts, such as the opportunity afforded to tourists to fire paint at an elephant. Paintball weapons also advanced in sophistication, resulting in controversy over the use of potentially harmful heavy automatic fire from machine-gun like paintball instruments.

Most paintballs were manufactured by pharmaceutical companies that already used the encapsulating equipment for the pellets in other items, such as vitamins and bath beads. These companies were making over three billion paintballs a year by the end of 1999. As the sport and the industry grew, many specialized manufacturers of paint pellets and other equipment sprang up, and these merged and consolidated in the 1990s. Industry leaders included the Brass Eagle Company and ZAP Paintballs, Inc.

Raw Materials

The paint used for paintballs is soluble in water, so that it washes easily out of players' clothes. It is nontoxic, as well, in case a player is hit in the mouth and accidentally swallows the paint. The basic materials for the paint are mineral oils, food coloring, calcium, ethylene glycol, and iodine. The paint is encapsulated in a bubble made from gelatin. This is the same material used in encapsulated medicines, such as many pain killers and cold treatments, and in liquid vitamins, such as vitamin E.

The Manufacturing Process

Making the paint

1 The paint for paintballs is a specialized product because it is both water-soluble and biodegradable, and has been developed for optimum characteristics in the encapsulating process. Typically, the paint is made at a specialty paint facility, then shipped to the encapsulating plant. A very large manufacturer may combine the two operations.

Encapsulation

2 Encapsulating the paint is done with specialized equipment. When the game of paintball was first getting started, manufacturing was done at pharmaceutical companies, which already had the equipment in place. As the industry evolved, paintball manufacturers furnished their own factories. The large machines cost millions of dollars. At a large facility, making paintballs is done as a continuous process, with the machines

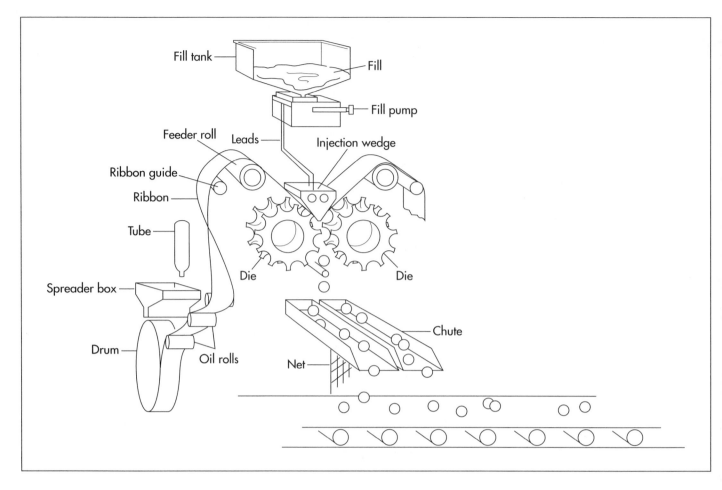

Paintballs are made using encapsulating machines.

active seven days a week, 24 hours a day. Several hundred workers staff the factory. Most are required to wear specialized clothing and footwear, as well as caps to cover their hair, in order to ensure a clean and relatively dustless work area. To make the capsules, workers load two wide strips of softened gelatin into the encapsulating machine. The strips move through two counter-rotating drums. These drums are lined with pockets or dimples that form the paintball casing. As the gelatin is pushed into the dimple, the machine automatically injects a precisely measured amount of paint into the cavity. It also automatically seals the two strips together, encapsulating the paint.

Tumbling and drying

3 The gelatin is soft and warm at this point. The balls must be cooled and hardened in a tumbling machine. This machine gently shakes the paintballs around. The rotating action of the tumbler spins the paintballs, so as they dry, they end up uniformly round.

Drying

4 Next, workers empty the tumblers and place the paintballs on shelves. The shelves are stacked on wheeled racks, and the paintballs are left to air dry. The amount of time the balls dry varies from factory to factory, and this, along with the exact formula of the gelatin, time in the tumbler, and many other aspects of paintball manufacturing, is regarded as a trade secret.

Inspection and packaging

5 When the balls are thoroughly dried, they are ready for packaging. Workers move the balls to the packaging area. They visually inspect them for an obvious flaws. A more rigorous quality check is performed on some of the batch. Workers load the balls into hoppers, and a machine automatically packages them by weight. Paintballs are sold by the case, which is supposed to hold 2,500 balls. But because the machine makes up the case by weight, the actual number in the case usually varies from approximately 2,490-2,510 balls.

Quality Control

A large paintball facility makes paintballs in a continuous process, but the process is still broken up into numbered lots, so that the manufacturers can perform an exact quality control process. A certain percentage of each lot is set aside for inspection and testing. After drying, a worker performs a visual check to find any obvious abnormalities. Then the balls are tested further. Workers place them in testing machines that measure the balls' weight and diameter. A drop test is done to test for brittleness. A properly manufactured paintball should burst on impact, but not sooner, so this is a very important step. After the paintballs have passed all these tests, some are taken to a target range and shot out of paintball guns as a final all-around field test.

Byproducts/Waste

Because paintballs are, for the most part, used outside in open areas, they are specifically manufactured to be biodegradable. Both the paint and the gelatin dissolve in water, so the waste from spent paintballs washes way in the rain.

Where to Learn More

Books

Barnes, Bill. *The Survival Game.* New Haven, CT: Mustang Publishing, 1989.

Periodicals

Bark, Kathleen Dombhart. "Paintball: Tactics Help Build Business Teamwork." *Memphis Business Journal* (May 18, 1992).

"Paintball Business Honors Charles J. Nelson for His Contributions to the Paintball Industry." *Paintball Business* (Winter 1999).

—*Angela Woodward*

Patent Leather

The process for making patent leather was invented in 1799 by an Englishman, Edmund Prior.

Background

Patent leather is leather that has been finished with chemicals that give it a shiny, reflective surface. It is usually black, and has long been popular for dress and dancing shoes. Most stages of the preparation of patent leather are the same as for other fine quality leathers. However, it is in the final finishing stage, when it is coated with a lacquer to give it its characteristic glossiness. All leather is derived from animal skins or hides. Most hides are a byproduct of the meat industry. The hides of cattle slaughtered for beef form the bulk of the leather industry. Other common leathers are made from the hides of sheep, goats, and pigs, and so-called novelty leathers are derived from reptile skins, such as alligator and snake, and even from the ostrich. Patent leather is usually light and thin, and usually derived from a calf or a kid. Today, however, patent leather can be made from any kind of hide, and need be of no finer quality than most shoe leathers.

Mammal hides are comprised of three layers: a hairy outer layer, a thick central layer, and fatty inner layer. The process of making leather, called tanning, involves removing the fat and the hair, and working a chemical change on the thick middle layer to preserve and strengthen it while giving it flexibility. A hide removed from a slaughtered animal begins to decompose within just a few hours. So the first step in tanning is to preserve the hide. Throughout history, this was usually done by salting. Then, the preserved hide is treated in any of a number of ways to remove the hair and dissolve the fat. It is then treated with chemicals that work on the collagen, a fibrous protein making up most of the middle layer of the skin. The word tanning derives from tannin, a chemical found in many plants that reacts with collagen to strengthen its molecular bonds. When tanned, the original hide becomes strong, elastic, and durable.

The treatment of animal hides to make leather is an ancient art. The basic technique of tanning leather dates back to prehistoric times, when primitive peoples apparently tanned hides with plant matter. The ancient Egyptians and the Hebrews tanned leather with plant products. The Hebrews used oak bark, and the Egyptians the pod of a plant called babul. The Romans had a thriving tanning industry, using certain tree barks, berries, and wood extracts. Tanning was lost in Europe during the Middle Ages, but the art was kept alive in the Arab world, and reintroduced to Europe later. By the eighteenth century, tanning was widespread in the Old World and the New. Though tanning was a relatively low-technology operation, it still required some specialized tools, such as fleshing knives, scrapers, and soaking vats. Up until the late nineteenth century, all tanning chemicals were plant derivatives, such as hemlock, oak, or sumac bark. Tanners salted hides, soaked them in lime to dehair them, delimed them in an acid solution, usually manure, and then soaked the hides in increasingly strong solutions of vegetable tannin.

At the end of the nineteenth century, chemical tanning became possible. In this method, the tanning agent is chromium sulfate. The process was discovered in 1858, and the first commercial production of chrome tanned leather was in New York in 1884. Though the initial method had some drawbacks, chrome tanning quickly replaced vegetable tanning. As the industry developed in the twentieth century, the tanning process was increasingly mechanized. Large machines made high volume possible. Earlier tanneries were usually situated near a source for vegetable tanning materials, such as the many that grew up in Virginia, Tennessee, and North Carolina in the United States because of the availability of chestnut wood. By the early twentieth century, vegetable tannins were being imported in large amounts from South America, and the ingredients for chrome tanning were not tied to any particular locality. Tanneries thus could be built anywhere, and centered in the Midwestern region of the United States, site of most beef slaughtering. Entering the twenty-first century, the tanning industry in the United States is declining as low labor costs in other parts of the world make imported leathers more economical.

Leather has many uses and comes in many forms, from thick, sturdy cow hide leather for straps and harnesses to soft kid leather for gloves. The most common shoe leather up through the nineteenth century would have been a very heavy sort to make sturdy boots. For practical purposes, both men and women in Europe also wore wooden shoes or iron-soled shoes called pattens to hoist the wearer above the mud and muck. From the time of Louis XIV up through the early nineteenth century, men's shoes were more subject to the whims of fashion than women's, as women's feet were usually covered by voluminous skirts. The exception was dancing shoes. Both sexes of the upper classes craved fancy, fashionable flat shoes for balls and parties. It was for this kind of shoe that patent leather first became popular. The process for making patent leather was invented in 1799 by an Englishman, Edmund Prior. Prior patented a process for painting leather with dyes and boiled oil, and finishing it with an oil varnish. In 1805 another patent was granted, this time to one Mollersten, for a leather finishing technique using linseed oil, whale oil, horse grease, and lamp black. The shiny, black, waterproof surface offered by this patent or "japanned" leather set off a craze for it in England and abroad. Patent leather first appeared commercially in 1822, and remained popular in cyclical fashion through the present day. The earliest patent leathers would have been made from fine leathers, such as calf or kid. The

leather was tanned by the usual process for making black shoe leather. From there, the tanner carefully coated the leather with a varnish imbued with dyes and other ingredients. A patent in 1854 described the varnish ingredients as "oil, amber, Prussian blue, litharge, white lead, ochre, whiting, asphalt, and sometimes copal." In practice, many tanners kept their varnish recipe secret, and even the ingredients listed in patent applications may have been falsified in order to throw off competitors. Linseed oil of sufficient purity and the dye known as Prussian blue seem to have been the basis of most patent leather finishes. Starting with a fine, black leather, the tanner built up layers of varnish, applying as many as 15 coats, drying the leather in the sun or in a stove in between. The trick was to get a smooth, hard finish that was also somewhat elastic, so the leather did not crack later. The modern process for producing patent leather is not very different, except in mechanization, from that used in the nineteenth century. The same problem exists of finding a balance between a hard finish and a flexible one, and manufacturers use varying recipes and techniques.

Raw Materials

The earliest patent leathers always started with a fine quality leather. Because the varnishes used today work better than the early linseed oil formulas, now almost any quality leather can be given a patent finish. Most patent leather today begins with cattle hide. The finish is a blend of polyurethane and acrylic. These two materials have different characteristics. Polyurethane gives a hard finish, shiny and durable, but acrylic results in a more flexible final product. So leather chemists combine the two for optimum qualities. The actual finish used thus will be different from tannery to tannery, and perhaps from batch to batch. The finishing material is also imbued with black dye. Dye formulas vary widely from plant to plant, as well. Other raw materials are common to leather manufacturing as a whole: salt for curing the hides; disinfectants; lime or other caustic chemicals for dehairing; various acids and salts for deliming the hides and getting them to the proper pH balance for tanning; chromium tanning salts, and water for various stages.

Animal hides must be cured in order to prevent decomposition.

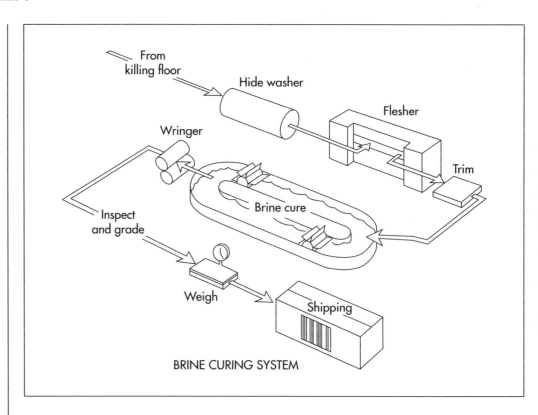

From killing floor

Hide washer

Flesher

Wringer

Trim

Brine cure

Inspect and grade

Weigh

Shipping

BRINE CURING SYSTEM

The Manufacturing Process

Preparing the hide

1 The hide used is usually cow, and it is produced as a byproduct of the meat industry in most cases. That is, cattle are principally slaughtered for their meat, and then the hide is sold to a tannery. The hide is removed by skilled workers who cut it carefully to preserve its integrity. Any stray cuts or marks can seriously affect the quality of the hide. Within hours after removal, the skin will begin to decay because of the large amount of organisms both on the hair side and the meat side. So the skin is immediately preserved in salt. The hides may be simply laid down, covered with salt on both sides, and the next hide stacked on top. Alternately, in a large commercial slaughterhouse, the hides are taken from the killing floor and sent through a chilling machine. This is a large tumble washer that both cleans off surface dirt and manure, and brings the temperature of the hide down so that the clinging fat solidifies. Next, workers pass the hides through another instrument called a fleshing machine. A pair of workers feed the hides one at a time through the cylinders of the fleshing machine, where the manure is knocked off into one container,

and the remaining fat and meat into another separate container. The fat and meat can be sold by the slaughterhouse. The cleaned hides are then loaded into a vat of brine.

At the warehouse

2 After the hides have cured in the brine for at least 24 hours, the slaughterhouse ships them to the tannery. In the United States, most tanneries maintain large warehouses for cured hides, and they could store hides for as long as a year before any further processing. This practice changed around the late 1970s, and now most domestic tanneries work on the "just in time" manufacturing principle, keeping very little hide in stock. So though the cured hides could be kept for quite some time before tanning, in present-day practices, they might proceed directly to the next step.

Soaking, liming, and bating

3 The cured hides undergo several steps at the tannery before they are ready for tanning. These collectively are called the "beamhouse" operations. Total time in the beamhouse takes 12-24 hours. The term beamhouse derives from ancient practice, when the hide was hung over a special curved log or table known as a beam for the

Patent Leather

To make patent leather, tanned leather is coated three times with a polyurethane/acrylic solution and then vacuum dried to seal the coating to the leather. Dye is added to the middle coat to give patent leather its shiny black look.

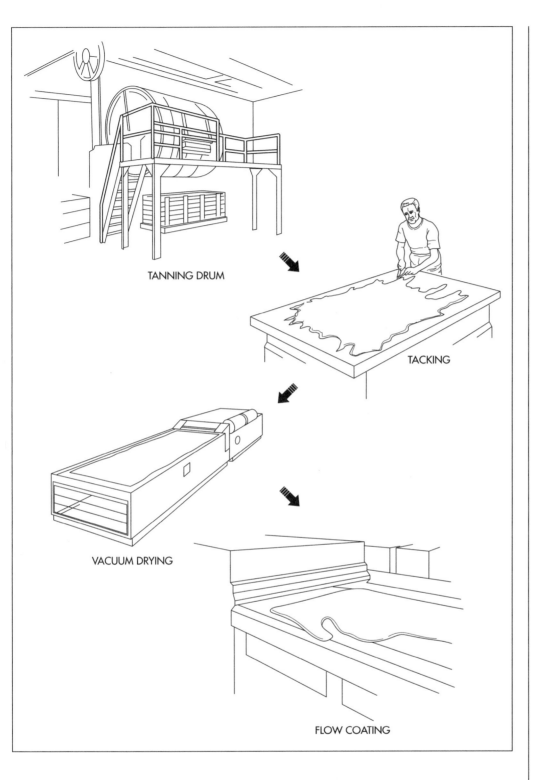

TANNING DRUM

TACKING

VACUUM DRYING

FLOW COATING

dehairing. First tannery workers soak the cured hides in cold water in a vat or drum. This removes the salt from the brine cure. Or if the hides have been cured in dry salt, it rehydrates them. Next lime or another caustic chemical is added to the soak, to loosen the hair. The hides swell up at this stage, becoming blue-white and rubbery. Then, the hides go through a step called bating. Bating gets rid of the hair and fat and other unwanted particles. It also slowly reduces the pH of the hides, from highly alkaline to neutral or slightly acidic. The hides are washed, then placed in a bath of warm water with some calcium salts and an enzyme. More warm water is run into the bath, gradually increasing the temperature. The action of the enzyme lowers the alkalinity of the hide. The fat also breaks down. Gradually, the water temperature is decreased. The hides are washed until all hair, fat, and chemicals have been removed.

Tanning

4 Now the hides are ready for tanning. Workers load them into a huge rotating drum. The drum is filled with the tanning solution, made of chromium salts in water. The hides soak in the tanning solution for eight to 12 hours. The chemical action of the chrome transforms the hide into leather. Dyes in the solution also give the leather its color. For patent leather, this is usually black. Workers remove the leather after the appropriate time in the bath, and send it to a drying area for at least 24 hours.

Finishing

5 All the previous steps apply to any leather. Only in the finishing is the leather transformed into the specific product of patent leather. In the United States, a common finishing technique used to be a spray application of the polyurethene and/or acrylic. But because of air pollution concerns, most patent leather finishing is done by some kind of so-called aqueous dispersion, that is, a liquid application. One common method is to use a machine called a flow coater. Workers load a tank above a conveyor belt with the liquid polyurethane/acrylic. The hides pass beneath the tank on a belt. A waterfall of overflowing liquid hits the traveling hide, and it becomes coated with the finish. Next, the finished hides are stretched on boards and pass through a heated tunnel to dry. Depending on the tannery and the particular application, the drying tunnel may use infrared lights or ultraviolet. The first coat of finish is formulated so it penetrates the leather completely. After drying, the hide is put through the flow coater again, this time for a middle coat that includes dye. Then it is dried as before, and put through for a third and final top coat. This top coat is clear, and dries hard, shiny, and waterproof.

Final steps

6 After the last coat dries, the leather is ready to move on to its buyer, most likely a shoe manufacturer. Despite all it has gone through, the leather is still in its original shape. It has not been cut except perhaps to trim some thick or damaged areas. The shoe manufacturer cuts it into many pieces, with as little waste as possible.

Quality Control

Quality control differs from tannery to tannery, and it depends mostly on for what the customer contracts. Good patent leather should not crack, the finish should be thoroughly dry and hard to the touch, not tacky, and it should not scuff easily. A fully equipped tannery might carry out tests for all these conditions, as well as chemical analyses of the finish. Other tanneries may just visually inspect the end product. Usually, the customer for the finished patent leather must agree with the tannery what tests should be carried out or what standards the leather should meet.

Byproducts/Waste

Tanning leather and finishing it into patent leather creates much waste water. And if a spray application of the finish is used, this creates air pollution. In the United States in the 1980s, the Environmental Protection Agency (EPA) stiffened its standards for air emissions from tanneries, and as a result, most now use water-based finish applications. Tanneries must find ways to deal with waste water, which is heavily polluted with chemicals. The water can be cleaned in a wastewater treatment facility. Then the cleaned water can be reused by the tannery. Some leather byproducts can also be reused. Rawhide scraps can be sold as dog chews. The waste hair, fat, and other animal solids can be collected and made into fertilizer. Though tanning is an industry that has a reputation for pollution and unpleasant smells, it is possible for a dedicated plant to recycle its waste for miminal environmental impact.

Where to Learn More

Books

McDowell, Colin. *Shoes: Fashion and Fantasy.* New York: Rizzoli, 1989.

Thorstensen, Thomas C. *Practical Leather Technology.* Huntington, NY: Robert E. Krieger Publishing Co., 1976.

Welsh, Peter C. *Tanning in the United States to 1850.* Washington, DC: United States National Museum, 1964.

Periodicals

McDowell, John. "Leather Company Creates Alternative to Landfilling." *BioCycle* (June 1998): 32.

—*Angela Woodward*

Pillow

Background

Americans usually have two or three pillows on their bed. Today, pillows are stuffed primarily with materials such as polyester (a synthetic), feathers, down, or a combination of the latter two. The least expensive pillows to manufacture are polyester, although they are the most durable, easily washed, and cause few allergic reactions. The most expensive is the pillow filled with goose down. Feathers are a moderately priced stuffing. Some higher-end pillows may be filled with a combination of goose feathers and down, and that ratio may be varied extensively according to price point (the more down, the more expensive). The pillow filling is distinguished by the tag on the pillow casing, which must be there by law in the event that the consumer may be allergic to the contents.

Pillows are still manufactured in great quantities in the United States. They are also produced outside the country, but pillows are generally not imported to the United States. Shipping is measured by volume and pillows are extraordinarily expensive to ship. Some manufacturers have tried to have pillows made out of the country—where labor is cheaper—and crush the pillows during transportation in order to save money. However, once the pillow is crushed, it is difficult for it to spring back to its original shape and much of its plushness is lost.

History

The shape and contents of pillows have varied little over time. The wealthier Greeks rested their heads and feet upon richly embroidered cushions and bolsters. The Egyptians, regarding the head as the seat of life, lavished much attention, detail, and money on pillows for the dead. The Chinese, however, thought that soft pillows robbed the body of vitality, and their pillows were made of wood, leather, and ceramic materials. Some were even filled with herbal remedies to cure disease, turn white hair black, restore lost teeth, and inspire sweet dreams.

For centuries, people slept fairly upright with not only a pair of pillows on the bed but a large, cylindrical bolster as well. These bolsters, sometimes nearly the width of the bed, were stuffed with down or some other type of batting and closed up. They were placed against the headboard and were the foundation for the pillows. Then, a pair of pillows was placed upright against the bolster. The sleeper would prop himself up against these pillows, resulting in a sleeping position that was closer to sitting than reclining. Until about the mid-1800s it was thought this position was better for the body.

Other fancy pillows were found on beds of the nineteenth and early twentieth centuries. Sometimes large, square pillows were placed within a decorative pillow cover and then placed against the pillows actually used for sleeping on a bed. These were often removed from the bed before sleep. Until **cotton** became easy to obtain around 1840, American women showed their needlework prowess by carefully hand weaving and sewing linen pillow cases and marking them with their initials and the number the case was within a set of pillow cases. As the American textile industry flourished throughout the 1800s, covers for pillows (which housed the stuffing) went from utilitarian linen to the sturdy cotton ticking, still seen on pillows and in fabric stores.

The traditional filler for pillows was, until recently, down and feather. However, as fabrics changed, so too did yarns. Synthetic polyester filling has replaced natural batts as it is has acceptable loft and shape retention, is relatively inexpensive, may be washed, and few people are allergic to it.

Raw Materials

The batting, or filling, itself is the most important part of the pillow. The most expensive filling is down. This is the light, fluffy undercoating of waterfowl, consisting of clusters of filaments growing from a central quill point. Down has a quill point but no quill shaft and is more resilient as a result. It is three-dimensional and therefore has more loft. Thousands of clusters are found in down that trap warm air to prevent heat loss. Duck down is smaller, more plentiful, and less expensive. It is important to note that not all down is the same. Down is rated by *fill power,* which is the volume of space in a calibrated cylinder that 1 oz (28 g) measures. The higher the number, the better the fill power.

Feathers are the principal covering of birds. They are flat and two-dimensional with a hard, tubular quill shaft that runs from one end to the other. Because they are flat, they are unable to effectively trap air and warmth. Feathers are strong, but not terribly soft. Duck feathers are the most common type of feather used in American pillows. Many manufacturers combine down and feather to make an affordable, comfortable pillow.

An other type of filling is polyester, a synthetic material. The cheapest polyester used for pillows is a continuous solid filament polyester which has good initial shape but loses loft fairly shortly. A better grade of polyester is called hollofill, which is also a continuous filament fiber but has a hollow core that gives the pillow more loft for a greater period of time than cheaper grade polyester.

The pillow filling determines the fabric chosen for the pillow casing. While the casing is generally cotton or cotton-polyester, the weight and closeness of the weave varies according to filling. The feather and down filling require a more expensive, very dense, tightly woven fabric that will keep the feather shafts from poking the sleeper and keep the fine down from working its way through the cloth. Polyester batts do not require such closely woven fabrics. These fabrics may have a starch placed on them during their manufacture to make them stiffer and more resistant to penetration. The only other material required for pillow manufacture is a sturdy thread for sewing the pillow itself.

The Manufacturing Process

The following process will describe the construction of a polyester-filled pillow, an inexpensive and commonly produced pillow. It is a small-medium size operation which produces between 2,000-3,500 pillows every day. The largest manufacturers of American-made pillows produce between 10,000-15,000 pillows each day.

1 The pillow covering must be constructed first. Sturdy cotton or cotton-polyester fabric is shipped to the factory in huge bolts. The fabric must be treated or *calendared* so that the sewn casing may be blown apart and easily separated during filling. Once calendared the fabric is taken to huge tables and cut apart—dozens of layers at a time—with either heavy shears or fabric cutting machines.

2 Stacks of rectangular-shaped fabric are taken to the sewing machines. In fully automated plants, automatic sewing machines are used to grab the fabric and sew them together. These machines are quite expensive and many plants still opt for people to do this work. In this case, a machine operator takes two pieces and sews them together around the edges, leaving a space of approximately 6 in (16 cm) open in order to stuff the pillow. As the operator sews the edges together, the tag that lists the pillow's contents is attached. The operator then turns the pillow covering inside out so that the seams are on the inside of the case.

3 The casings are moved to the pillow machine, which blows the polyester filling into the pillows. The machine has been loaded with polyester in one of two ways. More expensive machines need only to have an entire bale (about 600 lb [272 kg]) of polyester inserted into the machine and the machine unloads and combs it. Less expensive machines require an operator to unload

The blowing machine blows polyester filling into a pillow case.

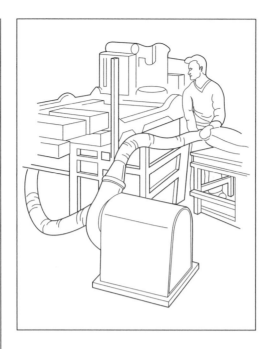

baled and tightly packed polyester by hand. Once the polyester is inside the machine, the blowing machine combs the polyester so it is fluffy and has some loft. The 6-in (15- cm) opening in the pillow covering is slipped onto a tube on the machine. Then, a blower pushes the polyester into the casing. Some machines can fill up to 100 pillows an hour.

4 When the pillow is filled, it is taken to another station and workers close the opening in the pillow case using an industrial sewing machine. The pillow is weighed at that time to ensure that it includes the requisite amount of batting.

5 The completed pillows are moved to machines for bagging them. The machine blows air into thin plastic bags to open them up and then inserts individual pillows. The bagged pillows are placed into boxes, ready for shipping.

Quality Control

The process described above is really quite labor-intensive as it includes many workers on the floor of the factory in order to fulfill the requirements of manufacturing pillows. As a result, these operators are able to scrutinize the quality of virtually all aspects of the manufacturing process. Operators are vigilant that the dozens of layers of fabric are carefully put down on the cutting table so that the fabric is not askew prior to cut-

ting. This guarantees that the casings can be quickly and evenly sewn. Sewing the casing of the pillow requires careful work and is monitored by supervisors as well as sewers. Polyester batting must be de-baled by hand (metal bands are removed from the large bales) and are examined to ensure that the quality is suitable for pillow inclusion. As operators load the blowing machine with polyester they examine the equipment for defects and inefficiencies. Workers are present as the pillows are blown with batting. Then the stuffed pillows are weighed to ensure that they are the weight represented in the labeling.

Byproducts/Waste

The primary byproducts of the polyester-filled pillows is the metal baling bands used to bundle and transport the polyester batt. This metal is desirable for recycling; one company has neighborhood recyclers pick this up free of charge and recycle it (they may be paid a fee for the scrap metal). Cartons are picked up for recycling a well. Because the fabric is generally produced and calendared elsewhere, bleaches and finishes are not a part of this operation. Polyester fibers themselves are not prone to becoming minute airborne fibers. The 2-in (5.8-cm) or 3-in (7.6-cm) cut lengths of fiber generally do not get into operators' lungs and are not the health-hazard that other fillings are. Down, with its very short filaments, creates a great deal of down dust, and it is imperative to use masks to protect the lungs when working with down and, to a certain extent, feathers.

The Future

Pillows have varied little since they were first used. They are now also made with blends of hypo-allergenic fibers so that even people with allergies or extremely sensitive skin are able to enjoy their comfort. In this age of therapeutic remedies, some pillows are reverting back to the Chinese method of including herbs to relieve aliments and give a better nights rest. Orthopedic pillows are also advancing rapidly. They are filled with or surrounded by foam (some even contain gel or water that can be heated or cooled) that is either already formed or forms around the head, to fully support the neck. These pillows help to relieve neck, back,

popular in the United States, and Gottlieb sold more than 50,000 games in its first year. Ray Moloney, one of Gottlieb's distributors, invented his own similar game, Ballyhoo, in 1932. This brightly colored game became even more popular than Baffle Ball. Moloney formed the Bally Manufacturing Company to manufacture his games, and sold 75,000 games in 1932. Bally continued as one of the major pinball manufacturers through the twentieth century. Both Bally and Gottlieb's company made money, even though the country was in the midst of the Great Depression, and their success set off a slew of imitators. More than a hundred companies began manufacturing similar games in the early 1930s, and pinball became a fixture of not only taverns, but also drugstores, barbershops, and gas stations. By 1935, the game design had changed so that the playing field had its own table. The games were electrified, so that parts of the playing field could light up, and the game could keep score and pay out prize money automatically. Harry Williams, whose Williams Manufacturing Company became one of the foremost pinball manufacturers in the United States, added significant thrill to the game by electrifying the playing field with a "kicker" that could shoot the ball out of a hole and back onto the field. Williams's addition made the game much more fast-paced.

Pinball in the 1930s was a more a game of chance than of skill. The movement of the balls was set by the plunger action at the beginning of the game, and players could only jostle the table to influence its course. The mathematics of the game had been studied by French philosopher Blaise Pascal and by Sir Francis Galton in England, who was a cousin of Charles Darwin and best known as the father of eugenics. These men had charted the probability of the ball reaching certain holes; skill at the game was close to that needed at roulette or dice. In the wild proliferation of Depression-era coin-operated games, some manufacturers made ones that emphasized the gambling aspect, such as mechanical dog races and horse races. This tarred the reputation of pinball. By the end of the 1930s, pinball shared other gambling games' bleak reputation, and many municipalities moved to ban it. Pinball was outlawed in Los Angeles and Chicago, centers of pinball manufacturing, and in New York,

Mayor Fiorello LaGuardia made opposition to pinball a focal point of his mayoral election campaign. Pinball was outlawed in New York City in 1941, and a mayoral commission declared that pinball could lead youths to a life of crime. Pinball still flourished, however, in suburban locations, particularly in roadhouses beyond city bounds.

The invention of flippers to push the ball on the playing field changed pinball decisively. Flippers were invented by an engineer for Gottlieb Manufacturing, and they were first put into the game Humpty Dumpty in 1947. By pushing buttons on the side of the game table, players could make the finger-like flippers bat the balls in play. Other manufacturers immediately copied Gottlieb, and flippers became standard on all pinball games ever since. Now the game definitely depended on the player's skill, and so pinball was able to disassociate itself from gambling games. A 1956 Federal court case made a definitive ruling that flipper-type pinball games were not gambling games and could not be regulated as such. However, municipalities still made distinctions between different types of pinball games. In some areas, machines that rewarded high-scoring players with a free game were illegal, and in other places, the automatic plungers that shoot the ball onto the playfield were the banned element. Such ordinances began to be repealed in the late 1960s, though it was not until January 1977 that pinball was finally legal again in Chicago. New York had legalized pinball the year before, and Los Angeles in 1972.

Games increased in artistry in the 1970s, particularly with the rise of colorful lighted backglasses. The backglass is the vertical portion of the game table that shows the score. These had always been eye-catching, but in the 1970s they became larger, and with evocative artwork, often with a fantastic theme. Pinball technology also advanced in the 1970s with the addition of microchips that allowed machines to tally scores and remember them. Games of the late 1970s had many new features, such as speech and enhanced sound, and more complicated scoring with bonus balls, mystery points, and triple-level playing fields. Nevertheless, by 1982, the pinball manufacturing industry arrived into a slump caused by the popularity of video games. To keep down costs, com-

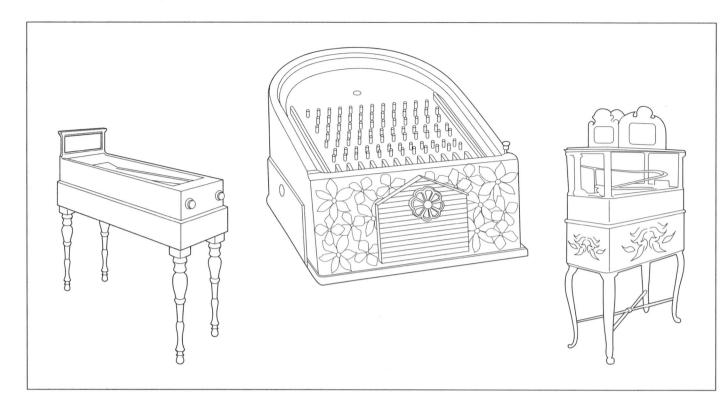

Examples of historic pinball machines

plicated games were abandoned for simpler versions, and new games were not nearly as innovative as the games of the previous decade. Major manufacturers went bankrupt or ceased production. By the end of the 1990s, only one manufacturer was left, Williams Electronics Games, Inc., in Chicago, which made both Williams and Bally brand games. Williams ceased production of traditional pinball games in October 1999. Old pinball games and pinball backglass art are avidly collected by fans.

Design

While a new pinball game took only 20-30 hours to manufacture, the design process was far more protracted. Designing a new game took a team of professionals from 12-18 months, and the investment cost could run up to a million dollars. Games were designed by teams, with members taking on specialized tasks. The team included a game designer, who came up with the game concept and generally organized the other members; a software developer; an artist to come up with the backglass art and playing field graphics; a mechanical engineer, responsible for the design of all the gadgets on the playing field; a sound designer to create the music and other sound effects for the game; a mechanical de-

signer, who brought together the drawings and designs from other team members; an animator to animate images on the scoring display; an electrical engineer and a cable designer to work out the circuitry and wiring; sculptors and model-makers to build the prototype devices for the playing field, and a publications writer to write the technical documents and manuals. The team might include other people, as well, such as someone who kept track of the cost of materials, and people working in marketing, sales, and licensing.

The design work starts with a concept, which might be an original idea or might come from a licensed product, such as a movie. After the concept is agreed upon, the design team works on the layout of the playing field. Designers fiddle with placements of any special elements the game will have, and try to work out how the player will make shots leading to the element. The placement of ramps, flippers, kickers, tilt mechanisms, and such is planned out, while the mechanical engineer produces prototypes for the special elements.

After the playfield design is more or less completed, the team then produces what is called a whitewood version of the game. This is a bare plywood prototype of the

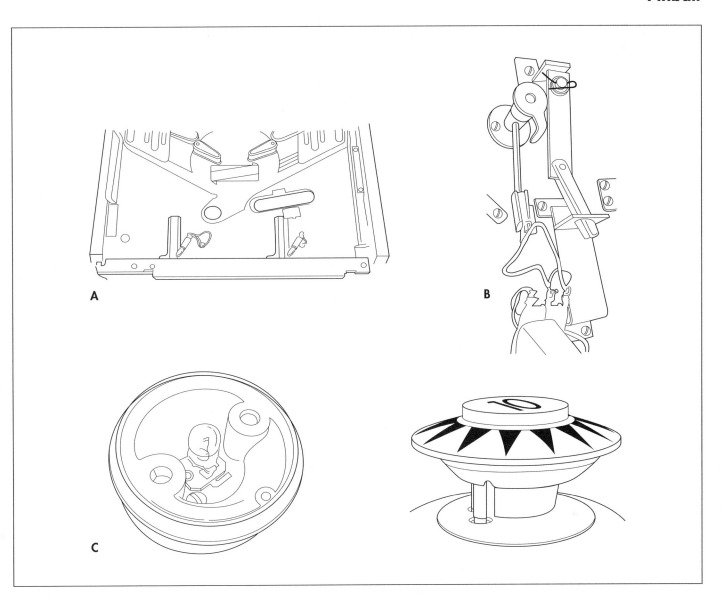

playfield. The designers use computer assisted design (CAD) software to record their plans, and this software is then used to control a computer-driven routing machine that cuts the board. The game elements for the whitewood are made by hand and glued on. Lights, ramps, and flippers are added. The whitewood may go through many revisions, as placements are changed or elements taken off or added. When the design seems satisfactory, the whitewood is wired to make it playable. Meanwhile the software developer has been creating the unique software to control the game. Williams, the last major American pinball manufacturer, had developed its own proprietary pinball operating system so that a basic framework could be used for each game, simplifying the design process. The software developer also works with the lighting and sound designers and animators to run all these effects. At the whitewood stage, the designers can actually play the game they have created, and work on developing rules.

Meanwhile the artist creates sketches for the backglass. If the game is based on a licensed theme, the artist needs to submit sketches for approval to the licensing agency. When the whitewood is near its final form, the artist makes sketches for the playfield. Pinball artists in the 1990s utilized traditional media such as pen and ink and paint, as well as computer graphics.

While the game is in development, team members keep constant track of how much the game might cost to make. If parts need to be ordered from an outside vendor, the team needs to retrieve estimates from the suppliers. The game has a total budget, and sometimes the cost of one spectacular effect means that other game elements, such as lighting, have to be redesigned to make them cheaper. The design team also creates a bill of materials, listing every part needed to make the game. Parts are ordered so that everything is in place when the game is ready to be mass produced.

After the design is finalized, workers at the plant assemble 10-20 prototypes. These are completely finished games, just like ones that will eventually be sold. The prototypes help the production plant determine that all the parts fit together, that the suppliers have sent the correct parts, and that mass production will go smoothly. Any glitches in the assembly are caught at this stage. And if any part of the manufacturing process needs to be modified, it is discovered now. The prototypes are also used for testing.

Raw Materials

The raw materials for pinball are not extraordinary. Inexpensive, sturdy plywood is used for the playfield, wiring boards, and cabinet. Metal parts, such as screws and springs, are made out of a variety of common materials, such as stainless steel and aluminum. Special elements on the playfield may be of molded plastic. Manufacturers require vendors to supply parts exactly to specifications, and to hold down production costs, which are considerable, the cheapest materials possible may be used.

The Manufacturing Process

Pressuring the playfield

1 Almost all the production of a pinball game is done by hand. The exception is the manufacture of the playfield. The design for the playfield is programmed into a special machine. Workers load plywood into the machine, and it squeezes each board under high pressure. The pressure incises each side of the board with marks, which show where all the assembly units go.

Screen printing

2 Pinball backglasses and playing field art are made by the silk-screen printing process, also called stencil printing. This is usually done by an outside vendor. From the artist's design, the printer makes a series of patterns called stencils that break down the design into four colors. That is, all the red parts of the design are isolated on one stencil, all the blue on another, and so on. Through combinations of four colors, every possible shade can be made. The stencils are made of a sieve-like fabric. Ink will pass through the open weave of the fabric and adhere to the material being printed. The printer lays the stencil, which might be a thin, plastic laminate, down on the material. An applicator sprays the color. Then the stencil is lifted, the material briefly dried, and the next stencil laid down. After the fourth color, the design is complete. For a backglass, the whole design could be screened as one piece. For the playfield, the artwork is broken up into small sections. At the pinball factory, workers punch out the sections and adhere them to the playfield. This way, the artwork is made to fit around the holes for the targets and other mechanical parts.

Hand assembly

3 Most of the remaining work is done by hand. Some of the small parts, such as spring-loaded bumper mechanisms, may have to be put together at the factory from components parts. Then these parts are assembled on the board. The wiring is very complicated. Workers painstakingly connect the wires, using screw drivers and soldering guns. The cabinets are made as rectangular hollow boxes. Workers screw rails on to hold the circuit boards and playfield. The screened art for the backglass is inserted in the backglass assembly, and that part attached to the main cabinet. Each game has close to 1,000 parts. The whole game takes from 20-30 hours to finish, and the factory itself requires about 1,000 workers.

Quality Control

Most of the quality control in pinball manufacturing is done before mass production begins. Individual parts from vendors are inspected and measured to make sure they are to specification. But the main check the manufacturer has on total product quality is in the prototypes made after the design process is complete. These 10-20 prototypes are tested in a variety of ways, including playability, general safety, and ability to withstand shipping. One is played on regularly, for field testing. One prototype is tested for sturdiness during shipping-this suffers a drop test. Another is sent to a private laboratory for a safety test, to make sure players could not be injured by the game. Other prototypes are used for publicity photographs, for the rules or manual writer, and perhaps for display at trade shows or promotions. Even after the prototypes pass inspection, the manufacturer might make a small run of some 50-100 games, just to make sure that everything is set in the production room for a bigger batch.

Where to Learn More

Books

Bueschel, Richard M. *Illustrated Historical Guide to Pinball Machines*. Vol. 1. Wheat Ridge, CO: Hoflin Publishing Ltd, 1988.

Eiden, Heribert, and Jurgen Lukas. *Pinball Machines*. West Chester, PA: Schiffer Publishing Ltd., 1992.

Sharpe, Roger C. *Pinball!* New York: E. P. Dutton, 1977.

—*Angela Woodward*

Polyurethane

Over the years, improved polyurethane polymers have been developed including Spandex fibers, polyurethane coatings, and thermoplastic elastomers.

Polyurethanes are linear polymers that have a molecular backbone containing carbamate groups ($-NHCO_2$). These groups, called urethane, are produced through a chemical reaction between a diisocyanate and a polyol. First developed in late 1930s, polyurethanes are some of the most versatile polymers. They are used in building insulation, surface coatings, adhesives, solid plastics, and athletic apparel.

Background

Polyurethanes, also known as polycarbamates, belong to a larger class of compounds called polymers. Polymers are macromolecules made up of smaller, repeating units known as monomers. Generally, they consist of a primary long-chain backbone molecule with attached side groups. Polyurethanes are characterized by carbamate groups ($-NHCO_2$) in their molecular backbone.

Synthetic polymers, like polyurethane, are produced by reacting monomers in a reaction vessel. In order to produce polyurethane, a step—also known as condensation—reaction is performed. In this type of chemical reaction, the monomers that are present contain reacting end groups. Specifically, a diisocyanate (OCN-R-NCO) is reacted with a diol (HO-R-OH). The first step of this reaction results in the chemical linking of the two molecules leaving a reactive alcohol (OH) on one side and a reactive isocyanate (NCO) on the other. These groups react further with other monomers to form a larger, longer molecule. This is a rapid process which yields high molecular weight materials even at room temperature. Polyurethanes that have important commercial uses typically contain other functional groups in the molecule including esters, ethers, amides, or urea groups.

History

Polyurethane chemistry was first studied by the German chemist, Friedrich Bayer in 1937. He produced early prototypes by reacting toluene diisocyanate reacted with dihydric alcohols. From this work one of the first crystalline polyurethane fibers, Perlon U, was developed. The development of elastic polyurethanes began as a program to find a replacement for rubber during the days of World War II. In 1940, the first polyurethane elastomers were produced. These compounds gave millable gums that could be used as an adequate alternative to rubber. When scientists found that polyurethanes could be made into fine threads, they were combined with nylon to make more lightweight, stretchable garments.

In 1953, the first commercial production of a flexible polyurethane foam was begun in the United States. This material was useful for foam insulation. In 1956, more flexible, less expensive foams were introduced. During the late 1950s, moldable polyurethanes were produced. Over the years, improved polyurethane polymers have been developed including Spandex fibers, polyurethane coatings, and thermoplastic elastomers.

Raw Materials

A variety of raw materials are used to produce polyurethanes. These include monomers, prepolymers, stabilizers which protect the integrity of the polymer, and colorants.

Isocyanates

One of the key reactive materials required to produce polyurethanes are diisocyanates. These compounds are characterized by a

(NCO) group, which are highly reactive alcohols. The most widely used isocyanates employed in polyurethane production are toluene diisocyanate (TDI) and polymeric isocyanate (PMDI). TDI is produced by chemically adding nitrogen groups on toluene, reacting these with hydrogen to produce a diamine, and separating the undesired isomers. PMDI is derived by a phosgenation reaction of aniline-formaldehyde polyamines. In addition to these isocyanates, higher end materials are also available. These include materials like 1,5-naphthalene diisocyanate and bitolylene diisocyanate. These more expensive materials can provide higher melting, harder segments in polyurethane elastomers.

Polyols

The other reacting species required to produce polyurethanes are compounds that contain multiple alcohol groups (OH), called polyols. Materials often used for this purpose are polyether polyols, which are polymers formed from cyclic ethers. They are typically produced through an alkylene oxide polymerization process. They are high molecular weight polymers that have a wide range of viscosity. Various polyether polyols that are used include polyethylene glycol, polypropylene glycol, and polytetramethylene glycol. These materials are generally utilized when the desired polyurethane is going to be used to make flexible foams or thermoset elastomers.

Polyester polyols may also be used as a reacting species in the production of polyurethanes. They can be obtained as a byproduct of terephthalic acid production. They are typically based on saturated aromatic carboxylic acids and diols. Branched polyester polyols are used for polyurethane foams and coatings. Polyester polyols were the most used reacting species for the production of polyurethanes. However, polyether polyols became significantly less expense and have supplanted polyester polyols.

Additives

Some polyurethane materials can be vulnerable to damage from heat, light, atmospheric contaminants, and chlorine. For this reason, stabilizers are added to protect the polymer. One type of stabilizer that protects against light degradation is a UV screener called hydroxybenzotriazole. To protect against oxidation reactions, antioxidants are used. Various antioxidants are available such as monomeric and polymeric hindered phenols. Compounds which inhibit discoloration caused by atmospheric pollutants may also be added. These are typically materials with tertiary amine functionality that can interact with the oxides of nitrogen in air pollution. For certain applications, antimildew additives are added to the polyurethane product.

After the polymers are formed and removed from the reaction vessels, they are naturally white. Therefore, colorants may be added to change their aesthetic appearance. Common covalent compounds for polyurethane fibers are dispersed and acid dyes.

Design

Polyurethanes can be produced in four different forms including elastomers, coatings, flexible foams, and cross-linked foams. Elastomers are materials that can be stretched but will eventually return to their original shape. They are useful in applications that require strength, flexibility, abrasion resistance, and shock absorbing qualities. Thermoplastic polyurethane elastomers can be molded and shaped into different parts. This makes them useful as base materials for automobile parts, ski boots, roller skate wheels, cable jackets, and other mechanical goods. When these elastomers are spun into fibers they produce a flexible material called spandex. Spandex is used to make sock tops, bras, support hose, swimsuits, and other athletic apparel.

Polyurethane coatings show a resistance to solvent degradation and have good impact resistance. These coatings are used on surfaces that require abrasion resistance, flexibility, fast curing, adhesion, and chemical resistance such as bowling alleys and dance floors. Water based polyurethane coatings are used for painting aircraft, automobiles, and other industrial equipment.

Flexible foams are the largest market for polyurethanes. These materials have high impact strength and are used for making most furniture cushioning. They also provide the material for mattresses and seat cushions in higher priced furniture. Semiflexible

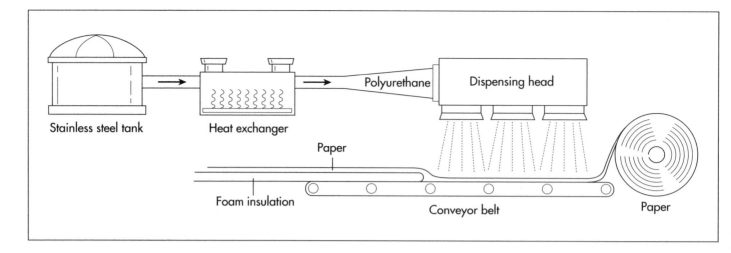

A diagram depicting the manufacturing processes used to create rigid polyurethane foam insulation.

polyurethane foams are used to make car dashboard and door liners. Other uses include carpet underlay, packaging, sponges, squeegees, and interior padding. Rigid, or cross-linked, polyurethane foams are used to produce insulation in the form of boards or laminate. Laminates are used extensively in the commercial roofing industry. Buildings are often sprayed with a polyurethane foam.

The Manufacturing Process

While polyurethane polymers are used for a vast array of applications, their production method can be broken into three distinct phases. First, the bulk polymer product is made. Next, the polymer is exposed to various processing steps. Finally, the polymer is transformed into its final product and shipped. This production process can be illustrated by looking at the continuous production of polyurethane foams.

Polymer reactions

1 At the start of polyurethane foam production, the reacting raw materials are held as liquids in large, stainless steel tanks. These tanks are equipped with agitators to keep the materials fluid. A metering device is attached to the tanks so that the appropriate amount of reactive material can be pumped out. A typical ratio of polyol to diisocyanate is 1:2. Since the ratio of the component materials produces polymers with varying characteristics, it is strictly controlled.

2 The reacting materials are passed through a heat exchanger as they are

pumped into pipes. The exchanger adjusts the temperature to the reactive level. Inside the pipes, the polymerization reaction occurs. By the time the polymerizing liquid gets to the end of the pipe, the polyurethane is already formed. On one end of the pipe is a dispensing head for the polymer.

Processing

3 The dispensing head is hooked up to the processing line. For the production of rigid polyurethane foam insulation, a roll of baking paper is spooled at the start of the processing line. This paper is moved along a conveyor and brought under the dispensing head.

4 As the paper passes under, polyurethane is blown onto it. As the polymer is dispensed, it is mixed with carbon dioxide which causes it to expand. It continues to rise as it moves along the conveyor. (The sheet of polyurethane is known as a bun because it "rises" like dough.)

5 After the expansion reaction begins, a second top layer of paper is rolled on. Additionally, side papers may also be rolled into the process. Each layer of paper contains the polyurethane foam giving it shape. The rigid foam is passed through a series of panels that control the width and height of the foam bun. As they travel through this section of the production line, they are typically dried.

6 At the end of the production line, the foam insulation is cut with an automatic saw to the desired length. The foam bun is then conveyored to the final processing

steps that include packaging, stacking, and shipping.

Quality Control

To ensure the quality of the polyurethane material, producers monitor the product during all phases of production. These inspections begin with an evaluation of the incoming raw materials by quality control chemists. They test various chemical and physical characteristics using established methods. Some of characteristics that are tested include the pH, specific gravity, and viscosity or thickness. Additionally, appearance, color, and odor may also be examined. Manufacturers have found that only by strictly controlling the quality at the start of production can they ensure that a consistent finished product will be achieved.

After production, the polyurethane product is tested. Polyurethane coating products are evaluated in the same way the initial raw materials are checked. Also, characteristics like dry time, film thickness, and hardness are tested. Polyurethane fibers are tested for things such as elasticity, resilience, and absorbency. Polyurethane foams are checked to ensure they have the proper density, resistance, and flexibility.

The Future

The quality of polyurethanes has steadily improved since they were first developed. Research in a variety of areas should continue to help make superior materials. For ex-

ample, scientists have found that by changing the starting prepolymers they can develop polyurethane fibers which have even better stretching characteristics. Other characteristics can be modified by incorporating different fillers, using better catalysts, and modifying the prepolymer ratios.

In addition to the polymers themselves, the future will likely bring improvements in the production process resulting in faster, less expensive, and more environmentally friendly polyurethanes. A recent trend in polyurethane production is the replacement of toluene diisocyanates with less-volatile polymeric isocyanates. Also, manufacturers have tried to eliminate chlorinated fluorocarbon blowing agents which are often used in the production of polyurethane foams.

Where to Learn More

Books

Kirk-Othmer *Encyclopedia of Chemical Technology*. John Wiley & Sons, 1997.

Oertel, G. *Polyurethane Handbook*. Second ed. Munich: Carl Hanser Publishers, 1993.

Seymour, Raymond, and Charles Carraher. *Polymer Chemistry*. New York: Marcel Dekker, 1992.

Ulrich, H. *The Chemistry and Technology of Isocyanates*. New York: John Wiley & Sons, 1996.

—*Perry Romanowski*

Pool Table

An estimated 36 million Americans play pool, one of the most popular participant sports in the country.

Background

In pool (the common American term for pocket billiards), a ball is struck with the end of a long, slender stick (cue), causing it to roll into other balls and knock them into holes (pockets) around the edges of the playing table. A short wall (rail) around the perimeter of the table keeps the balls on the playing surface. The rail is faced with a rubber cushion so balls that strike it rebound predictably and remain in play.

In the United States, the game's governing body is the Billiards Congress of America (BCA). One of the BCA's functions is to define specifications for equipment acceptable for sanctioned tournaments. Although it does not specify the exact size of an approved pool table, the BCA requires that its playing surface be twice as long as it is wide. BCA specifications for the table include maximum allowable surface deflections under a specified vertical force, surface flatness tolerances, size and shape requirements for the rubber cushion and the pockets, and composition requirements for the playing surface and its cloth covering.

History

The origin of billiard tables is uncertain. The most common theory is that around the fifteenth century, tables were used in France and England for an indoor version of a lawn game similar to croquet. A ball (*bille* in French) resting on the table was shoved with a stick (*billart* in French), in order to propel the ball through a wire gate to strike a wooden peg. The function of six pockets around the edges of the table is unclear. Vertical walls (banks in English) around the edges kept balls from falling off the table.

The first recorded billiard table was one sold to King Louis XI of France in 1470.

Billiards developed simultaneously in England. The rules varied from place to place, as described in *The Complete Gamester*, a book published in England in 1674. By this time, the club-shaped billart had evolved into a slender cue. It took another century for the wire gate and upright wooden peg to gradually disappear from billiard tables. Because playing surfaces were made of wood, they had a tendency to warp. As players began to purposely rebound balls off the table's edge walls, builders began to pad the banks with cloth stuffed with horsehair or rags.

During the 1700s, billiards remained popular in France and England, and caught on in the United States. Table sizes varied, but the 2:1 ratio of length to width became standard. Rails were padded with tightly rolled cloth, producing a somewhat more predictable ball rebound.

The Industrial Revolution contributed to a series of improvements in billiard tables. Between 1800 and 1850, chalk was first used on cue stick ends to increase friction, leather cue tips were invented, diamond-shaped sights were added to rails, slate was introduced as a superior table surface, and vulcanized rubber (which maintained its properties regardless of temperature fluctuations) was quickly adapted for rail cushions.

Subsequent refinements in pool tables have related primarily to construction techniques. For example, in older tables horizontal holes were drilled in the slate edges and filled with molten lead; screws running through the vertical edge of the rail were tightened into the lead-lined hole. In contrast, rails are

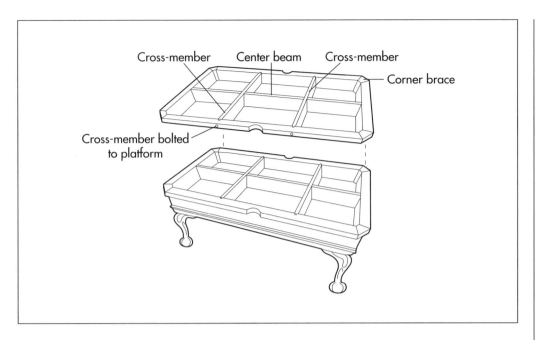

Cross-member   Center beam   Cross-member

Corner brace

Cross-member bolted
to platform

attached to modern tables by inserting a bolt vertically through a hole in the slate and tightening it into the bottom of the rail, pulling the rail and slate together snugly. Similarly, in older tables **brass** dowel pins were inserted into lead-lined horizontal holes drilled into the edges of the three slate sections where they would join to form the playing surface. In modern tables, the slate sections are held tightly together by screwing them to a wood frame, and joints are usually sealed with hot wax.

Raw Materials

Although some inexpensive pool tables use synthetic slate or plastic honeycomb sheets, the preferred playing surface (and the only one sanctioned by the BCA) is natural slate. It is quite dense, with the amount in a typical table weighing 450 lb (200 kg) or more. This mass helps keep the table stationary during play. Italian slate has long been the preferred type, but Brazilian slate now has some proponents.

Slate is prepared at the quarry, where computer-controlled, diamond-blade saws are used in conjunction with laser measuring devices to cut it into very flat sheets 0.75-1 in (1.9-2.5 cm) thick. The slate sheets are shipped in sets of three panels, which are certified as having been cut from the same slab. Three-section table surfaces are preferred because the smaller sections are

lighter and safer to lift, less prone to break, and easier to level during installation. Prior to shipment, properly sized holes are drilled in the slate for the pockets and for the bolts and screws that will be used to attach the slate to the table and rails.

The other major component of pool tables is wood. Usually at least two types are used. Poplar (tulipwood), a hardwood with superior self-healing properties that holds screws tightly and recovers well when staples are removed, is preferred for the structural framework of the table. Other hardwoods that provide a more attractive finish and are more resistant to nicks and scratches are used for the outer surfaces of the table. Examples are oak, maple, and mahogany.

Rails are usually produced by laminating two types of wood—an attractive, durable hardwood for the upper section and a functional softwood (like pine) or poplar for the lower section. Grade-A vulcanized rubber is preferred for the rail cushions, which are shaped to a particular triangular profile approved by the BCA. Canvas fabric is molded to the top and base of the cushion for proper rebound performance and secure attachment to the rail.

The cloth used to cover the slate and the rails is designed specifically for pool tables. By BCA edict, it must be primarily wool; available wool/nylon blends range from 100%/0% to 60%/40%. Although it is often

A diagram of pool table rail and blind.

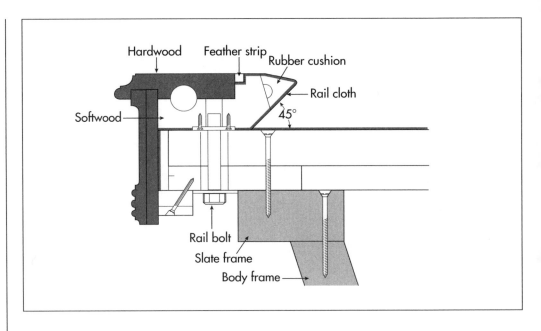

referred to as felt (a fabric formed by compressing fibers rather than weaving), it is actually a woven fabric with a nap (exposed, short, fuzzy fiber ends) on one surface.

Small components are made from various materials. Slate-sealing wax is specially formulated for this purpose, and is harder than beeswax. Diamond- or circular-shaped sights embedded in the rail tops are usually made of mother of pearl, abalone shell, or plastic. Pocket irons may be made of cast iron, zinc alloy, aluminum, rubber, or high-impact styrene plastic. Traditionally, pocket liners are made of leather (solid or net), but plastic or rubber is also used. Some tables use ball return ramps formed of materials such as polyethylene, aluminum, or heavy-gauge wire; they may be lined with rubber.

The Manufacturing Process

Construction techniques vary among manufacturers. The following description represents a generic process rather than an accurate account of a particular manufacturer's methods.

Preparation of components

1 Edge liners made of 0.75-in (2-cm) thick 1 x 8-in (1.9 x 18-cm, finished size) and 1 x 4-in (1.9 x 9-cm, finished) lumber are glued to the bottom of the slate around the edges. The wider strips are placed under the edges where pockets will be located. Until the wood glue is dry, the liners are clamped securely to the slate.

2 Following the pocket cutouts in the slate, pockets are sawed through the liner. Bolt and screw holes for attaching the rails and the table body are drilled in the liner to match the precut holes in the slate.

3 Sides for the body frame are cut from 2 x 12-in (4 x 28-cm, finished) lumber. Corners and top edges of the four sides must be carefully cut because once assembled, they will slope inward at a 15° angle from top to bottom. The frame is made smaller than the slate, so the slate overlaps the frame by 3.5 in (9 cm) on each side of the table. The sides are glued and nailed or screwed together.

4 Legs are prepared for the table. Solid wood pieces may be carved in decorative shapes, or hollow legs can be built by assembling wood sheets in a box shape. Wooden leg supports are glued and screwed into each corner of the frame.

5 A slate frame is built on the top edges of the body frame. Strips of 1.5 x 3-in (4 x 8-cm) wood are attached so they overhang the body frame by about 1 in (2.5 cm), except at the corners. Two cross members made from 2 x 6-in (4 x 14-cm, finished) lumber are glued and screwed between the long sides of the slate frame to support the

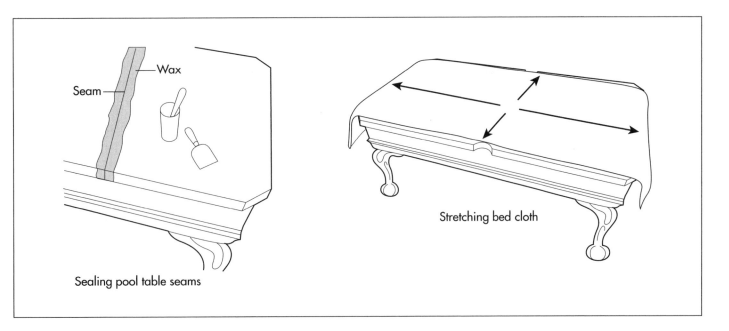

Sealing pool table seams

Stretching bed cloth

slate seams. A longitudinal support may be installed along the center of the frame, between the short sides. Pocket holes are cut in the corners and long sides of the slate frame.

6 Lower and upper rail components are sawed from appropriate woods. They are glued together to make six laminated sections 1.75 in (4.5 cm) thick and long enough to fit between each successive pair of pockets. The face angles on the rail are cut precisely for proper positioning of the rubber cushion. A groove is cut along the top edge to accept an anchoring strip for the cloth that will ultimately cover the rail.

7 Circular- or diamond-shaped sights are carefully placed at three locations on each rail section.

8 An apron (also called a blind) section about 4 in (10 cm) wide is cut to match the length of each rail section. This component will cover the ends of the slate and liner, slate frame, and body frame.

9 Each component is stained, fine sanded, and finished with catalyzed varnish, lacquer, and/or furniture wax. Rubber cushions are glued to the rail sections.

Assembly of components

Pool tables are usually shipped in pieces and assembled during installation of the table at the purchaser's location.

10 Legs are bolted to the underside of the table body. The structure is checked for level, and shims are inserted between the legs and the body if necessary.

11 The three sections of slate are screwed into place atop the table body. Shims may be placed as needed under the slate liners to ensure a flat, level surface. The seams between the three slate sections are sealed with hot wax, with any residue being carefully removed from the top surface.

12 Cloth is stretched tightly and uniformly across the slate, with the edges wrapped over the slate liner and stapled to its edges. Rail faces are also covered with cloth.

13 The rail sections are joined by inserting the pins of the pocket irons into holes drilled in the ends of each rail section. Bolts inserted upward through the slate are used to secure the rail atop the slate. Care must be taken to ensure that the sections are installed straight and tight. Pocket liners or ball return ramps are attached.

14 Blinds are glued and screwed to the bottoms of the rails and the edges of the slate liner.

Quality Control

Major manufacturers cut wooden components with computer-controlled equipment to ensure precision. They assemble the

Once the seams are sealed on the pool table frame, the bed cloth is stretched tightly over the frame and secured.

pieces of each table by hand at the factory, checking for proper fit before disassembling them for shipment. An installer, employed by the dealer rather than the manufacturer, reassembles the pieces at the purchaser's location. The quality of the manufacturing operation and of the installation process are both important for proper performance of the table.

Materials used in the table affect the quality, appearance, and cost of the table. An inexpensive table, for example, might have particle board components that do not hold screws or staples as well as solid wood. Tables vary widely in quality and cost; a casual player who wants a table for a few years of personal recreation can get one for around $600. So-called popularly priced tables, which are well-built, durable, and attractive, may cost $1,600-$3,000.

The Future

The use of alternative materials continues to be explored. Aiming for durability and stability, for example, one manufacturer recently introduced a pool table with a frame made of steel and rails formed of tempered aluminum. The metals are covered with a decorative synthetic veneer. The slate-topped table weighs 1,050 lb (480 kg), about the same as a wooden table. The manufacturer claims the table meets BCA specifications.

Where to Learn More

Periodicals

Bowman, Paul M. "Building a Pool Table." *Fine Woodworking* (March/April 1989). http://www.bestbilliard.com/resources/builtable.cfm. (May 31, 2000).

Other

BCA Equipment Specifications. Billiards Congress of America. http://www.bestbilliard.com/resources/specs.htm (January 6, 2000).

Gandy Pool Tables. http://gandys.com (May 31, 2000).

DiPaolo, Dennis. "Pool Table Buyer's Guide." http://www.seasonalstores.com/guide.html (May 31, 2000).

Best Billiard Sales and Service. http://www.bestbilliard.com (March 24, 2000).

SAM Billiards. http://www.billaressam.com (June 3, 2000).

—*Loretta Hall*

Popsicle

Background

The sound of an ice cream truck is a delight to the ears of children and adults alike on a hot summer day. That truck carries delicious concoctions that the industry calls collectively "frozen novelties." And it is guaranteed to carry a variety of frozen pops—ices, sherbets, pops in push tubes, and pops on sticks—in fruit flavors and colorful shapes and sizes. These are often called popsicles after the brand-named frozen pop.

History

The frozen pop has its origins in frozen desserts that are as old as civilization. The ancient Romans had blocks of ice carried down from the mountains in the summer. The blocks were ground into crushed ice that was flavored with fruit and syrup. The Chinese are credited with the same practice, and these sherbets, sorbets, and frozen ices were served in the thirteenth century court of Kublai Khan (1215–1294) when the Italian adventurer Marco Polo (1254–1324) visited. Other ancient cultures including those of Turkey, Persia (now Iran), the Arabian Peninsula, and India also knew of frozen, flavored ices. They were usually a privilege of the wealthy and were made of ice, fruit pulp, fruit syrup, and flowers for flavor, fragrance, and color. Frozen treats were served to honor guests at banquets or to cleanse the palate between courses, as sorbets are still used today.

In the sixteenth century, Catherine de Médicis (1519–1589), the Italian-born wife of French King Henry II (1519–1559), brought recipes for frozen ices from Italy and introduced the concept of freezing the ices artificially by plunging their containers into icy brine. The next major landmark in the development of frozen novelties occurred in three capital cities—Paris, London, and New York—in the 1820s. Street vendors sold ice cream and penny ices (frozen ice dollops, rather like small snow cones) that were held on tiny glass goblets. The customer bought an ice for a penny and ate it directly from the goblet (without any spoon or stick), and the vendor recycled the goblet for the next customer.

The little scoops of frozen ice or ice cream were called "hokeypokey" (probably derived from an Italian catchphrase), and the penny-ice men were also called hokeypokey men. The hokeypokey men were the original Good Humor men, named for the brand of ice cream on a stick that was introduced in 1922. Good Humor grew out of two popular inventions. The automobile (and the truck) helped vendors cover a larger area and carry more novelties than the street vendors with their push carts. And, in 1910, lollipops had become a popular candy for children in America and England. Ice cream on a stick and the Good Humor bar followed. Meanwhile, another icy revolution was underway.

The discovery of the frozen pop is attributed to an 11-year-old boy. In 1905, Frank Epperson mixed powdered soda pop and water but forgot about his preparation and left it outside on the porch of his home in Oakland, California. The concoction froze overnight, and the stick he had used to stir the powder in the water stuck in the frozen liquid. Frank pulled out the whole frozen mass and found that he had invented a new treat. He named it the "Epperson icicle." The following summer, he made his frozen treats in his family's icebox and sold them around his neighborhood under the shortened name of "Epsicle." Frank again rechristened his discovery the "popsicle" to show that they were made from soda pop. The selling price for the original Epsicles was five cents a piece. Epperson patented his treat in 1924 when he was 30 years old as the Popsicle.

From 1920 through the mid-1970s, Good Humor trucks sold all varieties of frozen treats until the cost of gasoline and insurance overwhelmed profits. Independent vendors still sell frozen treats in many neighborhoods, but the large variety of popsicles in grocery stores makes the household refrigerator the most familiar pop "vendor" today.

Raw Materials

The two principal kinds of non-dairy frozen treats manufactured are the popsicle and the juice bar. The popsicle is 90% water. Its other ingredients are sugar, corn syrup, gum, and stabilizers. These ingredients give the popsicle a texture called "mouth feel" that makes it pleasant to eaten. The sugars and stabilizers cause the pop to soften in the air so it is edible, instead of melting and dripping like an ice cube. Flavoring is highly concentrated and is usually some traditional fruit flavor and color. Specialty firms manufacture the flavorings and apply the chemistry needed for true flavor and color when the flavoring is reconstituted. Some flavors, particularly citrus, also have additives like citric acid that gives the flavor its citrus "bite." Juice bars have the same ingredients except that concentrated fruit juice provides the color and flavor. All other frozen treats (like fudge bars) are dairy products.

Popsicle makers also need sticks for the handles and printed wrappers. These are supplied by outside sources. The popsicle manufacturer will commission a design for the printing on the wrapper, which is etched on a plate and used to print the wrappers. Wrappers are purchased by the truckload, so a new product is a major cost commitment. Sticks are made of basswood by specialized subcontractors. Sometimes the sticks are also printed with jokes, funny sayings, or designs; they are bought by the pallet-load in quantities sufficient to last the manufacturing season.

Design

Design of a popsicle consists of a combination of color, flavor, and shape of the pop.

Flavors and colors tend to stay in a range of traditionally accepted fruit flavors. Molds can be made in almost any shape and can also be filled in stages to create layered colors or be coated with outer colors and flavors for contrast. The molds are made in a precision design process, however, and are very expensive—about $60,000 for each design. The decision to make a new mold is a major one for all popsicle makers. When the costs of modifying nutritional information, printing new wrappers, and advertising are added in, each new product may mean an investment of $100,000. Large manufacturers may produce up to 30 new items per year.

The Manufacturing Process

1 The popsicle begins with making water ice in sets of sterilized vats. The vats are refrigerated to 35-37°F (1.7-2.8°C). The base material, consisting of sugar, corn syrup, stabilizers, and gum, is mixed with water and then subdivided into several separate vats where the flavoring and coloring are added. The formulas for the mixes are constant, but trained inspectors flavor-test the batches and adjust flavorings and other ingredients to taste. In the refrigerated vats, the fluid becomes water ice—it does not quite reach the point of freezing and still can be pumped easily.

2 The water ice is pumped to a million-dollar machine called a Vita-Line. The machine is circular and about 15 ft (4.6 m) in diameter and transports sets of molds through several processes to form the pops. The machine is made of stainless steel and supports 200 to 300 strips of molds. Each strip may be 6, 8, or 12 molds wide, 4-8 ft (1.2-2.4 m) wide, and 30-40 ft (9-12 m) long. A typical popsicle machine will generate 4,320 pops per hour; larger machines produce twice as many, and a production line may include five machines.

A continuous chain across the top of the machine and around the wheel moves the molds through the process. All fluids are pumped by electrically powered pumps. Pneumatics (air pressure) control the movements of slave cylinders and the opening and closing of valves in a simple sequence of motions. In the first set of steps, the molds are pre-rinsed, washed, rinsed again,

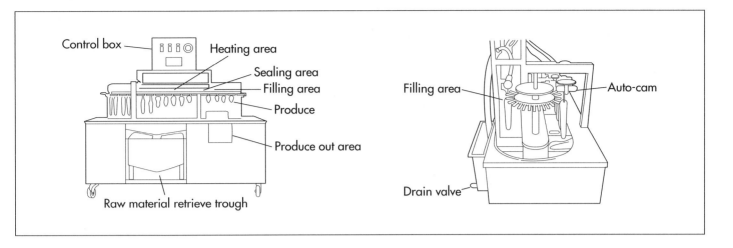

and sanitized (heated) in a minute-long process. The machine inverts the molds so that a spray bar can pre-rinse any materials out of the molds. Two wash bars, above and below the molds, wash and rinse the molds in a fresh water rinse. Another bar sanitizes the molds.

3 After washing, the molds are reinverted so they can be filled. The flavored water ice is pumped through sets of nipples into the molds, and the molds are pulled into a tank filled with water containing calcium chloride (salt brine) that is ammonia-cooled to -25--30°F (-32--34°C). As the molds are drawn along the 20-ft (6-m) length of the tank, their contents pass from the liquid to frozen state. If several different layers of flavors make up the particular pop, hoppers with different fluids are attached to the same machine that makes single-flavor pops. The first fill is injected, the mold is partially submerged in salt brine to freeze it, and a vacuum pump sucks out any remaining liquid on the frozen surface. During the second fill, the second flavor is added, flash-frozen, and sucked dry of standing water; finally, the third fill is added. Midway through the process, the water ice is partially frozen. At this stage, pops designed to have a twisted appearance are given that twist, and an injector pushes the sticks into the pops, which freeze from the outside toward the center so the frozen outside supports the stick. In the last 3 or 4 ft (0.9 or 1.2 m) of the machine, the centers of the pops freeze.

4 The molds then pass briefly through a tank containing 180°F (68°C) water that slightly heats the molds. An overhead extractor pulls the pops out of their molds by their sticks. If the pops are to be dipped to make an outside coating on them, the extractor lowers them into a dip tank, and the coating is flash frozen. The extractor bar then carries the popsicles to the bagging area where the bags are blown open by bursts of air as the pops are dropped by the extractor. They fall by gravity into the waiting bags.

The bagged pops ride on a conveyor along an assembly line where workers snap open boxes, fill the boxes with bagged pops, tape the boxes closed, and pack sets of boxes into a larger set called a master pack. The master packs are taken to the freezer where they are stored until shipment. In an alternative wrapping process, popsicles are dropped by the extractor bar on sheets of mylar or glassine paper that has been preprinted. The wrapping machine shapes the wrapper around each bar, seals the top and bottom, and cuts off the excess paper at both ends.

Quality Control

Quality control includes taste and visual tests. The water ice is "taste-tested" even though the mixture has been made precisely. The tester is usually an expert who knows when the flavor is not quite up to company standards despite the perfect measurements of the ingredients. The tester has the authority to adjust the mixture to the right taste.

The processes of the Vita-Line that are watched especially closely are the fill process, the extraction of the pops from the molds, and the dropping of the pops into the bags. As the pops are boxed, the assembly workers can reject any pops that look suspicious, so quality control in the popsicle factory is a "hands-on"

Two types of popsicle filling and sealing machines.

activity. Smaller companies benefit from this approach because they share a family atmosphere and pride in their product.

Byproducts/Waste

There are no byproducts or waste from the process of making water ice. Some waste water containing sugar may result from the steps in rinsing the molds, but the water is simply transported as waste water through the city's water treatment system—that is, no special treatment is required.

Major safety issues in the factory include the refrigeration process in which liquid ammonia must be handled. Low-pressure steam is used in cleaning and sanitizing the molds and machinery, and, of course, electricity supplies the power. Workers are trained in the operation of machines, so safe practices are ingrained and safety becomes a minimal concern.

The Future

The future of the popsicle will follow the demands of the public. Tastes lean toward traditional flavors and colors, so the refining and improving of traditional products is a primary interest among popsicle makers. Experts calculate that 100 ice pop flavors have been produced in the 75 years since the creation of the Epsicle, but cherry, orange, and grape are still the favorite flavors among the estimated 3 million popsicles sold every year. "Boutique" flavors such as daiquiri or huckleberry are made to catch the fancy of adventurous tastes, and combinations of types of pops like ice cream bars with an ice coating or popsicles with a small quantity of ice cream in the mixture are ways of producing new products that do not stray too far from tradition but appeal to consumers undecided between dairy and non-dairy novelties.

Manufacturers also watch their competitors, but small companies cannot match the investment needed for new sets of molds (for example) that may not be big sellers. Each new item also requires new labels with correct nutritional information that conforms to government labeling requirements and other details, and that can be very expensive. Cost does not seem to impede the popsicle's future, however; 98 new novelties were introduced in 1998, not counting the pop called "Frosty Paws," which was made for dogs. The future of pop-sicles relies on tradition rather than rapidly changing fads. A long, hot summer is the popsicle's best marketing tool.

Where to Learn More

Books

Dickson, Paul. *The Great American Ice Cream Book.* New York: Atheneum, 1972.

Liddell, Caroline, and Robin Weir. *Frozen Desserts.* New York: St. Martin's Press, 1995.

Wardlaw, Lee. *We All Scream for Ice Cream.* New York: Harper Trophy, Harper-Collins Publishers, Inc., 2000.

Wulffson, Don L. *The Kid Who Invented the Popsicle: And Other Surprising Stories About Inventions.* New York: Penguin Putnam Books for Young Readers, 1997.

Periodicals

Belleranti, Shirley W. "A treat from Marco Polo." *Hopscotch* 8, no. 2 (August-September 1996): 9.

Onoe, Phil. "At Large and At Small." *American Scholar* 67, no. 4 (1998).

Ward, Carol J. G. "Try frozen Popsicle treats to cool down this summer." *Knight-Ridder Newspapers* (August 10, 1999).

Other

Ching Tan Machinery Works, Taiwan. http://www.foodmachine.com.tw/e/d.htm (June 29, 2000).

Good Humor-Breyers Ice Cream. http://www.icecreamusa.com (June 29, 2000).

The Ice Screamer. P.O. Box 465, Warrington, PA 18976. http://www.icescreamers.com (June 29, 2000).

Perry's Ice Cream. http://perrysicecream.com. (June 29, 2000).

Popsicle Zone. http://www.popsicle.com (June 29, 2000).

Waukesha Cherry-Burrell Ice Cream. http://www.gowcb.com/products/IceCream/freezeindex.htm (June 29, 2000).

—*Gillian S. Holmes*

Raincoat

Background

Raincoats are jackets made of fabric that is specially treated to repel water. In 1836, Charles Macintosh invented a method for combining rubber with fabric, which was used in the first modern raincoats. Because of his inventions, all raincoats are called Mackintoshes or Macs by those in Great Britain. Most modern day raincoats are inspired in one way or another by Macintosh's brainchild.

Today there are many kinds of raincoats made of all types of fabric. An all-weather raincoat has a removable lining so it can be worn in any weather. Fold ups are foldable and usually made of vinyl. Vinyl raincoats are made of vinyl or of fabric that has a vinyl finish. Trenchcoats are worn by both men and women, and are often made of lightweight cotton/polyester fabric.

What is important to raincoat manufacture is efficient waterproofing. There are two important qualities: absorption (how much water can be soaked by the fabric) and penetration (the amount of water that can sink into the fabric). Raincoat fabrics are either absorbent or repellent. The best raincoats are made of tightly woven fabric.

History

People have been trying to make items of clothing waterproof for hundreds of years. As early as the thirteenth century, Amazonian Indians used a milky substance (rubber) extracted from rubber trees for this purpose. When European explorers came to the Americas in the sixteenth century, they observed the indigenous people using a crude procedure and rubber to waterproof items like footwear and capes.

By the eighteenth century, Europeans were experimenting with waterproofing fabric for clothing. François Fresneau devised an early idea for waterproofing fabric in 1748. Scotland's John Syme made further waterproofing advances in 1815. In 1821, the first raincoat was manufactured. Made by G. Fox of London, it was called the Fox's Aquatic. The raincoat was made of Gambroon, a twill-type fabric with mohair.

While these early attempts at waterproofing fabrics sometimes involved rubber, they were not particularly successful. When rubber was used in clothing, the articles involved were not easy to wear. If the weather was hot, the clothing became supple and tacky; if cold, the clothing was hard and inflexible. This problem was solved in the early nineteenth century by Macintosh.

The native of Scotland was a chemist and chemical manufacturer. Through experiments, Macintosh discovered a better way to use rubber in clothing. At the time, the gas industry was new. Coal-tar naphtha was one byproduct of the fractional distillation of petroleum, which was used in gasworks. This volatile oily liquid was a hydrocarbon mixture. Macintosh dissolved rubber in naphtha, making a liquid. This liquid was brushed on fabric making it waterproof.

In 1823, Macintosh patented his process for making waterproof fabric. This process involved sandwiching a layer of molded rubber between two layers of fabric treated with the rubber-naphtha liquid. It took some time to develop the industrial process for spreading the rubber-naphtha mixture on the cloth. The patented waterproof fabric was produced in factories beginning in 1824. The first customer was the British military. Mac-

Once designed and sized, the raincoats are sewn.

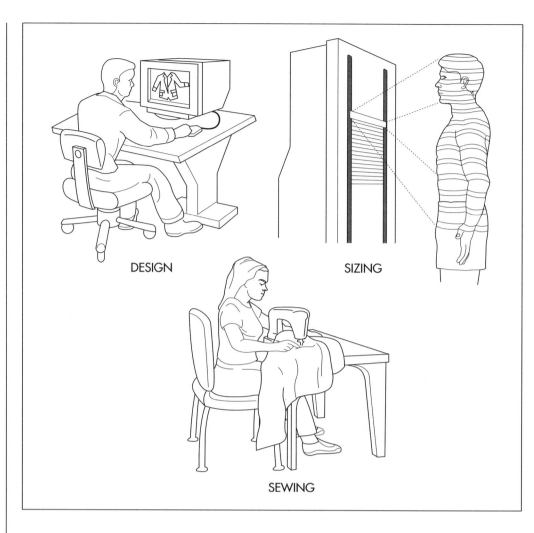

DESIGN

SIZING

SEWING

intosh's findings led to other innovative uses of rubber, including tires.

The process for vulcanizing rubber was developed by Charles Goodyear, a hardware merchant in Philadelphia, Pennsylvania, in 1839. Vulcanization means to heat rubber with sulfur, which made rubber more elastic and easier to meld. Four years later, Thomas Hancock took the waterproof fabric invented by Charles Macintosh and made it better using vulcanized rubber.

Americans continued to improve on Macintosh's process with the advent of the calendering process in 1849. Macintosh's cloth was passed between heated rollers to make it more pliable and waterproof. Another innovation involved the combination of only one layer of cloth with a layer of rubber. While such improvements made the cloth lighter than Macintosh's original, these raincoats were still rather hot even into the early twentieth century. Many raincoats were de-

signed with slits to make them cooler for their wearer.

Macintosh's fabric was not the only kind of waterproof fabric invented in the nineteenth century. In 1851, Bax & Company introduced Aquascutum. This was a woolen fabric that was chemically treated to shed water. This raincoat became popular at the end of the Crimean War (c. 1856).

Chemicially treated fabrics gradually began to predominate by the early twentieth century. For World War I, Thomas Burberry created the all-weather trench coat. The coat was made of a yarn-dyed fine twill cotton gabardine. The gabardine was chemically processed to repel rain. Though these trench coats were first made for soldiers, after the war ended in 1918, they spread in popularity. They were also much cooler than those made of Macintosh's fabric.

By 1920, raincoat design moved beyond the trenchcoat, though that coat remained a classic. Oil-treated fabrics, usually **cotton** and silk, became popular in the 1920s. Oil-skin was made by brushing linseed oil on fabric, which made it shed water. Car coats were introduced in the 1930s. These raincoats were shorter than trenchcoats and made for riding in automobiles. Rubber-covered and -backed raincoats, made of all kinds of fabrics, resurged in popularity between the wars as well.

After 1940, raincoats made of lightweight fabric became more popular. Military research led to the creation of raincoat fabrics that could be dry-cleaned. Vinyl was a preferred fabric in the 1950s for its impressive waterproofness, as was plastic (through the 1970s), though such raincoats retained heat. Innovations in fabrics continued to affect raincoats. Wool blends and synthetic blends were regularly used to make raincoats beginning in the 1950s. Such blends could be machine washed. There were also improved chemical treatments of cloth. Heat-welded seams were introduced as well, increasing how waterproof the fabric was.

In the 1960s, nylon was used to make raincoats, and in the 1970s, double-knit became a preferred fabric. Double-knit raincoats were not as water repellent as those made with other fabrics, but were designed differently to compensate. Still such raincoats were not as comfortable, and double-knit faded throughout the decade. Vinyl raincoats briefly had a renewed popularity, especially among women.

Modern day raincoats come in many fabrics, styles and colors. The gabardine trenchcoat remains a favorite. While natural and artificial blends, rubber and plastic are still used, plastic-coated artificial fibers used for Gore-Tex are very popular. Microfibers and other high-tech fabrics are taking over more of the raincoat material market.

Raw Materials

The primary material in a raincoat is fabric that has been specially treated to repel water. The fabric of many raincoats is made of a blend of two or more of the following materials: cotton, polyester, nylon, and/or rayon. Raincoats can also made of wool, wool gabardine, vinyl, microfibers and high tech fabrics. The fabric is treated with chemicals and chemical compounds, depending on the kind

of fabric. Waterproofing materials include resin, pyridinium or melamine complexes, **polyurethane**, acrylic, fluorine or Teflon.

Cotton, wool, nylon or other artificial fabrics are given a coating of resin to make them waterproof. Woolen and cheaper cotton fabrics are bathed in a paraffin emulsions and salts of metals like aluminum or zirconium. Higher quality cotton fabrics are bathed in complexes of pyridinium or melamine complexes. These complexes form a chemical link with the cotton and are extremely durable. Natural fibers, like cotton and linen, are bathed in wax. Synthetic fibers are treated by methyl siloxanes or silicones (hydrogen methyl siloxanes).

In addition to the fabric, most raincoats consist of buttons, thread, lining, seam tape, belts, trim, zippers, eyelets, and facings.

Most of these items, including the fabric, are created by outside suppliers for raincoat manufacturers. The manufacturers design and make the actual raincoat.

Design

To capture part of the market, raincoat design changes with the season and current fashion trends. Fabrics, lengths, cut and look are important to appeal to the consumer. Style is everything; the cut is important to distinguishing items on the market. For men, women, and children, there are many different styles of raincoats: short, commuter, car coats, sport, utility, and long raincoats. New fibers and finishes are regularly introduced.

Computer-aided design (CAD) gives designers the ability to combine fabrics, styles, and colors onscreen without having to make a sample. Designers create the patterns for the manufacturing process using CAD.

The Manufacturing Process

Much of the manufacturing process is done by Computer Aided manufacturing (CAM). Machines are run by computers, ensuring speed and efficiency.

Waterproofing the fabric

1 In this automated process, fabric passes through a series of rollers and into a tank

Waterproof material used for rain-coats is tested for its level of absorption or ability to repel moisture.

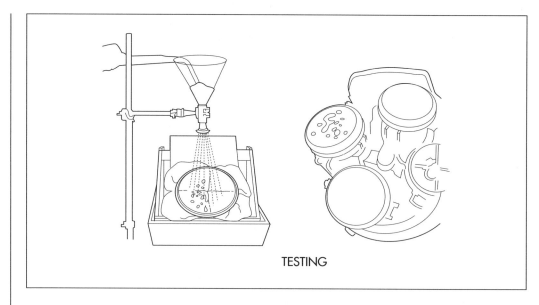

TESTING

containing a bath of the appropriate water-proofing materials.

2 Chemicals are allowed to soak into the fabric.

3 After the material leaves the bath, it is heated so the chemicals soak evenly into the fabric.

4 The fabric passes through another series of rollers and allowed to cool. This allows the chemicals to harden and stay in place on the fabric.

5 The treated fabric is re-rolled into bolts and readied for shipment to raincoat manufacturers.

Manufacturing the raincoat

1 The pieces that make the body of the raincoat are cut by a computer-operated cutter or large electrical cutter. The fabric is cut according to patterns that have been graded to each size the raincoat will be made in.

2 The interlining (which makes collars, cuffs, front facings, sleeve straps, belts and pocket welts stiff) is put inside the appropriate parts. Each part that contains interlining is fused by an automated fusing machine. The fusing process uses high heat and pressure to ensure the interlining is put in evenly.

3 An industrial sewing machine operator uses an industrial machine to sew all the

small parts of the raincoat, including the collar, belt, sleeve straps and pocket welts.

4 The facings (the underply of the coat that is sewn to the front edge) are set to the front edges of the fabric shell that makes up a coat's exterior. This is also done by a worker on an industrial sewing machine.

5 The large parts of the shell of the raincoat are assembled on an industrial sewing machine by a worker. Most come in pairs. First the right back panel and left back panel are joined together, creating a back seam. Then the rest of the large parts of the raincoat (the front of the raincoat, the sleeves etc.) are created in similar fashion.

6 On a separate line in the factory, a lining has been made. It is sewn into the appropriate pieces of the shell by a person.

7 All the large pieces of the raincoat are now assembled and sewn together by a worker using an industrial sewing machine.

8 An automated machine makes button-holes and sews the buttons on the raincoat.

9 Using an industrial sewing machine, the trim and sleeves are sewn on by hand. The belt is put on by hand as well.

10 The finished raincoat is examined by a inspector for quality control purposes.

11 A worker puts tags on the coat and puts it on the hanger.

12 A polybag is put over the coat, usually by a machine called an automatic bagger.

13 The bagged coats are loaded into the shipping container by a worker.

Quality Control

Quality control takes place at many steps of the manufacturing process. Before raincoats are even made, the quality of the fabric and dye are inspected. After the fabric is cut, the patterns must match and sizes must be right. As each piece is sewn together, the sewing is checked. The qualities inspectors look for include: stitch length, hem stitching, buttonhole stitching and alignment, and seam type. Anything defective is corrected. When the raincoat is completed, each part of it is inspected by hand.

Byproducts/Waste

In the actual production of raincoats, only scraps are created. They are usually thrown away.

The Future

The raincoat's fabric and its finishes will change. There will be improved water repellency, stain resistance, and wrinkle resistance. The fabrics themselves will have new weaves.

Another direction the future of raincoats might head is toward mass customization. This means the coat would be specially made to fit the consumer's body. Such a process would involve computers and the advent of certain kinds of software and scanning devices.

Where to Learn More

Books

Schoeffler, O. E., and William Gale. *Esquire's Encyclopedia of 20th Century Men's Fashions.* New York: McGraw-Hill Book Company, 1973.

Stone, Elaine. *The Dynamics of Fashion.* New York: Fairchild Publications, 1999.

Periodicals

Bober, Joanna. "Fashion 101: In the Trenches." *InStyle* (March 1, 1999): 141.

—*Annette Petrusso*

Roller Coaster

In May 2000, the Millenium Force opened at Cedar Point in Sandusky, Ohio. At 310 ft (94 m) tall and going 92 mph (148 kmp), it is the tallest and fastest roller coaster in the world.

Background

A roller coaster is an amusement park ride where passengers sit in a series of wheeled cars that are linked together. The cars move along a pair of rails supported by a wood or steel structure. In operation, the cars are carried up a steep incline by a linked chain. When the cars reach the top of the incline, they roll free of the chain and are propelled downward by gravity through a series of drops, rises, and turns. Finally the cars are braked to a stop at the starting point, where the passengers get out and new passengers get on. Roller coasters are considered by many to be the most exciting ride in any amusement park.

History

The origins of the roller coaster probably date back to Russia in the 1400s, where ice sledding was a popular winter activity. It became so popular that people in relatively flat areas constructed their own hills out of snow and ice. The tops of these artificial hills were reached by way of elevated wood towers with stairways from the ground. For a small charge, people could climb the stairway and take a quick, exciting ride down the hill on a sled.

By the 1700s, many owners of ice hills found a way to extend the profit potential of the ride beyond the winter months. They mounted wheels under small sleds and replaced the ice hills with ones constructed of wood. Brightly colored lanterns were hung along the slope to allow night operation.

Visitors from France saw these rides, which they called the Russian Mountains, and took the idea back with them. The first wheeled coaster opened in Paris in 1804, and the coaster craze quickly spread throughout France. As the popularity of the rides grew, operators vied for the public's patronage by building faster and more exciting coasters. Unfortunately, safety devices did not keep pace with the speed, and accidents were common. By the mid-1800s, the increasing number of injuries and a general loss of public interest took their toll. One-by-one the Russian Mountain coasters were dismantled.

The development of the roller coaster might have stopped there had it not been for a defunct coal-hauling railroad in the United States. The Mauch Chunk inclined railroad was built in Pennsylvania in the early 1800s to haul coal from a mine atop a mountain to barges in a canal below. Mules hauled the empty cars up the hill, and gravity brought the loaded cars, along with the mules, back down. In 1874 mining operations changed, and the railroad began hauling sightseers instead of coal. The one-and-a-half hour round trip cost one dollar and was an immediate success. The railroad continued to carry passengers until it closed in 1938.

The success of the Mauch Chunk inclined railroad as a tourist attraction provided the inspiration for several similar amusement park rides on a smaller scale. In the United States, LaMarcus Thompson built his Gravity Pleasure Switchback Railway ride at the beach on Coney Island, New York, in 1884. For a nickel, riders rode cars that coasted from one elevated station to another over a series of gentle hills supported on a wooden trestle. At the opposite end, the cars were switched onto a parallel track for the return trip.

The second roller coaster on Coney Island was built in late 1884 when Charles Alcoke opened his Serpentine Railway. Alcoke's coaster was the first to use an oval-track design. Riders sat sideways on open benches as they were whisked along at what was then considered to be a break-neck speed of 12 mph (19 kph). A third coaster was built on Coney Island in 1885 by Phillip Hinkle. Hinkle's coaster incorporated a chain lift to carry the cars up the first hill, thus allowing the passengers to board at ground level and saving them a climb.

Roller coaster development hit its peak in the 1920s when there were more than 1,500 wooden coasters in operation in the United States. The economic hardships of the 1930s and the wartime material shortages of the 1940s put an end to that era. Amusement parks closed by the hundreds, and their wooden roller coasters either fell into disrepair or were torn down. It wasn't until Walt Disney opened the Matterhorn Bobsled ride at Disneyland in 1959 that the era of modern steel roller coaster design began. Ironically, it took the construction of a new wooden coaster—the massive Racer at Kings Island near Cincinnati, Ohio, in 1972—before the coaster craze really caught on again.

By the late 1990s it was estimated there were over 200 major roller coasters in operation in the United States, with more being added every year. In May of 2000, the Millenium Force opened at Cedar Point in Sandusky, Ohio. At 310 ft (94 m) tall and going 92 mph (148 kmp), it is the tallest and fastest roller coaster in the world.

Raw Materials

Roller coasters are generally classified as either wooden coasters or steel coasters depending on the materials used for the support structure.

Wooden coasters use massive wooden trestle-style structures to support the track above the ground. The wood is generally a construction grade such as Douglas fir or southern yellow pine and is painted or otherwise treated to prevent deterioration. The wooden components are supported on concrete foundations and are joined with bolts and nails. Steel plates are used to reinforce critical joints. As an example of the immense number of parts required to build a wooden coaster, the American Eagle built for Six

Flags Great America in Gurnee, Illinois, used 2,000 concrete foundations; 1.6 million ft (487,680 m) of wood; 60,720 bolts; and 30,600 lb (13,910 kg) of nails. It was coated with 9,000 gal (34,065 L) of paint.

Steel coasters may use thin, trestle-style structures to support the track, or they may use thick tubular supports. The track is usually formed in sections from a pair of welded round steel tubes held in position by steel stanchions attached to rectangular box girder or thick round tubular track supports. All exposed steel surfaces are painted. Steel coasters can be just as complex as wooden ones. For example, the Pepsi Max Big One coaster at Blackpool Pleasure Beach in Blackpool, England, used 1,270 piles driven into the sandy soil for the foundation; 2,215 tons (2,010 metric tons) of steel, and 60,000 bolts. There were 42,000 sq. yd (35,087 sq. m) of painted surfaces.

The track and lift chain on both wooden and steel coasters are made of steel, and the cars usually have steel axles and substructures. The car bodies may be formed from aluminum or fiberglass, and the car wheels may be cast from urethane or some other long-wearing, quiet-running material.

Design

The design of a roller coaster ride is the first and most important part of the manufacturing process. Because each roller coaster is unique, every detail must be designed literally from the ground up.

To begin, roller coaster designers must consider what kind of riders will use the coaster. If the coaster is designed for small children, the hills and curves will be gentle, and the cars' speed will be relatively slow. Families usually want a somewhat faster ride with plenty of turns and moderate forces. Ultimate thrill seekers want extreme heights and speeds.

Designers must then consider the space available for the coaster. Roller coasters not only take a lot of ground space, but also a lot of air space. Designers look at the general terrain, other surrounding rides, power lines, access roads, lakes, trees, and other obstacles. Some amusement parks have added so many rides that a new roller coaster has to be designed to thread its way through existing rides and walkways.

The next objective for the designers is to achieve a unique "feel" for the coaster. Designers can draw on a number of techniques to provide a memorable ride. The initial incline can be made steeper or the speed of the lift chain can be made slower to heighten the apprehension of the passengers. Once up the incline, the first drop is usually designed to be the steepest, and therefore the fastest and scariest. Other drops can be designed with a brief flattened section in the middle, and are called double dips. Drops with very abrupt transitions to a flat or upturned section are called slammers because they slam the passengers down into their seats. Letting the cars run close to the ground, in what is called a gully coaster, gives the illusion of increased speed.

The advent of steel construction for coasters has allowed a number of variations on the basic roller coaster ride. In some modern coasters, the passengers sit suspended below the tracks rather than riding on top of them. In others, the passengers ride standing up rather than sitting down. Some coasters, known as bobsleds, have no track at all, and the cars roll free in a trough, like a bobsled run.

Most of the actual design and layout of a roller coaster is done on a computer. The height of the first incline must be calculated to give the cars enough energy to propel them all the way through the ride and back to the station. The horizontal and vertical forces that the loaded cars exert on the track must be calculated at every point to ensure that the support structure is adequate. Likewise, the forces exerted on the passengers must be calculated at every point. These forces are usually expressed as "g's," which are multiples of the force that gravity exerts on our bodies. For example, if a person weighs 100 lb (45.5 kg), then a 2 g force would exert 200 lb (91 kg) of force on that person. Coasters in the United States generally exert no more than about 3.5 g's, which is the limit that most people find tolerable. Three coasters outside the United States exert more than 6.5 g's and are considered ultra-extreme. Jet fighter pilots black out at about 10 g's.

Because each coaster usually incorporates one or more new and untried features, a working prototype of the new features may be built for testing and evaluation. The prototype is erected at the manufacturer's facility, and weighted test cars outfitted with instrumentation are propelled through the test section at the desired speed. Based on these tests, the designers may alter their original design before building the final product.

When the calculations, design, and testing are complete, a computer-aided drafting (CAD) program is used to prepare detailed drawings for each of the thousands of parts that will be used to build the new coaster.

The Manufacturing Process

The actual physical construction of a roller coaster may take place in a factory or on the amusement park site depending on the type and size of the coaster. Most steel coasters are built in sections in a factory, then trucked to the site and erected. Most wooden coasters are built piece-by-piece on the site. Here is the typical sequence of operations for manufacturing both modern steel coasters and classic wooden coasters.

Preparing the site

1 Before the roller coaster can be installed, the area where it is to be located needs to be cleared and prepared. This is usually done in the off season when the amusement park is closed. If it must be done while the park is still open, the area is fenced off to prevent the public from wandering onto the construction site.

2 If there are existing structures, vegetation, or utilities that need to be moved or demolished, this work is done first. If any of the surrounding terrain needs to be filled or excavated, that work is also done at this time.

3 Holes for the support structure foundations are surveyed and drilled or dug. Sturdy wooden forms are constructed to hold the concrete for each foundation point. In some areas where the soil is very sandy, large wooden piles may be driven into the ground as foundations rather than using poured concrete. If concrete is used, it is brought to the site in mixer trucks and pumped into place by a concrete pump with a long, articulating arm

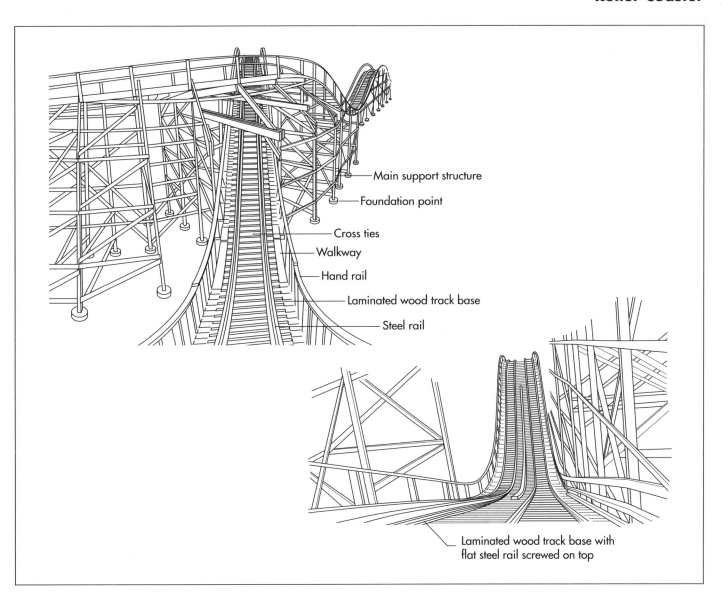

Main support structure

Foundation point

Cross ties

Walkway

Hand rail

Laminated wood track base

Steel rail

Laminated wood track base with
flat steel rail screwed on top

An example of a wood-construct-
ed roller coaster.

that can reach each foundation form. Connector plates are imbedded into the concrete on top of each foundation to allow attachment of the supports.

Erecting the main support structure

4 When the foundation is in place, work begins on the main support structure. The supports for steel coasters—in fact, almost all the parts for steel coasters—are made in a factory and shipped to the job site in sections on trucks. In the factory, the pieces for each support are cut and welded into the required shape using fixtures to hold them in the proper orientation to each other. If a complex three-dimensional bend is required, this may be done in a hydraulic tube bender that is controlled by information from the computer. On wooden coasters, the material for the supports is usually shipped to the site as unfinished lumber and the individual pieces are cut and assembled on site. In either case, the lower portions of the main supports are lifted by a crane and are attached to the connector plates protruding from the foundation points.

5 Once the lower supports are in place, they may be temporarily braced while the upper sections are lifted into place and connected. This work continues until the main support structure is complete.

Installing the track

6 With the main support structure in place, the track is installed. On steel coasters,

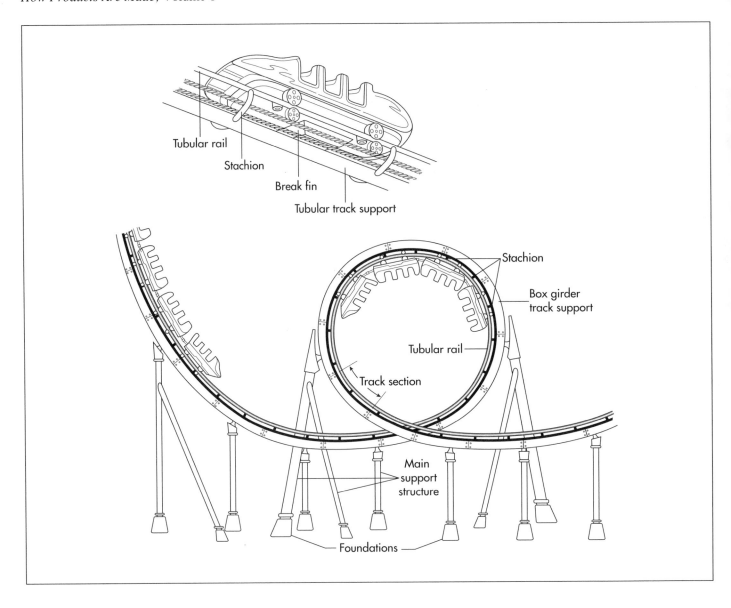

Tubular rail

Stachion

Break fin

Tubular track support

Stachion

Box girder
track support

Tubular rail

Track section

Main
support
structure

Foundations

An example of a steel-constructed
roller coaster and car.

sections of track are fabricated in the factory
with the stanchions and tubular tracks welded to the track supports. After the sections
are brought to the site, they are lifted into
place, and the track ends are slid together.
The sections are then bolted to the main support structure and to each other. On wooden
coasters, wood tie beams are installed across
the top of the main support structure along
the entire length of the ride. Six to eight layers of flat wood boards are installed length-
ways on top of the tie beams in two rows to
form a laminated base for the rails. The rails
themselves are formed from long, flat strips
of steel screwed into the wood base.

7 On steel coasters, walkways and
handrails are welded in place along the
outside of the track to allow maintenance

access and emergency evacuation of passengers. On wooden coasters, the portions of
the tie beams outside of the track are used as
walkways, and handrails are installed.

8 The lift chain and anti-rollback mechanisms are installed on the lift hill, and the
braking device is installed on the final approach to the station.

Fabricating the cars

9 The individual cars for the coaster are
fabricated in the factory. The subframe
pieces are cut and welded. The bodies are
stamped from aluminum or molded in fiberglass, then fastened to the subframe. Seat
cushions may be cut from foam, mounted on
a base, and covered with an upholstery.

Running wheels and guide wheels are bolted in place with locking fasteners. Brake fins, anti-rollback dogs, and other safety components are installed.

Finishing the ride

10 When the main construction is completed, electrical wiring is installed for the lighting, and the entire ride may be painted. The boarding station is constructed, signs are installed, and the landscaping is put in place.

Quality Control

The design and construction of roller coasters are covered by numerous governmental safety regulations. The materials used must meet certain strength requirements, and the actual construction is subject to periodic inspection. Every day, the coaster must be thoroughly inspected before it goes into operation.

Before the ride is open to the public, the cars are filled with weighted sandbags and sent through several circuits to ensure everything is operating properly. Government safety inspectors check make a final review before they give approval to operate.

The Future

The current trend to higher, longer, and faster coasters will probably continue for the near future. This is especially true now that roller coasters have become popular in Europe, Asia, and many other foreign countries. In the meantime, coaster designers will be looking for new ways to give riders a physical and visual thrill.

Where to Learn More

Books

Bennett, David. *Roller Coaster: Wooden and Steel Coasters, Twisters, and Corkscrews.* Edison, NJ: Chartwell Books, 1998.

Cook, Nick. *Roller Coasters, or, I Had So Much Fun, I Almost Puked.* Minneapolis, MN: Carolrhoda Books, Inc., 1998.

Periodicals

Lindsay, D. "Terror Bound." *American Heritage* (September 1998): 76-89.

Ruben, P. L. "Scream Machines." *Popular Mechanics* (August 1998): 80-83.

Other

World of Coasters. http://www.rollercoaster.com (November 29, 1999).

—*Chris Cavette*

Rubber Cement

In 1996, over 12 billion lb (5.4 billion kg) of adhesives were used in the United States.

Background

Rubber cement is a solution of unvulcanized (gum) rubber in a solvent, and is used as an adhesive. Ideally, it is meant to join two pieces of rubber together, which involves a chemical cohesion process. When joining two pieces of rubber, only one surface has to be coated with rubber cement since they are the same material. However, when joining paper together, both pieces need to be covered with rubber cement. When rubber cement dries, only the parts in contact with the paper remain, which holds the two pieces together. Despite this limitation, the household type of rubber cement finds wide use for applications such as mounting photographs. Unlike with white glue, the joined pieces of paper can be pulled apart without damaging either piece.

Adhesives are made from either natural animal or plant products or synthetic polymer. Natural adhesives are easy to apply and in general are water soluble. Synthetic adhesives are divided into four chemical categories: thermoplastic, thermosetting, elastomeric, and combinations thereof. Thermoplastic adhesives, such as polyvinyl alcohol and acrylics, can be resoftened since the materials do not crosslink upon curing. They require heat or a solvent to create a bond. Thermosetting adhesives, which include epoxies, cannot be heated and resoftened after curing because they do crosslink upon curing. Elastomeric adhesives are based on isoprene rubber or synthetic polymers that combine both elasticity and toughness. Silicone is a typical example.

The properties of the adhesive, the types of materials to be joined, and the condition of the surfaces all determine the performance of the joints and the service life of the bonded structure. The adhesives must be able to wet and spread properly on either surface to achieve molecular contact between the materials. Adhesives are used in a wide range of industries, including packaging, construction, electronics, transportation, furniture/woodworking, and medical.

In 1996, over 12 billion lb (5.4 billion kg) of adhesives were used in the United States. Construction applications had the highest share at 40%. This is also one of the largest applications for rubber cement. Other applications include heating, air conditioning, and automotive equipment. Manufacturers of rubber cement also sell a lot of their product to repackagers who market the product under their own name.

Some analysts put the global market for adhesives at $19.1 billion in 1997. Packaging, construction, and furniture/woodworking are the three largest segments of this global market, with over 65% of market revenues.

History

Natural adhesives have been around for at least several thousand years. Egyptian carvings show the gluing together of thin pieces of veneer to a wooden plank. Fibers in ancient fabric were joined together with flour paste, and gold leaf was bonded to paper with egg white. Animal glues improved during the eighteenth century and a century later, rubber- and nitrocellullose-based cements were introduced. During the 1900s, significant advances occurred, leading to the development of many synthetic adhesives that replaced some of the natural adhesives. Adhesives now had to be much stronger and more corrosion resistant.

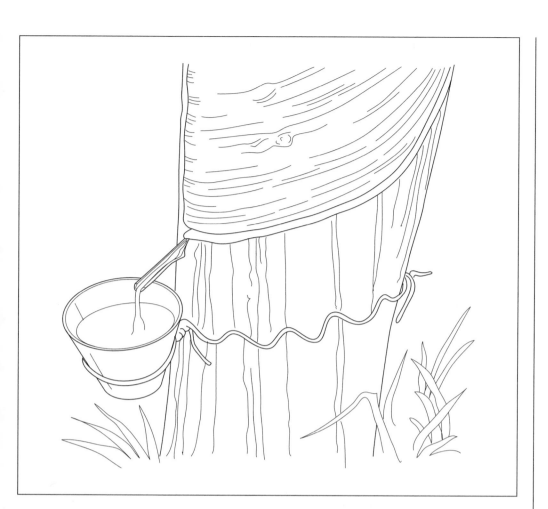

Natural rubber was first discovered by the Mayas and Aztecs over 2,500 years ago and used to make shoes and clothing waterproof. Centuries later in 1823, the Scottish chemist Charles Macintosh also investigated making waterproof textiles. He experimented with dissolving rubber in various chemicals and had the most success with naphtha. With this material he made a rubber paste and used it to join together two layers of cloth to make a **raincoat**.

Raw Materials

Rubber cement is an opaque liquid that contains pulverized natural or synthetic rubber and a solvent based on hexane or heptane. Grades of rubber cement may contain 70-90% heptane or hexane and 1-15% isopropyl alcohol (isopropanol) or ethyl alcohol (ethanol). The rubber is received in the form of large blocks or slabs, typically 100 lb (45 kg) in size. Thousands of gallons (liters) of liquid solvents are usually shipped by tank truck to the manufacturer.

Natural rubber comes from the *Hevea brasiliensis* tree originally found in Brazil. To make solid rubber, the tree is tapped and the latex is collected in a small cup, where it coagulates into a lump. This lump, together with the leftover flow and other pieces are collected together and processed at high temperature. This destroys most of the proteins and produces a solid material.

Synthetic rubbers include neoprene and latex. Synthetic rubbers are made using various chemical processes. The application determines what types of rubber and solvent are used.

Design

The properties and performance of the rubber cement are determined by the type and amount of ingredients. Typically, formulations are determined by the laboratory and then given to production.

The Manufacturing Process

The process to make rubber cement is relatively simple. After the rubber is broken down into smaller pieces, it is mixed with the hexane- or heptane-based solvent and then various sizes of containers are filled with the liquid. Most equipment is automated.

Mixing

1 First, the blocks or slabs of rubber are broken into smaller pieces. Rubber mills, equipped with two large rollers, are used. The rubber may be reduced in large high speed mixers equipped with sharp blades, which pulverize the rubber into a size similar to sawdust. The solvent ingredients are mixed in tanks, ranging from 40 to 6,000 gal (150 to 22,700 l), equipped with paddles. The rubber is added slowly until it is wetted by the solvent and is suspended or dissolved in the solution.

Packaging

2 Once the solution is thoroughly mixed it is fed into an automated filling line, which fills each container, caps it, and tightens the cap. Rubber cement is packaged in 4, 8, and 16 oz (118, 237, and 473 ml) bottles or quart (946 ml) and gallon (3.8 l) containers. Because the rubber cement is flammable, explosion proof equipment is used. The containers are then packaged in cardboard cases and properly labeled. For larger containers (tank trucks or drums), the solution is gravity fed or pumped into these containers from the bottom.

Quality Control

The raw materials are supplied according to the manufacturer's specifications. Each batch is checked for solids content, since better quality grades contain higher content. The percent solids content is obtained by weighing a sample, evaporating the liquid, and then weighing the remaining solid.

After mixing, each batch is also tested for viscosity, tackiness, heat load, and other properties before being packaged. Up to 20 different tests may be performed. After packaging, the containers are checked to make sure they are properly labeled.

Byproducts/Waste

Because of the tight controls and automated equipment, no waste is produced. The filling machine is programmed for the exact volume of each size of container. Any leftover material is recycled back into the process. Rubber cement is shipped with a material safety data sheet (MSDS) that outlines proper handling procedures since the ingredients are hazardous and flammable. The solvent is a volatile organic compound and is also subject to regulation as a hazardous air pollutant under the 1990 Clean Air Act Amendments. The U.S. Environmental Protection Agency also has included adhesives as one of the target categories that must comply with certain Maximum Achievable Technologies (MACT). A MACT sets a level of control designed to protect public health.

The Future

Demand for all adhesives in the United States is expected to reach over 15 billion lb (6.8 billion kg) by 2003, a 2.6% annual growth rate. The global market is expected to reach $26.2 billion in 2003, with a compound annual growth rate of 5.3%. Solvent-based adhesives will continue to be replaced by water-based adhesives since they are more environmentally friendly.

Where to Learn More

Other

Starkey Chemical Process Co. P.O. Box 10, 9600 W. Ogden Ave., LaGrange, IL 60525-2534. (708) 352-2565.

—Laurel M. Sheppard

Sailboat

Background

For people who like to be near the water, sailboats provide a means of skimming over its surface. Even when the water is frozen, iceboats (sailboats with runners or blades on the hull) can glide across the ice. The sailboat is a form of transportation, a type of recreation ranging from simple craft to the most elaborate racing yacht, and even a form of housing. A huge yacht with custom fittings and a crew is a symbol of wealth. A one-person boat with a tiny sail represents freedom. A majestic old clipper ship hearkens back to historic battles, the spice and tea trades, and sea legends like the "Flying Dutchman." And a bay or lake filled with sailboats and whitecaps with clouds rippling above makes a picture perfect for painting and a lasting memory.

Sailing was once a hobby of the rich, but the availability of free time and more cash to the "average" person has made sailing one of the most popular forms of recreation. Sailboats may be handcrafted or factory built in all sizes from day-sailers and other boats less than 11 ft (3.4 m) long, to the dinghy, larger single-masted sailboats, two-masted boats called yawls, and large yachts. Yachts are patterned after historic sailing vessels called brigantines, cutters, clipper ships, and schooners. Boats used for racing are specially designed for speed and maneuverability, while sailboats of all sizes that have onboard quarters for passengers and crew are sturdier in design with more details for comfort. Many sailboats also carry inboard or outboard diesel-powered motors in the event they are becalmed (motionless from lack of wind) or their sailors simply want a speedier return to port.

Of course, the sailboat is distinguished from other craft by its sails. A sail is simply a piece of fabric that is used to catch the wind to drive the boat across the water. Most modern sails are made of Dacron, a polyester fiber. Because the fabric is heated to meld the fibers together, the wind cannot escape through pores like those in woven cloth, and the surface has a very low friction factor. Polyester sails are also lightweight with little stretch.

Sails fall into two major categories and then into many subclasses. The two major categories are square and triangular sails. Square sails are mounted across the main axis of the boat to use the wind pressure to power the boat. Wind strikes only the back, or afterside, of square sails. Triangular sails follow the same axis as the boat, with fore sails at the front or bow of the ship and aft sails at the rear or stern. Both sides of triangular sails are used for forward motion, and they can be adjusted to make the best use of the wind's force.

The subclasses of sails are named for the pieces of rigging that support them or for neighboring gear. Masts are significant identifiers. A three-masted sailboat has a foremast, mainmast, and mizzenmast (toward the stern). Single-masted boats have a mainmast only. Two-masted vessels may have either a foremast and mainmast or a mainmast and mizzenmast, where the mizzenmast is the shorter of the two. Sails named for parts of the ship include gaff sails, jib-headed sails, spritsails, and lateen and lugsails. Sails are also named for specialized uses: summer sails are for tropical conditions, storm sails are used in bad weather, racing sails are needed only by racers, and cruising sails are the standard set

The Egyptians and other ancient people wove reeds together in mats to make sails, but the Egyptians were also the first to make cloth sails as early as 3300 B.C.

for everyday conditions. A three-masted square-rigger can be equipped with as many as 20 sails with unique names and purposes.

History

Animal skins were used as sails for the earliest boats and rafts. The Egyptians and other ancient people wove reeds together in mats to make sails, but the Egyptians were also the first to make cloth sails as early as 3300 B.C. Great sailors of the Mediterranean region like the Phoenicians sailed under cloth sails. Over the centuries, sails woven from a variety of fibers, such as hemp, flax, ramie, and jute, were sailmakers' favorites; but flax fiber was the primary material for sails throughout the age of exploration (approximately 1450–1650). **Cotton** gradually replaced flax as cultivation and processing of cotton increased. It was the victory of the racing yacht *America* in 1851 that crowned the cotton sail as supreme. This United States yacht defeated 14 British vessels in a sailing race around the Isle of Wight off the south coast of England and was the source of the name for the America's Cup Race, the greatest yacht race in the world.

Sailboats themselves began as single logs and simple rafts. More sophisticated shapes for hulls that would cut through the water grew out of military use, but also from merchant sailors who built extensive trading networks crisscrossing the Mediterranean Sea. When day-sailers were built for fishing and recreation, they were essentially miniature copies of naval ships like schooners and cutters. The elaborate yachts that were the playthings of royal families and the wealthy also copied naval sailing ships. By about 1850, a new engineering discipline called naval architecture was begun to design efficient hulls and other parts of sailboats according to the laws of physics and engineering and architectural principles. Sails and rigging and their effects on the speed of sailboats were essentially ignored until 1920. Since that time, aerodynamics have been used in their design. Today, modifications to complex craft like the boats that participate in the America's Cup Race are based on wind-tunnel testing and many other sophisticated analyses applied to boats, water, wind, and sails.

In parts of the world where waters are frozen for most of the year, iceboats were developed to skim the sailboat over the ice by mounting it on runners or blades. Archaeologists have found evidence of iceboats in Scandinavia dating back to 2000 B.C. Eyewitness accounts from Scandinavia, the Netherlands, and Baltic Coast countries like Latvia and Russia are much more recent, with the earliest from the seventeenth century. In the United States, the first known iceboat in the New World traveled up and down the Hudson River in New York in 1790. Like their warm-water counterparts, iceboats that race are called ice yachts, and ice yachting as an acknowledged sport dates from the nineteenth century.

Raw Materials

Sailboat manufacturers either fabricate their own parts or order them, depending on the intended volume of production. Items that are usually provided by specialty suppliers include masts, sails, engines, and metal fittings. Boatbuilders make their own fiberglass hulls, however, and the materials used to cast reinforced fiberglass include Gelcote polyester resin, a catalyst for the resin, woven fiberglass roving, and fiberglass. Manufacturers who build hulls from wood similarly order, age, and shape their own wood.

Roving is strand-like material that resembles burlap. It can be woven with biaxial, triaxial, or knit strands, and the designer specifies the type of roving depending on the planned design and weight of the finished sailboat; sailboats range from lightweight and very fast to strong and highly seaworthy.

Design

Manufacturers usually make several specific lines of sailboats. Their designs are drawn and printed on blueprints or drafted by computer design methods. When a design is new, a wooden plug is made from the blueprints to exactly match the configuration of the outside of the empty hull. The wooden plug is sanded, polished, and covered with a slick coating, something like boat-builder's Teflon, from which other materials can be removed.

A mold is built up on the outside of the plug; the inside of the mold is then the model for the outside of the sailboat-to-be. This mold is cast of fiberglass, and it must

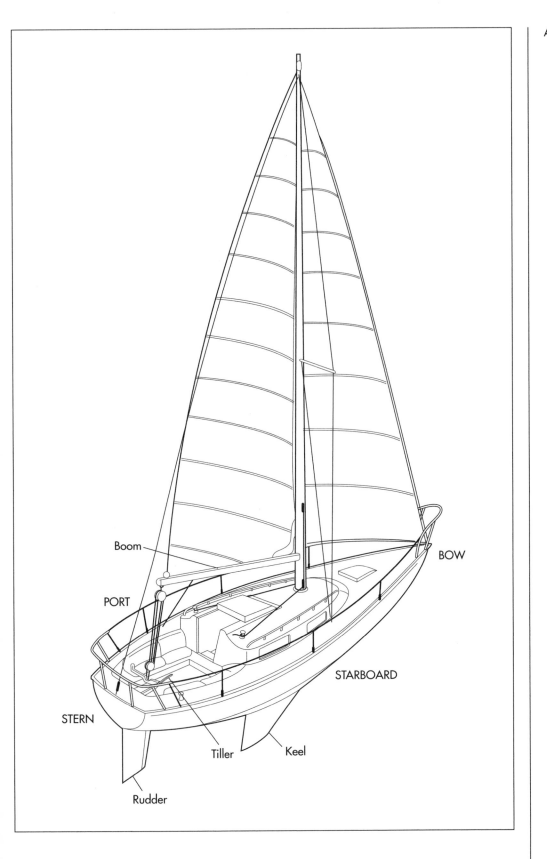

Boom

PORT

STERN

BOW

STARBOARD

Tiller

Keel

Rudder

be perfect. The process of making a plug and mold is very expensive, so the designers must be sure of the plan on paper before proceeding to plug construction.

After the mold is completed and approved, it is ready for use in duplicating sailboats of this design. The blueprint plan, plug, and mold are used to calculate the number and

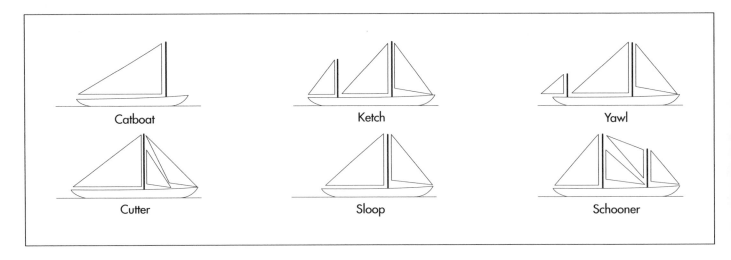

Catboat

Ketch

Yawl

Cutter

Sloop

Schooner

Examples of different types of rigs.

types of fittings on the sailboat and deck, engines, sails, and outfitting of any interior. These items are ordered from specialty suppliers in advance of production.

The Manufacturing Process

Manufactured sailboats typically range from 11-ft-long (3.4-m-long) day sailers to 28-ft-long (8.5-m-long) and sometimes longer luxury craft equipped with diesel motors and living quarters. The process described below is for the construction of small, fiberglass craft, but some remarks address larger varieties.

The hull

1 Building the hull of the sailboat begins from the outside in. The contact surface along the inside of the mold is the outside of the hull. The mold is lined with a parting agent—a non-stick coating that will help release the sailboat hull from the mold when the hull is complete. Color pigment is laid against the inside of the mold; effectively, the "paint job" is the first part of the sailboat to be made, although the pigment incorporates into the outermost fiberglass and resin. Sailboats can be colored in a rainbow of hues, thanks to the variety of pigments available.

The skin coat of the boat is made with 4-5 oz (124-156 g) of blown fiberglass that resembles cotton candy. Unlike the roving that will follow, the skin coat fiberglass is random-strand fiberglass that is blown in and then spread by hand. The skin coat is about 5 mils (0.005 in; 0.13 mm) thick, but

this is enough to keep the pattern of the roving from bleeding through to be detectable on the outer finish of the sailboat. Gelcote is applied to harden the fiberglass. This chemical resin reacts with a catalyst that causes a chemical reaction that converts the pliable fiberglass into a hard material.

2 Woven, 24-oz (746-g) fiberglass roving is placed against the skincoat. Layers of roving are rolled out by hand and are bound together for reinforcement and built up for thickness and strength. It is also pressed and molded into shape against the outer layers and the mold itself. Gelcote again is applied to begin the hardening of the fiberglass. The resin and catalyst are sprayed on with a carefully metered spray until the roving is saturated. After the Gelcote is applied, the fiberglass workers have 30-40 minutes to finish manipulating the fiberglass while it is in a state somewhat resembling cookie dough and before it begins to harden. The process of building up the fiberglass and spraying on the resin and catalyst is done in a temperature-controlled room in which the temperature is maintained at 72°F (22.2°C). Workers laminate the fiberglass for one-half of the hull (i.e., one side of the sailboat) at a time. Each half then cures for 24 hours to allow the catalyst to complete its chemical conversion.

3 The first layers of the remaining half of the sailboat are laminated and cured, and the workers then return to the first side to continue adding more layers to thicken the hull. Boats in the 18- to 24-ft-long (5.5- to 7.3-m-long) size range are typically made of three layers of roving. During the process of "glassing" the boat, stiffeners are also added

to the hull. Sections of precut and preformed marine plywood are laid in the bottom of the empty hull. As more roving thicknesses are added, the plywood is glassed into place; the resin and catalyst harden the plywood into place as structural components of the fiberglass. Depending on the design and size of the sailboat, the plywood components may consist of a system of stringers, ribs, and bulkheads that strengthen the boat against the impact and bending forces of waves.

4 In another part of the glassing room, the deck of the sailboat is being constructed. A corresponding deck mold is made. Like the lid of a shoebox, it fits over the top of the open hull and is mechanically fastened in place. The fiberglass deck is made by exactly the same process. A parting agent is applied to the underside of the deck mold (the upper side of the deck), pigment is placed, and a skin coat of blown fiberglass is sprayed on followed by application of the resin and catalyst. The layers of the deck are built up of fiberglass roving, stiffening sections of marine plywood are glassed into place, and special reinforcements are laminated into place wherever hardware or fasteners will pass through the deck. Reinforcement pieces are made of marine plywood or aluminum.

5 When fiberglass laminating has been completed on both the hull of the sailboat and its matching deck, the thoroughly cured sections are removed from their molds. A wooden wedge is driven between the mold and the fiberglass section, and the section pops out. Alternatively, the mold and section are submerged in water, and the water pressure between the two pieces pops out the fiberglass section. A chain hoist is used to lift the section out of its mold, and it is placed on an assembly cart so workers can move and access it easily.

6 The assembly phase begins with "green trimming," a process in which bits of flashing (fiberglass and resin excess) are trimmed off. The deck and hull sections of the sailboat are sanded and cleaned, all edges are checked for smoothness, and all joints are inspected for precise fit. Obviously, larger boats require many more assembly steps than day-sailers. If the boat is larger than 22 ft (6.7 m) long, it will be fitted with a cockpit liner, interior liner, cabins, sitting

An America's Cup contestant.

John Cox Stevens, a wealthy New Jersey real estate broker and sports promoter, spearheaded the organization of the New York Yacht Club in 1844. As commodore of the NYYC, in 1850 he organized a syndicate of five other club members that commissioned William H. Brown to construct a racing yacht. Following the design by George Steers, Brown finished *America* in 1851, in time for Stevens to accept an invitation from Britain's Royal Yacht Squadron to enter its race around the Isle of Wight. Pitted against 17 seasoned British boats, *America* started poorly but finished with a commanding lead and won the prize, a 100-guinea silver cup. Stevens accepted the **trophy** and kept it on display at his estate. After his death in 1857, the Cup became a trust of the NYYC.

The Cup was first challenged in 1870 by the 113-ft (34-m) English schooner *Cambria*. The 84-ft (26-m) United States defender, *Magic*, won the 35-mi (56-km) race. For the next 132 years, the *America*'s Cup remained in the United States. It was lost to Australia in 1983. The *Australia* II challenged the *Liberty* with Dennis Connor as skipper. At the next race in 1986, Connor won the Cup for the United States aboard the *Stars and Stripes*. The Cup was again lost in 1995 to New Zealand, who, in 2000, became the only country other than the United States to win the cup twice. The next challenge for the *America*'s Cup is set for 2003.

rooms and areas, toilets (called heads on boats), and sleeping areas. Usually, larger craft are custom-built, and these systems are made of wood.

7 Plumbing is installed next. Flexible pipe is used. Hoses are needed from fresh water tanks that store water for toilets and showers (again, on larger vessels), and return hoses are also required to return dirty water to the onboard sewage tank. Electrical lines are installed in accordance with electrical code requirements. Engines (if any) and me-

chanical equipment are fitted into the sailboat next. If the boat has an onboard engine, it is fitted to motor mounts that are installed on stringers—structural parts of the hull. A propeller shaft and propeller are fastened into place, and the engine controls are hooked up.

8 The green-trimmed and polished deck is inverted and fitted over the hull. A "dry fit" is done first to make sure the deck still fits the hull snugly. Mechanical fasteners made of stainless steel are passed through matching reinforced sections of the deck and hull. Before the deck is locked into place, a leakproof sealant especially made for marine construction is applied. The fasteners are then mechanically tightened. Deck hardware is installed. Hardware may include winches and turning blocks for winding lines and sails, rails, stays, and other features. Installation of hardware is an important indicator of the quality of the sailboat; the hardware itself, as well as the backing plates and locking nuts and bolts, must withstand heavy use and intense water action. Hardware is a very expensive part of sailboat construction; a single winch for a yacht made for the America's Cup Race may cost $20,000.

9 The mast is the last piece of major hardware added. Inside the hull, a wooden block or shoe called a tabernacle is installed as a seat for the mast. Masts are usually made of aluminum for light weight. When the mast is seated on the tabernacle, wires called stays are used to hold the mast in place. The stays are made from stainless steel wire and custom fit to each sailboat; manufacturers usually make their own stays in-house to assure the correct tension to stabilize the mast.

10 All wires (if any) for electrical connections and hoses for plumbing (if any) are hooked up and tested. If the sailboat is of the size and design to have windows in the cabin, templates are placed on the hull, and the outlines of the portholes are cut out. Sealant is added around the edges of the openings, and windows consisting of frames and glass are set in the openings. Additional sealant is applied to make secure seals. The painting of the exterior is checked for any mars or marks and detailed. Accents are painted on and feature tape is added for stripes and other decorations. Decals, numbering, and manufacturer's information is also added.

The sails

1 Sails are made by sailmakers who are highly skilled in design and material properties. A sailboat manufacturer may provide the sailmaker with a plan or the sailmaker may measure the boat's rigging and design the sails from that. Mathematics enter sail design in estimating the stretch of a sail with wind pressure applied, and the curvature of each sail's surface is calculated. This curvature is called the sail's draft, and it provides forward motion from the wind just like the curved wings of an airplane provide uplift. The sailmaker's shop (called a sail loft) is a critical tool because the plans for each sail are outlined in chalk on the floor and to full scale. The fabric, which is usually polyester, is laid over the plan, and the outlines are transferred onto the fabric. Each sail shape is numbered.

2 The sails are cut out according to the outlined dimensions drawn on them. Sails consist of several pieces, and these are sewn together with sail twine, a specialized fiber that is stronger than typical thread. When the sewing needle is threaded with sail twine, the twine is twisted to add strength, and wax is applied so the twine will hold that twist.

3 After all the pieces of each sail are sewn together, reinforcement is added to the parts of the sail that will take the most strain. Patches (reinforcements) are added to each corner and tabling (a thickened hem) is sewn along the forward edge (called the luff) and the foot of the sail. When the tabling is completed, ropes are also sewn inside the luff and foot edges of the sail to reinforce these parts of the sail against stretching.

4 Finally, hardware and fittings are attached to each sail. These may include metal slides, grommets, and reef points depending on the design of the boat and the purpose of the sail. The sails are shipped to the sailboat manufacturer, and sets are stored on the sailboats to which they belong.

Quality Control

Quality control is a continual process. All fittings and materials received from outside suppliers are checked upon receipt. Items like electrical wiring are bench-tested before installation in the boats, checked immediately after installation, and tested again during a final quality check.

Specialties such as glassing the fiberglass hull have critical requirements for temperature and placement. The Gelcote is applied against an indexing gauge because it cannot be too thick or too thin. Over catalyzing the resin produces intense internal temperatures in the material, and it cannot be worked. If catalyzing fails, the resin has to be chipped out—an expensive, time-consuming process. Errors in design of fiberglass thicknesses and placement of hardware as well as such errors in construction affect the weight and balance of the sailboat and how it will sit on the water; if the balance is off, the static list of the boat will be wrong.

Details are also important, and quality control checks are scheduled so they are corrected immediately. An uncorrected mistake within the hull may not be easy to reach after the deck is fastened in place. The final quality check should reveal only minor problems like tiny flaws in the exterior that are quickly repaired.

Byproducts/Waste

Sailboat manufacturers do not typically make byproducts, and they tend to specialize in several closely related lines. Only the largest companies have much wider ranges of products; their larger production allows an economy of scale in purchasing hardware and other supplies that appear to give them an advantage over smaller builders. Small builders know, however, that a lovingly crafted product is their boast, and they are willing to purchase smaller quantities (and waste less as well) in favor of custom quality.

Waste is a minor issue. Most materials can be recycled, and the business requires such strict quality control that waste is minimized by careful workmanship. Safety is a major concern, by contrast. Thirty years ago, the manufacture of sailboats was unregulated, and many boat builders ran "cottage industries" in their back yards. Regulations have made this almost impossible. Workers are trained in safety issues, particularly related to air quality, and they must wear respirators throughout the construction of fiberglass hulls because of the fumes generated by the catalyzing process. Air emitted from the

building must also be controlled in accordance with the regulations of the U.S. Environmental Protection Agency (EPA). Workers wear Tyvex suits to guard against splashing hazards, and chemicals are carefully stored and disposed according to regulatory requirements.

The Future

Sailboats are symbols of beauty and freedom. They are highly desirable forms of recreation and symbols of success, independence, and free time. Periods when the economy is strong lead to greater sailboat production and more sails visible on local bodies of water. A sailboat as a retirement home is the dream of many. The variety of sizes of sailboats suits them to a similar variety of lifestyles. All of these aspects seem to insure the future of the sailboat. They are a simple, elegant, and ancient form of transportation made modern; they also provide opportunities for people to enjoy that unique environment where the wind meets the water.

Where to Learn More

Books

Gustafson, Chuck. *How to Buy the Best Sailboat*. New York: Hearst Marine Books, 1991.

Hines, Vernon. *About the America's Cup*. Grand Junction, CO: Bookcliff Publishing Company, 1986.

Kentley, Eric. *Boat*. New York: Alfred A. Knopf, Inc., 1992.

Langone, John. *National Geographic's How Things Work: Everyday Technology Explained*. Washington, DC: National Geographic Society, 1999.

Marshall, Roger. *Designed to Win*. New York: W. W. Norton & Company, 1979.

Other

Boat Talk Boating Information Bureau. http://www.boattalk.com (October 1, 2000).

Glen-L Marine Designs. http://www.glen-l.com (September 29, 2000).

Martini Marine. http://www.martinimarine.com (September 29, 2000).

—*Gillian S. Holmes*

Salad Dressing

By the end of the twentieth century, over 60 million gal (227 million l) of salad dressings were sold in the United States.

Background

Salad dressing is a type of sauce used to bind and flavor greens and/or vegetables.

History

Using oil and vinegar to dress greens and vegetables dates to Babylonian times, some 2,000 years ago. The word salad can be traced to the ancient Romans who sprinkled salt on grasses and herbs, calling it *herba salata*. It was not long before Roman and Greek cooks experimented with combinations of olive oil, vinegar, and salt, then adding wine, honey, and a fermented fish sauce known as *garum*. The latter was made by soaking the intestines and other pieces of mackerel, salmon, sardines, and shad in brine and herbs.

The kings and queens of Europe were notably fond of salads with royal chefs tossing together as many as 35 ingredients. Henry IV of England was known to prefer a bowl of sliced new potatoes and sardines splashed with herb dressing. For Mary, Queen of Scots, it was lettuce, boiled celery root, truffles, chervil, and hard-cooked eggs in a mustard dressing.

Salad dressings were made from scratch in home kitchens until the turn of the nineteenth century when restaurant owners began packaging and selling their own dressings. One of the first was Joe Marzetti, proprietor of a Columbus, Ohio, restaurant. In 1919, Marzetti began to bottle a variety of dressings from old country recipes.

The Kraft Cheese Company entered the salad dressing industry in 1925; its first flavor was french, an oil and vinegar-based dressing flavored with tomato and paprika.

By the end of the twentieth century, over 60 million gal (227 million l) of salad dressings were sold in the United States. The most popular flavor by far was ranch dressing. The original brand, Hidden Valley Ranch, was created by Steve Henson who devised the recipe as a dry mix to be blended with **mayonnaise** and buttermilk. Henson and his wife Gayle served it at their California dude ranch, called the Hidden Valley Guest Ranch, in the late 1950s and early 1960s. Guests reported that the dressing was so popular, they often poured it on steaks and ice cream as well as on salads.

Guests began to request jars of the dressing to take home with them. When one man wanted 300 jars to take back to Hawaii, Henson offered to provide him with enough packages of the dried mixture instead. This led to a very lucrative mail-order business. The family eventually sold the business, which is now owned by Clorex.

Raw Materials

The primary ingredient in salad dressing is oil. In the United States, soybean oil is the most common type used in the production of salad dressings. Olive, peanut, and sunflower oils may also be used.

Stabilizers and thickeners, such as modified food starch, are mixed with the oil. The thickeners develop viscosity and protective colloid characteristics that help to prevent the breakdown of the blend during the various processing steps.

Other food ingredients are added depending on the type of dressing, including any or all of the following: eggs, vinegar, salt, honey,

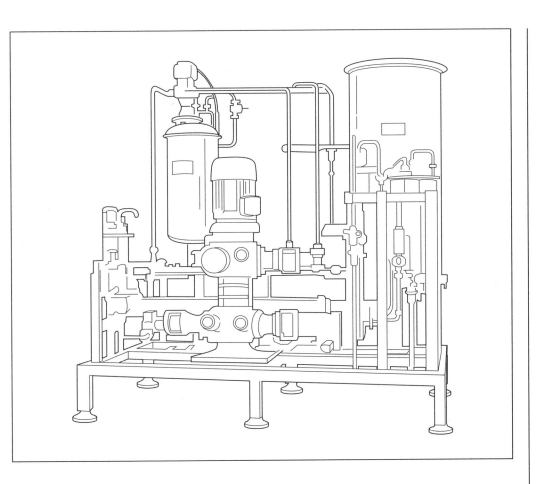

sugar, spices and herbs, tomato, vegetable bits, sherry, and lemon or lime juice. A large sub-industry is involved in the processing of these ingredients. Herbs and vegetables are usually blast-frozen and cut into pieces or reduced to flakes.

Monosodium glutamate (MSG) is a food additive developed in the early 1900s from seaweed. Today in the United States, MSG is extracted from the gluten of cereals. It is used in variety of foods, including salad dressings, to enhance flavor. However, the increase of allergic reactions experienced by consumers have caused some processors to cease using it.

Design

The salad dressing industry is constantly creating new and so-called improved flavors. Low fat and non-fat varieties are of particular interest to the consumer. The larger companies have food laboratories on site. Smaller companies and start-up companies often rely on research conducted by university food science institutes.

The Manufacturing Process

Creating the emulsion

1 The commercial salad dressing industry uses a continuous blending system to attain the correct degree of emulsification so that the mixture does not break down. An emulsion, or colloid, forms when the blending of two liquids, such as oil and water, causes one of the liquids to form small droplets that are evenly dispersed throughout the other liquid. In the production of commercial dressings, this basic blend moves continuously through a series of pumps and heat exchangers as the other ingredients are added.

Positive replacement pumps feature a cavity or set of cavities fitted with rotary impellers. A regulated pumping action causes the cavities to fill and empty. The impellers move the blended fluid from one cavity to another.

2 Manufacturers may use a rotational viscometer to test the viscosity, or consistency, of the dressing. This machine consists of a solid cylinder fitted inside a hollow

cylinder with a space of about (0.08 in) 2 mm in between. A sample of the dressing is poured into the space, the top is sealed and the cylinders are set spinning. As the dressing spins, it exerts torque on the center cylinder. A gauge then measures the consistency. Adjustments are made as necessary.

Adding ingredients

3 Pre-measured ingredients are piped through openings in the sides or from spigots up above.

Bottling the dressing

4 When the dressing is completely blended according to the recipe, it flows to the bottling station. Here, pre-sterizilized jars or bottles move along a conveyer belt as overhead spigots drop premeasured amounts of dressing into each container. The containers are immediately sealed with metal or plastic caps.

Labeling

5 Labels are mechanically affixed to each container. All ingredients and nutritional information must be printed on each label.

Quality Control

In the United States, the Food and Drug Administration (FDA) sets Standards of Identity for french dressing, and the modified mayonnaise called salad dressing that is used as the basis for most creamy dressings. French dressing must contain at least 35% vegetable oil by weight, vinegar, and tomato and/or pa-

prika products. Salad dressing must contain a minimum of 30% vegetable oil, 4% egg yolk ingredient, vinegar or lemon juice, and spices.

At the processing plant, each shipment of raw and processed ingredients are tested upon arrival. Daily samplings are taken of all stored materials to insure their freshness.

Where to Learn More

Books

Coyle, L. Patrick, Jr. *The World Encyclopedia of Food.* New York: Facts on File, 1982.

Schlesinger, Chris, and John Willoughby. *Lettuce in Your Kitchen.* New York: William Morrow, 1996.

Trager, James. *The Food Chronology.* New York: Henry Holt, 1995.

Periodicals

Barr, Susan Learner. "Well-dressed." *Shape* (August 1997).

Other

Association for Dressings and Sauces. http://www.dressings-sauces.org (December 2000).

Kraft Unit Operations. http://www.kraftunitops.com (December 2000).

Hidden Valley Ranch. http://www.hiddenvalleyranch.com (December 2000).

—*Mary McNulty*

Saxophone

Background

A saxophone is a single reed, woodwind instrument first developed in the mid-1800s by Adolphe Sax. It is composed of a mouthpiece, conical metal tube, and finger keys. Sound is produced when air is blown through the instrument causing the reed to vibrate. This sound is amplified as it travels through the instrument's main body. Saxophones consist of numerous parts and pieces which are made separately and then assembled.

History

Most instruments have steadily evolved over many years. In fact, no one person can be said to have invented common instruments like the flute or the oboe. The saxophone however, can be directly credited to Adolphe Sax who invented it during the 1800s. Sax was born in Belgium in 1814 and learned to make instruments from his father who was a musical instrument maker. By the age of 16, Sax was already an accomplished instrument maker himself. Some of his achievements included improving the clarinet's design and adding piston valves to the cornet. During his time, he produced some of the highest quality clarinets, flutes, and other instruments.

When he set out to develop the saxophone, he wanted to create an instrument that could blend the orchestral sounds of the woodwinds with the brass instruments. His new instrument would have the tone quality of a woodwind and the power of a brass. The first saxophone he built was a large, bass saxophone. Since a conical shape was needed, it was easier to make the instrument out of brass than wood. On March 20, 1846, Sax patented this instrument. Smaller saxophones such as the alto and tenor were created a short time later.

In addition to his instrument-making prowess, Sax was also an entrepreneur. To promote his new instrument he staged a "battle of the bands" between the traditional French infantry band and one that used his saxophone. Sax's group won the contest, and the military officially adopted the saxophone into their bands. This caused a significant level of resentment toward Sax and many instrument manufacturers and musicians rejected the saxophone as an acceptable instrument, refusing to produce or play it. This prevented the saxophone from being used for its original purpose in the orchestra.

However, many composers were impressed with the sound of the saxophone and steadily incorporated it into their pieces. This versatile instrument was used in many musical styles. For example, it has been used in opera such as *Bizet's l'Arlesienne* and also worked into Ravel's orchestral piece, *Bolero*. In the United States, the instrument was made famous by J. P. Sousa who used it extensively in his marching band compositions.

The true potential of the sax was realized by jazz musicians during the early 1900s. Artists like Charlie Parker and John Coltrane helped make it the most popular woodwind solo instrument for jazz. Both of these musicians had distinctly different sounds. The individualized sound is a result of various mouthpiece materials and structures, reed hardness, and the musicians mouth position. For jazz musicians, the mouthpiece was modified so the instrument would be louder.

The typical saxophone is a single reed instrument constructed from brass with a curved bottom.

335

Raw Materials

Saxophones are primarily made from brass. Brass is a composite alloy made up of metals including copper, tin, nickel, and zinc. The most common type used for instruments is yellow brass which contains 70% copper and 30% zinc. Other types include gold brass and silver brass which have different ratios. The zinc in brass makes the alloy workable at lower temperatures. Some custom manufacturers use special blends of brass for different saxophone parts. A small amount of arsenic or phosphorous also may be added to make the brass more useful in tubing applications.

Other materials are used to make the saxophone. Most of the screws are composed of stainless steel. Cork is used to line the joints and water keys. In some cases, a wax is applied to these joints. Mouthpieces can be made from various materials, however, the material has little effect on the sound. The most common material is black, hard rubber or ebonite. Metal or glass mouthpieces are also available. Plastic resonators are made and the instrument is often coated with a lacquer. Nickel plating on the keys helps strengthen them and keeps them attractive.

Design

The typical saxophone is a single reed instrument constructed from brass with a curved bottom. Originally available in 14 different sizes and keys, today that number has been reduced to six. This includes—in order of pitch from highest to lowest—the sopranino, soprano, alto, tenor, baritone, and bass saxophones. Generally, the smallest instrument is the sopranino and the largest is the bass.

The saxophone mouthpiece is the part of the instrument that the musician blows in to produce the sound. The construction of the mouthpiece has an important effect on the final sound of the instrument. It makes the difference between the sax player in a symphony orchestra and one in a rock band. There are two main parts of the mouthpiece that affect tone: the tone chamber and the lay (or the facing) which is the opening between the mouthpiece's reed and its tip. Mouthpieces are typically marked with a letter or number to denote the width of the lay.

The reed is attached to the saxophone and vibrates to create the sound. Saxophone reeds are made from bamboo (*Arundo donax*) which is grown in southern France. The reed can be made soft or hard depending on the desire of the musician. The ligature is the part that holds the reed on the mouthpiece. It attaches to the mouthpiece with screws. They can be made from innumerable materials such as leather, metal, or plastic.

The crook is the part that joins the mouthpiece and the main instrument body. At the top of it is a cork which is important for tuning the instrument. The tone changes depending on where the mouthpiece is positioned on the cork. The other end of the crook is a metal joint that fits into the main body of the saxophone. It connects with a screw to keep the crook in place.

Saxophone keys are of two types, closed standing and open standing. Closed standing keys are those that are held closed by a **spring** when the instrument is not being played. When the key is pressed, the hole it covers is opened. Open standing keys are held open by a spring and close when the key is pressed. Each key has a pad on its end which provides an airtight seal on the hole.

The saxophone tube is a long, metal tube which steadily gets wider at one end. It has holes drilled in the side at specific spots to create notes. When all the holes are closed, the instrument works much like a bugle amplifying the sound of the vibrating reed. When a hole is opened, the sound is modified producing a different note. The conical shape of the saxophone makes the overtones octaves. This makes fingering easier because the higher pitched notes are produced with the same fingering as lower pitched ones.

The Manufacturing Process

Since saxophone demand is relatively high, their manufacture is largely an automated process. The primary production steps include piece formation, assembly, and final polishing.

Parts production

Production of the various saxophone parts is a specialized operation and often done by contract manufacturers. They produce the pieces and send them to the saxophone producers for assembly.

1 The main body of the saxophone is produced from brass. This is made by first putting a brass tube on a long, tapered mandrel and then lubricating it. The brass is reshaped and made a consistent thickness by a doughnut-shaped die that is drawn down the mandrel. The tube is then heated to make it more malleable. Since heating creates an oxide residue on the tube's surface, it is soaked in a sulfuric acid bath.

2 Depending on the type of saxophone, the modified tube is taken to a shaping station where it is bent to give it a curl if needed. Two types of bending methods may be employed. In one case, the tube is put in a die which matches the desired curve. Highly pressurized water is then forced through the tube causing it to expand and conform to the walls of the die.

3 The tone holes can be produced by hand or machine. In the traditional method, workers would slip the brass tubes on a steel mandrel that was loaded with pulling balls. A drill press was then lowered and threaded into the pulling ball. The drill press was then raised, pulling the ball through and creating a hole with a rim or chimney. This same process was done with each tone hole in the shaft. In more modern production operations, several tubes are loaded into a machine which automatically creates the tone holes. Computer controls ensure that the instruments are perfect each time. After the tubes are formed, the body is coated with a clear lacquer finish.

Key construction

4 Saxophone keys were first forged by hand but today they are die-cast or stamped. In the die-casting method, a molten alloy is forced into a steel die. When it cools, the metal takes on the desired key shape. The stamping method involves a large stamping machine which cuts the keys from a sheet of metal. Depending on the use, the keys may be soldered together and polished. Polishing is done in a tumbling machine or by hand. The keys may also be metal plated, improving the appearance and durability.

John Coltrane

John William Coltrane Jr., was born on September 23, 1926, in rural North Carolina. Discovering jazz through the recordings of Count Basie and Lester Young, he persuaded his mother to buy him a saxophone, settling for an alto because it was supposedly easier. Coltrane showed a proficiency on the saxophone almost immediately. After studying at the Granoff Studios and at the Ornstein School of Music in Philadelphia, he joined a cocktail lounge band. He played for a year with a Navy band in Hawaii before landing a spot in the Eddie Vinson ensemble in 1947. For Vinson's band, Coltrane performed on the tenor sax. After a year with Vinson, Coltrane joined Dizzy Gillespie's group for four years. By then he was experimenting with composition and technical innovation.

In the 1950s Coltrane played horn for Miles Davis and Thelonious Monk; the latter showed him tricks of phrasing and harmony that deepened instrumental control. Coltrane devoted himself to rapid runs in which individual notes were virtually indistinguishable, a style quickly labeled "sheets of sound." This music was not easily understood, but it represented an evolution welcomed by various musicians and composers.

By 1965 Coltrane was legendary. He continued to experiment, even at the risk of alienating his growing audience. His work grew more complex, ametric, and improvisatorial. Coltrane continued to perform and record even as liver cancer left him racked with pain. He died at 40, only months after he cut his last album *Expression*.

5 Most keys have pads attached to them. These pads are made of layers of cardboard, felt, or leather. They are typically stamped or cut and then glued to each key by line workers. The keys are finished by being drilled and fitted with springs and

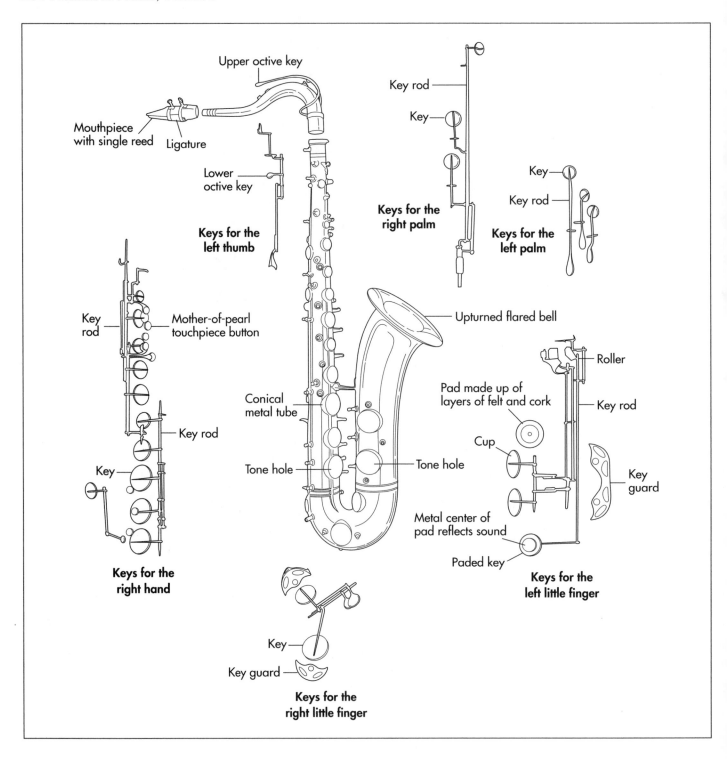

Upper octive key

Mouthpiece
with single reed

Ligature

Lower
octive key

**Keys for the
left thumb**

Key rod

Key

**Keys for the
right palm**

Key

Key rod

**Keys for the
left palm**

Key
rod

Mother-of-pearl
touchpiece button

Key rod

Key

**Keys for the
right hand**

Conical
metal tube

Tone hole

Upturned flared bell

Roller

Key rod

Pad made up of
layers of felt and cork

Cup

Tone hole

Metal center of
pad reflects sound

Paded key

Key
guard

**Keys for the
left little finger**

Key

Key guard

**Keys for the
right little finger**

A tenor saxophone and its parts.

screws that allow them to be attached to the instruments.

Final assembly

6 When all the pieces are formed, they are assembled into a complete instrument. The keys are mounted to the main tube on small posts. These posts are first screwed onto the main body. Holes are drilled into the posts to hold the key springs. The keys are then screwed onto the posts and seated on the tone holes. Since an airtight seal is needed for the instrument to perform correctly, the seal is tested and adjusted if necessary.

7 The crook is attached to the main tube, as is the mouthpiece. Typically, the mouthpiece is manufactured separately out of hard rubber. Other things such as the strap ring are attached at this point. The joints are usually lined with cork and waxed so they fit together smoothly. The main body is stamped with the manufacturers name and all other finishing steps are done.

8 When the instrument is assembled, it is played to ensure that it produces a quality sound. After this step, the saxophone is disassembled and put into a cushion lined case. The case is then shipped to the retailer.

Quality Control

Each saxophone piece is checked during the various phases of manufacture. This is done typically through visual inspection by trained workers. Inspectors check for things such as deformed parts, inadequate soldering, and other unacceptable variations. Additionally, more rigorous evaluations can be performed. Measuring devices like a vernier caliper or micrometer are used to check the physical dimensions each part.

Sound quality is also tested prior to shipment. Manufacturers employ professional musicians who can verify that tone quality, intonation, and playability are within acceptable limits set for the specific model of instrument. The instrument sound may also be checked under different acoustical settings. In general, if the saxophone is produced according to specifications no adjustments are needed. However, the tone holes can be drilled further to make the instrument sound less sharp or filled in with shellac to make it sound less flat.

The Future

Saxophone manufacturing and design is still changing. Since popularity has grown within the last few years, saxophones with differing bow radiuses and bell flares have been produced. More parts are now removable and enable easier cleanup. The Selmer Series III alto even has an additional tone hole to improve pitch. As developments continue with this instrument, quality and sound continue to improve.

Where to Learn More

Books

Harvey, Paul. *Saxophone*. London: Kahn & Averill, 1995.

Kernfeld, Berry. *Saxophone. The New Grove Dictionary of Jazz*. London: Macmillan, 1988.

Othmer-Kirk. *Encyclopedia of Chemical Technology*. Vol. 22. Wiley-Interscience, 1992.

Other

Classical Saxophone Online. http://www.classicsax.com (January 2001).

International Saxophone Home Page. http://www.saxophone.org (January 2001).

—Perry Romanowski

Seedless Fruits and Vegetables

Seedless fruits and vegetables are produced by meticulous cross-breeding, and it can take decades to bring a new strain to commercial viability.

Background

The fruits that are grown, sold, and eaten are essentially the ripened ovary of a plant. In the wild, fruit-bearing plants spread their seeds either by dropping their ripe fruits to the ground or by being eaten by animals, who then excrete the seeds. The tasty fruit is merely the mechanism by which the plant passes its seeds along. But from the human consumer's point of view, the seeds can be a nuisance. Spitting out hard, bitter seeds lessens the pleasure of eating grapes, for example. As a result, horticulturists have developed seedless varieties of popular fruits and vegetables. Seedless varieties make up over half the United States grape market, the seedless navel orange is a mainstay of the orange industry, and the seedless watermelon saw increasing popularity since its introduction in the 1990s. Seedless fruits and vegetables are produced by meticulous cross-breeding, and it can take decades to bring a new strain to commercial viability.

History

Careful breeding of plants to yield desired results, such as small seeds or bigger fruits, has been done since the dawn of agriculture. The scientific underpinning of plant breeding began to be understood in the mid-nineteenth century, with the work of Gregor Mendel. In 1856, Mendel, the father of genetics, was the first to publish his findings on the statistical laws governing the transmission of plant traits between generations. Mendel studied how specific traits in the pea plants in his garden were passed down to succeeding generations, and he formulated the idea of some sort of unit within plants that was responsible for heredity. His work

lay fallow for some time, then was rapidly extended in the early twentieth century. By the middle of the twentieth century, researchers had established that inheritance is transmitted by genes, which express chemical information resulting in characteristic traits. For seedless fruits, it is important to understand more of the details of genetic transmission. Genes in plants and animals are usually deployed in pairs, called an allele. One gene in the allele is usually dominant, and the other recessive. This means that typically only one trait is expressed in the biological makeup of the organism, though there is still a second gene for that trait. This is important because every cell in an organism carries a complete genetic map of itself, called the chromosomes, in its nucleus. When a cell divides, the chromosomes double, and then a copy goes into the new cell. The exception is the sex cells, the ovum and sperm. These cells only carry half the genetic material, which is one chromosome, or one half of each gene pair. When the ovum and sperm meet, the gene pairs recombine, and the new individual created through sexual reproduction has a new full set of genetic material, with half inherited from each parent. In traditional plant breeding, the horticulturist tries to optimize a trait by breeding together plants that both have, for example, small seeds. If the new generation of plants has inherited the small seed gene from both parents, it should also have small seeds, and be able to pass this trait to its offspring in turn. Many factors complicate the· picture, so that in real circumstances only a small percentage of the offspring may show the desired trait.

Seedless oranges and seedless grapes are the result of cultivation of naturally occurring seedless plants. The navel orange is descended from a seedless orange tree found on a plantation in Brazil in the nineteenth century. This tree was a mutation, that is, something in its genetic material had spontaneously changed, resulting in this unique plant. Orange growers propagated new trees from the original navel, so that all the navel oranges available in markets today are descended from that Brazilian tree. The common supermarket green seedless grapes are descended from a European seedless grape strain that probably originated between the Black and Caucasus Seas. Grape growers spread this variety all over the world, and the same species exists under many different names. It has been grown in the United States since at least 1872 under the name Thompson. Other seedless grape varieties, even red and black varieties, are also descended from the Thompson. The Thompson has a genetic abnormality that causes the seeds to arrest development. Though the flower is pollinated and the ovum fertilized, the seeds stop growing after a few weeks. So, the grape is not entirely seedless; rather, the seeds are aborted, and exist as tiny specks inside the fruit. Commercial growers treat the plants with a growth hormone called gibberillin, which is normally secreted by developing seeds. The flowers are dipped or sprayed with the hormone so that the grapes grow big and juicy despite the arrested seeds.

Seedless watermelons began to be a big seller in United States markets in the 1990s. Besides the convenience of having few or no hard black seeds when consuming the fruit, the new variety has a hard shell, making it easy to ship and giving it longer shelf life. Seedless watermelons are sterile, that is seedless, because they have three sets of chromosomes. This condition is called triploid. Standard watermelons, like Thompson grapes and most other organisms, have two sets of chromosomes, and are called diploid. To produce triploid watermelons, a diploid parent is pollinated by tetraploid watermelon, which has four chromosomes. During sexual reproduction, the new organism inherits half of each parent's genetic material. As a result, the new watermelon gets one chromosome from the diploid parent, and two from the tetraploid, making it triploid. The triploid hybrid is virtually seedless. It produces a very few seeds, and these can be planted to grow new water-

melons. But the new plants must be pollinated by standard diploid watermelons in order to produce fruit.

Research & development

The development of a new strain of seedless fruits or vegetables is a painstaking process. Research is typically carried out by horticulturists working at an agricultural development laboratory or government research station, where they can devote years to the work. A researcher studies thousands of seedlings to find ones with the desired characteristics. In searching for a seedless variety, other factors have to be taken into account, as well. The seedless fruit will not be commercially viable if it does not have good flavor, if it is prone to disease, if it is misshapen, etc. The fruit must be as good as seeded varieties, with seedless as an added advantage. So the researcher breeds likely plants, studies the offspring, and breeds these with other likely plants. The developer of the Flame Seedless, a red seedless grape, experimented with over 100,000 seedlings in the course of the quest. The plant that produced the Flame was a cross of five different varieties.

The traditional process for breeding seedless fruits was to cross a seeded female plant with a strain of seedless male. The offspring were seedless about 15% of the time. Then a successive generation could be produced from this 15%. Starting in the 1980s, horticulturists found ways to speed up the process by culturing the tissue of seedless plants. With grapes, the aborted seeds of the seedless strain are grown in a petri dish or test tube. Then these seedless strains can be crossed with other seedless strains, resulting in offspring that is 50-100% seedless. This technique has been used with great success with grapes, speeding up the time it takes to bring a new seedless variety to market. With watermelons, the sprouting tip of a seedless plant is placed in a petri dish filled with growth regulators and nutrients, and the one tip will sprout as many as 15 clone plants. This technique has also been used to produce seedless tomatoes.

Cultivation

Scaling up

1 After a new strain is developed, commercial seed growers license the stock and

Mendel's Second Law.

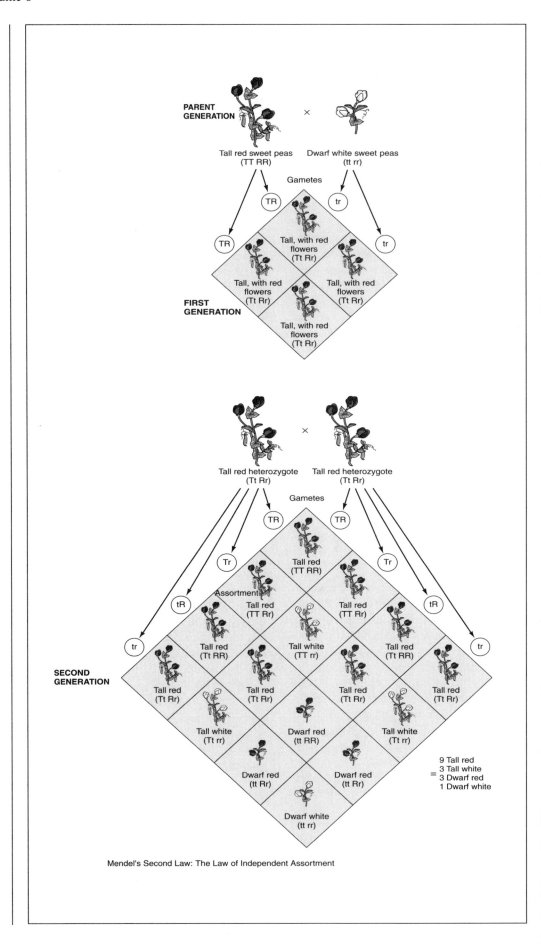

Mendel's Second Law: The Law of Independent Assortment

begin producing enough seed on a large enough scale to sell to farmers. The following process given is for seedless watermelons. The commercial seed grower obtains seed or seedlings from the developer, and plants them in an isolation field. These are the tetraploid (four chromosomal) plants. The isolation field must be three miles from any other watermelon field, so that bees do not pollinate the seedless plants with other varieties. Or, the isolation field may be surrounded by another plant, such as corn. The isolation field is typically one to five acres. The grower carefully maintains the plants, probably treating them with a growth hormone. Bees pollinate the plants, which produces possibly hundreds of pounds of stock seed. Next, this stock seed is planted in a seed production field along with a variety of regular diploid watermelons. When the plants flower, the cultivators pollinate them by hand. Pollen from the male diploid plant is transferred to the female tetraploid flower. The fruit that results has triploid seeds. These seeds are sold to farmers.

Germination

2 Because the triploid plants are generally fragile and easily killed by improper temperatures, they are usually germinated in a greenhouse and transferred to the field as seedlings. Triploid seeds also have an unusually thick coating, so the farmer may nick the rounded end of the seed before planting, to speed germination. The farmer buries the seeds in light potting soil and adds a little water. The greenhouse then needs to be maintained at around 85°F (29°C). After most of the seeds have sprouted, the temperature is lowered to between 70-80°F (21-27°C).

In the field

3 When the seedlings are three to four weeks old and have developed two or three leaves, the farmer transplants them into the field. The farmer also must plant a standard diploid watermelon in the field, one row for every two or three of the triploid. These may have been directly seeded into the field earlier, or transplanted at the same time. The diploid variety is chosen to mature at the same time or earlier than the triploid. To avoid confusion at harvest time, the diploid should also have a different rind

color than the triploid, so it is clear which one is seedless. Both varieties grow and mature, and bees pollinate the flowers. The triploid plants produce seedless watermelons. These are harvested and sold. To grow more triploid watermelons the next year, the farmer must start over with new seed.

The Future

Because of the success of sophisticated tissue culture methods, the time it takes to develop seedless fruits and vegetables is lessening. This means horticulturists can plan varieties to fill specific market gaps, such as a seedless black grape that matures in August, when few black grapes are available. Another technique that may quicken the production of seedless varieties is gene transfer. Biologists can fuse a new gene into a fruit plant that tells the plant to produce a growth hormone. The growth hormone stimulates the growth of the fruit even without pollination. The unpollinated plants produce no seeds. In the late 1990s, this method was successfully conducted on tomatoes and watermelons. This kind of biotechnology is one of the fastest growing areas of plant science. So the future may produce many more seedless fruit and vegetable varieties, without the long testing and development time needed in the past.

Where to Learn More

Periodicals

Mlot, Christine. "Seedless Wonders for Winter Markets." *Science News* (December 6, 1997): 359.

"No More Seeds in Watermelons?" *USA Today Magazine* (June 1993):7.

Tracy, Eleanore Johnson. "Flame Seedless: the Hottest Thing in Grapes." *Fortune* (July 23, 1984): 81.

Other

Access Excellence: About Biotech. http://www.accessexcellence.org (January 2001).

Nebraska Cooperative Extension. http://www.ianr.unl.edu/pubs/horticulture (January 2001).

—*Angela Woodward*

Semiconductor Laser

The concept behind lasers was first proposed by Albert Einstein, who showed that light consists of wave energies called photons.

Background

A laser, which is an acronym for Light Amplification by Stimulated Emission of Radiation, is a device that converts energy into light. Electrical or optical energy is used to excite atoms or molecules, which then emit light. A laser consists of a cavity, with plane or spherical mirrors at the ends, that is filled with lasable material. This material can be excited to a semistable state by light or an electric discharge. The material can be a crystal, glass, liquid, dye, or gas as long as it can be excited in this way.

The simplest cavity has two mirrors, one that totally reflects and one that reflects between 50 and 99%. As the light bounces between these mirrors, the intensity increases. Since the laser light travels as an intense beam, the laser produces very bright light. Laser beams can also be projected over great distances, and can be focused to a very small spot.

The type of mirror determines the type of beam. A very bright, highly monochromatic (one wavelength or one color) and coherent beam is produced when one mirror transmits only 1-2% of the light. If plane mirrors are used, the beam is highly collimated (made parallel). The beam comes out near one end of the cavity when concave mirrors are used. The type of beam in the first case makes lasers very useful in medicine since these properties allow the doctor to target the desired area more accurately, avoiding damage to surrounding tissue.

A semiconductor laser converts electrical energy into light. This is made possible by using a semiconductor material, whose ability to conduct electricity is between that of conductors and insulators. By doping a semiconductor with specific amounts of impurities, the number of negatively charged electrons or positively charged holes can be changed.

Compared to other laser types, semiconductor lasers are compact, reliable and last a long time. Such lasers consist of two basic components, an optical amplifier and a resonator. The amplifier is made from a direct-bandgap semiconductor material based on either gallium arsenide (GaAs) or InP substrates. These are compounds based on the Group III and Group V elements in the periodic table. Alloys of these materials are formed onto the substrates as layered structures containing precise amounts of other materials.

The resonator continuously recirculates light through the amplifier and helps to focus it. This component usually consists of a waveguide and two plane-parallel mirrors. These mirrors are coated with a material to increase or decrease reflectivity and to improve resistance to damage from the high power densities.

The performance and cost of a semiconductor depends on its output power, brightness, and operating lifetime. Power is important because it determines the maximum throughput or feed rate of a process. High brightness, or the ability to focus laser output to a small spot, determines power efficiency. Lifetime is important because the longer a laser lasts, the less it costs to operate, which is especially critical in industrial applications.

The simplest semiconductor lasers consist of a single emitter that produces over one watt of continuous wave power. To increase power, bars and multibar modules or stacks have been developed. A bar is an array of 10 to 50 side-by-side individual semiconductor lasers integrated into a single chip and a stack is a two-dimensional array of multiple bars. Bars can produce 50 watts of output power and last over 5,000 hours. Because such high powers produce a lot of heat, cooling systems must be incorporated into the design.

History

The concept behind lasers was first proposed by Albert Einstein, who showed that light consists of wave energies called photons. Each photon has an energy that corresponds to the frequency of the waves. The higher the frequency, the greater the energy carried by the waves. Einstein and another scientist named S. N. Bose then developed the theory behind the phenomenon of photons' tendency to travel together.

Laser action was first demonstrated in the microwave region in 1954 by Nobel Prize winner Charles Townes and his co-workers. They projected a beam of ammonia molecules through a system of focusing electrodes. When microwave power of appropriate frequency was passed through the cavity, amplification occurred and the term Microwave Amplification by Stimulated Emission of Radiation (M.A.S.E.R.) was born. The term laser was first coined in 1957 by physicist Gordon Gould.

Townes also worked with Arthur Schawlow and the two proposed the laser in 1958, receiving a patent in 1960. The first practical laser was invented that same year by a physicist named Theodore Maiman, while he was employed at Hughes Research Laboratories. This laser used a pink ruby crystal surrounded by a flash tube enclosed within a polished aluminum cylindrical cavity cooled by forced air. Two years later, a continuous lasing ruby was made by replacing the flash lamp with an arc lamp.

In 1962, laser action in a semiconductor material was demonstrated by Robert Hall and researchers at General Electric, with other United States researchers soon following. It took about another decade for the first semiconductor diode laser to be developed that could operate at room temperature, which

was first demonstrated by Russian researchers. Bell Labs followed the Russian researchers' success, while also improving laser lifetimes. In 1975, Diode Laser Labs of New Jersey introduced the first commercial room-temperature semiconductor laser.

Despite this progress, these lasers still were inadequate for telecommunications applications. Instead they found wide use (after other performance and lifetime improvements) in audio compact disks after Philips (The Netherlands) and Sony (Japan) developed a CD in 1980 using a diode laser. By the end of the decade, tens of millions of CD players were being sold every year. More recently, digital video disks have become available for optical storage, which are also based on diode lasers.

As power has increased, semiconductor lasers have expanded into other applications. Since 1995, performance of high-power diode lasers has jumped by a factor of 25. With this higher reliability, large groups of diode lasers can now be combined to create "stacks" of up to 25 individual diode lasers.

In 1999, laser-diode revenues represented 64% of all lasers sold, up from 57% in 1996, and were projected to reach 69% in 2000. In terms of units sold, semiconductor lasers have accounted for about 99% of the total (over 400 million units), which means most laser light is now produced directly or indirectly (via diode pumping) by semiconductor lasers. In addition to industrial applications, semiconductor lasers are being used as pump sources for solid state lasers and fiber lasers, in graphics applications such as color proofing and digital direct-to-plate printing, and for various medical and military applications (target illumination and ranging). In 2000, *Laser Focus World* estimated that about 34% of medical therapy lasers were of the semiconductor type.

Raw Materials

The conventional semiconductor laser consists of a compound semiconductor, gallium arsenide. This material comes in the form of ingots that are then further processed into substrates to which layers of other materials are added. The materials used to form these layers are precisely weighed according to a specific formula. Other materials that are

A double heterostructure laser.

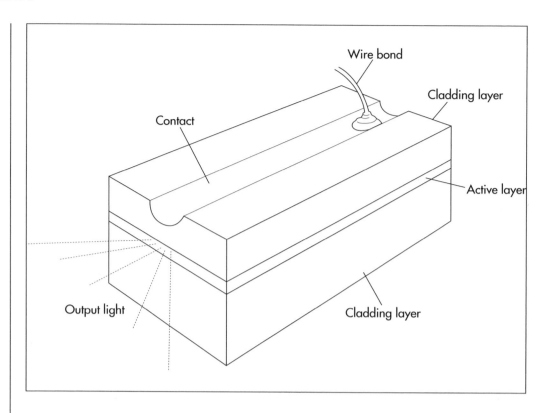

Contact

Wire bond

Cladding layer

Active layer

Output light

Cladding layer

used to make this type of laser include certain metals (zinc, gold, and copper) as additives (dopants) or electrodes, and silicon dioxide as an insulator.

Design

The basic design of a semiconductor laser consists of a "double heterostructure." This consists of several layers that have different functions. An active or light amplification layer is sandwiched between two cladding layers. These cladding layers provide injection of electrons into the active layer. Because the active layer has a refractive index larger than those of the cladding layers, light is confined in the active layer.

The performance of the laser can be improved by changing the junction design so that diffraction loss in the optical cavity is reduced. This is made possible by modifying the laser material to control the index of refraction of the cavity and the width of the junction. The index of refraction of the material depends upon the type and quantity of impurity. For instance, if part of the gallium in the positively-charged layer is replaced by aluminum, the index of refraction is reduced and the laser light is better confined to the optical cavity.

The width of the junction can also affect the performance. A narrow dimension confines the current to a single line along the length of the laser, increasing the current density. Peak power output must be limited to no more than 400 watts per cm (0.4 in) length of the junction and current density to less than 6,500 amperes per centimeter squared at the junction to extend the life of the laser.

The Manufacturing Process

Making the substrate

1 The substrates are made using a crystal pulling technique called the Czochralski method, where a crystal is grown from a melt. The elements are first mixed together and then heated to form a solution. The solution is then cooled, which solidifies the material. A seed crystal is attached to the bottom of a vertical arm so that the seed barely contacts the material at the surface of the melt. The arm is raised slowly, and a crystal grows underneath at the interface between the crystal and the melt. Usually the crystal is rotated slowly in order to avoid producing impurities in the crystal. By measuring the weight of the crystal during the pulling process, computer controls can vary the pulling rate to produce any desired diameter.

Growing the layers

2 The most common method for growing the layers onto the substrate is called liquid-phase epitaxy (LPE). Layers that have the same or fixed crystal-growth direction as that of the substrate can be grown on the substrate when the substrate comes into contact with a solution of the desired composition. As the temperature is decreased, the semiconductor compound (such as GaAs) comes out of the solution in crystalline form and is deposited onto the substrate.

An LPE system consists of a reactor (where the layers are grown), a substrate loading system, a pump and exhaust system (for removing air or impure gases after the materials are put in or taken out), a gas flow system (to move hydrogen gas through the reactor to remove impure gases) and a temperature control system. Pure materials are used for making the reactor so that the layers are not contaminated. The loading box is usually filled with nitrogen gas to purge the air while opening the reactor. The reactor typically consists of a quartz tube, in which a graphite boat and boat holder are placed. The graphite boat consists of an outer frame, a substrate holder, a spacer and a melt box.

3 The source materials for the layers are first rinsed and etched in order to clean the surface. After drying the etched materials, they are loaded into each melt box of the graphite boat. To grow each layer, the materials are first melted by heating to a specific temperature and then the substrate holder is pulled along with the substrate from the first melt to the next. The substrate is kept at each melt for a certain time under a fixed cooling rate, usually 33°F (0.5°C) per minute, according to a specific program designed for each composition. The temperature is automatically controlled using thermocouple sensors.

Fabricating the laser device

4 After the layered structure is grown, several other processes are completed to form the laser device. First, the substrate is mechanically polished until the thickness decreases to 70-100 microns in preparation for cleaving. Next, a very thin **silicon** dioxide film is formed on the substrate surface. Stripes are formed by photolithography and chemical etching. Contact electrodes are applied using an evaporation method. Next, a

laser resonator is formed by cleaving the wafer along parallel crystal planes. The completed laser devices are then attached to a copper heat sink on one side and a small electrical contact on the other.

Quality Control

The substrate onto which the semiconductor structure is grown must meet certain requirements regarding crystal direction, etch pit density (EPD), impurity concentration, substrate thickness and wafer size. The crystal direction must be within several degrees. Etch pits, which are rectangular hills or holes, are revealed by selectively etching the substrate with some type of acid solution. The etch pit density (number of etch pits per square centimeter) is used for estimating dislocation density, which affects the laser lifetime. An EPD of 10^3 per centimeter squared or less is required. Impurity concentrations are around 10^{18} per cubic centimeter. Substrates can range in size up to 3 in (7.6 cm) in diameter and typically are sliced into 350-micron thick pieces.

After the growth process, the surface of the semiconductor wafer is examined by an optical microscope. To examine the layered structure, a ground or cleaved cross section of wafer is stained and etched to increase the contrast of the layers using a scanning electron microscope. X-ray diffraction is used to determine the compositions of the layers and to measure the lattice patterns of the structure. The impurity concentration and refractive index of the layers is also measured using several analytical methods. After the laser device is fabricated, such operating parameters as voltage/current curves, threshold current density and spectral characteristics are measured.

The Future

Industry analysts from Frost & Sullivan predict that the diode laser systems market will reach nearly $4.6 billion by 2005. This growth is partially due to expanding applications in materials processing as high-power diode lasers become less expensive than solid state lasers. The compact size and electrical efficiency also make high-power semiconductor lasers attractive for industrial applications such as heat treating and welding. New material compositions and processing methods are also being developed to expand the applications.

Where to Learn More

Books

Iga, Kenichi, and Susumu Kinoshita. *Process Technology for Semiconductor Lasers: Crystal Growth and Microprocesses.* New York: Springer-Verlag, 1996.

The Photonics Dictionary, 4th International Edition. Pittsfield, MA: Laurin Publishing, 2000

Periodicals

Anderson, Stephen. "Diodes Dominate Laser Applications." *Laser Focus World* (April 2000). http://lfw.pennnet.com (January 2001).

Lang, Robert. "Semiconductor Lasers: What's Vital to Commercial Devices." *The Photonics Design and Applications Handboook 2000* Pittsfield, MA: Laurin Publishing, 2000.

"Making Photons: Lasers and Light Sources." *Photonics Spectra* (January 2000): 90-92.

Matthews, Steve. "Out of the Lab and into the Home." *Laser Focus World* (May 2000). http://lfw.pennnet.com (January 2001).

Matthews, Steve. "Semiconductor Lasers 2000: The Early Years: Promise and Problems." *Laser Focus World* (April 2000). http://lfw.pennnet.com (January 2001).

McComb, Stephen, and Michael Atchley. "High-Power Laser Diodes: Reliable, Multikilowatt Semiconductor Lasers Mature." *Laser Focus World* (December 1999). http://lfw.pennnet.com (January 2001).

—*Laurel M. Sheppard*

Shoelace

Background

It is understand how important shoes are for protecting feet from hazards and weather, and proper fit is necessary to maximize protection and comfort for the shoe's wearer. The shoelace is one way to ensure the proper fit, and a simple pair of laces, costing less than two dollars, can make all the difference in the world to the look and fit of a shoe. The lace is just a simple, woven band that pulls the shoe together to hold it to the foot. A shoelace consists of only two components: the woven tape that pulls the shoe tightly together, and the *aglet*, the hardened, taped end that fits through the eyelets on a shoe or boot.

Shoelaces have surprising importance in our lives. A real watershed in a child's life is when he has finally learned to tie his shoes. Athletes are vehement about checking and double-checking shoelaces before races or other fast-moving events lest they trip on them or take time on the field to retie them. American championship skater Tonya Harding nearly forfeited time on the ice during championship competition in 1994 because her skate lace had broke and her skate fit improperly.

Shoelaces are still manufactured in the United States although they are also made overseas where labor is cheaper. They are made in one of two ways in the United States The more common method includes the old-fashioned braiding of the shoelace using bobbins on machines that may be decades old. It is a simple process but still effective in producing significant numbers of laces a day. This method permits extensive variation within lace manufacture—one may vary fibers used, color, the number of ends or yarns, and design as desired to produce an array of laces. Recently, some new machinery has been developed for the completely computerized weaving of a shoelace on a narrow fabric loom. Much of this machinery is European in manufacture and it is not universally adopted in the United States (wholesale replacement of older machinery by an established company would be quite an investment).

History

The history of shoelaces is inextricably bound with the history of the shoe and how it was secured and designed in different eras and cultures. In 2000 B.C., ancient Mesopotamians wore simple pieces of leather that fit beneath the foot and were bound to the foot and ankle with laces that were likely of rawhide. Without these laces, the soles were useless. The footwear of ancient Greeks included sandals with rawhide lacing, and the ancient Etruscans donned high-laced shoes with turned-up toes. Roman soldiers spread the use of shoes to western Europe, particularly the utilitarian footcovering of the marching soldier. Ancient Britons adopted the Romans' simple sole with a thong between the large toe and second toe, with rawhide straps securing the sandal to the foot. There was a fair variety of laced shoes during this period, including shoes of more luxurious fabrics and furs that have not survived. Many of the more expensive shoes were secured with pins rather than laces. Both ancient Romans and Greeks deplored carelessness in appearance, and those with haphazardly tied laces were ridiculed.

Shoes of the Middle Ages are less frequently found in excavations than those of the Romans, perhaps because the sturdy leather

Eyelets for shoelaces, which guided shoelaces as they passed through the shoe, were hand-sewn of sturdy thread until about the 1840s, when metal eyelets were developed.

A. Shoelace tipping machine. B. Tipping mechanism. C. The frog that wraps the lace and secures the ends in preparation for tipping.

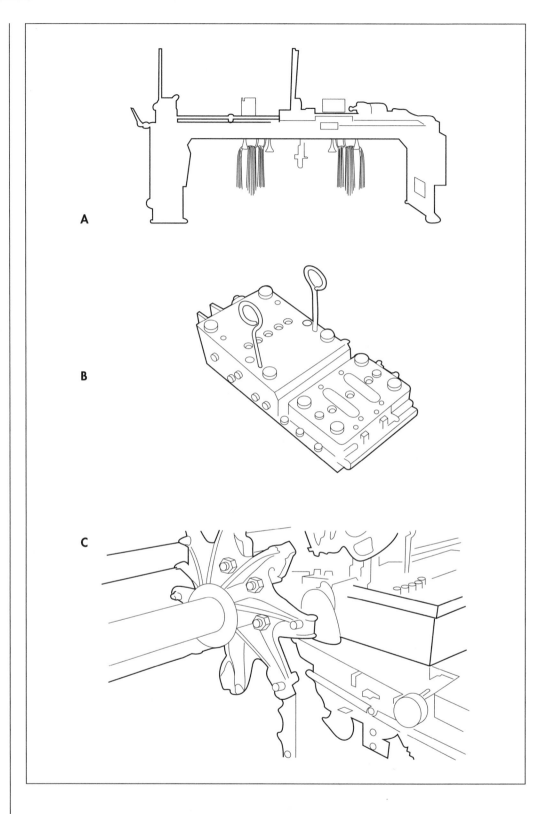

of the Roman sandal seems to defy deterioration. Nevertheless, there was a variety of footwear during this era as well. The poor went without shoes. Other shoes were actually breeches of linen that were laced to the leg and down over the foot, leaving the toes exposed. Other shoes were moving toward boots, covering the upper portion of the foot, and strapped together. Later in the Middle Ages, cordwainers (shoemakers) were able to fit boots and shoes of the wealthy more precisely and some of these shoes were secured with buttons or just a few short laces.

In the Elizabethan era, shoes were often secured with straps, laces, buckles and pins—both plain and jeweled. Prosperous American colonists generally secured their colored leather or fabric shoes with buckles or ribbons, while those with less money wore sturdy, simple leather shoes and boots tied with short rawhide laces. Eyelets for shoelaces, which guided shoelaces as they passed through the shoe, were hand-sewn of sturdy thread until about the 1840s, when metal eyelets were developed. Victorian boots for women were secured in a variety of ways. Popular gaiters actually used elastic gores that made it easy to pull the boots on and off, and that held the boots tight to the feet. Button boots were popular, as were ankle-high boots with metal eyelets for fabric laces. Shoes and laces were mass-produced in New England, particularly Massachusetts and New Hampshire, around the time of the Civil War. Textile braiding machinery, constructed by New England firms for the purpose of making shoelace braid, was readily available. Some of it is still in use in this country as the basic process of weaving the shoelace braid has changed little, even though fibers and colors have changed.

Raw Materials

The materials used for shoelaces vary according to the type of lace under construction. Typical fibers used for laces today include **cotton**, textured polyester, spun polyester, nylon, and polypropylene. The aglet, the hard plastic end of the shoelace that pushes through the eyelet in the shoe, is made of clear plastic. Acetone is used in the process of securing the aglet to the braided lace.

The Manufacturing Process

This essay will discuss the method of shoelace manufacture that uses a braiding machine to make shoelaces. It should be noted that the tipping and cutting of the woven braid described below is fairly standard across American factories.

1 First, there is a large room filled with dozens of shoelace braiding machines. Each machine resembles a horizontal circle and is equipped with 44 bobbins that all contribute to the manufacture of a single shoelace. Attached to the machine is a basket that will catch the shoelace as it is woven.

2 Next, the braiding begins. Electric motors start the braiding machinery. The bobbins start weaving the thin shoelace, with the braiding action resembling a maypole as the bobbin thread is deployed in a specific order. Side gears on the machine tell the machine how fast to take up the yarn to vary the tightness of the weave. The more quickly the yarn is taken up, the looser the braid. As the braid becomes longer and longer, it falls directly into a can that holds it there until the braid is finished. Each braiding machine can produce about a gross (144 shoelaces) every thirty minutes. Generally, the can holds about 13 lb (5.9 kg) of braid. The amount of braid this basket may hold varies according to thickness of the braid. Heavy sport laces, such as hockey laces, quickly add up to 13 lb (5.9 kg).

3 The cans of braids are moved to the tipping department. Each piece of braid is put into a machine that performs a variety of functions. First, the automatic tipping machine immerses the braid in acetone. (The acetone will allow the braid to hold the plastic tip tightly.) Then, the braid is automatically inserted into a die that holds acetate tape. The die is heated and presses the acetate tape at specific intervals (the length of the shoelace). Together, the acetone, the heat and the die pressure ensure that the shoelace will accept the acetate permanently. So, the braid is one long piece of shoelace that has a clear, 1-in (2-cm) wide band of acetate every 30 in (76 in) or so. Now the shoelaces are hung on the machine to dry (the acetone must evaporate) for about 20 minutes before the long braid is cut apart.

4 The laces move along and another die advances and cuts each band of acetate in half. Thus, each aglet is now about 0.5 in (1.3 cm) wide (the acetate band was about 1 in (2 cm) wide but was cut in half). The process of cutting the aglet in half cuts apart the shoelaces as well. The laces fall into a basket as they are cut. The basket counts the laces as they are dropped in. One basket can hold a half a gross of laces at a time.

5 Now the lace must be paired up. An operator takes two baskets, or one gross, of shoelaces, and places the baskets on his or

her lap. Then, the operator takes a lace in the right hand and one in the left, and feeds them into a pairing machine. The laces are sucked up into this elevator at right and left, are wound around a cardboard cylinder as a pair, and are pushed through a chute for packaging.

6 The pair of laces wound on a cardboard cylinder are sent to the blister packaging machine. Cardboard is put behind the laces, polystyrene in front, and the package is heat-sealed. The pairs are sent to a cardboard shipping box for movement out of the factory.

Quality Control

Control of product varies by factory and method of production. However, one company that utilizes braiding machines controls quality in three ways. First, the most important way that the employees control quality is in monitoring the bobbins as they weave the braid. When the bobbin is empty it drops down and the action stops until a new bobbin is put on. However, the new yarn must be knotted into the braid to continue the weaving. The operator ties a large knot into the shoelace so that the flaw is noticeable. The operator must remove that shoelace with a knot before it moves into the automatic tipping machine, or the big knot will explode the heated metal die (it is too bulky for the die).

Also, operators inspect each and every shoelace that is produced. Some laces are dirty and if so, they are removed and considered flawed. Inspection occurs when the laces are hanging to dry and as the operator is sending them into the pairing machine.

Maintaining machinery is essential to the successful operation of the business. The older braiding machines are decades old and are still efficient and precise because they are thoroughly cleaned and oiled each week.

Byproducts/Waste

Those companies that dye their own yarns must deal with the liquid effluvia ejected from their factory and reclaim the water. Bleaches and dyes are an environmental concern, and governmental authorities monitor their disposal. Acetone disposal, should there be any, is also a concern. Some shoelace companies even produce their own cardboard packaging (others purchase it). One such company has been cited for using solvent-based coatings for the cardboard and causing environmental problems as a result. The company has been urged to move to water-based cardboard coatings.

Where to Learn More

Books

Swann, June. *Shoes.* New York: Drama Books Inc., 1982.

Wilson, Eunice. *A History of Shoe Fashions.* London: Pitman Publishing, 1974.

Other

Artur Mueller Company. http:// www.artur-mueller.com (January 2001).

St. Louis Braid Company. http:// www.st-louisbraid.com (January 2001).

—*Nancy E.V. Bryk*

Shortbread

Background

Shortbread is a traditional Scottish baked good with a relatively simple recipe that consists of three basic ingredients (flour, butter, and sugar). Like most baked goods, it is produced in three steps consisting of ingredient mixing, product forming, and baking. Although shortbread is frequently eaten around Christmas and the New Year, it is consumed year round in many countries. In essence, shortbread—with its centuries' old history—is the granddaddy of all butter cookies and a mainstay in Europe. Today, there are several companies in the United States that manufacture shortbread exclusively.

History

Shortbread's namesake is a bit of a conversation piece. The majority classify shortbread as a cookie, but there are some who consider it a biscuit or even a cake. One would be hard pressed, though, to find someone willing to classify it by its given name—a bread. This confusion surrounding the classification of shortbread is further complicated by why shortbread is called short. A review of the literature on shortbread turns up at least two reasons. First, shortbread calls for a large percentage of shortening thus the name shortbread. Second, short refers to the desired crispness or "shortness" of the final product. Historically, the namesake shortbread was defended by early Scottish bakers who fought to prevent shortbread from being classified as a biscuit to avoid paying a government tax on biscuits.

However, one thing that is not typically contested about shortbread is its origination. Scotland is credited as the birthplace of shortbread. In Scotland one can find regional shortbread variations. For example, in Shetland and Orkney the people add caraway seeds and call it "Bride's Bonn." At holiday time in Edinburgh, shortbread is commonly adorned with pieces of citrus peel and almonds. Shortbread has a reputation as being a tea-time accompaniment, but it is also enjoyed with milk, coffee, wine, or champagne.

Raw Materials

The main ingredient in shortbread is white flour. Flour is made from wheat seeds, which in turn are made of three main parts: the outer coat or bran, the germ, and the endosperm. In white flour the bran and germ are removed leaving only the endosperm. The endosperm is made primarily of starch and protein, and it enables the dough to be stretched and rolled without breaking.

Most "authentic" shortbread recipes rely on real butter for their fat. Vegetable shortening tends to give the cookie an undesirable texture and flavor. In fact, in 1921 the British government proclaimed that in order to be called shortbread a product must get at least 51% of its fat from real butter. Cookies marketed as shortbread outside Britain, however, do not have such a requirement. Typically unsalted or sweet butter is recommended in shortbread recipes so as not to affect the taste of the cookie.

Shortbread recipes usually call for granulated or confectioners' sugar. All refined sugar is made from sugar cane or sugar beet. Hot water is used to draw out the sugar in a process called diffusion. The resulting juice is purified and concentrated by evaporation. It is then cystallized out of solution. Different types of sugar can be made from this point based on the size of the sugar crystal.

Once the ingredients are mixed, the dough is extruded into the desired shapes and baked in a tunnel oven.

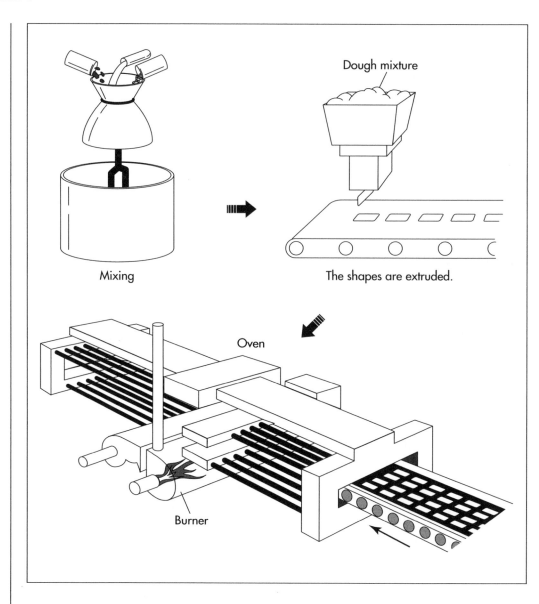

Mixing

Dough mixture

The shapes are extruded.

Oven

Burner

The more screening or refining that the crystals undergo, the smaller the particle. For example, granulated sugar has a much larger crystal size than confectioners' sugar.

Beside the three primary ingredients, many shortbread recipes call for salt, which helps to enhance the taste. One of the distinguishing features of traditional shortbread is its simple almost bland taste. In order to improve the subtle flavor, some manufacturers add eggs, cream, dried fruits, nuts, and even chocolate. Some larger commercial bakeries take additional liberties with their shortbread. For example, they may use vegetable shortening instead of butter t o keep costs down. Baking soda (sodium bicarbonate) is also used in some recipes to help the dough to rise.

Design

Today, you can find shortbread in a variety of shapes and sizes. One specialty shortbread manufacturer will even put customers' own photographs on their cookies. Traditionally, however, shortbread was one large round cookie with notched edges made by pinching the dough between the finger and the thumb.

The Manufacturing Process

Ingredient handling/warehousing

1 Most of the major ingredients are delivered to the bakeries in large quantities and stored in bulk tanks or silos made of stainless-clad steel. It is important that the

internal environment of the tanks is controlled so that the temperature and humidity are just right. Some tanks are refrigerated to keep the raw ingredients from spoiling.

Ingredient assembly

2 Keeping track of numerous bulk ingredients is a big task. In most large bakeries there is a department responsible for gathering the exact quantities of each ingredient and delivering it to the mixing area. Scales are used to ensure the correct amount of each ingredient. Additionally, some ingredients require processing such as milling or grinding before they are sent to be mixed and formed.

Mixing and forming

3 There are a variety of factors that are important when it comes to mixing the dough: the temperature of the ingredients, the mixing time, and the order in which ingredients are added. Shortbread dough requires that the shortening be at room temperature for proper mixing. Over mixing shortbread dough causes it to become tough and oily. Additionally, the sugar and shortening are typically mixed first with the flour being folded in later.

4 Once the dough is mixed, it must be formed into individual cookies. Shortbread comes in a variety of shapes, but it is typically molded by a rotary molding machine. The thickness of the dough is crucial. If it is too thick it will be too doughy. On the other hand, if it is too thin it may be too crispy or even burn. The molding machine ensures a uniform shape for baking.

Baking and cooling

5 Once the dough is properly mixed and formed, it must be baked. Ovens in commercial bakeries are 300-ft (91-m) long tunnels with adjustable speed conveyor belts and extremely sensitive temperature controls. Many chemical and physical changes take place during baking so it is important that the process is closely monitored. Each 18-30 ft (5.5-9.1 m) section of the oven has its own temperature controls and doors that allow employees to observe the cookies and vent the oven if necessary.

6 After the shortbread is baked, it must be cooled on a cooling conveyor. Controlled cooling helps the shortbread to retain appearance, taste, and texture. Controlled cooling also prevents condensation when the shortbread is packaged.

Packaging

7 The last step in the manufacturing process is packaging. Because one of shortbread's desired properties is crispness, it is important that the manufacturer package shortbread in a rigid and airtight container to prevent the cookies from breaking or getting soggy from moisture. Often manufacturers choose a tin box or canister to hold the shortbread. Other manufacturers house the shortbread in plastic trays surrounded by some type of outer paper packaging. The outer package, whether it is tin or paper, is decorated to make it appealing to the consumer. Oftentimes shortbread manufacturers choose a traditional Scottish plaid design for their packages. Individual shortbread packages are then put into case boxes that are stacked on pallets for shipment to stores. All individual boxes and cases are coded so that they can be traced back to the time and place of manufacture.

Quality Control

Quality control begins with the ingredient assembly stage of production during the measuring, weighing, and processing of the raw ingredients. Additionally, most large manufacturers have quality control (QC) labs responsible for making sure the materials meet determined specifications. Characteristics such as appearance, color, odor, and flavor are checked. QC technicians also test for particle size, viscosity of oils, and pH of raw materials.

In addition, there are a variety of non-technical measures that bakeries take to prevent product contamination. For instance, throughout the bakery hairnets are worn. Also, most bakery personnel wear special uniforms with no pockets and are forbidden from wearing jewelry. These precautions will keep personal items from accidentally falling into the raw ingredients or dough.

The finished product is also carefully monitored. Like the inspection of the raw material, the finished products must be examined for appearance, flavor, texture, and odor. The product is compared to a standard established during product development. Specially trained testers are responsible for detecting subtle differences that deviate from the norm.

The Future

As a product with a lot of history behind it, shortbread is shrouded in tradition. With its high fat content shortbread does not pretend to be a health food, although food industry trends are for low-fat and organic products. In an attempt to keep up with these trends, some shortbread manufacturers have started making low-fat and organic varieties. Like many products nowadays, shortbread is also beginning to be marketed and sold via the Internet. This online option makes it easy for new and repeat customers to enjoy authentic shortbread made by small to medium sized bakeries in Scotland and England. Shortbread has stood the test of time and will continue to be manufactured and con-

sumed in both its traditional format and "healthy" varieties.

Where to Learn More

Books

Karoff, Barbara. *The Best 50 Shortbreads.* San Leandro, CA: Bristol Publishing, 1995.

Other

ABC Official Girl Scout Cookie Bakers. ABC Bakery and Marketing Consultants. (1997). http://www.girlscoutcookiesabc.com/ pages (January 2000).

Irish Sugar. Sugar Educational Website. http://www.irish-sugar.ie/pages/product/ prodtext/prodi/mprodi.htm (January 2000).

The Scots Kitchen. Scotweb Ltd. http:// www.scotweb.co.uk/kitchen/BAK/short bread.html (January 2000).

—*Sandy Delisle & Perry Romanowski*

Silicon

Background

Second only to oxygen, silicon is the most abundant element in Earth's crust. It is found in rocks, sand, clays and soils, combined with either oxygen as silicon dioxide, or with oxygen and other elements as silicates. Silicon's compounds are also found in water, in the atmosphere, in many plants, and even in certain animals.

Silicon is the fourteenth element of the periodic table and is a Group IVA element, along with carbon germanium, tin and lead. Pure silicon is a dark gray solid with the same crystalline structure as diamond. Its chemical and physical properties are similar to this material. Silicon has a melting point of 2570° F (1410° C), a boiling point of 4271° F (2355° C), and a density of 2.33 g/cm^3.

When silicon is heated it reacts with the halogens (fluorine, chlorine, bromine, and iodine) to form halides. It reacts with certain metals to form silicides and when heated in an electric furnace with carbon, a wear resistant ceramic called silicon carbide is produced. Hydrofluoric acid is the only acid that affects silicon. At higher temperatures, silicon is attacked by water vapor or by oxygen to form a surface layer of silicon dioxide.

When silicon is purified and doped with such elements as boron, phosphorus and arsenic, it is used as a semiconductor in various applications. For maximum purity, a chemical process is used that reduces silicon tetrachloride or trichlorosilane to silicon. Single crystals are grown by slowly drawing seed crystals from molten silicon.

Silicon of lower purity is used in metallurgy as a reducing agent and as an alloying element in steel, brass, aluminum, and bronze. When small amounts of silicon are added to aluminum, aluminum becomes easier to cast and also has improved strength, hardness, and other properties. In its oxide or silicate form, silicon is used to make concrete, bricks, glass, ceramics, and soap. Silicon metal is also the base material for making silicones used in such products as synthetic oils, caulks and sealers, and anti-foaming agents.

In 1999, world production was around 640,000 metric tons (excluding China), with Brazil, France, Norway and the United States major producers. This is a continued decline compared to the last several years (653,000 tons in 1998 and 664,000 in 1997). Though data is not available, China is believed to be the largest producer, followed by the United States. One estimate puts China's production capacity as high as 400,000 metric tons per year, with over 400 producers. Exports from this country have increased in recent years.

Consumption of silicon metal in the United States was roughly 262,000 metric tons, at a cost of 57 cents per pound. The annual growth rate during 1980–1995 was about 3.5% for silicon demand by the aluminum industry and about 8% by the chemical industry. Demand by the chemical industry (mainly silicones) was affected by the Asian economic crisis of the late 1990s.

History

Silicon was first isolated and described as an element in 1824 by a Swedish chemist, Jons Jacob Berzelius. An impure form was obtained in 1811. Crystalline silicon was first produced in 1854 using electrolysis.

In 1999, world production of silicon was around 705,467 short tons (640,000 tonnes)—excluding China—with Brazil, France, Norway, and the United States major producers.

The reaction between silica and carbon within an electric arc furnace produces silicon.

$$2C + SiO_2 \rightarrow 2CO + Si$$

The type of furnace now used to make silicon, the electric arc furnace, was first invented in 1899 by French inventor Paul Louis Toussaint Heroult to make steel. The first electric arc furnace in the United States was installed in Syracuse, New York in 1905. In recent years, furnace technology, including the electrodes used for heating elements, has improved.

Raw Materials

Silicon metal is made from the reaction of silica (silicon dioxide, SiO_2) and carbon materials like coke, coal and wood chips. Silica is typically received in the form of metallurgical grade gravel. This gravel is 99.5% silica, and is 3 x 1 or 6 x 1 in (8 x 3 cm or 15 x 3 cm) in size. The coal is usually of low ash content (1-3% to minimize calcium, aluminum, and iron impurities), contains around 60% carbon, and is sized to match that of the gravel. Wood chips are usually hardwood of 1/2 x 1/8 inch size (1 x .3 cm size). All materials are received as specified by the manufacturer.

The Manufacturing Process

The basic process heats silica and coke in a submerged electric arc furnace to high temperatures. High temperatures are required to produce a reaction where the oxygen is removed, leaving behind silicon. This is known as a reduction process. In this process, metal carbides usually form first at the lower temperatures. As silicon is formed, it displaces the carbon. Refining processes are used to improve purity.

The Reduction Process

1 The raw materials are weighed and then placed into the furnace through the top using the fume hood, buckets, or cars. A typical batch contains 1000 lb (453 kg) each of gravel and chips, and 550 lb (250 kg) of coal. The lid of the furnace, which contains electrodes, is placed into position. Electric current is passed through the electrodes to form an arc. The heat generated by this arc (a temperature of 4000° F or 2350 ° C) melts the material and results in the reaction of sand with carbon to form silicon and carbon monoxide. This process takes about six to eight hours. The furnace is continuously charged with the batches of raw materials.

2 While the metal is in the molten state, it is treated with oxygen and air to reduce the amount of calcium and aluminum impurities. Depending on the grade, silicon metal contains 98.5-99.99% silicon with trace amounts of iron, calcium and aluminum.

Cooling/Crushing

3 Oxidized material, called slag, is poured off into pots and cooled. The silicon metal is cooled in large cast iron trays about 8 ft (2.4 m) across and 8 in (20 cm) deep. After cooling, the metal is dumped from the mold into a truck, weighed and then dumped in the storage pile. Dumping the metal from the mold to the truck breaks it up sufficiently for storage. Before shipping, the metal is sized according to customer specifications, which may require a crushing process using jaw or cone crushers.

Packaging

4 Silicon metal is usually packaged in large sacks or wooden boxes weighing up to 3,000 lb (1,361 kg). In powder form, silicon is packaged in 50-lb (23-kg) plastic pails or paper bags, 500-lb (227-kg) steel drums or 3,000-lb (1,361-kg) large sacks or boxes.

Quality Control

Statistical process control is used to ensure quality. Computer-controlled systems are used to manage the overall process and evaluate statistical data. The two major process parameters that must be controlled are amounts of raw materials used and furnace temperatures. Laboratory testing is used to monitor the chemical composition of the final product and to research methods to improve the composition by adjusting the manufacturing process. Quality audits and regular assessments of suppliers also ensure that quality is maintained from extraction of raw materials through shipping of the final product.

Byproducts/Waste

With statistical process control, waste is kept to a minimum. A byproduct of the process, silica fume, is sold to the refractory and cement industries to improve strength of their products. Silica fume also is used for heat insulation, filler for rubber, polymers, grouts and other applications. The cooled slag is broken down into smaller pieces and sold to other companies for further processing. Some companies crush it into sandblasting material. Because electric arc furnaces emit particulate emissions, manufacturers must also comply with the Environmental Protection Agency's (EPA) regulations.

The Future

Though industry analysts predicted demand for chemical-grade silicon by Western countries would increase at an annual average rate of about 7% until 2003, this growth may be slower due to recent economic declines in Asia and Japan. If supplies continue to outpace demand, prices may continue to drop. The outlook for the automotive market is positive, as more car makers switch to an aluminum-silicon alloy for various components.

Other methods for making silicon are being investigated, including supercooling liquid to form bulk amorphous silicon and a hydrothermal method for making porous silicon powder for optical applications.

Where to Learn More

Books

Kirk-Othmer. *Encyclopedia of Chemical Technology*. New York: John Wiley & Sons, Inc. 1985.

Periodicals

Bendix, Jeffrey. "The Heart of Globe is in Cleveland." *Cleveland Enterprise* (Fall 1991).

Ward, Patti. "Heroult Electric Arc Furnace Stands the Test of Time." *Iron and Steelmaker* 26, no. 11 (November 1999). http://www.issource.org/magazine/Web/9911/Ward-9911.htm.

Other

Annual Minerals Review: Silicon. U.S. Geological Survey, 1998.

Mineral Commodity Summaries: Silicon. U.S. Geological Survey, February 2000.

Mineral Industry Surveys: Silicon in February 2000. U.S. Geological Survey, May 2000.

—*Laurel M. Sheppard*

Skateboard

In 1997, there were 8.2 million skateboarders and around 48,186 reported injuries, 0.006% of which resulted in hospitalization.

Background

A skateboard is a small piece of wood in the shape of a surfboard with four wheels attached to it. A single person rides the skateboard, guiding the movement with his feet. While some use skateboards as transportation over short distances, most are used to perform stunts.

Skateboards consist of three parts: the deck (the actual board), the truck (a component usually made of metal that holds the wheels to the deck), and the wheels. The average skateboard deck is about 32 in (81.3 cm) long, 8 in (20.3 cm) wide, and is a little less than 0.5 in (1.3 cm) thick. The deck has a defined nose and tail with a concave in the middle. Skateboard wheels are usually made of **polyurethane** and range in width from about 1.3-1.5 in (3.3-3.8 cm). While nearly all skateboards have similar shapes and characteristics, their dimensions vary slightly based on use. There are skateboards built for speed, slalom, and freestyle.

Since skateboards first came into widespread use in the 1960s, their popularity has come in waves. Newfound interest is usually related to technical innovation, though a core constituency of skateboard enthusiasts has always remained.

History

Though there is unconfirmed evidence that a skateboard-like apparatus existed as early as 1904, the more commonly accepted predecessor to the skateboard was created in the 1930s. In Southern California, a skatescooter was made out of fruit crates with wheels attached to the bottom. This evolved into an early skateboard that was made out of 2x4 ft (61x121.9 cm) piece of wood and four metal wheels taken from a scooter or roller skates. This version of the skateboard featured rigid axles which cut down on the board's maneuverability.

Recognizable skateboards were first manufactured in the late 1950s. These were still made of wood and a few were decorated with decals and artwork. Skateboards became especially popular among surfing enthusiasts, primarily in California. Surfers practiced on skateboards when the ocean was to rough, and they soon became known as "sidewalk surfers." One of the first competitions was held for skateboarders in 1965. While skateboards were popular through most of the 1960s, riders were not respected and the activity was banned in some cities. The first wave of skateboard popularity was over by 1967.

Five years later, in 1973, there was a renewed interest in skateboards when wheels made of polyurethane were introduced. These early polyurethane wheels were composites of sand-like material that was formed into a wheel with an adhesive binder under extreme pressure. With the advent of polyurethane wheels, boards became easier to control and more stunts were possible.

Also in the 1970s, skateparks were introduced. Skateparks were specially designed places that catered to skateboarders. They had obstacle courses, pools (empty bowls, usually below ground level like an empty pool), and pipes (large, circular type) to challenge skateboard riders. With skateparks also came more competition, recognition, and sponsorship. Skateboarders sometimes decorated the bottom of their boards with logos of their sponsors. By the end of the 1970s, skateboarding again became controversial after it became identified antisocial behavior. Due to the amount and severity of the injuries, skateparks closed in fear of lawsuits and the sport returned underground.

When popular interest in skateboarding briefly re-emerged in the mid-1980s, it was not due to any particular technical innovation, though skateboard manufacturers were always experimenting with different materials in the production of decks. Instead, skateboarding videos featuring skateboarders performing extremely difficult and dangerous stunts using ramps, stairs, and even handrails generated new interest in the sport. At the same time skateboard art had also emerged. The bottom of skateboard decks were now elaborately decorated with logos and other designs. Continued resistance to skateboarders led to another downturn in popularity at the end of the 1980s, though not as severe as previous years.

By the middle of the 1990s, skateboarding again became popular mainly due to high-profile exposure like ESPN and MTV's X-Games competitions. These televised events of "extreme sports" showed the best of many kinds of skateboarding. Skateboarding was regarded as the first extreme sport. Though skateboarding was still banned or regulated in many communities, such exposure gave the sport an air of legitimacy. It is not as dangerous a sport as many think. In 1997 there were 8.2 million skateboards and around 48,186 reported injuries, 0.006% of which resulted in hospitalization. Compared to a more commonly accepted sport like basketball—which had 4.5 million participants in 1997 and 644,921 reported injuries (0.124% resulting in hospitalization)—the fear seems misplaced.

Skateboard art also continued to evolve. Art was based on street trends and whatever was hot at the moment: comics, bands, logos, and original art. In the mid-1990s, deck manufacturers would introduce an average of six board designs per month, making only 1,000 of each. While skateboard manufacturers experimented with different thicknesses of veneers that made up decks, little changed in the actual manufacture of skateboards at the beginning of the twenty-first century.

Raw Materials

Most skateboard decks are made of glue and wood (usually maple), but some are made of composites, aluminum, nylon, Plexiglas, fiberglass, foam, and other artificial materials. They are usually decorated by screenprinting. Skateboard trucks are usually made of aluminum or other metal (steel, **brass**, or another alloy), though a few are made of nylon. Skateboard wheels are made of polyurethane (a synthetic rubber polymer).

While some low-end skateboards are assembled by manufacturers, most components are sold separately to consumers who put them together on their own. To assemble a skateboard, the consumer also needs ball bearings (usually full precision and made of metal) and a piece of grip tape. Grip tape comes in a large piece bigger than the deck and looks like a piece of sandpaper. It is put on the top of the deck to provide traction.

Design

Skateboard decks, trucks, and wheels have different designs depending on how the skateboard will be used. Decks differ in their angle of concavity and the shape of the nose and tail. Manufacturers design their own boards with their own signature styling. They use templates to impose their design on the shape of the board. Companies that manufacture decks and wheels also make their products stand out by their individual art designs. While some of this artwork is created on computer, some is also done by hand.

The Manufacturing Process

Decks

1 A piece of maple wood undergoes a treatment that allows it to be peeled into veneers (thin sheets of wood) that are then delivered to the deck factory. They are stored in a climate-controlled environment to ensure the moisture content is optimized. Too much moisture is not good for the manufacturing process.

2 Each veneer is then put into a glue machine by hand. This machine evenly coats each veneer with a water-based glue specially designed for wood.

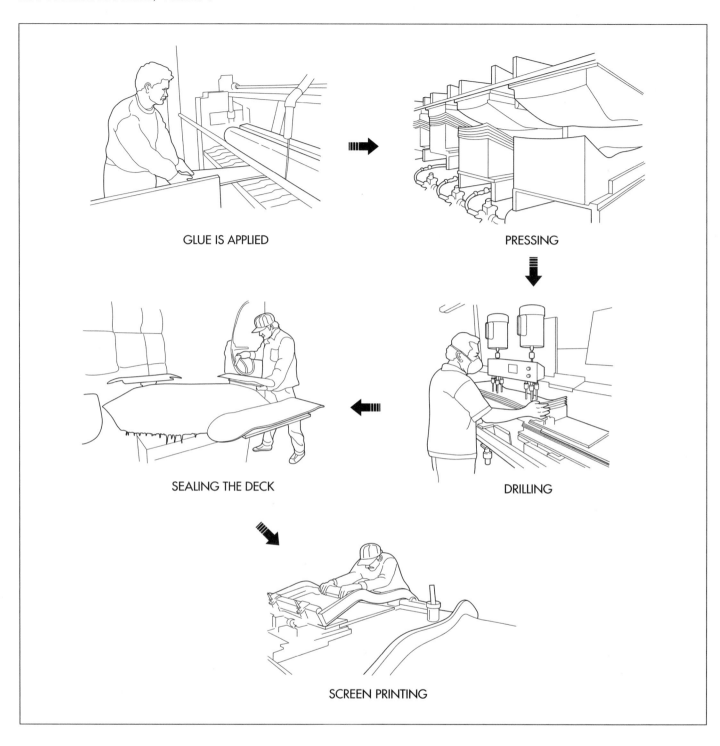

GLUE IS APPLIED

PRESSING

SEALING THE DECK

DRILLING

SCREEN PRINTING

The manufacturing steps to make the skateboard deck.

3 After being coated, the veneers are numbered and stacked according to grain and level of use. Each skateboard is made of seven layers of veneer. The first, second, fourth, sixth, and seventh layers have the grain running from the nose to the tail of the board. The third and fifth have the grain running from side to side. These stacks are put into a two-part mold inside a hydraulic press. The mold creates the nose, concave,

and tail of each skateboard. Each press makes five to 15 decks at one time. The resulting laminate sits in the press for anywhere from a few minutes to a few hours. The longer the time, the more naturally the wood and glue set.

4 After the laminates are removed from the press, eight holes for the truck mount are drilled by hand with a drilling rig.

5 A worker—called the shaper—takes the newly drilled board and, with a previously made template, hand-shapes each deck with a band saw. The deck is hand sanded and coated with a paint or sealant.

6 After the deck is dry, a decorative design is imposed by screenprinting. Each color is hand painted separately on a custom screenprinting machine. The decks are then dried and readied for shipment.

Trucks

7 With one of three materials (wood, plastic, or clay) a master truck pattern is hand tooled. This is used to make a match plate. With the plate, a sand mold is made for making the actual truck. A sand mold uses sand as its primary mold material, usually with clay and water. The material is packed around the plate then removed.

8 Aluminum ingots are heated to 1,300°F (706.7°C) in a furnace, reducing them to a liquid. This liquid aluminum is poured by hand into the sand mold's pouring basin sprue hole and through runners into the gate (the actual opening of the mold's cavity). The sand mold has the truck's axles in place before the aluminum is poured in. The mold is allowed to cool, then broken by hand and the parts removed. These pieces include the kingpin knob, pivot cup, baseplate, and riser pad. Using machines, a worker heat-treats each part. The parts are then grinded, polished, and drilled.

9 Finally, each truck is hand-assembled with kingpins, brushings, grommets, washers, and nuts and prepared for shipment.

Wheels

10 In metering machines, two polyurethane components are heated and mixed together in a certain ratio. High-quality polyurethane wheels are mixed together at elevated temperatures (lesser quality wheels are mixed at room temperature). This step creates a liquid. If the wheels are to be colored, the pigment is now added and the resulting mixture is poured into aluminum molds via a mix chamber (if the urethane is high quality it is heated again) and allowed to harden into a solid.

11 The wheel is removed by hand and cured on trays. Many wheel molds are running at the same time on a conveyor system and 300 wheels can be made per hour.

12 The resulting wheel slug is cut to shape by hand on a lathe. With a blade, the sidewalls (also known as the radius) and tread (riding surface) are cut into the wheel.

13 If the wheel is to be decorated, this semi-automated process is next. Digital artwork is converted to film to make a photo-etched print plate. The image on the plate is printed on the wheel with a pad printing machine. The silicone pad is on an inked printing plate and transfers the images to the wheels. Wheel printing that incorporates more than one color goes through one pad for each color. The wheels are then packaged for shipment.

Assembling the skateboard

14 After purchasing/manufacturing the three separate components, the consumer or manufacturer must put them together. Grip tape is needed to provide traction on the board. Grip tape comes in a large rectangular sheet, bigger than the actual deck. It is smoothed over by hand to get rid of any air bubbles. Using a file or other flat-edged object, the edge of the board under the grip tape is defined. With a safety knife or scissors, the extra parts of the grip tape are removed.

15 With an awl or an other sharp, pointed object, the eight truck holes are exposed through the grip tape, and the mounting bolts are placed. The truck is then installed over the bolts and tightened with the locknuts.

16 One set of bearings and a spacer are placed on each of the four truck axles. The wheel is put on next, flush with the bearings and spacer. The other set of bearings is put in the wheel. The wheels are secured with washers and a lugnut. The skateboard is now ready to be ridden.

Quality Control

When the components are purchased separately, the consumer must follow all instructions for his own safety. All screws must be

An example of a skateboard truck.

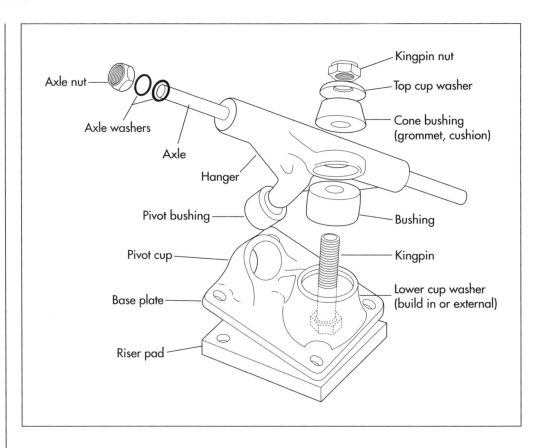

Axle nut

Axle washers

Axle

Hanger

Pivot bushing

Pivot cup

Base plate

Riser pad

Kingpin nut

Top cup washer

Cone bushing
(grommet, cushion)

Bushing

Kingpin

Lower cup washer
(build in or external)

tightly secured so that they will continue to hold the trucks in place while stunts are being preformed. Manufacturers continually check the finished boards to see that they are secure and meet safety requirements.

Byproducts/Waste

In the production of wheels, any polyurethane left over is sent to a landfill. At the present time, it is too costly to recycle.

The Future

Decks might be made of more artificial materials inside a wood exterior. One deck of the future has Nomex honeycomb at its core, with Kelver as one of the structural materials. Even with traditional wood decks the number of veneer layers may increase or decrease. The most noticeable difference might be the art on the bottom of the skateboard. Instead of being applied with a screenprinting process, decks might use a sublimation printing process.

Wheels may change in their shape, color or decoration, but not much will improve on

polyurethane itself. If a new material comes on the market, this may affect how wheels are manufactured.

Where to Learn More

Books

Cassorla, Albert. *The Skateboarder's Bible: Technique, Equipment, Stunts, Terms, Etc.* Philadelphia: Running Press, 1976.

Periodicals

Brower, Steven, and John Gall. "Skateboard Art." *Print* 50, no. 6 (November-December 1996): 52.

Stoughton, Stephanie. "A Wheel Challenge to Succeed: Manufacturer Finds Momentum is Critical." *Washington Post* 5 (May 28, 1998).

Other

Skateboard.com: Frontside. http://www.skateboard.com (June 10, 2000).

—*Annette Petruso*

Skyscraper

Background

There is no precise definition of how many stories or what height makes a building a skyscraper. "I don't think it is how many floors you have. I think it is attitude," architect T. J. Gottesdiener told the *Christian Science Monitor*. Gottesdiener, a partner in the firm of Skidmore, Owings & Merrill, designers of numerous tall buildings including the Sears Tower in Chicago, Illinois, continued, "What is a skyscraper? It is anything that makes you stop, stand, crane your neck back, and look up."

Some observers apply the word "skyscraper" to buildings of at least 20 stories. Others reserve the term for structures of at least 50 stories. But it is widely accepted that a skyscraper fits buildings with 100 or more stories. At 102 stories, the Empire State Building's in New York occupied height reaches 1,224 ft (373 m), and its spire, which is the tapered portion atop a building's roof, rises another 230 ft (70 m). Only 25 buildings around the world stand taller than 1,000 ft (300 m), counting their spires, but not antennas rising above them.

The tallest freestanding structure in the world is the CN Tower in Toronto, Canada, which rises to a height of 1,815 ft (553 m); constructed to support a television antenna, the tower is not designed for human occupation, except for a restaurant and observation deck perched at 1,100 ft (335 m). The world's tallest occupied structure is the Petronas Twin Towers in Kuala Lumpur, Malaysia, which reach a height of 1,483 ft (452 m), including spires. The Sears Tower in Chicago boasts the highest occupied level; the roof of its 110th story stands at 1,453 ft (443 m).

In some ways, super–tall buildings are not practical. It is cheaper to build two half-height buildings than one very tall one. Developers must find tenants for huge amounts of space at one location; for example, the Sears Tower encloses 4.5 million square feet (415,000 square meters). On the other hand, developers in crowded cities must make the fullest possible use of limited amounts of available land. Nonetheless, the decision to build a dramatically tall building is usually based not on economics, but on the desire to attract attention and gain prestige.

History

Several technological advances occurred in the late nineteenth century that combined to make skyscraper design and construction possible. Among them were the ability to mass produce steel, the invention of safe and efficient elevators, and the development of improved techniques for measuring and analyzing structural loads and stresses. During the 1920s and 1930s, skyscraper development was further spurred by invention of electric arc welding and fluorescent light bulbs (their bright light allowed people to work farther from windows and generated less heat than incandescent bulbs).

Traditionally, the walls of a building supported the structure; the taller the structure, the thicker the walls had to be. A 16–story building constructed in Chicago in 1891 had walls 6 ft (1.8 m) thick at the base. The need for very thick walls was eliminated with the invention of steel-frame construction, in which a rigid steel skeleton supports the building's weight, and the outer walls are merely hung from the frame almost like curtains. The first building to use this design

The world's tallest occupied structure is the Petronas Twin Towers in Kuala Lumpur, Malaysia, which reach a height of 1,483 ft (452 m), including spires.

was the 10-story Home Insurance Company Building, which was constructed in Chicago in 1885.

The 792-ft (242–m) tall Woolworth Building, erected in New York City in 1913, first combined all of the components of a true skyscraper. Its steel skeleton rose from a foundation supported on concrete pillars that extended down to bedrock (a layer of solid rock strong enough to support the building), its frame was braced to resist expected wind forces, and its high–speed elevators provided both local and express service to its 60 floors.

In 1931, the Empire State Building rose in New York City like a 1,250-ft (381-m) exclamation point. It would remain the world's tallest office building for 41 years. By 2000, only six other buildings in the world would surpass its height.

Raw Materials

Reinforced concrete is one important component of skyscrapers. It consists of concrete (a mixture of water, cement powder, and aggregate consisting of gravel or sand) poured around a gridwork of steel rods (called rebar) that will strengthen the dried concrete against bending motion caused by the wind. Concrete is inherently strong under compressive forces; however, the enormous projected weight of the Petronas Towers led designers to specify a new type of concrete that was more than twice as strong as usual. This high-strength material was achieved by adding very fine particles to the usual concrete ingredients; the increased surface area of these tiny particles produced a stronger bond.

The other primary raw material for skyscraper construction is steel, which is an alloy of iron and carbon. Nearby buildings often limit the amount of space available for construction activity and supply storage, so steel beams of specified sizes and shapes are delivered to the site just as they are needed for placement. Before delivery, the beams are coated with a mixture of plaster and **vermiculite** (mica that has been heat-expanded to form sponge-like particles) to protect them from corrosion and heat. After each beam is welded into place, the fresh joints are sprayed with the same coating material.

An additional layer of insulation, such as fiberglass batting covered with aluminum foil, may then be wrapped around the beams.

To maximize the best qualities of concrete and steel, they are often used together in skyscraper construction. For example, a support column may be formed by pouring concrete around a steel beam.

A variety of materials are used to cover the skyscraper's frame. Known as "cladding," the sheets that form the exterior walls may consist of glass, metals, such as aluminum or stainless steel, or masonry materials, such as granite, marble, or limestone.

Design

Design engineers translate the architect's vision of the building into a detailed plan that will be structurally sound and possible to construct.

Designing a low-rise building involves creating a structure that will support its own weight (called the dead load) and the weight of the people and furniture that it will contain (the live load). For a skyscraper, the sideways force of wind affects the structure more than the weight of the building and its contents. The designer must ensure that the building will not be toppled by a strong wind, and also that it will not sway enough to cause the occupants physical or emotional discomfort.

Each skyscraper design is unique. Major structural elements that may be used alone or in combination include a steel skeleton hidden behind non-load-bearing curtain walls, a reinforced concrete skeleton that is in-filled with cladding panels to form the exterior walls, a central concrete core (open column) large enough to contain elevator shafts and other mechanical components, and an array of support columns around the perimeter of the building that are connected by horizontal beams to one another and to the core.

Because each design is innovative, models of proposed super tall buildings are tested in wind tunnels to determine the effect of high wind on them, and also the effect on surrounding buildings of wind patterns caused by the new building. If tests show the building will sway excessively in strong winds,

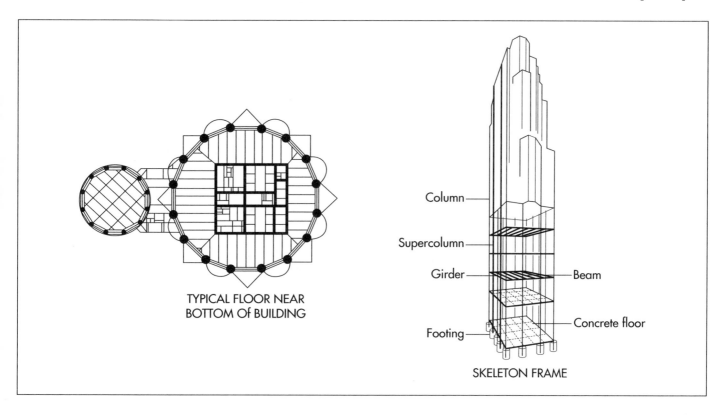

TYPICAL FLOOR NEAR
BOTTOM Of BUILDING

Column
Supercolumn
Girder — Beam
Footing
Concrete floor

SKELETON FRAME

An example of a skyscraper ground floor design and building frame.

designers may add mechanical devices that counteract or restrict motion.

In addition to the superstructure, designers must also plan appropriate mechanical systems such as elevators that move people quickly and comfortably, air circulation systems, and plumbing.

The Construction Process

Each skyscraper is a unique structure designed to conform to physical constraints imposed by factors like geology and climate, meet the needs of the tenants, and satisfy the aesthetic objectives of the owner and the architect. The construction process for each building is also unique. The following steps give a general idea of the most common construction techniques.

The substructure

1 Construction usually begins with digging a pit that will hold the foundation. The depth of the pit depends on how far down the bedrock lies and how many basement levels the building will have. To prevent movement of the surrounding soil and to seal out water from around the foundation site, a diaphragm wall may be constructed before the pit is dug. This is done by digging a deep, narrow trench around the perimeter of the planned pit; as the trench is dug, it is filled with slurry (watery clay) to keep its walls from collapsing. When a section of trench reaches the desired depth, a cage of reinforcing steel is lowered into it. Concrete is then pumped into the trench, displacing the lighter slurry. The slurry is recovered and used again in other sections of the trench.

2 In some cases, bedrock lies close to the surface. The soil on top of the bedrock is removed, and enough of the bedrock surface is removed to form a smooth, level platform on which to construct the building's foundation. Footings (holes into which the building's support columns can be anchored) are blasted or drilled in the bedrock. Steel or reinforced concrete columns are placed in the footings.

3 If the bedrock lies very deep, piles (vertical beams) are sunk through the soil until they are embedded in the bedrock. One technique involves driving steel piles into place by repeatedly dropping a heavy weight on their tops. Another technique involves drilling shafts through the soil and into the bedrock, inserting steel reinforcing rods, and then filling the shafts with concrete.

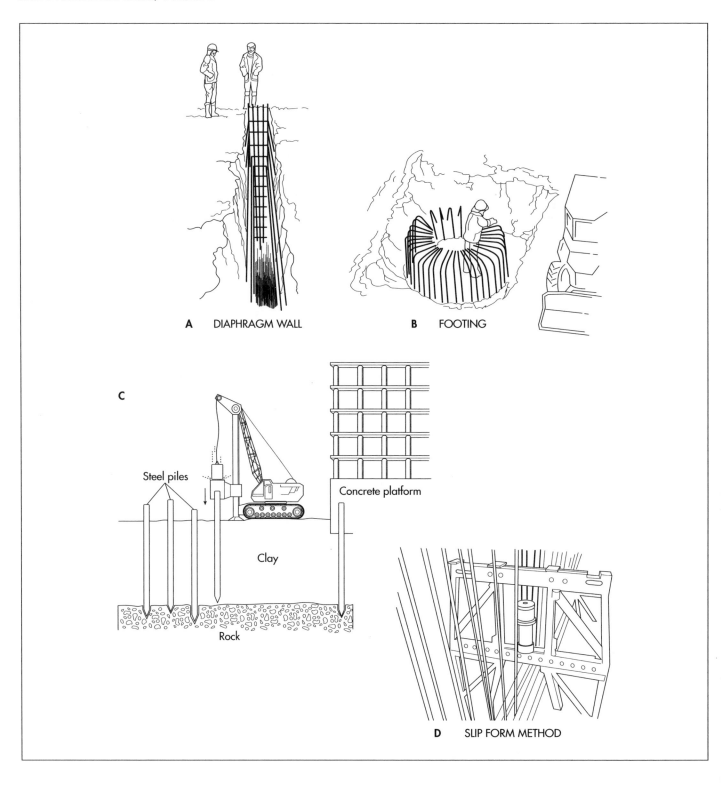

A DIAPHRAGM WALL

B FOOTING

C

Steel piles

Concrete platform

Clay

Rock

D SLIP FORM METHOD

A. Diaphragm wall. B. Footing. C. One type of foundation for a skyscraper uses steel piles to secure the foundation to the ground. D. The slip form method of pouring concrete.

4 A foundation platform of reinforced concrete is poured on top of the support columns.

The superstructure and core

Once construction of a skyscraper is underway, work on several phases of the structure proceeds simultaneously. For example, by the time the support columns are several stories high, workers begin building floors for the lower stories. As the columns reach higher, the flooring crews move to higher stories, as well, and finishing crews begin working on the lowest levels. Overlapping these phases not only makes the most efficient use of time, but it also ensures that the structure remains stable during construction.

5 If steel columns and cross-bracing are used in the building, each beam is lifted into place by a crane. Initially, the crane sits on the ground; later it may be positioned on the highest existing level of the steel skeleton itself. Skilled workers either bolt or weld the end of the beam into place (rivets have not been used since the 1950s). The beam is then wrapped with an insulating jacket to keep it from overheating and being weakened in the event of a fire. As an alternative heat-protection measure in some buildings, the steel beams consist of hollow tubes; when the superstructure is completed, the tubes are filled with water, which is circulated continuously throughout the lifetime of the building.

6 Concrete is often used for constructing a building's core, and it may also be used to construct support columns. A technique called "slip forming" is commonly used. Wooden forms of the desired shape are attached to a steel frame, which is connected to a climbing jack that grips a vertical rod. Workers prepare a section of reinforcing steel that is taller than the wooden forms. Then they begin pouring concrete into the forms. As the concrete is poured, the climbing jack slowly and continuously raises the formwork. The composition of the concrete mixture and the rate of climbing are coordinated so that the concrete at the lower range of the form has set before the form rises above it. As the process continues, workers extend the reinforcing steel grid that extends above the formwork and add extensions to the vertical rod that the climbing jack grips. In this way, the entire concrete column is built as a continuous vertical element without joints.

7 In a steel-skeleton building, floors are constructed on the layers of horizontal bracing. In other building designs, floors are supported by horizontal steel beams attached to the building's core and/or support columns. Steel decking (panels of thin, corrugated steel) is laid on the beams and welded in place. A layer of concrete, about 2-4 in (5-10 cm) thick, is poured on the decking to complete the floor.

The Empire State Building.

The Empire State Building was intended to end the competition for tallest building. It was to tower 102 stories, 1,250 ft (381 m) above Manhattan's streets. Its developers, John J. Raskob and Pierre Samuel Du Pont, along with former New York Governor Alfred E. Smith, announced in August 1929 their intention to build the world's tallest building. They chose the construction firm Starrett Brothers and Eken, and the architectural firm Shreve, Lamb, and Harmon for the project with William F. Lamb as the chief designer. It is set back from the street above the fifth floor and then soars uninterrupted for more than 1,000 ft (305 m) to the 86th floor. The exterior is limestone and granite and vertical chrome-nickel-steel alloy columns extend from the sixth floor to the top. The building contained 67 elevators and 6,500 glass windows, topped with a 200-ft (61-m) mooring mast for dirigibles.

The Empire State Building was completed on April 11, 1931, 12 days ahead of schedule and officially opened on May 1, 1931. The building took its place in history as the tallest building ever built, holding this title for more than 40 years. It was not until 1972, when the 1,348-ft- (411-m-) tall twin towers of the World Trade Center were completed that the Empire State Building was surpassed in height. The World Trade Center in turn was surpassed in 1974 by the Sears Tower in Chicago, which at 1,453 ft (443 m) became the tallest building in the world.

The exterior

8 In most tall buildings, the weight of the structure and its contents is borne by the support columns and the building's core. The exterior walls themselves merely en-

close the structure. They are constructed by attaching panels of such materials as glass, metal, and stone to the building's framework. A common technique is to bolt them to angle brackets secured to floor slabs or support columns.

Finishing

9 When a story of the building has been enclosed by exterior walls, it is ready for interior finishing. This includes installation of such elements as electrical wires, telephone wires, plumbing pipes, interior walls, ceiling panels, bathroom fixtures, lighting fixtures, and sprinkler systems for fire control. It also includes installation of mechanical components like elevators and systems for air circulation, cooling, and heating.

10 When the entire superstructure has been completed, the top of the building is finished by installing a roof. This may be built much like a floor, and then waterproofed with a layer of rubber or plastic before being covered with an attractive, weather–resistant layer of tiles or metal.

Quality Control

Various factors are taken into consideration when assuring quality control. Because of the huge scale of skyscrapers, a small positioning error at the base will be magnified when extended to the roof. In addition to normal surveying instruments, unusual devices like global positioning system (GPS) sensors and aircraft bombsights may be used to verify the placement and alignment of structural members.

Soil sensors around the building site are used to detect any unexpected earth movement caused by the construction activity.

Byproducts/Waste

Excavation of the foundation pit and basement levels require the removal of enormous amounts of dirt. When the 110-story World Trade Center towers were built in New York in the early 1970s, more than 1 million cubic yards (765,000 cubic meters) of soil and rock were removed and dumped in the Hudson River to create 23.5 acres (95,100 square meters) of new land, on which another skyscraper was later constructed.

The Future

Plans have been developed for several new skyscrapers that would break existing height records. For example, a 108-story building at 7 South Dearborn Street in Chicago, expected to be completed by 2004, will be 1,550 ft (473 m) tall. It will provide 43 acres (174,000 square meters) of enclosed space on a lot only 200 ft (61 m) square.

In 1956, American architect Frank Lloyd Wright announced plans for a mile-high (1.6-km tall) skyscraper in which 100,000 people could work. In 1991, another American architect, Dr. Eugene Tsui, designed a 2-mile (3,220-m) tall building that would provide space for living, working, and recreation for 1,000,000 people. Although such buildings may be theoretically constructable, they are currently impractical. For example, human comfort levels limit elevator speeds to no more than 3,000 ft/min (915 m/min). To accommodate the 100,000 people working in Wright's proposed structure, the number of elevator shafts would have taken up too large a portion of the building's area.

Improvements in elevator technology will be important for future skyscraper designs. Self-propelled, cableless elevator cars that move horizontally, as well as vertically, have been proposed, but are still under development. Computerized car dispatching systems using fuzzy logic could be refined to carry people more efficiently by grouping passengers whose destinations are near each other.

Where to Learn More

Books

Books Dunn, Andrew. *Structures: Skyscrapers.* New York: Thomson Learning, 1993.

Michael, Duncan. *How Skyscrapers Are Made.* New York: Facts on File Publications, 1987.

Periodicals

Hayashi, Alden M. "The Sky's the Limit." *Scientific American Presents: Extreme Engineering* (Winter 1999): 66 ff.

Richey, Warren. "New Rush of Buildings Reaching for the Clouds." *The Christian Science Monitor* (July 8, 1998): 1.

Other

Dankwa, E. T. *New York Skyscrapers.* http://mx3.xoom.com/iNetwork/NYC (March 2000).

"Ultima's Tower, Two-Mile High Sky City." *Tsui Design & Research.* http://www.tdrinc.com/ultima.html (March 2000).

—*Loretta Hall*

Slime

Background

Slime is a unique play material composed of a cross-linked polymer. It is classified as a liquid and is typically made by combining polyvinyl alcohol solutions with borate ions in a large mixing container. It often has an unpleasant odor, a green color, and is cold and slimy to the touch.

In scientific terms, slime is classified as a non-Newtonian fluid. These are thick liquids that have a variable viscosity. Viscosity is a measurement of the resistance to flow when a shearing force is applied. Newtonian fluids have a constant viscosity depending on their composition. For example, water is always a thin liquid with a low viscosity. Molasses is thick and has a high viscosity. Non-Newtonian fluids, like slime, have a different viscosity based on the amount of force put on them. If a small amount of force is applied, such as stirring them slowly with your fingers, they feel thin and water-like. If a high force is applied, like throwing it against a wall, the resistance is very strong. They are called non-Newtonian fluids because they do not behave as predicted by Newton's laws. Other materials that also behave like this include ketchup, gelatin, glue, and quicksand.

The molecular structure of slime is the factor responsible for its interesting behavior. Toy slime is typically composed of tangled, long-chain polymer molecules. These polymer molecules can be thought of as spaghetti strands. When put together on a plate, the strands are mixed together making a tangled mess. If the strands are rubbed together, they line up and become smoother. This motion gives the mass its slimy, slippery feel.

While the intermixing of the polymer strands will give some built-in viscosity, a cross-linking agent is also present in slime to give it the non-Newtonian fluid behavior. Cross-linking agents are ions that help temporarily connect polymer strands with relatively weak ionic bonds. These bonds are strong enough to hold the polymer strands together but not strong enough to make the mass a solid.

History

The story of toy slime's development dates back to the beginning of the twentieth century when the science of synthetic polymers was being determined. During the 1920s, Nobel laureate Hermann Staudinger laid the groundwork for our modern understanding of polymer science. He suggested a new molecular model for polymers; one of long, chain-like molecules and not aggregates or cyclic compounds as previously thought. In 1928, his models were confirmed by Meyer and Mark. These two scientists studied the dimensions of natural rubber using x-ray techniques. By the 1930s, Staudinger's models were widely accepted and extensive development of synthetic polymers began in earnest.

Manufacturers have sold polymeric play materials like slime for years. They are known to not only amuse children and adults, but also help in the development of dexterity and creativity. The earliest of these toys were moldable materials like modeling clay. The need for improved and varied play materials led to the development of silly putty in the 1950s. Glow-in-the-dark silly putty was subsequently introduced. During the 1980s, various slime-type toys were introduced. These products were made from such materials as polyvinyl alcohol, guar gums, or even fortified milk.

Design

Although a wide variety of slime variants are sold, they have many common characteristics. In general, slime is a gooey liquid available in small tubs. It can be sold by itself or as part of a toy set, such as an accessory for an **action figure**. Slime has a slightly unpleasant odor and is cold and slimy to the touch. Colors vary but the most common are green, blue, and red. Some manufacturers add fragrances to improve the odor.

One of the key aspects of slime formulation is that the materials must be safe for young children. In general, this means that the raw materials used to make slime must be non-irritating to the skin or eyes and non-toxic in case of ingestion. Additionally, consumers demand that slime (and toys like it) will not damage things like clothing, upholstery, fabric, or carpeting.

Raw Materials

Slime formulas are initially produced in laboratories by chemists. These scientists begin by determining what aesthetic features the slime will have. For example, they decide what the consistency will be, what color it will have and what it will smell like. Consumer testing is often used to help in making these decisions. After the features are determined, small test batches are made in the laboratory using the primary raw materials. The most common ingredients used in the production of toy slime are water, polymeric materials, gelling agents, colorants, fillers, and preservatives.

The most abundant material in slime is water, typically making up over 90% of the formula. Generally, specially treated deionized water is used. Water is a diluent that gives the slime its liquid consistency. The source of the water can be from underground wells, lakes, and rivers.

The polymeric materials are responsible for the important characteristics of slime. The most commonly used material is polyvinyl alcohol (PVA). PVA is a long chain polymer that has a backbone of carbon molecules with numerous hydroxyl (OH) groups

attached. In a typical slime formula, about 2% PVA is needed. Another similar polymer is polyvinylacetate (PVAC). This has a slightly different chemical makeup but it behaves in the same way as PVA when a gelling agent is added. Certain "natural" polymers can also be used to produce slime. Common examples include guar gum, which is derived from the bean of the guar plant, methylcellulose, which comes from plants, and cornstarch.

While the polymeric materials give slime its substance, a gelling agent is needed to give it the non-Newtonian liquid behavior. In classic slime formulations, sodium borate (Borax) or sodium tetraborate is used. When dissolved in water, sodium borate dissociates into sodium ions and borate ions. If a polymer is present like PVA, the borate ions interact with the polymer chains and form weak ionic bonds that make the solution thicker. These bonds also give the ability to stretch when a force is applied. Typically, sodium borate makes up about 2% of the final product. The ratio of the polymer to the gelling agent is one factor in determining the consistency of the slime.

A variety of additional ingredients are added to slime products to improve color and odor. In general, the polymeric solutions used in slime manufacturing are colorless. Therefore, a variety of dyes are added to give the product color. These are typically government-certified food colorings, such as FD&C Blue and FD&C Yellow. To improve the odor of slime, a fragrance is often added. Fragrances are made up of volatile oil materials. Since contamination from mold and bacteria is possible, preservatives like formaldehyde or methylparaben are also added to prevent microbial growth. Additionally, various acids or bases may be included to control the pH of the slime.

The Manufacturing Process

The production of slime occurs in two steps. First, a batch of slime is made. Then it is filled into its final packaging. This process can vary however, as filling may be easier if the cross-linking agent is not added until the polymeric solution is in the final package. The following description outlines a batch process.

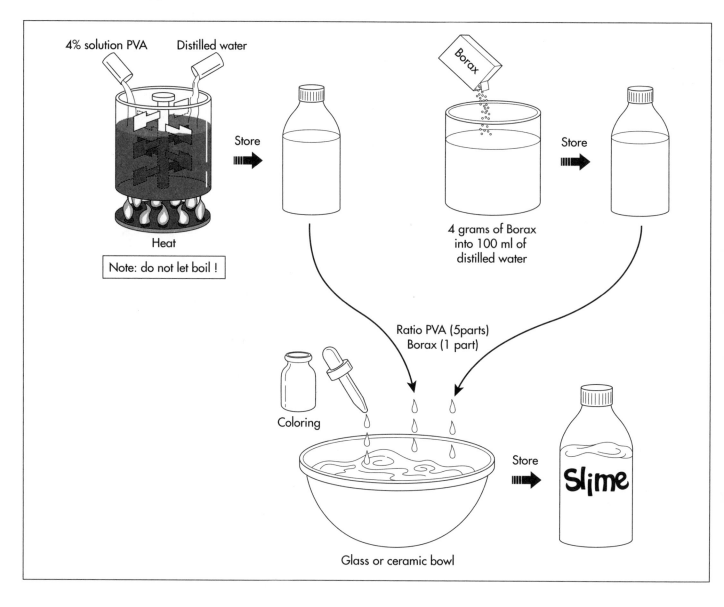

4% solution PVA   Distilled water

Store

Heat

Note: do not let boil !

Borax

Store

4 grams of Borax
into 100 ml of
distilled water

Ratio PVA (5parts)
Borax (1 part)

Coloring

Store

Slime

Glass or ceramic bowl

Slime is made by combining a 5:1 mixture of polyvinyl alcohol (PVA) solution and a Borax solution.

Compounding the batch

1 The batches of slime are made in large, stainless steel tanks fitted with mixers and a temperature control system. Starting with water, compounders add the rest of the raw materials according to the formula instructions at specific times and temperatures. Using computer controls, the mixing speed and temperature are regulated. Slime is typically heated to get the PVA to mix into the solution. Colorants, preservatives, and other additives are added last.

Quality control check

2 After the compounding phase, a batch sample is taken to the quality control lab for approval. In this area, physical and chem-ical evaluations are performed to ensure that the batch meets the minimum specifications outlined in the formula instructions. Tests, such as pH determination, viscosity checks, and appearance and odor evaluations, may be conducted. Adjustments can be made at this point if needed. After the batch is approved, it is pumped to a holding tank, where it is stored prior to filling.

Filling and packing

3 The filling stage depends on the type of packaging in which the slime will be sold. For the typical tub of slime, the process occurs on a filling line. At one end of the filling line is a hopper containing the empty tubs. These tubs are physically manipulated by a rotating machine until they

are standing upright. They are then moved along a conveyor belt to the filling heads, which contain the slime solution.

4 The filling heads are a series of piston tubes connected together on a circular carousel. As the tubs pass under the filling heads, product is injected into them.

5 Once filled, the tubs are moved along to a capping machine. This machine sorts the lids, places them on the tubs, and tightens them down.

6 If necessary, the tubs pass through a labeling machine. Sometimes this is not required because pre-decorated packaging is used. Further packaging steps might occur at this point, such as shrink wrapping or adding a cardboard backing. The packages are then put into boxes and stacked onto pallets. The pallets are hauled to storage and eventually shipped to local retail outlets.

Quality Control

Quality control is a continuous process occurring during the entire production process. Prior to batching, the raw materials are inspected and tested to ensure they meet minimum specifications. This inspection process typically involves work by quality control chemists, who check various physical and chemical characteristics. They measure characteristics of the incoming raw materials such as particle size, pH, viscosity, appearance, and odor. If a raw material does not meet the predetermined specifications, it is rejected. Batches are similarly tested.

If a batch does not meet all of the specifications, an adjustment may be made. For example, if the color of the batch is off more dye can be added. Preservation tests help ensure that the toys will be safe from microbial growth upon long-term usage. Stability tests also are performed. These studies monitor the performance of a finished batch under a variety of environmental conditions. On the filling lines, line inspectors inspect the bottles appearance, cap sealing, and filling weights.

The Future

Toys like slime have an intrinsic quality that makes them both fascinating and fun. While current materials enjoy commercial success, there remains a need for improved and varied play material formulations. Already some variations on the classic slime product are available. For example, slime kits are sold that allow the user to create their own slime products. In the future, toy manufacturers will focus on finding new ways to sell current slime and also create new, novel materials.

Where to Learn More

Books

Kirk-Othmer. *Encyclopedia of Chemical Technology.* New York: Wiley Interscience, 1997.

Shibayama, M. "Poly(Vinyl Alcohol)-Ion Complex Gels." In *Polymeric Materials Encyclopedia..* Edited by J. Salamone. New York: CRC Press, 1996.

Periodicals

Casassa, E. Z., A. M. Sarquis, and C. H. Dyke "The Gelation of Polyvinyl Alcohol with Borax." *Journal of Chemical Education* 63 (1986).

Sarquid, A.M. "Dramatization of Polymeric Bonding Using Slime." *Journal of Chemical Education* 63 (1986).

—*Perry Romanowski*

Snowshoe

Apparently when the English began settling the Americas, they did not take to snowshoes and were defeated by their more agile foes in the French and Indian War, culminating in a loss in the so-called Battle on Snowshoes near Lake George in New York in 1758.

Background

Snowshoes allow people to walk across the top of deep snow. They distribute weight so that the walker does not sink into soft drifts, and enable people to roam through landscapes that are usually impassible with only ordinary footwear. Snowshoes are of ancient origin, and until roughly 1950, they were made of wood, with rawhide bindings. After 1950, manufacturers came up with new materials and designs. Snowshoes of the twenty-first century are most often made of lightweight metal and other manmade materials. The design too has altered somewhat from the traditional snowshoe. Modern snowshoes, also called Western snowshoes, are often an asymmetrical shape. A large shoe might be 30 in (76 cm) long by 10 in (25 cm) wide, and a small shoe, such as the type used for racing, may be slightly narrower, and only 25 in (64 cm) long. With the development of lightweight snowshoes that require no maintenance and little specialized equipment, the sport grew. Snowshoeing became very popular in the United States in the 1990s, and in many areas, it rivals cross-country skiing.

History

The first snowshoes originated in Central Asia in about 4000 B.C. Probably people crossing the Bering Straits land bridge into North America came on snowshoes or brought the technology with them. Native North Americans used snowshoes extensively. They made hundreds of different shapes and varieties, suitable for many different terrains. Many wood-framed, or traditional, snowshoes are named for the Native American tribe that used that particular de-

sign, such as the paddle-shaped Huron snowshoe, and the front-pointed Ojibwa model. The tribes that relied most on snowshoes were the Athabascans on the West coast, who made snowshoes with upturned toes, and the Algonquins of the upper Midwest and Canada. Tribes living on the plains, such as the Sioux and Blackfoot, also used snowshoes in winter, though later they became more dependent on horses for winter travel. The snowshoe designs perfected by the Algonquins and other woodland tribes remained in use through most of the twentieth century. Native American snowshoes were made of a hard wood, such as ash. The wood was soaked or steamed to make it pliable, then bent into shape. The frame was laced with rawhide, and the lacing was often beautifully intricate.

French trappers and traders who worked and lived in the St. Lawrence River valley adopted snowshoes from the Native American people. Apparently when English people began settling in the same region, they did not take to snowshoes, and eventually this became a serious military disadvantage. English troops were defeated by their more agile foes in the French and Indian War, culminating in a loss in the so-called Battle on Snowshoes near Lake George in New York in 1758. Only then did snowshoes become standard winter gear for the British in North America. Later, as European settlers pushed west across North America, they brought snowshoes with them as essential tools of winter travel.

Snowshoeing purely for recreation also has a long history. Snowshoe clubs were formed in Quebec in the late eighteenth century. Groups met for hikes or for competitive races. The races were sometimes for long distances, sometimes for speed over short distances, and some even required the participants to jump hurdles. In French Canada this grew into a serious sport, and fostered an intense rivalry between Native American and European participants. Some native snowshoe makers developed lightweight racing shoes with a small, narrow design, until the Montreal snowshoe club set a limit in 1871, requiring that snowshoes had to weigh at least 1.5 lb (0.68 kg). More relaxed snowshoe hikes, including women and children, were also a fixed part of the social scene in French Canada and the Eastern United States up through the 1920s and 1930s.

Snowshoeing seemed to have fallen out of popularity across most of North America until advances in materials attracted a new generation to the sport. In the 1950s, snowshoe designers experimented with a variety of materials to make durable, lightweight shoes. A Canadian company, Magline, developed a magnesium snowshoe in the late 1950s that used webbing made from steel airplane cable coated with nylon. Aluminum snowshoes arrived around the same time, as well as snowshoes made of new materials, such as Lexan, which is the trade name for polycarbonate, the material used in astronauts' helmets. Even manufacturers of traditional wood-framed snowshoes experimented with manmade materials in the lacings, for example switching from rawhide to nylon coated neoprene. The new materials surpassed the old, because they did not require extensive care and maintenance. Wood and rawhide shoes had to be revarnished periodically and the webbing waxed, or else they would lose their water resistance. And like all wood products, they were subject to shrinking and swelling with changes in humidity. New materials, however, eliminated these problems.

Raw Materials

Traditional snowshoes are made of a hard wood, usually ash. The webbing material is rawhide, which is strips of denuded animal skin. The animal from which the webbing material was retrieved was traditionally moose, deer, or caribou. But in the twentieth century, most manufacturers switched to cow hide. At least one United States manufacturer imported water buffalo hide for an extra-tough webbing, but cow hide is generally the cheapest source for acceptable quality rawhide. Some manufacturers continued to make wood frame snowshoes even after new materials grew popular, but they switched to neoprene for the webbing. Most snowshoes today are framed in aluminum, usually in an alloy form that is both ultralight and very strong. Some manufacturers coat the aluminum with powdered plastic. The deck, or section on which the foot rests, is made of various materials, usually neoprene, polyurethane, or a composite material such as **polyurethane** coated with nylon. The material for the binding, which holds the snowshoe to the foot, is usually similar to the decking material. Other materials used may be plastic for some straps, rivets, or eyelets, steel for rivets, and aluminum or other metal for the cleat fitted on the bottom of the snowshoe. Some manufacturers use graphite for the snowshoe frame. This is the ultralight material popular in tennis racket manufacturing.

The Manufacturing Process

Traditional snowshoes

Traditional wooden snowshoes are still manufactured very much like they were thousands of years ago. The wood used is usually ash, which is strong and straight wood with an appropriate moisture content.

Forming the frame

1 For the very finest snowshoes, the wood is carefully split by hand to get the straightest grain. However, most manufacturers rely on wood sawn at a sawmill, for reasons of cost. The manufacturer checks the wood for poor grain and large knots, and discards any wood that is not of the right quality. Then the wood is steamed, to make it pliable. When the wood is soft enough, the manufacturer bends the wood into the shape of the snowshoe frame. The frames are then dried, usually in a kiln or hot room. This takes two to seven days.

Preparing the frame

2 After the frames have dried sufficiently, the manufacturer sands them and coats them with varnish. Holes are drilled for the lacing, and two wooden cross pieces are inserted.

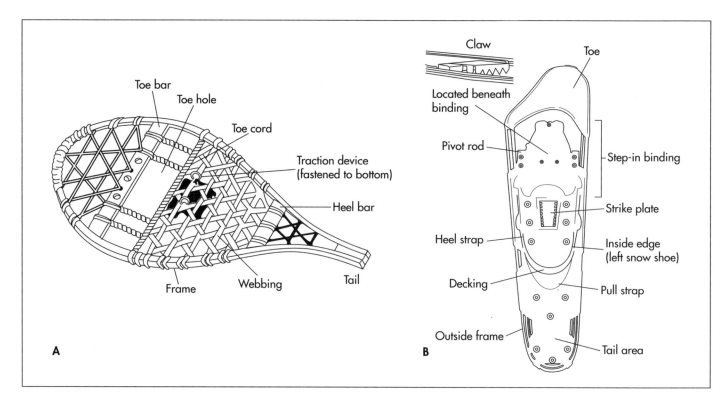

A. Traditional snowshoe. B. Aluminum snowshoe.

Lacing

4 Lacing is usually done by highly skilled, specialized workers. Lacing with genuine rawhide is usually done on site in the snowshoe factory, since the hide can spoil and has to be kept under optimum conditions. Neoprene does not spoil, and many manufacturers hire workers to do neoprene lacing in their homes. An expert lacer can finish up to ten pairs of snowshoes in a day. The laces are woven in an intricate pattern and drawn tightly through the holes in the rim of the frame. A heavier gauge lacing is usually used for the center portion of the shoe, directly under the foot.

Finishing

5 The worker who laces the webbing usually also affixes the binding, which is the arrangement of straps that holds the foot to the snowshoe. Then, the snowshoes are returned to the factory and inspected. They undergo a further period of drying. Next, they are boxed and shipped to retailers.

Aluminum snowshoes

1 For aluminum snowshoes, the metal arrives at the snowshoe factory in precut lengths of seamless tubing. After inspecting the tubing for any obvious flaws, workers insert each piece into a machine called a bender. This is a machine that has been designed for the specific finished shape and diameter of the snowshoe frame. It locks around the tubing tightly, and then a worker depresses a lever, and force is exerted to make the bend. Some machines work manually, with the worker providing the muscle power. Other machines are hydraulic. When the worker removes the tubing from the bender, it is shaped exactly as needed for the snowshoe frame. The frame may be ready to for the decking at this point, or it may next be powder coated.

Powder coating

2 Powder coating is usually done by an outside contractor. The aluminum frames are electrically charged, and then sprayed with a fine mist of dry, powdered plastic. The powder clings to the charged metal. Then the frames are heated, and the plastic melts onto them. Then the frames are returned to the snowshoe factory for the remaining steps.

Cutting the deck and binding

3 The deck is the main part of the snowshoe, taking the place of the webbing in

a traditional shoe. The decking material arrives at the factory in a wide roll. Workers unroll the material and feed it into a die stamper. This is a machine mounted with razors in the outline of the deck shape. The stamper lowers and presses through the material, cutting it in one motion. The manufacturer arranges the dies so that the decks can be cut very close together. As little as a quarter inch of material may be lost. This material is not recyclable, and represents the biggest waste in the manufacturing process. The binding straps are cut with different dies, but in the same process, and the material used is very similar.

Printing

4 Next, the cut deck is imprinted with the manufacturer's logo. The deck is passed under a heated die stamper. The stamper is carved with the logo, and affixed with a piece of plastic that bears heat-sensitive ink. When the heated stamper presses the plastic onto the deck, it leaves an inked impression of the logo. The ink dries almost immediately.

Clipping the deck to the frame

5 Now workers rivet the deck to the frame. This may be done by hand, or in a large facility, the process may be automated. Small plastic straps fit around the frame and hold the deck in a sandwich between the ends. The rivet is punched through. Rivets are placed at many points along the frame, to hold the deck securely.

Top and bottom

6 To finish the snowshoe, a cleat needs to be affixed to the bottom, and the binding and a plastic pivot strap need to be attached to the top of the deck. The plastic pivot strap is die-cut. The cleat is often supplied by an outside contractor. The binding and strap are fitted to the deck and the cleat fixed underneath. Then, these three parts are riveted together through the deck, so they are all held in place by the same part. At this point, the snowshoes are finished. They do not need any drying or curing, but are ready to be boxed and sent to retailers.

Quality Control

A conscientious manufacturer inspects all the raw materials for the snowshoes as they come into the factory. The workers check for problems with each step of the process. Because the parts fit very precisely, any fault in the process is usually immediately apparent. For example, if the decks were not cut correctly, they could not be clipped to the frames. The manufacturing process is also slow. Most makers are small, and produce a quality product. Snowshoes are not rushed through a highly automated assembly line, so visual inspection along the way is usually adequate quality control. The worker who tags the finished snowshoes and prepares them for shipping may act as a final inspector. For design problems, most manufacturers have relationships with avid snowshoers who can take a new pair through several hundred miles in a few weeks, so the manufacturer can get relatively rapid feedback from users about design flaws or successes.

The Future

Though nonwood snowshoes are described as modern or new-style, they have nevertheless been around for about 50 years. In other words, they are not terribly new. Increased popularity of the sport in the 1990s led to more marketing of the equipment and opening of more resorts and trails to snowshoers. But new technology does not seem imminent. Snowshoes are appearing in gaudier colors and in styles aimed at new segments of the market, such as women and children. This may ultimately confuse the consumer, since men, women and children can essentially wear the same snowshoe.

Where to Learn More

Books

Edwards, Sally, and Melissa McKenzie. *Snowshoeing*. Champaign, Illinois: Human Kinetics, Inc., 1995.

Prater, Gene. *Showshoeing*. Seattle: The Mountaineers, 1997.

Wolfram, Gerry. *Walk into Winter* New York: Charles Scribner's Sons, 1977.

—*Angela Woodward*

Sodium Chlorite

Today, sodium chlorite is an important specialty chemical with sales over $18 million annually.

Background

Sodium chlorite is a compound used for water disinfection and purification. It is produced in large quantities as flakes or a solution from chlorine dioxide and sodium hydroxide. Its use as a bleach for textiles was first discovered during the 1920s. Today, sodium chlorite is an important specialty chemical with sales over $18 million annually.

In its dried state, sodium chlorite ($NaClO_2$) is a white or light yellow-green solid. The greenish tint comes from trace amounts of ClO_2 or iron, which are production residuals. Sodium chlorite has a molecular weight of 90.44 and decomposes at about 392°F (200°C). It is generally soluble in water, but its solubility increases as the temperature of the water rises. Sodium chlorite is a powerful oxidizer that will not explode on percussion. The anhydrous salt does not absorb water and is stable for up to ten years.

Sodium chlorite is used for a variety of applications. It is used as a disinfectant and purification chemical for water. It is also employed as a textile-bleaching and water anti-fouling agent. Additionally, it is used in the paper and electronics manufacturing industries as a bleaching agent.

When put in an acid solution, sodium chlorite breaks down into chlorine dioxide. When added to a municipal water supply, chlorine dioxide helps control unwanted tastes and odors. It also aids in the removal of ions like iron and manganese. One added benefit is that it helps eliminate trihalomethanes in drinking water.

As a textile-bleaching agent, sodium chlorite is effective with various fibers. It can be used on **cotton**, bast fibers, and man-made fibers like nylon, Perlon, Dralon, and Rhovyl. It has an oxidizing effect on many of the natural waxes and pectins found in cellulose fibers. It helps solubilize them and makes the fiber more even and workable. It has the added benefit of destroying natural color matter without attacking the fibers themselves. This makes it useful for making permanent white fabrics without compromising tensile strength.

Sodium chlorite is also used for various industrial applications. It controls microbial contamination in industrial cooling systems and towers. It is used in place of chlorine in industrial ammonia plants because it does not react negatively with ammonia. Since it is an oxidizer, it is often a part of flue gas scrubber systems. Food-processing companies use it for washing fruits and vegetables because it is a fungicide. Meat and poultry are also washed with a solution, as is food processing equipment. Finally, it is an antimildew agent in detergent compositions and has been used in toothpaste and contact lens solutions.

History

The development of sodium chlorite as an industrial chemical began in 1921 when E. Schmidt found that cellulosic fibers could be purified with chlorine dioxide without being appreciably damaged. Unfortunately, chlorine dioxide gas is extremely explosive at high concentrations. These discoveries prompted researchers to look for safe and economical ways to deliver chlorine dioxide for bleaching purposes. The first company to introduce sodium chlorite for this purpose was the Mathieson Chemical Corporation.

In 1960, sodium chlorite became the standard material for continuous bleaching operations in the United States, replacing hydrogen. In subsequent years, other uses for sodium chlorite were discovered.

Raw Materials

The primary raw materials used in the production of sodium chlorite are chlorine dioxide, sodium hydroxide, and hydrogen peroxide. Chlorine dioxide is a gas at room temperature. Its color is intensely greenish-yellow. Chlorine dioxide provides the source of chlorine that is converted to sodium chlorite. In production, it is stored as a liquid solution in glass-lined steel containers.

Sodium hydroxide is a fused solid with a crystalline structure. Also known as caustic soda, it is corrosive to skin and vegetable tissue, causing severe burns. It is typically produced through the electrolysis of sodium chloride solutions. Hydrogen peroxide is a colorless liquid that is caustic and bitter to taste. Pure H_2O_2 is a thick, syrupy liquid that rapidly decomposes into oxygen and water. In nature, it occurs only in trace amounts in snow or rain. It is naturally generated during lightening storms. It is typically used in dilute solutions during the manufacture of sodium chlorite.

Other materials are typically added to sodium chlorite powders or solutions before they are sold. Commercial sodium chlorite bleaching solutions contain special ingredients including anticorrosive agents, buffering agents, chlorine dioxide fume controllers, and surfactants. Anticorrosive agents are used to prevent the corrosion of stainless steel bleaching equipment. Buffer salts help liberate the chlorine dioxide that is produced during the bleaching process. Surfactants help stabilize solutions and allow for cleaning and penetration effects. Stabilized sodium chlorite solution can be stored for long periods without loss of activity. When the sodium chlorite is sold as a solid, sodium chloride is often included to make it safer to handle and store.

The Manufacturing Process

While a variety of chlorites are available, sodium chlorite is the only one produced commercially. It is sold in solution or as a solid. The technical grade is made up of about 80% sodium chlorite and the rest is sodium chloride. Large scale production is based on a reaction of chlorine dioxide in a sodium hydroxide solution. Hydrogen peroxide is also present as the reducing agent. Sodium chlorite is manufactured in three phases, chlorine dioxide production, sodium chlorite generation, and recovery.

Chlorine dioxide production

1 While there are five principal methods for generating chlorine dioxide, the most common is the Hooker R-2 process, which generates chlorine dioxide from sodium chlorate. During production, solutions of both sodium chlorate and sodium chloride are pumped into a reaction vessel in approximately equal ratios. Concentrated sulfuric acid is also added to the reaction. Next, air is bubbled into the bottom of the container to create rapid agitation and dilution of the chlorine dioxide that is produced. During this process, both chlorine dioxide and chlorine gas are created.

2 These gases are separated out from the reaction vessel. The chlorine dioxide is separated by being absorbed in a conventional, water chilled tower. The chlorine gas is passed through separation towers and is picked up as sodium or calcium hypochlorite. This process produces about a 95% yield of chlorine dioxide.

Sodium chlorite generation

3 The chlorine dioxide gas is pumped into a vessel containing a cooled, circulating solution of sodium hydroxide. These compounds react to form sodium chlorite and sodium chlorate in approximately equal amounts. Water and oxygen are also generated. To minimize sodium chlorate production, a reducing agent is added. Typically, hydrogen peroxide is used, although sodium peroxide and sodium amalgam may also be employed. This step is closely monitored because sodium chlorate is highly undesirable in the final product.

Isolation and purification

4 Even though steps are taken to minimize its production, sodium chlorate must still be reduced before the sodium chlorite can

A chemical reaction that creates sodium chlorite.

$$2\ ClO_3 + 2\ NaOH \dashrightarrow NaClO_2 + NaClO_3 + H_2O$$

be isolated. This is accomplished by adding extra hydrogen peroxide.

5 The spent reactive solution is then pumped through a fractional crystallization tower to purify the sodium chlorite. This method takes advantage of the large solubility differences between the chlorite and other related salts that can be formed. After purification, the sodium chlorite solution is evaporated and tumble dried. If an anhydrous (devoid of water) product is desired, the evaporated powder is mixed with water at 100°F (38°C). The solution is saturated and cooled to 77°F (25°C). When this happens, the anhydrous salt spontaneously crystallizes out of the solution. A rotary drum, steam heated dryer is used to isolated the crystals, resulting in flakes or a fine powder. Occasionally, multiple drying steps are required.

6 The anhydrous salt can then be converted into powder, granules, or a solution. Granules are used more often because they are safer, with lower toxic risks and fire hazards, and a homogeneous composition can be created. Using typical methods, the particle size of the granules can be tightly controlled. Prior to packaging, solid sodium chlorite is mixed with sodium chloride to make it safer to handle.

7 Solutions are prepared by mixing powdered sodium chlorite with various anticorrosive agents, buffering agents, and surfactants in a mixing vessel. These solutions are used for commercial bleaching processes and can be formulated to be extremely stable.

8 Depending on the final use, sodium chlorite solution is packaged in plastic containers, drums, tote tanks, and tanker trucks. In the United States, powdered or flaked sodium chlorite is shipped in lined drums. Bulk transportation of the solid is not allowed because of safety concerns.

Quality Control

To ensure the quality of the sodium chlorite that is produced, the production process is monitored at each stage. The starting raw materials and the final product are all subjected to a variety of chemical and physical tests to determine that they meet the required specifications. Some of the commonly tested characteristics include appearance, odor, pH, density, specific gravity, and melting point. If the final product is a solution, its chemical activity is tested to make sure it has the correct concentration. For solid granules, particle size is determined and modified if necessary.

Byproducts/Waste

Manufacturing sodium chlorite produces some undesirable byproducts, such as chlorine dioxide, that cannot be released into the immediate environment. Concentrated fumes of chlorine dioxide are toxic, and cause sickness, appetite loss, and nausea in line operators. In the production plant, circulation of fresh air is essential. The chlorine dioxide gas is also highly corrosive. For this reason, sodium chlorite solutions must be stored in specially coated containers. Materials such as glass, porcelain, some plastics, or earthenware are typically used. Titanium is the most resistant metal used today. In the textile industry, molybdenum alloy stainless steels are used to store the sodium chlorite bleach solutions.

The Future

With increased applications for chlorine dioxide, improvements in sodium chlorite production are currently being studied. Sodium chlorite research is focused on reducing the environmental impact of bleaching systems and finding quicker, less expensive production methods. New bleaching formulations are constantly being developed by formulating chemists.

Where to Learn More

Books

Kirk-Othmer. "Carbon & Graphite Fibers to Chlorocarbons & Chlorohydrocarbons." In *Encyclopedia of Chemical Technology.*

Vol. 5, edited by Jacqueline I. Kroschwitz and Mary Howe-Grant. New York: John Wiley & Sons, Inc., 1993.

Periodicals

Busch, Gretchen. "Vulcan to Expand Treatment Business with Chlorite Buy." *Chemical Marketing Reporter* (June 15, 1992).

Rittmann, Douglas, and Joel Tenney. "Generating chlorine dioxide gas: Chlorate vs. chlorite." *Water Engineering & Management* (Sept. 1998).

—*Perry Romanowski*

Solid State Laser

The world laser system market is expected to increase from $4.7 billion in 2000 to $8 billion in 2005, with the solid state laser market reaching over $1.1 billion, compared to $4.6 billion for diode lasers.

Background

A laser, which is an acronym for Light Amplification by Stimulated Emission of Radiation, is a device that converts electrical or optical energy into light. Electrical or optical energy is used to excite atoms or molecules, which then emit monochromatic (single wavelength) light. A laser consists of a cavity, with plane or spherical mirrors at the ends, that is filled with lasable material. This material can be excited to a semi-stable state by light or an electric discharge. The material can be a crystal, glass, liquid, dye, or gas as long as it can be excited in this way. A solid state laser is one that uses a crystal, whose atoms are rigidly bonded, unlike a gas. The crystal produces laser light after light is pumped into it by either a lamp or another laser.

The simplest cavity has two mirrors, one that totally reflects and one that reflects between 50 and 99%. As the light bounces between these mirrors, the intensity increases. Since the laser light travels in the same direction as an intense beam, the laser produces very bright light. Laser beams can also be projected over great distances, and can be focused on a very small spot.

The type of mirror determines the type of beam. A very bright, highly monochromatic and coherent beam is produced when one mirror transmits only 1-2% of the light. If plane mirrors are used, the beam is highly collimated (made parallel). The beam comes out near one end of the cavity when concave mirrors are used. The type of beam in the first case makes lasers very useful in medicine since these properties allow the doctor to target the desired area more accurately, avoiding damage to surrounding tissue.

One way to excite the atoms to a higher energy level is to illuminate the laser material with light of a higher frequency than the laser light. Otherwise known as optical pumping, these solid state lasers use a rod of solid crystalline material with its ends polished flat and parallel and coated with mirrors to reflect the laser light. Ions are suspended in the crystalline matrix and emit electrons when excited.

The sides of the rod are left clear to admit the light from the pumping lamp, which may be a pulsed gas discharge producing flashing light. The first solid-state laser used a rod of pink ruby and an artificial crystal of sapphire. Two common solid state lasers used today are Nd:YAG (neodymium:yttrium aluminum garnet) and Nd:glass. Both use krypton or xenon flash lamps for optical pumping. Brilliant flashes of light up to thousands of watts can be obtained and operating lifetimes are near 10,000 hours.

Since laser light can be focused to a precise spot of great intensity, enough heat can be generated by a small pulsed laser to vaporize different materials. Thus, lasers are used in various material removal processes, including machining. For instance, ruby lasers are used to drill holes in diamonds for wire drawing dies and in sapphires for watch bearings.

History

The concept behind lasers was first proposed by Albert Einstein, who showed that light consists of mass-less particles called photons. Each photon has an energy that corresponds to the frequency of the waves. The higher the frequency, the greater the energy carried by the waves. Einstein and another scientist named S. N. Bose then developed the theory for the phenomenon where photons tend to travel together. This is the principle behind the laser.

Laser action was first demonstrated in the microwave region in 1954 by Nobel Prize winner Charles Townes and co-workers. They projected a beam of ammonia molecules through a system of focusing electrodes. When microwave power of appropriate frequency was passed through the cavity, amplification occurred and the term microwave amplification by stimulated emission of radiation (M.A.S.E.R.) was born. The term laser was first coined in 1957 by physicist Gordon Gould.

A year later, Townes worked with Arthur Schawlow and the two proposed the laser, receiving a patent in 1960. That same year, Theodore Maiman, a physicist at Hughes Research Laboratories, invented the first practical laser. This laser was a solid state type, using a pink ruby crystal surrounded by a flash tube enclosed within a polished aluminum cylindrical cavity cooled by forced air. The ruby cylinder was polished on both ends to be parallel to within a third of a wavelength of light. Each end was coated with evaporated silver. This laser operated in pulsed mode. Two years later, a continuous ruby laser was made by replacing the flash lamp with an arc lamp.

After Maiman's laser was successfully demonstrated, other researchers tried a variety of other substrates and rare earths, including erbium, neodymium, and even uranium. Yttrium aluminum garnet, glass, and calcium fluoride substrates were tested. The development of powerful laser diodes (a device that forms a coherent light output using electrodes or semiconductors) in the 1980s led to all-solid-state lasers in the continuous-wave regime that were more efficient, compact and reliable. Diode technology improved during the 1990s, eventually increasing output powers of solid state lasers to the multikilowattt level.

Nd:YAG and ruby lasers are now used in many industrial, scientific and medical applications, along with other solid state lasers that use different type of crystals. Nd:YAG lasers are also being used for monitoring pollution, welding and other

uses. This type of crystal is the most widely used—more than two-thirds of crystals grown are this type. Other crystals being grown include Nd:YVO4 (yttrium orthovanadate), Nd:glass, and Er:YAG.

Raw Materials

Optical, mechanical, and electronic components made of various materials (crystals, metals, semiconductors, etc.) are usually supplied by other manufacturers. Outsourcing varies from laser manufacturer to manufacturer. A solid state laser consists of two major components, or "boxes." One component contains the optics (lasing crystal and mirrors), and the other contains the electronics (power supply, internal controls). Sometimes these two components are integrated into one box.

Design

The design of the laser cavity is determined by the application. Typically, the research and development group develops the design. This design determines the operating characteristics, including power, wavelength, and other beam properties. The designers also incorporate safety features as required by the Food and Drug Administration(FDA).

The Manufacturing Process

1 Usually, all or most of the components are manufactured elsewhere. For instance, crystal growers provide the lasing material. To grow an Nd:YAG crystal, a high-purity oxide powder compound of the desired elements is placed in a crucible and melted in a radio frequency furnace at high temperatures. A seed crystal is then brought into contact with the liquid surface. When the seed crystal is slowly lifted, rotated, and cooled slightly, a single crystal of the desired composition emerges at the rate of about 0.02 in (0.5 mm) per hour.

Typical Nd:YAG crystals range from 2.4-3.1 in (60-80 mm) in diameter by 6.9-8.9 in (175-225 mm) in length. Rods, wafers and slabs in various geometries are extracted from the grown crystal, then fabricated, polished, and coated to customer specifications. Finished products range from rods as small as 0.02 in (0.5 mm) in diameter by 1 in (25 mm)

A solid state laser consists of a cavity with plane or spherical mirrors at each end that is filled with a crystal, whose atoms are rigidly bonded. After light is pumped into it by either a lamp or another laser, the crystal produces light that bounces between the mirrors, increasing intensity and producing a very bright light.

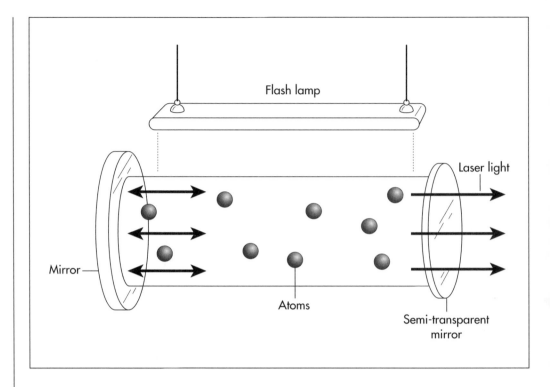

long to slab geometries as large as 0.3 x 1.5 in (8 x 37 mm) in cross section by 9.2 in (235 mm) long. The most common Nd:YAG rod geometry is a right circular cylinder.

Assembly

2 Once the laser is designed and the components received, the optics are integrated with the mechanical components. A technician follows a blueprint, placing the optical components in the desired positions, using metal holders or mounting devices. This procedure is performed in a clean room environment to avoid contamination of the optical components.

Alignment

3 Next, the lasing cavity is aligned so it operates at the desired specifications. This is performed on a test table by another technician, using another laser to help with the alignment.

Final testing

4 Before shipping the laser to the customer, it goes through a step called end testing, which basically checks the laser for proper operation, including output power, beam quality and other characteristics. The laser is operated for a number of hours to make sure it passes inspection.

Quality Control

Most laser manufacturers follow international quality standards that provide feedback loops throughout the manufacturing process. The laser also goes through several major testing procedures as previously described.

All laser devices distributed in the United States must be certified as complying with the federal laser product performance standard and reported to the Center for Devices and Radiological Health (CDRH) Office of Compliance prior to distribution to end users. This performance standard specifies the safety features and labeling that all lasers must have in order to provide adequate safety to users. Each laser must be certified that it complies with the standard before being introduced to the market. Certification means that each unit has passed a quality assurance test that complies with the performance standard. Those that certify lasers assume responsibility for reporting and notification of any problems with the laser.

Byproducts/Waste

Since suppliers of the various components usually follow total quality management procedures, the laser manufacturer does not test the components for defects and there is little waste. If defective components are found, they are sometimes sent back to the manufacturer.

The Future

Solid state lasers are being designed that have higher power, are faster, have shorter wavelengths, and better beam quality, which will expand their applications. For instance, lasing materials are being developed that will be able to squeeze many billions of pulses into one second, resulting in femtosecond lasers delivering dozens of pulses in each nanosecond. Solid state lasers that can provide power on the terawatt or petawatt level are also being tested for producing nuclear reactions, with the potential of being used in nuclear medicine applications such as CAT scanning. Nd:YAG lasers are expanding into the electronics industry for drilling, soldering and trimming applications. Lasing crystals continue to be made to last longer.

The world laser system market is expected to increase from $4.7 billion in 2000 to $8 billion in 2005, with the solid state laser market reaching over $1.1 billion, compared to $4.6 billion for diode lasers. Solid state lasers are replacing dye, ion and HeNe type lasers in certain markets. Other analysts predict flashlamp-pumped solid state lasers will grow to $660 million and diode- pumped solid state lasers to $312 million by 2003. The latter type of laser will become more popular for such industrial applications as general-purpose marking and materials processing, as costs come down and higher powers become available. These lasers are also being designed with minimal maintenance.

Where to Learn More

Books

Ambroseo, John. "Lasers: Understanding the Basics." In *The Photonics Design and Applications Handbook 2000*. Pittsfield, MA: Laurin Publishing, 2000.

Craig, Bruce, and Mark Keirstead. "Diode-Pumped Lasers: Big Choices, Small Package." In *The Photonics Design and Applications Handbook 2000*. Pittsfield, MA: Laurin Publishing, 2000.

The Photonics Dictionary, 4th International Edition. Pittsfield, MA: Laurin Publishing, 2000.

Teppo, Edward. "Nd:YAG Lasers: Standing the Test of Time." In *The Photonics Design and Applications Handbook 2000*. Pittsfield, MA: Laurin Publishing, 2000.

Periodicals

Hand, Aaron. "Lasers Squeeze into Tighter Board Assemblies." *Photonics Spectra* (July 1999): 96-101

"Lasers and Light Sources: Making Photons." *Photonics Spectra* (January 2000): 90-94.

Moody, Stephen. "From Earth to Space, Lasers Take on Pollution." *Photonics Spectra* (October 1999): 96-103.

Smith, James. "Lasers Continue Across Nuclear Fission Threshold." *Photonics Spectra* (April 2000): 42.

Steinmeyer, G., et al. "Ultrafast Lasers." *Photonics Spectra* (February 2000): 100-104.

—*Laurel M. Sheppard*

Spam

Background

Spam is a brand name for a canned meat product containing ham, pork, salt, flavorings, and preservatives that are mixed and cooked under vacuum pressure. There are other brands of similar canned pork meat products, but Spam—made by Hormel Foods Corporation—is the original and the best-selling of the brands.

The standard Spam can is brick-shaped and holds 7 oz (198 g) of meat. A 2-oz (57-g) serving contains 170 calories, provides 7 g of protein, 140 calories of fat, and has 0.75 g of sodium. It contains small amounts of cholesterol and iron. Americans eat approximately 3.8 cans per second. Two American plants produce 44,000 cans of Spam every hour. Hawaii consumes the most Spam in the world—about four million cans yearly (it is particularly popular in sushi).

Spam is an important protein source and economical as well. Unopened cans require no refrigeration and Spam has an indefinite shelf life because it is heat-sealed within the tin. It can, therefore, be shipped all over the world without spoiling. Thus, it is an important food source in many places where fresh meat is difficult to obtain or expensive (such as Hawaii and Guam). Spam has become a kitschy favorite with Spam t-shirts and cookbooks selling quite well. Spam has also made it onto the worldwide web with several websites dedicated to the product. The term spam has also come to mean unwanted junk e-mails received on personal computers.

History

Spam was first released onto the American market in 1937. Jay Hormel, the son of a successful Minnesota meat-packing house owner, was an energetic young man with big plans for his father's company. Hormel brought out canned ham in 1926. When his product was imitated, Hormel added spices to make it distinct. In the early 1930s, many companies were producing canned pork in large containers. Hormel's competition included lips, snouts, even ears in their meats but Hormel refused to use these refuse parts. Instead, he used the shoulder of the pig (a cut of meat rarely used because of its time-consuming removal from the bone). Hormel's meat was superior and more expensive than the competition's, but once opened it was indistinguishable. Hormel sought a way to seperate his product from the rest, and he decided to try two things: reduce the size of the can so it was family-sized and design a distinctive label.

Hormel's first experimental 12-oz (340-g) cans of this pork luncheon meat turned out to be 8 oz (227 g) of meat and 4 oz (113 g) of useless juice. As the heat cooked the meat in the sealed can, cells broke down and released an excessive amount of juice. Hormel tried many things to reduce the juice. Ultimately he discovering that it was not enough to put it in a can that was vacuum sealed, but the meat must also be mixed in a vacuum in order to minimize the juice released while cooking.

The new luncheon meat was not available for a while, awaiting a marketable name and an iconic label. After much dispute, the name Spam seemed perfect. Most believe it to be a combination of the words spiced and ham, but the original product contained no ham. (Hormel later added ham to the mixture because so many thought it was already in the product.) Upon release the meat was not an instant seller, but Spam was touted for its value and convenience.

By 1941, 40 million cans of Spam had been sold. During World War II, Spam was sent overseas to feed American G.I.s. Hormel supplied Allied troops with 15 million cans of Spam per week throughout the war. World leaders—including Eisenhower, Margaret Thatcher, and Nikita Khrushchev—credited Spam for its effectiveness. After the war, Hormel actively advertised the product, getting big names to sing its praises. Plants overseas also began producing Spam. By 1959, Hormel had manufactured its billionth can. By 1962, the 12-oz (340-g) can was joined by a 7-oz (198-g) can for single people and small families. Other innovations included Spam with cheese chunks and smoke-flavored product (1972) and Spam-Lite (1992). A major re-design of the label occurred in 1997, and both the old and new version entered the Smithsonian.

Raw Materials

The primary ingredient in Spam is chopped pork shoulder meat mixed with ham. About 90% of Spam is pork from a pig's shoulders. The remaining 10% (or so) comes from the pig's buttock and thigh, better known as ham. This ratio varies according to ham and pork prices. The U.S. Department of Agriculture does not permit any nonmeat fillers in lunchmeat, nor does it allow pig snouts, lips, or ears. The second ingredient is salt, added for flavor and for use as a preservative. Also, a small amount of water is used to bind all ingredients together. Sugar is also included for flavor. Finally, sodium nitrate is added to prevent botulism and acts as a preservative as well. It is the sodium nitrite that gives Spam its bright pink color—without it, Spam would discolor and become brown.

The Manufacturing Process

1 Pigs are no longer butchered by the Hormel Company, so meat is purchased from dealers and brought into the plant. Pork shoulders and ham are brought into the plant and cut apart. The pork shoulders are put into a powerful hydraulic press that literally squeezes the meat off the bone. The deboned meat is put into a large gondola or basket. Ham, however, must be cut away from the bone by hand. The meat-cutters remove and sort the meat from the shanks in the ham trim lines. The whitest, fattiest pieces are put into a large gondola marked "white," while meatier pieces are hand-sorted into one marked "red." The gondolas remain in a refrigerated area until they are needed.

2 Next, the gondolas are wheeled from the cold storage area and onto the main floor. The meat is transferred to a crane-like machine and then dumped into a large metal trough equipped with a drill bit. There, the drill bit thoroughly grinds the red and white pieces dumped in the trough. The batch is weighed (usually about 8,000 lb [3,628 kg] at this point) and passed under a metal detector (to catch a stray knife or mixing component). A small sample of Spam is analyzed to ensure it has the right combination of pork to ham and white to red pieces.

3 The ground meat is then distributed by the gondolas into several vacuum mixers. When these mixers are in the open position, they look like giant gas grills, but they are equipped with a refrigerated ammonia outer core that brings the meat temperature down to below freezing (32°F [0°C]). Then, the other ingredients in Spam—salt, sugar, water, and sodium nitrite—are added. The mixer lid is closed, creating an airtight seal, and the batch is mixed. The reason the vacuum is induced, the meat chilled, and the salt added is to reduce the amount of juice released by the meat when it is cooked. If too much liquid is released during cooking, the can would contain a large amount of gelatin.

4 While the Spam is being mixed, machines elsewhere are pushing empty, upside-down Spam cans off storage pallets one layer at a time. The plain silver cans are pushed onto a conveyor belt and sent toward the filler.

5 Nearly 1,000 lb (454 kg) of Spam is manually unloaded from the first mixer, dumped into receivers, and fed through pipes. The mixture moves through the pipes until it reaches the cone-shaped can fillers. As the cans travel underneath the fillers, a device picks each one up and deposits the raw, ground Spam into the can (from the

Spam is a mixture of ground pork meats, seasonings, and other ingredients that have been cooked under vacuum pressure.

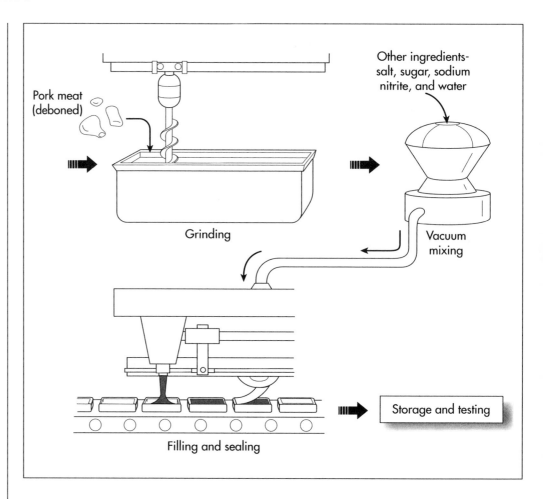

bottom) in one motion. The can is filled as the machine lifts it.

6 The can is sealed at a closing machine. They are then stamped with an identifying code so that the product can be traced back to the manufacturer.

7 Now, the closed cans head to the six-story-tall hydrostatic cooker. Spam is cooked in the can by very hot water within the cooker. The cans approach to cooker in a line, an arm swings out and pushes 24 cans onto a shelf. The shelf moves upward, and an arm swings out an pushes another group of cans onto a shelf. In two hours, 66,000 thousand cans will travel up and down 11 chambers in this huge cooker as they are heated, sterilized, washed, and cooled.

8 As the cans leave the hydrostatic cooker, they are now cool and ready for labeling. The labels sit at the end of the cooker in long rolls. An automatic labeler attaches a polypropelene film label on each can, and the labeler cuts the label to the correct length.

9 The cans are now ready for boxing. Twenty-four cans are fed onto flat pieces of cardboard, and a box is formed around the cans using the cardboard. The boxes are moved, and when a palette is filled with boxes, the entire pallet is shrink-wrapped. The cans are stamped with a date and other identifying numbers. A huge robot crane, driven by computer, transfers the pallet to a rack of shelving in the building. When the pallets get to the loading dock, then they are hoisted into the shelves by machines.

10 The Spam cans cannot be shipped out for 10 days. One of every 1,000 cans produced must undergo extensive testing to make certain the meat was properly cooked. If there are no problems, the cans may be sold.

Quality Control

Hormel would likely agree that Spam begins with quality pork and ham. Hormel no longer supplies its own meat for Spam, but the company chooses the meat carefully. Meat-cutters who cut the meat from the ham carefully perform their tasks and throw the pieces into the appropriate gondola. Also, the huge hydrostatic cooker has an alarm that trips if the computer detects there is any problem with the batch. The workers must fix that problem within three minutes. If they don't, the entire batch's viability is in question.

Portions of each batch are examined to make sure the batch has the right amount of pork shoulder to ham. The U.S. Department of Agriculture does not permit any Spam cans to leave the processing plant for 10 days. One out of every 1,000 cans be subjected to a 100°F (38°C) test to see if the can bulges or shows any other signs of improper cooking. The bacteria content is also tested. Finally, taste tests are routine at Hormel Foods Corporation. Every Friday all executives involved in Spam production meet to visually inspect (and sometimes taste) several different batches of Spam produced during the week.

The Future

Since Spam was first released it has undergone many transformations. From plain Spam to Turkey Spam to Spam-Lite. People are coming up with endless recipes that call for Spam, and Hormel is trying to incorporate every consumer's need into their product development. Spam with less sodium is now available. The launch of Hormel's website dedicated to Spam now provides consumers with a catalog devoted to Spam and Spam labeled products.

Where to Learn More

Books

Wyman, Carol. *Spam: A Biography.* San Diego: Harcourt, Brace & Co., 1999.

Other

"Spam: An Authorized Biography." A Manual for Public Relations. Hormel Foods, 2000.

—*Nancy E.V. Bryk*

Springs

During the third century B.C., Greek engineer Ctesibius of Alexandria developed a process for making "springy bronze" by increasing the proportion of tin in the copper alloy, casting the part, and hardening it with hammer blows. He attempted to use a combination of leaf springs to operate a military catapult, but they were not powerful enough.

A spring is a device that changes its shape in response to an external force, returning to its original shape when the force is removed. The energy expended in deforming the spring is stored in it and can be recovered when the spring returns to its original shape. Generally, the amount of the shape change is directly related to the amount of force exerted. If too large a force is applied, however, the spring will permanently deform and never return to its original shape.

Background

There are several types of springs. One of the most common consists of wire wound into a cylindrical or conical shape. An extension spring is a coiled spring whose coils normally touch each other; as a force is applied to stretch the spring, the coils separate. In contrast, a compression spring is a coiled spring with space between successive coils; when a force is applied to shorten the spring, the coils are pushed closer together. A third type of coiled spring, called a torsion spring, is designed so the applied force twists the coil into a tighter spiral. Common examples of torsion springs are found in clipboards and butterfly hair clips.

Still another variation of coiled springs is the watch spring, which is coiled into a flat spiral rather than a cylinder or cone. One end of the spring is at the center of the spiral, and the other is at its outer edge.

Some springs are fashioned without coils. The most common example is the leaf spring, which is shaped like a shallow arch; it is commonly used for automobile suspension systems. Another type is a disc spring, a washer-like device that is shaped like a truncated cone. Open-core cylinders of solid, elastic material can also act as springs. Non-coil springs generally function as compression springs.

History

Very simple, non-coil springs have been used throughout history. Even a resilient tree branch can be used as a spring. More sophisticated spring devices date to the Bronze Age, when eyebrow tweezers were common in several cultures. During the third century B.C., Greek engineer Ctesibius of Alexandria developed a process for making "springy bronze" by increasing the proportion of tin in the copper alloy, casting the part, and hardening it with hammer blows. He attempted to use a combination of leaf springs to operate a military catapult, but they were not powerful enough. During the second century B.C., Philo of Byzantium, another catapult engineer, built a similar device, apparently with some success. Padlocks were widely used in the ancient Roman empire, and at least one type used bowed metal leaves to keep the devices closed until the leaves were compressed with keys.

The next significant development in the history of springs came in the Middle Ages. A power saw devised by Villard de Honnecourt about 1250 used a water wheel to push the saw blade in one direction, simultaneously bending a pole; as the pole returned to its unbent state, it pulled the saw blade in the opposite direction.

Coiled springs were developed in the early fifteenth century. By replacing the system of weights that commonly powered clocks with a wound spring mechanism, clockmak-

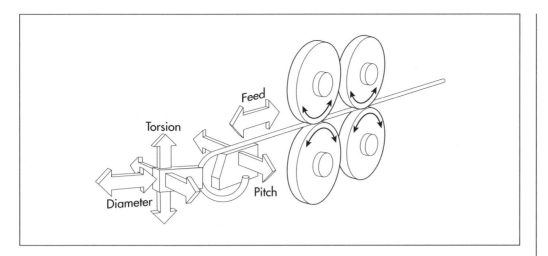

ers were able to fashion reliable, portable timekeeping devices. This advance made precise celestial navigation possible for ocean-going ships.

In the eighteenth century, the Industrial Revolution spurred the development of mass-production techniques for making springs. During the 1780s, British locksmith Joseph Bramah used a spring winding machine in his factory. Apparently an adaptation of a lathe, the machine carried a reel of wire in place of a cutting head. Wire from the reel was wrapped around a rod secured in the lathe. The speed of the lead screw, which carried the reel parallel to the spinning rod, could be adjusted to vary the spacing of the spring's coils.

Common examples of current spring usage range from tiny coils that support keys on cellular phone touchpads to enormous coils that support entire buildings and protect them from earthquake vibration.

Raw Materials

Steel alloys are the most commonly used spring materials. The most popular alloys include high-carbon (such as the music wire used for guitar strings), oil-tempered low-carbon, chrome **silicon**, chrome vanadium, and stainless steel.

Other metals that are sometimes used to make springs are beryllium copper alloy, phosphor bronze, and titanium. Rubber or urethane may be used for cylindrical, non-coil springs. Ceramic material has been developed for coiled springs in very high-tem-

perature environments. One-directional glass fiber composite materials are being tested for possible use in springs.

Design

Various mathematical equations have been developed to describe the properties of springs, based on such factors as wire composition and size, spring coil diameter, the number of coils, and the amount of expected external force. These equations have been incorporated into computer software to simplify the design process.

The Manufacturing Process

The following description focuses on the manufacture of steel-alloy, coiled springs.

Coiling

1 Cold winding. Wire up to 0.75 in (18 mm) in diameter can be coiled at room temperature using one of two basic techniques. One consists of winding the wire around a shaft called an arbor or mandrel. This may be done on a dedicated spring-winding machine, a lathe, an electric hand drill with the mandrel secured in the chuck, or a winding machine operated by hand cranking. A guiding mechanism, such as the lead screw on a lathe, must be used to align the wire into the desired pitch (distance between successive coils) as it wraps around the mandrel.

Alternatively, the wire may be coiled without a mandrel. This is generally done with a central navigation computer (CNC) ma-

Examples of different types of springs.

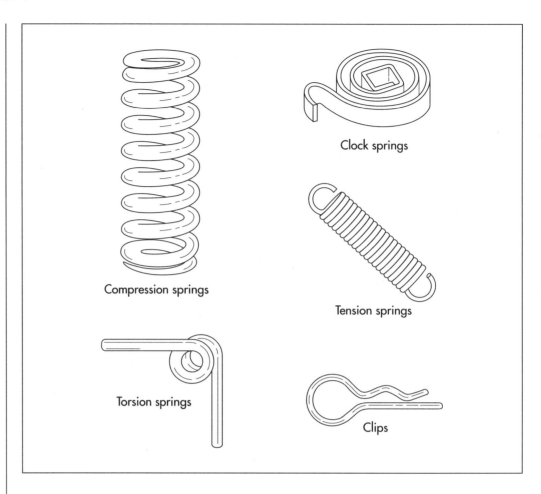

Compression springs

Clock springs

Tension springs

Torsion springs

Clips

chine. The wire is pushed forward over a support block toward a grooved head that deflects the wire, forcing it to bend. The head and support block can be moved relative to each other in as many as five directions to control the diameter and pitch of the spring that is being formed.

For extension or torsion springs, the ends are bent into the desired loops, hooks, or straight sections after the coiling operation is completed.

2 Hot winding. Thicker wire or bar stock can be coiled into springs if the metal is heated to make it flexible. Standard industrial coiling machines can handle steel bar up to 3 in (75 mm) in diameter, and custom springs have reportedly been made from bars as much as 6 in (150 mm) thick. The steel is coiled around a mandrel while red hot. Then it is immediately removed from the coiling machine and plunged into oil to cool it quickly and harden it. At this stage, the steel is too brittle to function as a spring, and it must subsequently be tempered.

Hardening

3 Heat treating. Whether the steel has been coiled hot or cold, the process has created stress within the material. To relieve this stress and allow the steel to maintain its characteristic resilience, the spring must be tempered by heat treating it. The spring is heated in an oven, held at the appropriate temperature for a predetermined time, and then allowed to cool slowly. For example, a spring made of music wire is heated to 500°F (260°C) for one hour.

Finishing

4 Grinding. If the design calls for flat ends on the spring, the ends are ground at this stage of the manufacturing process. The spring is mounted in a jig to ensure the correct orientation during grinding, and it is held against a rotating abrasive wheel until the desired degree of flatness is obtained. When highly automated equipment is used, the spring is held in a sleeve while both ends are ground simultaneously, first by coarse wheels and then by finer wheels. An appro-

priate fluid (water or an oil-based substance) may be used to cool the spring, lubricate the grinding wheel, and carry away particles during the grinding.

5 Shot peening. This process strengthens the steel to resist metal fatigue and cracking during its lifetime of repeated flexings. The entire surface of the spring is exposed to a barrage of tiny steel balls that hammer it smooth and compress the steel that lies just below the surface.

6 Setting. To permanently fix the desired length and pitch of the spring, it is fully compressed so that all the coils touch each other. Some manufacturers repeat this process several times.

7 Coating. To prevent corrosion, the entire surface of the spring is protected by painting it, dipping it in liquid rubber, or plating it with another metal such as zinc or chromium. One process, called mechanical plating, involves tumbling the spring in a container with metallic powder, water, accelerant chemicals, and tiny glass beads that pound the metallic powder onto the spring surface.

Alternatively, in electroplating, the spring is immersed in an electrically conductive liquid that will corrode the plating metal but not the spring. A negative electrical charge is applied to the spring. Also immersed in the liquid is a supply of the plating metal, and it is given a positive electrical charge. As the plating metal dissolves in the liquid, it releases positively charged molecules that are attracted to the negatively charged spring, where they bond chemically. Electroplating makes carbon steel springs brittle, so shortly after plating (less than four hours) they must be baked at 325-375°F (160-190°C) for four hours to counteract the embrittlement.

8 Packaging. Desired quantities of springs may simply be bulk packaged in boxes or plastic bags. However, other forms of packaging have been developed to minimize damage or tangling of springs. For example, they may be individually bagged, strung onto wires or rods, enclosed in tubes, or affixed to sticky paper.

Quality Control

Various testing devices are used to check completed springs for compliance with specifications. The testing devices measure such properties as the hardness of the metal and the amount of the spring's deformation under a known force. Springs that do not meet the specifications are discarded. Statistical analysis of the test results can help manufacturers identify production problems and improve processes so fewer defective springs are produced.

Approximately one-third of defective springs result from production problems. The other two-thirds are caused by deficiencies in the wire used to form the springs. In 1998, researchers reported the development of a wire coilability test (called FRACMAT) that could screen out inadequate wire prior to manufacturing springs.

Computer-operated coiling machines improve quality in two ways. First, they control the diameter and pitch of the spring more precisely than manual operations can. Second, through the use of piezoelectric materials, whose size varies with electrical input, CNC coiling heads can precisely adjust in real time to measurements of spring characteristics. As a result, these intelligent machines produce fewer springs that must be rejected for not meeting specifications.

The Future

Demands of the rapidly growing computer and cellular phone industries are pushing spring manufacturers to develop reliable, cost-effective techniques for making very small springs. Springs that support keys on touchpads and keyboards are important, but there are less apparent applications as well. For instance, a manufacturer of test equipment used in semiconductor production has developed a microspring contact technology. Thousands of tiny springs, only 40 mils (0.040 in or 1 mm) high, are bonded to individual contact points of a semiconductor wafer. When this wafer is pressed against a test instrument, the springs compress, establishing highly reliable electrical connections.

Medical devices also use very small springs. A coiled spring has been developed for use in the insertion end of a catheter or an endoscope. Made of wire 0.0012 in (30 micrometers or 0.030 mm) in diameter, the spring is 0.0036 in (0.092 mm) thick—about the same as a human hair. The Japanese company that developed this spring is attempting to make it even smaller.

The ultimate miniaturization accomplished so far was accomplished in 1997 by an Austrian chemist named Bernard Krautler. He built a molecular spring by stringing 12 carbon atoms together and attaching a vitamin B_{12} molecule to each end of the chain by means of a cobalt atom. In the relaxed state the chain has a zigzag shape; when it is wetted with water, however, it kinks tightly together. Adding cyclodextrin causes the chain to return to its relaxed state. No practical application of this spring has yet been found, but research continues.

Where to Learn More

Other

"Coil Spring Making Process—Automotive." Industrial Engineers and Spring Makers. http://www.ozemail.com (November 2000).

"H & R Spring Overview." http://www.hr springs.com/abouthr.html (November 2000).

Silberstein, Dave. "How to Make Springs." http://home.earthlink.net/~bazillion/intro.ht ml (November 2000).

—*Loretta Hall*

Steel Wool

Background

Steel wool is the name given to fine metal wire that are bundled together to form a cluster of abrasive, sharp-edged metal strips. The metal strips are massed together in a sheet, folded, and turned into pads that are easily held in the hand. These steel wool pads are used for a variety of purposes, but primarily as an abrasive material, sometimes replacing sandpaper. Steel wool may be best known to consumers as the pink-colored abrasive pads that have soap added so that they may be used to scrub pots and pans. Steel wool comes in a variety of grades, or thicknesses, from coarse to extra fine. The coarser the wire, the more abrasive the steel wool is against the surface. Fine sanding is always done with the finest steel wool grade (generally referred to as extra fine). Steel wool is made by a few manufacturers in the United States, but a fair amount of it is made overseas as well as in Mexico.

Steel wool gets its name from the fact that the fuzzy, grey mass of metal strings resembles wool before it has been carded and in some ways does resemble a fiber. It is not, however, truly spun as is a fiber. Instead, steel wool is produced by pulling metal rods through a series of metal dies that slice into the rods and cut away unnecessary metal—a process known as drawing. The rod is thus reduced to a fine strand, with the *swarf* or metal that is peeled away utilized in other products.

The production of steel wool generates heat as the cutting tool slices into these metal rods. Fires are a hazard during the production process and necessitate careful watch. Oil minimizes this fire hazard by reducing friction. However, the product does contain some oil, and manufactures and purchasers of steel wool need to be aware of the oil content as the product can spontaneously combust even beyond the factory. Steel wool must be stored away from electrical outlets or other sources of electricity or flame.

History

For many years the properties of small pieces or circles of metal were recognized for their ability to clean and cut through grease and grime, particularly those embedded in metal. The Victorians used peculiar pot-scrubbers that had a metal wire handle to which was attached many dozens of small circles of steel intertwined. Referred to as wire dish cloths, these scrubbers were touted as "the most convenient and most popular utensil extant." The scrubber was submerged in soap and water, then pressed against cast iron or aluminum pots, cleaning the surface easily.

However, mechanics who ran metal lathes noticed that the metal shavings resulting from peeling away metal from a part or tool was an interesting bit of waste. It is said that well before 1900, mechanics gathered up this swarf and used it to polish metal surfaces.

Steel wool was mass-produced sometime in the early part of the twentieth century. Its use infiltrated the American home when steel wool pads soaked with soap became a kitchen necessity. Throughout the late nineteenth century and early twentieth century, enterprising mechanics gathered up these leftover steel turnings and mixed them with soft soap. There is some contention as to who decided to manufacture and market these soap-soaked steel wool pads first. It is known that by World War I some entrepreneurs real-

Steel wool gets its name from the fact that the fuzzy, grey mass of metal strings resembles wool before it has been carded and in some ways does resemble a fiber. It is not, however, truly spun as is a fiber.

A steel wool cutting machine.

ized that these pads were necessary for cleaning the newly invented aluminum cookware that had started to overtake cast iron pots. One pot salesman had so many complaints from housewives about the difficulty of cleaning their new aluminum pots that these pads were a gimmick to get the women to try the new pots. They worked wonders in cleaning and still do, although some are unhappy with the fact that these steel wool pads can rust if left on the sink wet. A replacement for these pads has been a sponge, resembling steel wool, made of a mass of synthetic fibers and is rust-proof. But the cutting edges of the steel wool cannot be duplicated in these colorful, synthetic pads.

Raw Materials

Raw materials used in the manufacture of steel wool includes the metal rod that is to be thinly shaved and made into wool. These metals may vary and can include low-grade carbon steel wire, bronze, aluminum, and stainless steel. The only other raw material used in the process is the oil that is put on the cutting tools to lessen the friction generated between metal rod and cutting tool.

The Manufacturing Process

1 The entire cutting of steel wires into finely shaved steel wool happens on an approximately 50-ft (15.2-m) long machine called a steel wool cutting machine. The raw material is received at the factory and transported to the cutting machine with a forklift and readied for loading onto the machine. Skilled workers then take an end of a metal rod on the huge spools and wrap the end around the circular spool visible on one side of the cut-

ting machine. Each spool has 15 grooves on the surface. Thus, each spool can accommodate the shaving of 15 spools of wire. The operators who thread this machine work very quickly and it takes them only a few minutes to thread the wire into the machine.

2 The wire rod move down one side of the machine, going from spool to spool, to the other side of the machine. As it moves through the spools the wire rod moves against a metal cutting tool resembling a large sawtooth blade. The attached blades move against the wires, shaving the wire to generate thinner fiber-like wires of steel. The cutting tool has many consecutive sawtooth edges that are set very closely. The closer the teeth are placed together, the less waste there is as the cutting tool runs across the surface of these metal rods. These cutting tools press against these steel wires, producing a very fine wire that is pyramidal in shape due to the shape of the cutting tool. This results in production of a steel wool strand that is quite sharp on two sides of the pyramid. (That is why it is easy to be cut with steel wool and gloves should always be worn when using the product.)

3 As one spool runs out of wire, another spool is simply wound onto the spool, and the shaving process continues. As the cutting tool slices into metal, a tremendous amount of heat is generated. The tool must be kept cool with oil to reduce the chance of fire. Fire is a serious hazard in the production of steel wool; however, machine operators are fully trained to put out the fires. The cutting tool also dulls quickly and has to be re-shaped and sharpened approximately every three hours. Thickness of the product is varied by the size of the razor-like edges. The thicker the steel wool, the more slowly the product moves through the machine.

4 After the wire has moved up one side and down the other, the cutting tools have fully formed the steel wool. The usable steel wool product is wound up into big rolls underneath the machine that weigh approximately 40 lb (18.1 kg) each. These large rolls are run through a machine that cuts the steel wool in length (perhaps 2 ft [61 cm]) and width, then rolls the strip and presses it into steel wool pads (if one examines a steel wool pad it is essentially a roll of steel wool that has been rolled to form a pad and can be eas-

ily unrolled, exposing all sides of the product to the surface being scraped or sanded). These steel wool pads are then hand-packaged in packaging that reflects the grade of steel that ranges from extra-fine to coarse. Large steel wool manufacturers cut over 2,000 short tons (1,814 t) of steel each year.

Quality Control

Quality of steel wool is measured for fiber thickness, oil content, and weight. Perhaps the most important factor in steel wool production is the consistent thickness of the metal rods used to make the thin metal strips. In order for the grades to be considered uniform and reliable, the raw materials must be of absolutely consistent thickness, ensuring that the product will be shaved at the correct thickness each and every time. Similarly, the cutting tool must be regularly checked for sharpness. Approximately every three hours, the cutting tool must be sharpened. If it is not, it may snag or the rods may not be cut consistently in the pyramidal shape and grade desired. Some manufacturers easily and quickly change out those blades and re-grind them using a blade grinding machine. It is essential that a grade is consistent in its quality. If a woodworker requires extra fine steel wool to complete final finish sanding before staining and coarse steel cuts into the finish, the surface is ruined. Too much oil in the pad is also detrimental. Excessive oil can prevent the pad from soaking up the product (stain or wood stripper) and can mar the surface with oil. In addition, excessive amounts of oil in steel wool can make the product combustible.

Byproducts/Waste

The left-over wire (the metal that is cut away from the metal rod and is not usable steel wool) is collected and sent out the back of the machine via conveyor belt and moves to the hammermill. Here, the hammermill chops the scrap metal into metal dust that is sold to the automotive industry and used in the formation of brake pads. The small, left-over pieces that remain after the metal rod is cut are rolled onto a spool and cut into smaller pieces. This scrap is sold to concrete companies and is increasingly replacing rebar as it is significantly stronger than the reinforcing bars currently used in concrete construction. Lint and steel wool dust, as well as fumes, are generally collected with a cyclone dust collector, thus keeping these particulates out of the circulation within the plant.

The Future

Since the advent of steel wool, the product has undergone few changes. Soap has been added for use with pots and pans, and these steel wool types come in a handful of sizes and colors that are appealing to the consumer. In the future, consumers will see different types of grease fighting agents applied to the steel wool pads.

Where to Learn More

Other

New Scientist. http://www.newscientist.com (January 2001).

SFI Steel Wool Machines. http://sfisteel-wool.com (January 2001).

—*Nancy E. V. Bryk*

Storm Shelter

Background

More than half of the United States lie in a broad strip between the Appalachian and Rocky Mountains that is commonly called Tornado Alley. It has more tornado activity than any other area of the world. The rest of the country is not immune; tornadoes have occurred in every state.

Each year, United States residents spend more than 3 billion person-hours under tornado watches (official alerts issued in areas experiencing severe weather that might generate tornadoes). According to the Wind Engineering Research Center at Texas Tech University, a tornado actually forms during more than half of these watches. During 1999, 13 states suffered a total of 30 fatal tornadoes causing 95 deaths. Although there have been no studies of the number of lives saved by storm shelters, there is broad agreement that properly constructed shelters are highly effective, not only in tornadoes—the most violent of wind storms, with winds sometimes exceeding 300 mph (485 kmph)—but also in hurricanes and other severe weather events.

A family that wants to have their own storm shelter has three basic choices. They can build a monolithic dome home (a seamless concrete structure) that looks unconventional but is strong enough to withstand tornadic winds and debris impacts. Alternatively, they can heavily reinforce a room or closet in their conventional house, producing a "safe room" (in-residence shelter). At least one company sells an epoxy-coated steel panel kit for this purpose, and the Federal Emergency Management Agency (FEMA) offers free instructions for a do-it-yourself, reinforced concrete version in its book *Taking Shelter from the Storm: Building a Safe Room Inside Your House.*

A third choice is for the family to build a shelter underground, beneath their house or nearby. These can be purchased as prefabricated units and installed in a few hours or they can be custom built on site. Manufactured shelter costs begin at less than $3,000, with installation adding on about $300-$500. Custom-built shelters can be as extensive as desired; one company that builds survival shelters for nuclear explosions as well as natural disasters suggests adding amenities such as a billiard room, bowling alley, or pistol shooting range. In general, however, storm shelters are small—often around 50 ft^2 (4.6 m^2) and about 6 ft (1.8 m) high—and intended to accommodate six to10 people for only a few hours.

History

The history of storm shelters overlaps with that of fallout shelters designed for protection from nuclear warfare. Threats to the security of United States territory during World War II, the uncertainty of the nuclear weapons era, and the Cold War between communism and democracy combined to launch the United States' first civil defense programs in 1949. In 1950, Congress authorized both a nationwide system of nuclear bomb shelters and a relief effort for victims of natural disasters. People were encouraged to build underground bomb shelters for their families and stock them with survival provisions. In tornado-prone areas, these shelters served a dual purpose.

In 1971, during the presidency of Richard M. Nixon (1969–1974) and its philosophy of de-

During 1999, 13 states suffered a total of 30 fatal tornadoes causing 95 deaths.

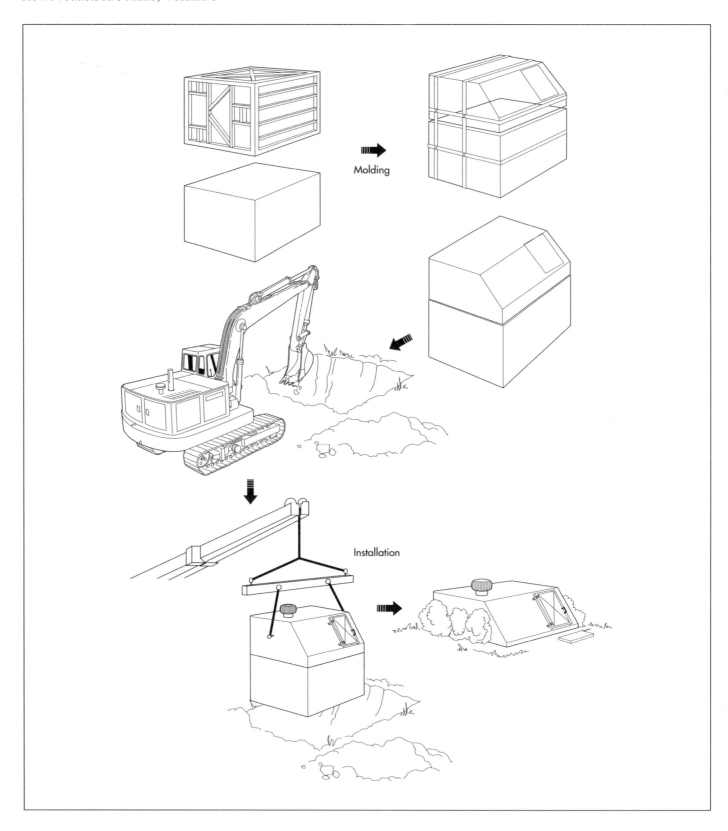

Molding

Installation

An underground storm shelter is a molded concrete structure that is buried in the ground and used as a protective shelter against tornadoes.

tente (peaceful coexistence with communist governments), the focus of government-supported programs for civil defense shifted from military attacks to natural disasters. After a brief reversal during Gerald Ford's presidency (1974–1977), the focus returned to protection from natural disasters and peacetime accidents like that at the Three Mile Island nuclear power plant in 1979. Following widespread devastation from Hurricane Hugo and California's Loma Prieta earthquake in 1989 and Hurricane Andrew in 1992, FEMA's mission broadened from disaster relief and reconstruction to loss prevention. In accord with this philosophy, the federal government now offers financial incentives to homeowners who construct in-residence or underground storm shelters.

Raw Materials

Safe rooms are constructed of steel panels or reinforced concrete. Site-constructed underground storm shelters are usually made of reinforced concrete. Prefabricated underground shelters are made from various materials, including corrugated steel culverts, steel plate (perhaps galvanized or epoxy coated), reinforced fiberglass, high-density polyethylene, and concrete reinforced with rebar (steel rods) or fibermesh (fine, polypropylene fibers distributed uniformly through the wet concrete).

Shelter doors or hatches are usually made from steel, fiberglass, or steel-plated plywood or aluminum. Stairs or ladders in underground shelters may be made of wood, aluminum, steel, or fiberglass. Stainless steel and zinc are used for other hardware items like bolts and anchor chains.

Design

Underground shelters come in various shapes including spheres, domes, horizontal tubes, and rectangular boxes. Most manufacturers offer their own patented designs, some of which are available in several sizes. They usually come with battery-powered lights, wall-mounted benches, and a ladder or stairs. Some manufacturers offer optional accessories like a chemical toilet, indoor-outdoor carpeting, a telephone jack, and a weather-band radio.

To minimize the risk of occupants being trapped by debris covering the door, some shelters are equipped with a hydraulic or screw jack for forcing open the door. Other models feature a second, emergency exit hatch at the end of the chamber opposite the entrance.

Safe rooms are readily accessible to the handicapped. Most underground shelters are not designed for entrance by mobility-im-

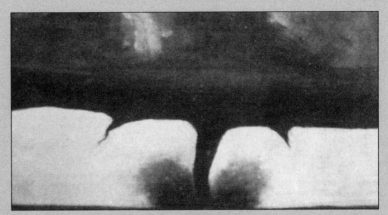

The oldest known photograph of a tornado sighted on August 28, 1884, near Howard, South Dakota.

A tornado is a rapidly spinning column of air formed in severe thunderstorms. The rotating column, or vortex, forms inside the storm cloud then spirals downward until it touches the ground. Although a tornado is not as large as a thunderstorm, it is capable of extreme damage because it packs very high wind speeds into a compact area. Tornadoes have been known to shatter buildings, drive straws through solid wood, lift locomotives from their tracks, and pull the water out of small streams. An average of 800 tornadoes strike the United States each year. Based on statistics kept since 1953, Texas, Oklahoma, and Kansas are the top three tornado states. Tornadoes are responsible for about 80 deaths, 1,500 injuries, and millions of dollars in property damage annually. While it is still impossible to predict exactly when and where tornadoes will strike, progress has been made in predicting tornado development and detecting tornadoes with Doppler radar.

paired persons, but a few manufacturers offer some models that are. Canton Enterprises' horizontal-entry version, for example, can be recessed into a hillside, or it can be installed at ground level and have dirt banked around its walls.

The Manufacturing Process

The following description is representative of the manufacturing process for a prefabricated, reinforced concrete storm shelter. Certain details may vary among manufacturers.

Shell construction

1 Steel molds are used to cast the upper and lower sections of the shelter. Walls are 3-4 in (7.5-10 cm) thick, while floors and ceilings are 5-6 in (13-15 cm) thick.

2 Metal reinforcement is placed in each mold. For example, one manufacturer uses 0.5-in (1.3 cm) diameter steel rebar placed at 12-in (30-cm) spacing.

3 Concrete is poured into the mold. The mold is vibrated to minimize air bubbles.

4 When the concrete has set, the shell is removed from the mold and inspected for flaws.

5 The top and bottom halves of the shell are joined. Their edges are designed to interlock, and a tar sealant or waterproof mastic is applied prior to mating the two pieces. Steel straps are placed across the seams and bolted in place to reinforce the joint.

6 Steps or a ladder are installed. The door is mounted using multiple hinges, a piano hinge, or a frame that allows the door to slide open and closed. One or two air vents are installed. Each consists of a vertical pipe topped with either a wind turbine or a screen-covered 180° elbow.

Installation

The installation procedure is similar for all types of in-ground shelters. The following description mentions there refinements necessary for each type.

7 A **backhoe** is used to dig a hole for the shelter. A worker with a shovel hand-finishes the hole, leveling the bottom and clearing the corners. The hole should be about 2 ft (61 cm) larger than the shelter in each dimension.

8 The backhoe or a crane-equipped tow truck is used to lift the shelter and lower it into the hole. Some shelters are equipped with fittings that the lifting device can grasp. Others are lifted by chains attached to brackets on the outside of the shelter or by a pair of nylon straps wrapped vertically around the shelter.

9 A prefabricated reinforced concrete shelter weighs at least 12,000 lb (5,500 kg). The smallest sizes of steel and fiberglass shelters weigh about 3,500 lb (1,600 kg) and 1,400 lb (640 kg), respectively. Because they could float if the ground became saturated, steel and fiberglass shelters must be anchored during installation. This is usually done by bolting or chaining them to mounts embedded in a concrete foundation. Some manufacturers also recommend pouring concrete around the shelter's walls.

10 If concrete is not poured around the shelter, soil excavated from the installation site is generally used to backfill the remaining void around the shelter. If the shelter is also intended to function as a bomb shelter, backfilling consists of gravel or crushed rock.

11 Some underground shelters are installed with their tops at ground level, but most are designed to be covered with 2-3 ft (61-91 cm) of soil. The surface may then be planted with grass or topped with patio decking.

Quality Control

FEMA bases its acceptability standards for storm shelters on research conducted at the Wind Engineering Research Center. Since 1970, the center has studied the damage caused by 100 windstorms and tornadoes. It found that the greatest threat to people is flying debris that can penetrate windows, walls, and roofs of buildings, forcibly striking occupants. Shelter designs can be tested in the center's laboratory to determine their ability to withstand the impact of a 15-lb (6.8-kg) section of 2x4-in (5x10-cm) lumber fired by a compressed-air cannon at speeds of 100 mph (160 kmph) for walls and doors or 67 mph (120 kmph) for roofs. A FEMA-acceptable shelter must also be able to withstand winds of 250 mph (400 kmph); very few tornadoes exceed this wind speed.

In addition to the standard tests, some manufacturers tout their product's ability to withstand somewhat more creative trials. For example, one advertises that its shelter withstood the impact of a minivan, and another reports its unit stopped a 38-caliber bullet. As part of its inspection of each shelter produced, one manufacturer fills the completed shell with water to verify that it is water-tight.

The Future

Storm shelters are usually made out of concrete or steel, but recent technological developments have shown that fiberglass is becoming very popular. Older shelters tend to rust or corrode after many decades of being exposed to the worst of the elements. Mildew is also a large factor in underground shelters. Fiberglass is mildew resistant and usually guaranteed not to rust or corrode. Advances have also been made with new steel and concrete models. Many companies sell warranties with their shelters promising that they will not leak, rust, or float.

Where to Learn More

Other

E-Z Kit Safe Room. http://www.twisterpit.com/ez_kit.htm (April 15, 2000).

"Inresidence Shelters for Protection from Extreme Winds." Texas Tech University Wind Engineering Research Center. http://www.wise.ttu.edu/inshelter/inshelte.htm (April 12, 2000).

Jarrell Storm Shelters. http://www.jarrellstormshelters.com (April 13, 2000).

Storm Shelters USA. http://www.shelter-susa.com (April 14, 2000).

"Taking Shelter from the Storm: Building a Safe Room Inside Your House." Federal Emergency Management Agency (FEMA). http://www.fema.gov/mit/tsfs01.htm (April 11, 2000).

—*Loretta Hall*

Teeth Whitener

In the late 1990s, sales for teeth whiteners soared to $33.7 million. By comparison, consumers spent $2.43 billion on the entire oral hygiene category which include toothpastes and mouthwashes.

Background

Teeth whiteners are products designed to enhance the appearance of teeth by removing stains and improving brightness. These whiteners typically contain bleaching agents, such as hydrogen peroxide or other peroxygen-type chemicals, that remove organic residue and oxidize stains so they are less visible. While teeth whiteners have been used by dentists for many years, they have only been commercially available since the mid-1990s. These products are available in two primary forms: as toothpastes that whiten teeth as they clean and as specialty liquids applied to teeth separate from the brushing process.

History

Teeth-whitening formulations were first developed for use in denture cleaning compounds. From there, they gained popularity among dentists for general teeth whitening. The first clinical treatments administered by dentists required several lengthy appointments that involved etching the teeth with an abrasive or an acid and then bleaching them with a 30-33% solution of hydrogen peroxide and applying heat. Finally, the teeth were polished to restore a lustrous surface. Dentists typically charged $50-150 for each of these clinical procedures.

In the 1990s, commercial tooth whiteners began appearing on the market for the general public. These products contained some of the same active ingredients as the professional products but at lower concentrations. To achieve significant whitening, they must be used for several minutes each day for four to six weeks. While they can improve the general appearance of teeth, these products are not effective in removing deeper stains such as those caused by tetracycline, fluorosis, jaundice, or internal bleeding. In the late 1990s, sales for teeth whiteners soared to $33.7 million. By comparison, consumers spent $2.43 billion on the entire oral hygiene category which include toothpastes and mouthwashes.

While the products currently on the market are generally recognized to be less effective than professional treatments, there is still concern that they may damage oral tissue. Therefore teeth whiteners are the source of some controversy. The U.S. Food and Drug Administration (FDA) has expressed concerns that these products should be considered drugs rather than cosmetics and the American Dental Association (ADA) says that overuse of these products may damage tissue, cause cell changes, or harm dentin and enamel. Although the products remain on the market, their future is not yet clear, and an extreme shift could occur if the FDA eventually rules that whiteners can no longer be sold as cosmetics.

Raw Materials

Water constitutes the largest portion of the formula and is used as diluent for the other ingredients. Dionized or demineralized water is used because the metal ions found in hard water can interfere with the action of the other raw materials.

Stain removal can be achieved chemically with bleaching agents such as hydrogen peroxide and physically with abrasives such as carbonates. Chemical bleaching agents include hydrogen peroxide, sodium percarbonate, and sodium perborate. Abrasive ma-

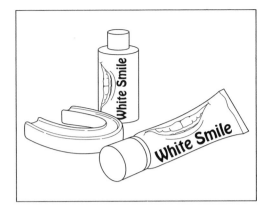

terials used in whiteners include calcium or magnesium carbonate, calcium phosphates, insoluble sodium metaphosphate, silica xerogels or aerogels, and hydrated aluminas.

Detergents are used in teeth whiteners to provide foam and help cleanse. These chemicals are surfactants such as sodium lauryl sulfate, sodium lauryl sarcosinate, sodium lauryl sulfoacetate, or dioctyl sodium sulfosuccinate.

Binders and thickeners increase the viscosity of the product. Some that are typically used are carboxymethylcellulose, carrageenan, gum tragacanth, gum karaya, Irish moss, sodium alginate, Carbopol resins, magnesium aluminum silicates, and block copolymers.

Therapeutic agents may also be added to the product. They include sodium citrate, which interferes with the metabolic activity of pathogenic bacteria and helps reduce gingivitis; and pyrophosphate, which is an effective tartar-control agent.

Humectants such as sorbitol, glycerin, and propylene glycol are used in the formulas to reduce moisture loss from the product; this prevents "crusting" if the cap is left off of the package for an extended time. Flavors are included to improve consumer appeal. Common flavors are peppermint, spearmint, wintergreen, sassafras, and anise. Additionally, sweeteners, such as saccharin, are added to further improve the taste of the product. Preservatives such as methyl and propyl paraben and sodium benzoate are used to prevent bacterial growth in the product. Finally, titanium dioxide, which contributes to teeth whitening; sodium bicarbonate, which controls pH; and certified colors are other common ingredients.

Design

Teeth whiteners for home use are carefully formulated to be efficacious and safe. The most significant challenge when designing these products is to eliminate potential incompatibility between the peroxygen bleaching agents (e.g., hydrogen peroxide) and the other ingredients in the formulation. Initially, this challenge was overcome by the development of a dual-delivery system where gelled hydrogen peroxide was kept separate from the rest of the formulation with a dual chambered package. With this method, both portions of the product were co-extruded onto the toothbrush at the time of use. Eventually, stable formulations were designed which replaced hydrogen peroxide with a solid peroxygen bleaching agent (either sodium percarbonate or calcium peroxide) in an anhydrous formulation. This approach also allows the incorporation of other active ingredients such as baking soda and tartar-control agents. If care is taken to select compatible thickeners, and if contamination from other sources is controlled, these formulations can maintain shelf stability for the life of the product.

Aesthetic considerations also impact formula design. For example, taste, appearance, and consistency must be designed to the consumer's liking before these products can be commercially acceptable. Proper packaging must also be considered during the design phase of product development. Teeth whiteners may be designed as pastes that can be applied using a simple toothbrush or as liquids used in conjunction with plastic or rubber dental dams that hold the solution close to the tooth surface. This application process helps the product perform better. In addition, some teeth whiteners are designed with a pre-rinse treatment that eliminates some of the residue found on the tooth surface.

A number of manufacturers have marketed successful teeth whitening products. For example, Proctor and Gamble's Crest introduced its MultiCare Plus Extra Whitening toothpaste in the late 1990s. Other popular brands include Aquafresh Whitening Advanced Freshness by Smith Beacham; Rembrandt Daily Whitening Gel with Safe Peroxide by the Den-mat Corporation; and Mentadent Advanced Whitening by Cheseborough Ponds to name just a few.

The Manufacturing Process

Raw material staging

1 Raw materials are first analyzed to make sure they comply with all relevant specifications. Once they have been approved they are pre-weighed and staged in the manufacturing area in preparation for production.

Charging the batch tank

2 Toothpaste-like teeth whiteners are typically made in stainless steel batching tanks equipped with planetary-style mixers that stir the batch without whipping in excessive amounts of air. These tanks are sealed and equipped with an apparatus that uses a vacuum to pull out trapped air. This prevents the formation of bubbles in the finished product. Batching tanks range in size from 25-625 gal (100-2,500 L)

Mixing

3 The humectants (glycerine and sorbitol) are added first and the powdered thickening agents are dispersed in the liquid. By dispersing the powders in the non-aqueous solvents they will not clump when they come in contact with water.

4 After the powders are dispersed, the water is added to dilute the batch. The batch is mixed for about 20 minutes, allowing the water soluble ingredients to fully hydrate.

5 Then, a vacuum is applied to draw excess air from the batch. The sweeteners, surfactants, and other ingredients are mixed in and the batched is mixed for another 15 minutes.

6 Finally, the flavor chemicals are added and the batch is stirred for an additional 10 minutes until it reaches a smooth, bubble-free consistency.

Filling and packaging

7 The finished product is transferred to a filling machine via high pressure pumps. The filling machine consists of nozzles connected to a metered dispensing system. The packaging, usually plastic tubes, is fed under the filling nozzles via a conveyor belt. Each tube moves down the conveyor and as it passes under the filling head the nozzle dispenses a pre-set amount of product into the open end of the tube. The tube then moves down the conveyor to a sealing machine that clamps the tube shut and seals the plastic with heat or ultrasonic vibrations. Each tube imprinted with a batch code to allow the tracking of each lot. The sealed tubes are then transferred to cartons for shipping.

Quality Control

As with other personal care products, the quality of teeth whiteners are carefully monitored during production. Before manufacture begins, all raw materials are assessed to ensure they meet the established specifications. After the batch is complete, the finished product is checked for quality including basic chemical parameters such as pH and viscosity. Both these factors can affect the stability of the product as well as its aesthetic appeal. For example, if the pH is too low, the solubility and efficacy of the whitening agents may be affected. The batch is also checked to ensure it is free of microbial contamination.

The quality these products may be impacted by actions of the FDA. In February 1994, the FDA issued a Notice of Proposed Rule Making (NPRM) in the form of a tentative final monograph for oral antiseptic drug products. This monograph lists active ingredients and establishes testing and efficacy requirements for teeth whiteners. Active ingredients are classified Category II (not generally recognized as safe or effective) or Category III (more data needed to establish safety and effectiveness). As of the end of 2000, the FDA has not established when it will issue the final monograph on this category and it continues to regulate these products as cosmetics, not as drugs. The National Tooth Whitener Coalition, a trade organization composed of teeth whitener manufacturers, is fighting to ensure that the FDA does not over regulate this product category.

The Future

The primary factor affecting the future teeth whitener products is the regulatory environment. Depending on future actions by the FDA, these products may continue to be sold as cosmetics or they may be regulated as drugs. The fact that most of these products contain hydrogen peroxide continues to be of concern. In addition, chemists are continually developing improved formulations. While changes in the regulatory environment will impact the future of teeth whitening products, it is likely that advances in chemistry will result in products with improved performance, taste, and stability.

Where to Learn More

Periodicals

"Guidelines for the Acceptance of Peroxide-Containing Oral Hygiene Products, Council on Dental Therapeutics." *Journal of the American Dental Association* 125 (August 1994):1141-1142.

Rosendahl, Iris. "Tooth Whiteners Wite Hot Despite Heat from FDA." *Drug Topics* 136, no. 6 (March 23, 1992): 96(2).

"Striped Gel Toothpaste." *Soap Cosmetics Chemical Specialties* 72, no. 12 (December 1996): 42(4).

Thorsen, Eric. "Oral Care Cosmetics Brighten Category Sales." *Discount Store News* 37, no. 16 (August 24, 1998): 44(1).

—*Randy Schueller*

Thompson Submachine Gun

Tommy guys were used in Chicago's notorious St. Valentine's Day Massacre in 1929 and carried by renegade killers Bonnie and Clyde in the 1930s.

Background

A machine gun is a weapon that fires a continuous stream of bullets as long as the trigger is held down. Many inventors worked to come up with such a gun, and early models are the well-known Gatling gun, used prominently in the American Civil War, and Hiram Maxim's fully automatic weapon, patented in 1883. Machine guns of various makes were instrumental in the trench battles of World War I. After World War II, the machine gun was for the most part replaced by different types of more powerful automatic assault rifles. The lightweight machine gun known as the "Tommy gun," or Thompson submachine gun, was developed for use in World War I, and then marketed to law enforcement personnel. It became notorious as the gun of choice of gangsters in the 1920s and 1930s. It is still manufactured, finding a market primarily with gun collectors.

History

The gun invented by Richard Jordan Gatling in 1862 was the first widely used weapon of the machine gun type. The Gatling gun was not strictly a machine gun, as it was not completely automatic. Its rotating barrel had to be cranked by hand. Ammunition was fed into the Gatling through a top-mounted hopper. It could fire a thousand rounds a minute. American arms inventor Benjamin Berkeley Hotchkiss came up with an improved Gatling-type gun in 1872. Both the Hotchkiss and the Gatling were made obsolete by the invention of the Maxim machine gun in 1883. The Maxim was fully automatic, firing continuous rounds powered by the recoil energy of the exploding shell. Other early machine guns were John Browning's Browning Automatic Rifle of 1892, and an improved version of the Browning developed by an American army colonel Isaac Newton Lewis in 1911. By the Boer War of 1899–1902, the effectiveness of the machine gun was well demonstrated, and European countries adopted various weapons of Maxim, Hotchkiss, and Lewis in the years leading up to World War I. These machine guns were heavy, needed to be supported by a block or tripod, and they tended to overheat quickly, requiring some sort of cooling system.

The inventor of the Thompson submachine gun was Kentucky-born Army officer John Taliaferro Thompson. He was born into a military family, and spent his youth on military bases across the United States. He graduated from the military academy West Point in 1882 and then entered the army. By 1890 Thompson was working in the Ordnance Department, where he remained for the rest of his career. Thompson became a specialist in small arms, and by 1903 he was working on modernizing many of the Army's weapons designs. He developed a new model rifle based on the German Mauser in 1903, and in 1907 he was put in charge of small arms design, development, and production in the Ordnance Department in Washington. Thompson's dream was to convince the United States Army to adopt or develop an automatic rifle, but his ideas were considered radical. The machine gun's several inventors had all gone to Europe to market their weapons, and the U.S. Army remained uninterested. Thompson eventually retired from the army in 1914, and went to work for the Remington Arms Corporation, one of the leading American weapons manufacturers. At Remington he pursued plans to design his own automatic rifle.

Through personal contacts, Thompson met business magnate Thomas Fortune Ryan, and the financier agreed to provide the inventor with capital. In 1916, Thompson launched a new company, the Auto-Ordnance Corporation, to develop, manufacture, and market a new automatic rifle. This firm, based in New York, contracted with a Cleveland machine-tool firm, Warner & Swasey, to build and test its prototypes. Auto-Ordnance's first attempts at an automatic rifle failed. In 1917, with the European countries engaged in trench warfare in World War I, John Thompson decided to opt for a new design entirely. This was to be a small, hand-held machine gun. The Maxims and other machine guns in use in World War I were large, relatively immobile weapons that were used primarily defensively. Thompson envisioned a gun of similar swift firepower, that soldiers could run with, and so use in offensive assaults.

Auto-Ordnance began working feverishly on this "miniature" machine gun. The first workable designs were done in 1918, and the company made several prototypes and got them ready to ship to American troops overseas. The prototypes reached the dock in New York the day the Armistice was signed, and Auto-Ordnance thus lost out on its intended market. The company went back to work, trying to modify the gun for use other than in trench warfare. In 1919 the company unveiled its Thompson submachine gun, the "sub" indicating that it was much smaller than the massive machine guns used in Europe. The premier United States gunmaker Colt agreed to manufacture the Thompson, and the first guns were ready in March, 1921. Though Auto-Ordnance hoped to get a large order from the U.S. Army, it instead found eager takers in countries like Honduras and Panama, where the guns were used to solve labor disputes. Within months of the gun's introduction, the Thompson found its way to underground fighters of the Irish Republican Army (IRA). Auto-Ordnance marketed the gun heavily to police departments, touting the "pocket machine gun" as a great way to stop bank robbers and other motorized bandits. Unfortunately, it was these criminals who seized on the merits of the Tommy gun. In 1925 gangsters in Chicago used Thompsons in vendettas, finding them ideal for quick killing from a safe distance. The submachine guns were apparently easily and legally available at sporting goods stores. Notorious gangster Al Capone supposedly stopped at a Chicago sporting goods store to get a gun, and Capone's first known Tommy-gun killing followed on April 27, 1926. The guns spread through the underworld, first in other parts of the Midwest, and then to New York. They were used in Chicago's notorious St. Valentine's Day Massacre in 1929 and carried by renegade killers Bonnie and Clyde in the 1930s.

During the 1930s, the Tommy gun continued to be identified with desperados, gangsters, and bank robbers. In 1932, Auto-Ordnance at last convinced the United States Army to buy its guns, but the Army bought only small quantities. However, on the eve of World War II, the company suddenly received an order from France for 3,000 Tommy guns. The French order was soon followed by a British one, and the U.S. Army too ordered over 20,000 Thompsons in 1940. Colt refused to manufacture more of the submachine guns because of the bad press the weapon had received, and the Thompson was redesigned and somewhat simplified to fill the World War II orders. The Thompsons of the 1940s were manufactured by a company in Bridgeport, Connecticut, where they were the only light machine guns being mass-produced by any of the Allied countries. But even the new, improved design was soon obsolete. By the end of the war, the Thompson had been surpassed by the cheaper, lighter British STEN gun and its United States counterpart, the M3. The M3 was known as the "grease gun," an inelegant thing that was made of stamped metal, welded together. Ugly as it was, it could be mass-produced for a fraction of the cost of the Thompson.

After the war, demand for the Thompson was practically gone. Auto-Ordnance Co. changed hands several times, always on the verge of bankruptcy. In the 1970s, the company was acquired by a former employee, Ira Trast, who redesigned the classic Thompson as a semi-automatic weapon. The intended market was mostly gun enthusiasts who wanted a working gun that looked like the infamous gangster weapon. In 1999 the company changed hands again. It was bought by Kahr Arms in Blauvelt, New York. In order to produce a historically ac-

Richard Jordan Gatling.

Richard Jordan Gatling was born in 1818 in Hertford County, North Carolina. Gatling helped his father develop machines for sowing and thinning **cotton**. In 1839 Gatling invented a screw propeller for ships and went on to develop agricultural machines, such as a hemp-breaking device and a steam plow.

When the Civil War began in 1861, Gatling focused his efforts on armaments. In 1862 he invented the weapon that has bore his name ever since, the Gatling gun. Considered the first practical machine gun, the Gatling gun was capable of firing 250 shots per minute. It consisted of 10 breach-loading rifle barrels—cranked by hand—rotating around a central axis. Each individual barrel was loaded by gravity feed and fired while the entire assembly evolved. Cartridges were automatically ejected as the other barrels were fired. It was operated by two people: one fed the ammunition that entered from the top, and the other turned the crank that rotated the barrels. At first, the Union Army was uninterested in Gatling's invention, but General Benjamin Butler (1818–1893) eventually bought several Gatling guns. They worked so well on the battlefield that the government finally agreed to adopt them in 1866, but by then the war was over.

After the war, Gatling continued to improve his gun. Eventually, it was capable of firing 1,200 shots per minute at all degrees of elevation and depression. Gatling's gun was used all over the world and remained in the United States military arsenal until 1911.

guns are now an interesting blend of old and new technology. Parts are machined by precision instruments controlled by computers, and then the guns are carefully assembled by hand by trained artisans.

Raw Materials

The raw materials for Thompson submachine guns are mostly steel, with lighter alloys for small and flexible parts such as springs. The stocks are made of walnut, a traditional hardwood for gun manufacturing.

Design

The original design process for the Thompson was quite lengthy, and involved numerous drawings and prototypes. The gun was redesigned for use in World War II to make a simpler model that was easier to mass-produce. The Thompsons produced after World War II were assembled out of surplus parts by a company that had bought Auto-Ordnance's inventory. When the parts inventory began to run low, Auto-Ordnance was sold to Kahr Arms, a manufacturer of guns, other weapons, and parts, as well as many other metal products. At this point, Kahr wished to make complete Thompson guns out of new parts. Kahr's engineers consulted the scores of original drawings for historical accuracy, and also went through a process known as reverse engineering.

In reverse engineering, engineers take apart a finished product and figure out how it was made. Drawings are made from already available parts, instead of new parts being made from engineers' drawings. To make the Thompson according to modern methods, a drawing for each part was produced using computer software known as computer aided design, or CAD. Next, a separate set of drawings were made, called machine or shop drawings. These are blueprints that show exactly how each part needs to be cut. These drawings are converted to computer codes that can be read by the actual cutting machines.

The Manufacturing Process

Cutting the steel

1 The manufacturer first receives its raw material at the factory as steel bars. These

curate gun, Kahr researched the original engineering drawings for the Thompson, digging through records going back to 1919. Kahr then used modern computer design and drafting techniques to produce completely new engineering drawings based on the old designs. Thompson submachine

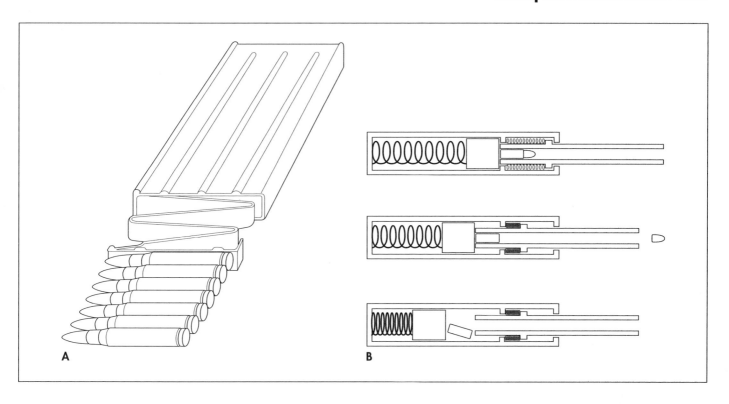

A

B

are cut by a number of specialized machines. The machines read the computer-generated blueprints and cut exactly to specification. The component parts of the gun have been designed so that they can be produced in only one or two operations, thus cutting down the possibility of deviation from the desired specifications. Some of the main parts of the Thompson that are cut from solid steel are the barrel, the receiver, the bolt, and the frame.

Other metal parts

2 Not every part needs to be cut from solid steel. Some smaller parts are stamped. These are done by a sub-contractor who specializes in stamping. Large stamping machines press down on sheets of metal, working something like a cookie cutter. Springs are also purchased from a sub-contractor who specializes in **spring** manufacturing.

The stock

3 The stock is made of walnut. This is made by a sub-contractor according to the gun manufacturer's design specifications. Workers use wood-cutting tools to cut and shape the stock from walnut boards, and ship them to the machine gun maker.

Subassembly

4 The Thompson gun has a total of between 60 and 70 parts total. Rather than workers assembling the whole gun at once, the process is broken down into five main subassemblies. Workers at the factory are divided into different subassembly stations. Parts belonging to a particular subassembly are set out, and workers fit parts together by mating surfaces and/or securing parts with screws. Workers are selected for jobs because they have a background in firearms, and they go through a three-month internship before they are fully qualified. Workers are paid by the piece, and so they strive to be fast as well as accurate.

Final assembly

5 Other workers put the entire gun together from the finished subassemblies. Because of the high precision of the machining, parts fit neatly into each other. They are snapped into place and secured with screws. The wooden stock is screwed on last, and the gun is cleaned and polished. Then the finished Thompson moves to a quality control area for a final check.

Quality Control

The maker of Thompson submachine guns works under international standards for

A. A magazine feed system. B. Using a recoil, the barrel and breechblock lock together before firing, move backwards after firing, and propel the bullet forward to be discharged.

manufacturing quality. These are standards that apply to the metal machining techniques used, whether the end product is a gun or an exercise machine. To list itself as a factory following these standards, the manufacturer submits to random audits of its facilities several times a year. So the entire facility follows strict guidelines for quality control. As far as specific quality control tests for the Thompson, the guns undergo tests for function and for cosmetics. Each finished gun is carefully inspected for obvious outward flaws such as scratches or blemishes on the stock. And each gun is test-fired. Quality control workers at a test firing range shoot off six or seven rounds from each gun. Then the guns are wrapped, boxed, and distributed to wholesalers.

The Future

Modern guns have taken the place of Thompson submachine guns for warfare and other uses. But they have historical significance, and may be collected by gun enthusiasts for that reason. Although the design and inner workings of the Thompson has changed for modern manufacturing, it is the distinctive outward appearance of the gun that will surely remain unchanged in the future. Manufacture of the Thompson will likely continue as long as our fascination with history and the underworld is alive.

Where to Learn More

Books

Helmer, William J. *The Gun that Made the Twenties Roar.* London: Macmillan, 1969.

Hosley, William. *Colt: The Making of an American Legend.* Amherst, MA: University of Massachusetts Press, 1996.

—*Angela Woodward*

Toilet Paper

Background

Most of us can't imagine living without toilet paper. The average American uses over 100 single rolls—about 21,000 sheets—each year. It's used not only for bathroom hygiene, but for nose care, wiping up spills, removing makeup, and small bathroom cleaning chores. Manufacturers estimate that an average single roll lasts five days.

Toilet paper, paper towels, napkins, and facial tissues are sanitary papers, personal products that need to be clean and hygenic. They're made from various proportions of bleached kraft pulps with relatively little refining of the stock, rendering them soft, bulky, and absorbent. Sanitary papers are further distinguished from other papers in that they are creped, a process in which the paper is dried on a cylinder then scraped off with a metal blade, slightly crimping it. This softens the paper but makes it fairly weak, allowing it to disintegrate in water.

Toilet paper can be one- or two-ply, meaning that it's either a single sheet or two sheets placed back-to-back to make it bulkier and more absorbent. Color, scents, and embossing may also be added, but fragrances sometimes cause problems for consumers who are allergic to perfumes. The biggest difference between toilet papers is the distinction between virgin paper products, which are formed directly from chipped wood, and those made from recycled paper. Most toilet paper, however, whether virgin or recycled, is wrapped around recycled cardboard cylinders.

History

Before paper was widely available, a variety of materials were employed. The Romans used an L-shaped stick (like a hockey stick) made of wood or precious metal; at public toilets people used sponges on sticks that were kept in saltwater between uses. In arid climates, sand, powdered brick, or earth was used. Until the late nineteenth century, Muslims were advised to use three stones to clean up. One favorite tool was a mussel shell, used for centuries. Until the early twentieth century, corn cobs were used.

In the late fifteenth century, when paper became widely available, it began to replace other traditional materials. Sometimes old correspondence was pressed into service, as were pages from old books, magazines, newspapers, and catalogs. People also used old paper bags, envelopes, and other bits of scrap paper,which were cut into pieces and threaded onto a string that was kept in the privy.

Toilet paper is a fairly modern invention, making its debut around 1880 when it was developed by the British Perforated Paper Company. Made of a coarser paper than its modern incarnation, it was sold in boxes of individual squares. In America, the Scott Paper Company made its Waldorf brand toilet paper in rolls as early as 1890. The first rolls were not perforated, and lavatory dispensers had serrated teeth to cut the paper as needed. It was a nearly "unmentionable" product for years, and consumers were often embarrassed to ask for it by name or even be seen buying it. Timid shoppers simply asked for "Two, please," and the clerk presumably knew what they wanted. To keep things discreet, toilet paper was packaged and sold in brown paper wrappers.

During the 120 years since its introduction, toilet paper has changed little, although it's

The average American uses over 100 single rolls of toilet paper—about 21,000 sheets— each year.

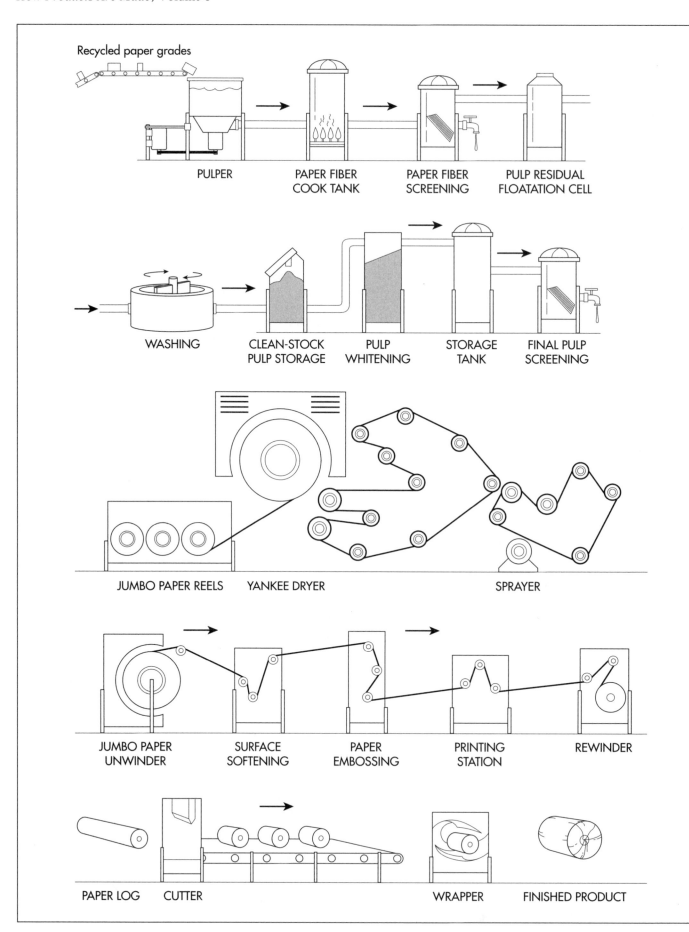

Recycled paper grades

PULPER PAPER FIBER COOK TANK PAPER FIBER SCREENING PULP RESIDUAL FLOATATION CELL

WASHING CLEAN-STOCK PULP STORAGE PULP WHITENING STORAGE TANK FINAL PULP SCREENING

JUMBO PAPER REELS YANKEE DRYER SPRAYER

JUMBO PAPER UNWINDER SURFACE SOFTENING PAPER EMBOSSING PRINTING STATION REWINDER

PAPER LOG CUTTER WRAPPER FINISHED PRODUCT

now perforated, and may be scented, embossed, or colored. Recently, toilet paper manufacturers increased the number of sheets on a roll, allowing consumers to replace the roll less frequently.

Raw Materials

Toilet paper is generally made from new or "virgin" paper, using a combination of softwood and hardwood trees. Softwood trees such as Southern pines and Douglas firs have long fibers that wrap around each other; this gives paper strength. Hardwood trees like gum, maple and oak have shorter fibers that make a softer paper. Toilet paper is generally a combination of approximately 70% hardwood and 30% softwood.

Other materials used in manufacture include water, chemicals for breaking down the trees into usable fiber, and bleaches. Companies that make paper from recycled products use oxygen, ozone, sodium hydroxide, or peroxide to whiten the paper. Virgin-paper manufacturers, however, often use chlorine-based bleaches (chlorine dioxide), which have been identified as a threat to the environment.

The Manufacturing Process

1 Trees arrive at the mill and are debarked, a process that removes the tree's outer layer while leaving as much wood on the tree as possible.

2 The debarked logs are chipped into a uniform size approximately 1 in x 1/4 in. These small pieces make it easier to pulp the wood.

3 The batch of wood chips—about 50 tons—is then mixed with 10,000 gallons of cooking chemicals; the resultant slurry is sent to a 60-ft (18.3-m) -tall pressure cooker called a digester.

4 During the cooking, which can last up to three hours, much of the moisture in the wood is evaporated (wood chips contain about 50% moisture). The mixture is reduced to about 25 tons of cellulose fibers, lignin (which binds the wood fibers together) and other substances. Out of this, about

15 tons of usable fiber, called pulp, result from each cooked batch.

5 The pulp goes through a multistage washer system that removes most of the lignin and the cooking chemicals. This fluid, called black liquor, is separated from the pulp, which goes on to the next stage of production.

6 The washed pulp is sent to the bleach plant where a multistage chemical process removes color from the fiber. Residual lignin, the adhesive that binds fibers together, will yellow paper over time and must be bleached to make paper white.

7 The pulp is mixed with water again to produce paper stock, a mixture that is 99.5% water and 0.5% fiber. The paper stock is sprayed between moving mesh screens, which allow much of the water to drain. This produces an 18-ft (5.5-m) wide sheet of matted fiber at a rate of up to 6,500 ft (1981 m) per minute.

8 The mat is then transferred to a huge heated cylinder called a Yankee Dryer that presses and dries the paper to a final moisture content of about 5%.

9 Next, the paper is creped, a process that makes it very soft and gives it a slightly wrinkled look. During creping, the paper is scraped off the Yankee Dryer with a metal blade. This makes the sheets somewhat flexible but lowers their strength and thickness so that they virtually disintegrate when wet. The paper, which is produced at speeds over a mile a minute, is then wound on jumbo reels that can weigh as much as five tons.

10 The paper is then loaded onto converting machines that unwind, slit, and rewind it onto long thin cardboard tubing, making a paper log. The paper logs are then cut into rolls and wrapped packages.

Recycled toilet paper

Toilet tissue made from recycled paper is made from both colored and white stock, with staples and pins removed. The paper goes into a huge vat called a pulper that combines it with hot water and detergents to turn it into a liquid slurry. The recycled pulp then goes through a series of screens and rinses to remove paper coatings and inks.

Opposite page:
A diagram of toilet paper manufacturing process.

The pulp is whitened somewhat and sanitized with oxygen-based products like peroxide. It then goes through steps 7 through 10 like virgin paper products, producing a cheaper, less-white paper.

Quality Control

Paper companies often maintain their own tree stands in order to ensure the quality of the paper they manufacture. The chemicals used in the pulping process are also carefully tested and monitored. Temperatures at which a slurry is cooked is ensured, too, by checking gauges, machinery, and processes. Completed paper may be tested for a variety of qualities, including stretch, opacity, moisture content, smoothness, and color.

Byproducts/Waste

The first waste product produced in the papermaking process, the bark removed from tree trunks, burns easily and is used to help power the paper mills. In addition, black liquor, the fluid removed from the pulp after cooking, is further evaporated to a thick combustible liquid that is also used to power the mill. This reduction process, in turn, yields a byproduct called tall oil that is widely used many household products. About 95% of the cooking chemicals are recovered and reused.

But other problems associated with the industry are less easily solved. The production of virgin toilet paper has spawned two current controversies: the destruction of trees, and the use of chlorine dioxide to bleach the paper. While virgin paper processing does necessitate the destruction of trees, they are a readily renewable resource and paper companies maintain large forests to feed their supply. Despite this, some activists have proposed that toilet paper be manufac-

tured only from recycled products and suggest that consumers boycott toilet paper made of new materials.

These activists object to new paper processing because it often uses chlorine bleaching, which produces dioxins, a family of chemicals considered environmental hazards, as a byproduct. Paper and pulp mills are the primary producers of dioxins, and manufacturers must carefully assess their effluvia to counteract the emission of dioxins. Increasingly, virgin paper makers use alternative bleaching methods that substitute oxygen, peroxide, and sodium hydroxide for chlorine. Some simply reduce the amount of chlorine used in the process. Others experiment with cooking the wood chips longer, removing more lignin earlier in the process, which requires less bleach. Better pulp washing also removes more lignin, and reduces the amount of bleach needed for whitening.

Where to Learn More

Books

Ierley, Merritt. *The Comforts of Home.* New York: Clarkson Potter, 1999.

Muir, Frank. *An Irreverant and Almost Complete Social History of the Bathroom.* New York: Stein and Day Publishers, 1983.

Other

Charmin Bathroom Tissue. http://www.charmin.com (January 2001).

Georgia Pacific. Student Resources. http://www.gp.com/resourcecenter/process.html (January 2001).

Marcal Paper Products. http://www.marcalpaper.com (January 2001).

Toy Model Kit

Background

Scale models or model kits are produced by the millions and give hours of pleasure to hobbyists. They're sold in specialized hobby and craft stores as well as toy, department, and drug stores—even supermarkets may carry them. "Scale" indicates that the model is a miniaturized version of an actual object, like an automobile, made to a specific fraction of the real thing. Scales typically range from 1:24 to 1:100; a 1:24-scale model is 1/24th of the size of the real object. Common scales are 1/24, 1/48, 1/72, and 1/100. Many of the same engineering skills go into producing scale models as into the real machines represented.

Model kits are made in approximately five skill levels. Snap-together models provide all the pieces with tiny tabs that fit in specially shaped holes. The plastic pieces are made in the same colors as the original; and virtually no tools, adhesives, paint, or other equipment are needed to build them. The plastic parts simply snap together in a sequence depicted on an instruction sheet; these models are an excellent introduction to the kinds of pieces in a model kit and prepare a novice builder for the next step. The next three levels are beginner, intermediate, and advanced. The level is described on the box containing the kit, and, as the names suggest, the kits become increasingly complicated with more steps in assembling and detailing the models. The most advanced kits are called customized kits and provide the model builder with a variety of styles of engines, bumpers, hubcaps, and other details of an automobile (for example), as well as options for making a truly unique model suiting the builder's imagination.

History

Model-making is as old as civilization. Scale models of buildings, boats, and furniture were buried in tombs of ancient Egyptians to represent possessions the dead took into the next world. Many ancient models survived in the tombs while the original objects did not; these have given historians an understanding of what life in ancient civilizations was like. During the Napoleonic Wars (1799–1815), French prisoners of war carved beautiful model warships from wood scraps; these models are so detailed that they have become documents of warfare and ships lost at sea. They are also highly prized today among antique collectors.

During the Industrial Revolution of the eighteenth and nineteenth centuries, inventors of new tools, machines, artworks, and other objects began by building models of their ideas. Industrial technology was found to have its own beauty, and large machines like locomotives were admired and copied in miniature. Early in the twentieth century model ships and airplanes were sold in kits. Balsa wood pieces were machine-cut to fit together easily and could be painted and rigged like the originals, although they were fragile. In the 1920s, some firms produced scale models of their products out of metal and wood as promotional models. Citroën, the French car manufacturer, produced delightful models that are now valued collectibles.

World War II moved modeling into a full-scale industry and hobby for two reasons. First, plastics were invented and perfected during the years before and during the war. Their versatility made them ideal for mass-produced model kits. Second, the machines of the war stimulated the public's interest in

In 1951, Revell introduced its first all-plastic model kit of an early automobile: a classic 1910 Maxwell, in which the driver was a scaled-down version of radio comedian Jack Benny.

modeling. Slim fighters and heavy bombers intrigued many hobbyists who saw the real aircraft flying overhead on the way to war. The exploits of navy ships, both small and large, in the Pacific also fired imaginations. When soldiers and sailors returned home, they had more time and money for recreation, including building models of the machines they knew so well. Monogram Models introduced its first kits of warships in 1945.

The returning war veterans were also able to afford the automobiles that rolled off Detroit assembly lines; they also built models of the cars they owned—and the ones the dreamed of. In 1951, Revell introduced its first all-plastic model kit of an early automobile: a classic 1910 Maxwell, in which the driver was a scaled-down version of radio comedian Jack Benny. By the mid-1950s, more detailed kits and models that could be customized appeared. By the 1960s scale modeling was a full-fledged hobby with thousands of models covering hundreds of subjects. By 2000, the scale-model industry had produced more automobiles than all the automotive giants of Detroit combined.

In the 1960s, the scale-model industry expanded into ancient history and science fiction. Models of dinosaurs that once roamed the earth, monsters like Godzilla, superheroes like Superman, television characters like the Lone Ranger, and celebrities like Elvis Presley were mass-produced in scale form. The scale-model manufacturers also provided outlets for the public's interest in the Space Race during the 1960s, and models of the newest spacecraft were often on the hobby store shelves before the real-life vessel had taken flight. Fantasy followed here, too, with models of starships and intergalactic craft that have flown on television and in the movies. As techniques for precision casting of true-color parts continued to improve, scale models became important teaching tools. Detailed anatomical figures that can be snapped apart and reassembled are members of many classrooms, as just one example.

Raw Materials

Plastic is the essential raw material for the majority of scale models made today. The plastic used to mold the parts is purchased in bulk quantities by the manufacturer. These pellets are purchased in common colors, but additional pigments can be added to the plastic while it is being melted. Printed items, including decals, the instruction sheet, and the box are also important parts of model kit. The box front usually bears a full-color photo or an artist's detailed depiction of the completed model, or a photo of the original object (like an automobile). Designs, layouts, photos, and artwork are prepared by artists in the model-maker's design studio; they are printed by specialized printers.

Paints and glue for assembling and detailing the model are not part of the kits. These are manufactured and sold separately because they have shorter shelf-lives, might leak inside the box, and could damage model parts as they move. Major model manufacturers produce or market their own brands of paint, glue, and assembly tools.

Other raw materials for models include balsa wood. Before the development of plastics, this was the prime material for model building, and it was sold in lengths and widths like small boards. Sometimes the balsa wood had outlines stamped on it that the builder could follow when cutting out pieces; other times an appropriate assortment of wood was packaged together with a set of patterns and instructions printed on paper for the model builder to follow. More often, the model builder used imagination and skill to scale, cut, assemble, and detail wood models. Balsa wood is still sold for model building, but it is far outclassed in volume by the plastic model industry.

Design

From inception to production, creating a new scale model kit can take a full year and cost the manufacturer an investment of up to $250,000. For the scale model of a classic automobile, for example, the design process begins with pure research: taking hundreds of photographs of a working example of the car. The car is photographed inside and out; pictures are taken of every detail and from many angles, and measurements are photographed along with the object for the designer to use in the studio to reproduce the car exactly.

For a new car design, car manufacturers give model makers computerized information on part specifications—sometimes even before the first actual automobile has been assembled—in a highly confidential process. The model designer uses computer-aided drafting and design (CADD) software to sort this information and create the measurements and configuration that will be used to make a scale design. The designer transfers this information to a set of drawings that will be used to make the molds for the model. This process can consume several hundred hours of engineering time.

From photographs, computer data, and paper drawings, the design moves to pattern-making phase. Skilled artists follow the designer's drawings and carve out a pattern model from balsa or other soft wood. The pattern model is made at two to three times larger than the scale of the model kit, allowing additional details to be added to the pattern. This also proves the accuracy of design and provides a basis for all of the molds that will be made of the car parts. As they carve the pieces, the pattern makers fit them together. Accuracy in pattern-making is within several ten-thousandths of an inch (fractions of a millimeter).

When the wooden pattern model is complete, each part is coated with an epoxy resin, a plastic material that hardens as it cures. The wooden piece is removed from the resin, and the resin has trapped the shape of the piece in a cavity mold. A core mold of resin is made from the cavity mold; the two fit closely together, but there is a small space between them. The plastic model part will be formed in this space. Preparation of the wooden pattern and the resin molds takes over 1,000 hours.

Meanwhile, other design engineers use the design drawings to lay out the "tool," the metal cast of a number of parts that will be molded on a single form of plastic called a tree. The tree is usually roughly rectangular along its outer edge so it will fit in a box. Several standard box sizes are used. From the central "trunk" of the tree, a number of plastic "branches" or arms protrude. The end of each branch narrows to a node where it joins a piece of the model. The model piece can be snapped off for assembly. The tool designers also use CAD to map out the tool layout. They must design the orientation of each model piece on the tree to pre-

cise angles so that, when the plastic is injected in the tool, it fills all the cavities. The trees are also designed to release quickly from the mold.

The resin molds are used to make the individual tools for each of the model parts, using a pantograph to copy the exact shape of each piece and draw it at the smaller scale of the actual model. The pantograph has two needle-like parts: one is run over the surface of the resin mold while the second, a cutting blade, carves the steel to the same shape at the correct scale. When the tool maker has completed the scaled tool, he polishes it to a high sheen and adds more details by hand. Some of these details are too fine to be seen by the naked eye.

Another set of designers works on the paper portions of the kit. To develop the instructions, the designers take the pattern model (and sometimes samples of the first production run of the plastic model) apart and reassemble it. They describe the steps as they go, writing and drawing them as instructions. Other artists look at the photographs of the real automobile and design decals for the model. These may be copies of real decals on the car, or they may be other design features like racing stripes. Sometimes, more research is needed to capture these details. The illustrations on the box lid are also created. These serve the model builder as a color reference guide, so they must be true to both the original automobile and the decals made for the model. The illustrations on the lid may be photographs of the real car or artists' impressions. The box lid for a model of a Fokker triplane, for example, may show the Red Baron's famous airplane in a dogfight. Elsewhere on the lid, the artists describe the kit, its level of difficulty, the parts enclosed, and the manufacturer's details.

The Manufacturing Process

1 With the many design steps now complete, the individual steel tools for the separate parts of the model are taken into the factory where they are placed in a larger frame called the die, which can weigh more than 1/2 tons (1.4 metric tons). With the two halves fitted together, the die looks like a large steel box from the outside. It is lifted

An example of a toy model kit.

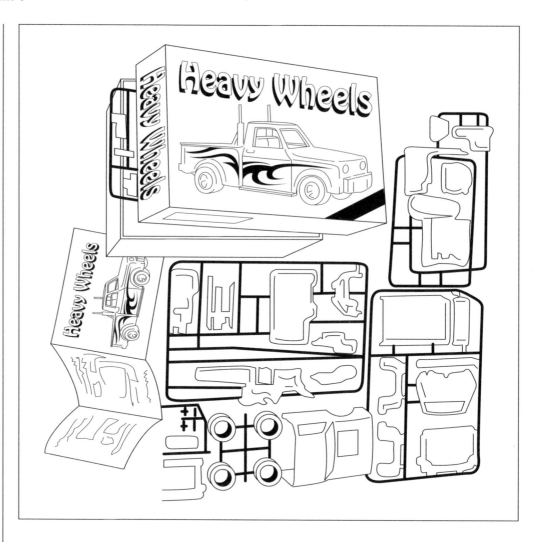

with a hoist into the cavity of the injection molding machine.

2 In a separate part of the injection-molding machine, the fine, confetti-like pellets of plastic are poured into a storage hopper above the machine. When molding is about to begin, the pellets are fed into a heating chamber where they are melted at a temperature of 500° F (260° C). The liquid plastic is pumped into the cavity of the injection molding machine where the tool is contained. The pressurized liquid plastic is injected into the tool at a pressure of 1,000 lbs per square in (58 kg per square cm). Aided by the careful layout within the tool, the combination of high pressure and high heat forces the liquid plastic into every crevice of the tool. In several seconds, the tree forms, the plastic cools, and the tool opens slightly allowing the plastic tree to fall out onto a conveyor belt. The injection machine closes the tool again, and more plastic is injected for the next tree of parts. The trees are injection-molded at a rate of about two trees per minute.

Any gaps around the edges of each tree will also fill with plastic. These appear on the trees as paper-thin excess and are called flash. When the model builder begins making a model, the flash must be trimmed off.

3 At another machine, the clear plastic parts like the automobile windshields, are molded. Rubber tires are produced by a third machine, and other specially colored parts may be produced by yet another machine in the plant. Chrome parts, like the radiator grille, are first injection-molded then placed on the conveyor where they move to a dipping machine that picks individual trees off the conveyor and dips them in a metallic coating. The dip dries quickly, the machine redeposits the chrome-colored trees on the conveyor, and they move toward the package assembly area.

4 At the package assembly area, assembly-line workers sit along one side of a conveyor belt that feeds box bottoms to them. They select one of each of the trees of model parts from a series of bins behind them, slip the trees into plastic bags, add sheets of decals and folded instructions, close the bag, and put it in the bottom of the box. The bottoms travel along the conveyor to a machine that caps them with lids and seals the box ends or covers the boxes with shrink wrap. Boxes are multipacked in cartons and taken to the shipping department.

Quality Control

Making scale models is design intensive, with much of the process taking place in the design studio rather than the factory. Designers are extremely knowledgeable about their subjects, whether warships or cartoon heroes, and are highly attentive to the details in their designs. Designers can reject work at any point in the process. A review board approves the initial design concept, the design drawings, pattern and tool making, and samples from the first production run of the plastic model. In the factory, quality-control personnel observe the injection molding and dipping processes. Quality-control personnel observe the assembly line, and when kits have been assembled, sample boxes are pulled from the line and checked to confirm that all parts are present.

Byproducts/Waste

Toy model manufacturers are able to recycle much of the waste associated with manufacturing plastic models. Plastic waste can be melted and remixed in future batches. Steel from tool making is collected and sent to a metal recycler. Paper products are made from recycled paper.

Safety Concerns

Safety for factory workers is particularly important for the pattern and tool makers and personnel that control the injection-molding process. Pattern and tool makers wear gloves and safety glasses as they carve out patterns, make resin molds, and etch steel tools with the pantograph. The area where resin molds are made is equipped with ventilation hoods. Operators of the injection-molding machines wear heavy gloves to protect them from the heat generated by the machines. The machines themselves are fully automated and require observation only.

The Future

The health of the toy model kit industry has a number of positive indicators. In the 1990s, modeling giants Revell and Monogram merged, and, as of 2000, made 10 million kits a year. Model-making clubs like the International Plastic Modelers' Association have branches in many cities, in large businesses, and on the Internet. Fairs and other displays often have model-building contests as well as exhibitions of models and dioramas; these include not only "scratch-built" models (the highest level in which the hobbyist makes all the parts and details by hand to complete a model) but manufactured kits from snap-together through customized models.

The fascination of the toy model kit has several elements. It is a construction hobby that requires handwork and skill to shape a collection of small pieces into a completed work of art. The investment required can begin with only pocket money for a few tools, finishes, and a kit. Toy model kits are outstanding teaching tools. Research has gone into their design, and students can learn from personal research about the actual object such as when the automobile was manufactured, which airlines fly particular plane designs, and how a comic strip artist created his character. With the physical parts provided in the model kit, the model builder can use patience, thought, handwork, and imagination to transform history, science, and popular culture into three-dimensional models of objects that were or that may be in the future.

Where to Learn More

Books

Bowen, John, ed. *Scale Model Sailing Ships*. New York: Mayflower Books, Inc., 1978.

Ellis, Chris. *How to Make Model Aircraft*. New York: Arco Publishing Company, Inc., 1974.

Ellis, Chris. *The Scale Modeler's Handbook*. Secaucus, NJ: Chartwell Books, Inc., 1979.

Gordon, Theron L. *How to Build, Customize & Design Plastic Models.* Blue Ridge Summit, PA: TAB Books, Inc., 1982.

Harris, Jack C. *Plastic Model Kits.* New York: Crestwood House, Macmillan Publishing Company, 1993.

Marmo, Richard. *Building Plastic Model Aircraft.* Blue Ridge Summit, PA: TAB Books, Inc., 1990.

Price, Brick. *The Model Shipbuilding Handbook.* Radnor, PA: Chilton Book Company, 1983.

Other

Revell-Monogram. http://www.revell-monogram.com (January 2001).

Testor Corporation. http://www.testors.com (January 2001).

—*Gillian S. Holmes*

Trophy

Background

Trophies are a category of awards given primarily for academic, work, and sport contests or events. They are physical evidence that one person or group has bested another in some contest. Imposing and sculptural, trophies often include a figure, sports equipment, or animal (for agricultural fairs) associated with the contest in which the winner has excelled. They are tangible evidence of prowess and have extraordinary meaning for their recipients. They're proudly displayed in homes and schools.

Trophies range from inexpensive to almost priceless. They can be unique, like the one-of-a-kind Stanley Cup awarded to each year's National Hockey League's champion, or mass-produced, molded plastic figures costing less than a dollar. Each is considered a treasure regardless of its monetary value.

Once of metal atop a wooden or metal base, many trophies are now made of molded plastic colored to resemble gold, silver, or **brass**. Trophy parts are manufactured within a factory but assembled by award dealers (retailers) who sell to the public. As a result, there is an astonishing array of parts that may be purchased, allowing the retailer to build an award to meet every customer's needs. For example, trophies may be purchased with silvertone, goldtone, or clear plastic figurines; and with marble-like or wood-like bases. The figures atop trophies are currently available in hundreds of forms, including suit-attired saleswomen, Irish step dancers, bait casters, pistol marksmen, and women's lacrosse, to name just a few.

History

Since ancient times trophies have marked victories. In fact, the word "trophy" is derived from the Greek *tropaion*, which comes from the verb *trope*, meaning "to rout." In ancient Greece, trophies reflected victory in war, and were created on the battlefield at the place where the enemy had been defeated. These trophies included captured arms and standards, and were hung upon a tree or a large stake made to resemble the figure of a warrior. They were inscribed with details of the battle and were dedicated to a god or gods. Naval trophies consisted of entire ships (or what remained of them) laid out on the nearest beach to represent the conquest. The deliberate destruction of a trophy was considered a sacrilege since it was given in thanks and tribute to a god.

The ancient Romans wanted to keep their trophies closer to home. Instead of a tribute to the victors and the gods on the spot of the victory, the Romans had special trophies constructed in Rome. These magnificent trophies often included columns and arches atop a foundation. There remain, still, outside of Rome, huge stone memorials that were originally crowned with sculpted stone trophies, now gone.

Little is known about awards or trophies given during the Middle Ages. Chalices, or two-handled cups, were given to winners of sporting events as early as at least the very late 1600s in the New World. An exquisite, small two-handled sterling cup in the Henry Ford Museum in Dearborn, Michigan, was given to the winner of a short horse race between two towns in New England about 1699 and is called the Kyp Cup (made by sil-

In ancient Greece, trophies reflected victory in war, and were created on the battlefield at the place where the enemy had been defeated.

versmith Jesse Kyp). Chalices, particularly, are associated with sport events, and were typically made in silver and given in horse racing, and later, boating and early automobile racing (which became popular over a century with the birth of the automobile). Sporting awards often take the form of a cup, including the Davis Cup, a major tennis trophy first awarded in 1900, the Stanley Cup, given to National Hockey League champs (1894), and the World Cup, given since 1967 to top male and female alpine skiers, to name a few.

Trophies are less expensive and awarded more frequently, thanks to manufactured plastic trophies. In addition to having a variety of figures from which to choose, trophies are of three primary forms: the clear plastic **action figure** crowning the base, the gold-colored or silver-colored action figure atop the base, or a rectangular plinth of plastic that is holographically decorated by computer with words or graphic of the event, equipment, or sport scene and placed on the plastic base.

Raw Materials

Trophies are produced almost exclusively from plastic; one trophy may include several different types. Hot-stamp metallic foils are pressed into the columnar shafts to impart to give the figurine metallic color (unless it's clear plastic). Gypsum is inserted into the base and metal studs are molded into the trophies to give them strength.

Design

A trophy is not designed as an entire piece; instead, it is broken into several components that are designed and redesigned; an assembler then chooses individual components to make a trophy. First there is the base upon which the entire trophy sits. This is often plastic made to look like marble or wood. Bases are generally categorized as crescent, sculpted, tiered, or a specialty form. Next comes the column or columns. These are the vertical piece or pieces on which the figurine may rest, or they may hold yet another tier upon which the figurine sits. These are often plain metal, imitation marble, or holographs. Next comes the riser, a small, decorative element that sits upon the base and between the columns. The riser may indicate the recipient's position: first, second or third. Some trophies have another tier atop the columns upon. Finally, on top of everything, comes the figurine.

One large trophy manufacturer describes their design process as four-part: talking to customers, brainstorming new ideas, producing the idea visually or physically, and then executing some models for testing. First, the new products division talks to consumers. They gather ideas on what components are popular, changes that might be made, figurine needs not reflected in the current catalog, etc. Customer opinion is gathered through market research and focus groups. Then, new product developers articulate these needs and also brainstorm possibilities for redesigned components. The most promising ideas are discussed with a team of developers who can help make the idea real. This group includes sculptors, graphic artists, conceptual designers, and design engineers. These artists and engineers produce either a drawing or a basic sculpture of a riser or figurine that gives the idea visual or three-dimensional form.

Once the concept is approved by a committee, the steel tool-and-die department creates the die for the new plastic part. If the part is a not tooled, like a Mylar plinth or plaque, then a flat die is used. The appropriate material is chosen for the component based upon durability and its intended function (e.g., support or decoration) on the trophy. The new design is then carefully assessed by a committee that scrutinizes it for design aesthetic, manufacturing difficulty, tooling needs, durability, and decoration. If there is a problem in any of these areas, the part is returned for reassessment. Of particular concern is ease of manufacturing. Reassessment and adjustments may take up to four weeks alone. Once final approval is received, the new part is ready for mass-production.

The Manfacturing Process

1 Different parts of the trophy are produced in different ways. The components are either molded using steel dies, or extruded through a die using pressure and some heat. Most of the parts, including the risers, base, and figurine, are injection molded. To mold a plastic part, an automatic feeder system is fed a continuous stream of plastic pellets.

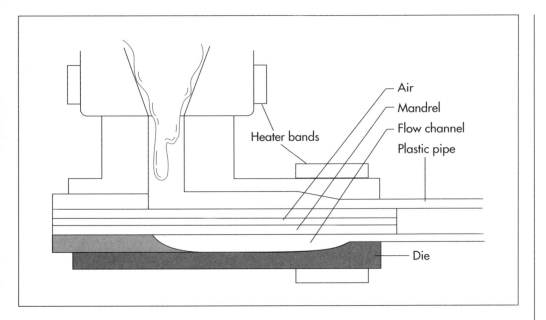

Heater bands

Air
Mandrel
Flow channel
Plastic pipe

Die

The machines are loaded with many millions of pounds of plastic pellets each year to make various parts of the trophy.

2 The molding machine is fitted with very expensive specially designed steel dies. The machines melt the pellets into a liquid and, using extraordinary pressure against the dies, form the trophy bases, risers, and figures. The dies form between up to 12 components of a single design per cycle (the number varies on the size of the component). The machines are operated by a worker who oversees the production. A metal stud is inserted into the body of each of these figures to ensure the strength of this component as parts of the figure (ankle, wrist, neck) may be of thin plastic and subject to breakage. One of the largest manufactories of trophy parts runs 40 such molding machines six days a week, three shifts a day.

3 Bases must then be filled with gypsum to give the base a proper weight without using too much brittle plastic. Figurines that are not to be given a silver- or gold-tone finish are essentially done and are pushed through the machine and out, ready to be assembled at the retailers.

4 Figures that are to receive a metallized finish are put operator onto the hot stamp foil machine. The plastic figures are washed with a top coat, then metal foil (which comes in 54-in (1.4-m) wide rolls and comes off in linear feet) is heated and

pressed onto them. Those parts that are to be colored as well as metallized are decorated in a similar fashion. The figures are now moved away from the machine, ready to be boxed and shipped to assemblers.

5 The columns are form from plastic pellets that go into a machine, melted, then forced (extruded) through a formed die. As they go through the die they're cut to the required length. These, too, then move away from the machine ready for boxing. made of extruded plastic

6 The components are automatically bagged and boxed by a machine and ready for shipment to the assemblers.

Quality Control

Precision machinery, including the steel dies, are inspected, maintained, and cleaned regularly to insure the production of accurate, high-quality parts. All materials, from the rolls of metal foils to the various plastics used in production, are approved upon arrival at the factory. Once the trophies have been assembled, a series of inspectors check them as they come off the line, looking for unacceptable variations in the molded parts or improper color or foil application.

Byproducts/Waste

Parts that do not pass quality inspection are sorted by plastic type and applied color, then reground for use at a later time in another product or component. Important parts like the figures and clear vinyl or acrylic plinths for specialty trophies, however, are made of new materials only.

Plastic molding and extruding machines do kick up a certain amount of ambient plastic materials. Air scrubbers are used inside such factories. Water ejected from such plants may contain effluvia and plastic molding companies reclaim and clean the water from the plant.

Where to Learn More

Other

PDU (Plastic Dress Up) Millenium Edition, c.2000 Trophy Parts Catalog.

Rousseau's Sporting Goods and Awards. http://www.r-sports.com (January 2001).

STAMPFOIL. http://www.stampfoil.it/english.htm (January 2001).

—Nancy E.V. Bryk

Tunnel

Background

A tunnel is an underground or underwater passage that is primarily horizontal. Relatively small-diameter ones carry utility lines or function as pipelines. Tunnels that transport people by rail or by automobile often comprise two or three large, parallel passages for opposite-direction traffic, service vehicles, and emergency exit routes.

The world's longest tunnel carries water 105 mi (170 km) to New York City from the Delaware River. The lengthiest person-carrying tunnel is the Seikan Railroad Tunnel. It is a 33-mi (53-km) long, 32-ft (9.7-m) diameter railroad connection between Japan's two largest islands, Honshu and Hokkaido.

One of the most anticipated tunnels was the Channel Tunnel. Completed in 1994, this tunnel connects Great Britain to Europe through three, 31-mi (50-km) long tunnels (two one-way and one service tunnel). Twenty-three miles (37 km) of this tunnel are underwater.

History

Tunnels were hand-dug by several ancient civilizations in the Indian and Mediterranean regions. In addition to digging tools and copper rock saws, fire was sometimes used to heat a rock obstruction before dousing it with water to crack it apart. The cut-and-cover method—digging a deep trench, constructing a roof at an appropriate height within the trench, and covering the trench above the roof (a tunneling technique still employed today)—was used in Babylon 4,000 years ago.

The first advance beyond hand-digging was the use of gunpowder to blast a 515-ft (160-m) long canal tunnel in France in 1681. The next two major advances came about 1850. Nitroglycerine (stabilized in the form of dynamite) replaced the less powerful black powder in tunnel blasting. Steam and compressed air were used to power drills to create holes for the explosive charges. This mechanization eventually replaced the manual process made famous by John Henry, the "steel-driving man," who swung a 10-lb (4.4-kg) sledge hammer with each hand for 12 hours a day, pounding steel chisels as deep as 14 ft (4.2 m) into solid rock.

Between 1820 and 1865, British engineers Marc Brunel and James Greathead developed several models of a tunneling shield that enabled them to construct two tunnels under the Thames River. A rectangular or circular enclosure (the shield) was divided horizontally and vertically into several compartments. A man working in each compartment could remove one plank at a time from the face of the shield, dig ahead a few inches, and replace the plank. When space had been dug away from the entire front surface, the shield was pushed forward, and the digging process was repeated. Workers at the rear of the shield lined the tunnel with bricks or cast iron rings.

In 1873, American tunneler Clinton Haskins kept water from seeping into a railroad tunnel under construction below the Hudson River by filling it with compressed air. The technique is still used today, although it presents several dangers. Workers must spend time in decompression chambers at the end of their shift—a requirement that limits emergency exits from the tunnel. The pressure within the tunnel must be carefully balanced with the surrounding earth and water pressure; an imbalance causes the tunnel either to collapse or burst (which subsequently allows flooding).

Soft soil is prone to collapse and it can clog digging equipment. One way to stabilize the soil is to freeze it by circulating coolant through pipes embedded at intervals throughout the area. This technique has been used in the United States since the early 1900s. Another stabilization and waterproofing technique—widely used since the 1970s—is to inject grout (liquid bonding agent) into soil or fractured rock surrounding the tunnel route.

Shotcrete is a liquid concrete that is sprayed on surfaces. Invented in 1907, it has been used as both a preliminary and a final lining for tunnels since the 1920s.

In 1931, the first drilling jumbos were devised to dig tunnels that would divert the Colorado River around the construction site for Hoover Dam. These jumbos consisted of 24-30 pneumatic drills mounted on a frame welded to the bed of a truck. Modern jumbos allow a single operator to control several drills mounted on hydraulically controlled arms. In 1954, while building diversion tunnels for construction of a dam in South Dakota, James Robbins invented the tunnel boring machine (TBM), a cylindrical device with digging or cutting heads mounted on a rotating front face that grinds away rock and soil as the machine creeps forward. Modern TBMs are customized for each project by matching the types and arrangement of the cutting heads to the site geology; also, the diameter of TBM must be equal to the diameter of the designed tunnel (including its lining).

Raw Materials

Materials used in tunnels vary with the design and construction methods chosen for each project. Grout used to stabilize soil or fill voids behind the tunnel lining may contain various materials, including sodium silicate, lime, silica fume, cement, and bentonite (a highly absorbent volcanic clay). Bentonite-and-water slurry is also used as a suspension and transportation medium for muck (debris excavated from the tunnel) and as a lubricant for objects being pushed through the tunnel (e.g., TBMs, shields). Water is used to control dust during drilling and after blasting, which is often done with

a low-freezing gelatine explosive. Water-and-salt brine or liquid nitrogen are common refrigerants for stabilizing soft ground by freezing. The most common modern lining material, concrete reinforced by either steel or fiber, may be sprayed on, cast in place, or prefabricated in panels.

Choice of method

A tunnel's construction method is determined by several factors, including geology, cost, and potential disruption of other activities. Different methods may be used on individual tunnels that are part of the same larger project; for example, four separate methods are being used on portions of Boston's Central Artery/Tunnel project.

The Manufacturing Process

Preparing

1 Site geology is evaluated by examining surface features and subsurface core samples. A pilot tunnel about one-third the diameter of the planned main tunnel may be constructed along the entire route to further evaluate the geology and to test the selected construction method. The pilot tunnel may run alongside the main tunnel's path and eventually be connected to it at intervals to provide ventilation, service access, and an escape route. Or the pilot tunnel may be enlarged to produce the main tunnel.

2 If soil stabilization is required, it may be done by injecting grout through small pipes placed in the ground at intervals. Alternatively, a refrigerant may be circulated through pipes embedded in the ground to freeze the soil.

Mining

3 There are seven different methods used to remove material from the tunnel path. The first is the immersed tube method. Workers prepare an underwater tunnel site by digging a trench at the bottom of the waterway. Steel or reinforced concrete sections of tunnel shell are constructed on dry land. Each section may be several hundred feet (100 m or more) long. The ends of the section are sealed, and the section is floated to the tunnel site. The section is tied to anchors adjacent to the trench, and ballast tanks built

into the section are flooded. As the section sinks, it is guided into place in the trench. The section is connected to the adjoining, previously placed section, and the plates sealing that end of each section are removed. A rubber seal between the two sections ensures a watertight connection.

In the cut-and-cover method workers dig a trench large enough to contain the tunnel and its shell. A box-shaped tube is constructed, often by in-place casting of reinforced concrete. In certain types of soil or in close proximity to other structures, tunnel walls may be built before digging begins in order to keep the trench from collapsing during excavation. This may be done by driving steel sheets into the ground or building a slurry wall (a deep trench that is filled with watery clay as dirt is removed). When the desired size is attained for a section of wall, a cage of steel reinforcing rods is lowered into it and concrete is pumped in to displace the wet-clay slurry. As digging progresses enough for the excavation machinery to be below grade, temporary surface panels may be laid across the trench to allow traffic to move across it. When the tunnel shell has been completed, it is covered by replacing excavated soil.

The third method is the top-down method. A parallel pair of walls are embedded into the ground along the tunnel's route by driving steel sheet piles or constructing slurry walls. A trench is dug between the walls to a depth equal to the planned distance from the surface to the inside of the tunnel roof. The tunnel roof is formed between the walls by framing and pouring reinforced concrete on the bottom of the shallow trench. After the tunnel roof has cured, it is covered with a waterproofing membrane and excavated soil is replaced above it. Conventional excavating machinery, such as a front-end loader, is used to dig out the soil between the diaphragm walls and under the tunnel roof. When sufficient depth has been reached, a reinforced concrete floor is poured to complete the tunnel shell.

With the drill-and-blast method a drilling jumbo is used to drill a predetermined pattern of holes in the rock along the tunnel's path. Carefully planned charges of dynamite are inserted in the drilled holes. The charges are detonated in a sequence designed to

The Eurotunnel.

Construction on the English Channel Tunnel between England and France, a dream for centuries envisioned and encouraged by Napoleon, was begun in 1987. Originally referred to as the Chunnel and now known as the Eurotunnel, it was completed in 1994 at a cost of $13 billion. The two rail tunnels (one for northbound and one for southbound traffic) and one service tunnel are each 31 mi (50 km) in length and have an average depth of 150 ft (46 m) under the seabed. It is the first physical link between Britain and the European continent. Passenger rail service is provided, as well as the ferrying of automobiles and trucks. Travel times from London to Paris have been reduced from more than five hours (over sea) to three hours via the Eurotunnel.

The Seikan Tunnel in Japan was placed in service in 1988. The 33mi-(53km-)long tunnel connects the northern tip of Japan's main island of Honshu with the island of Hokkaido, passing under the Tsugaru Strait. The Seikan Tunnel is the world's longest submarine tunnel, involving excavation 330 ft (100 m) below the seabed across a strait where the sea is up to 460 ft (140 m) in depth.

break away material from the tunnel's path without unduly damaging the surrounding rock. Air is circulated through the blast area to remove explosion gases and dust. Rubble dislodged by the blast is hauled away. Pneumatic drills and hand tools are used to smooth the surface of the blasted section and remove loose pieces of rock.

It is usually necessary to stabilize and reinforce the surface of the newly blasted section with a preliminary lining. One technique involves inserting a series of steel ribs connected by wood or steel braces. Another technique, called the new Austrian tunneling method (NATM), involves spraying the surface with a few inches (several centimeters) of concrete. In appropriate geologic condi-

Shield tunneling.

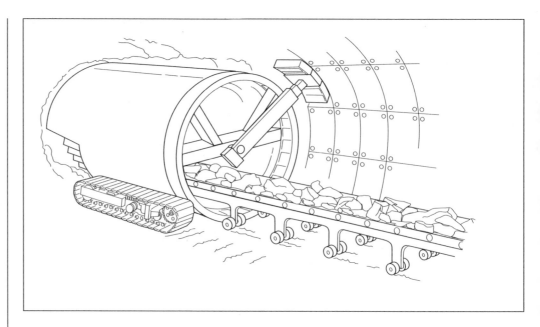

tions, this "shotcrete" lining may be supplemented by inserting long steel rods (rock bolts) into the rock and tightening nuts against steel plates surrounding the head of each bolt.

A fifth method to remove material from the tunnel is the shield driving or tunnel jacking method. Some tunnels are still dug using a Greathead-style shield. The top of the shield extends beyond the sides and bottom, providing a protective roof for workers digging in advance of the shield. The leading edge of the shield top is sharp so it can cut through the soil. Excavation may be done by hand or with power tools. Excess material is passed back through the shield on a conveyor belt, loaded into carts, and hauled out of the tunnel. When workers have dug out material in front of the shield as far as the top extends, jacks at the rear of the shield are braced against the most recently installed section of tunnel lining. Activating the jacks pushes the shield forward so workers can begin digging another section. After the shield has moved forward, the jacks are retracted, and steel or reinforced concrete ring segments are bolted into place to form a section of permanent lining for the tunnel.

Tunnel jacking is a similar technique, but the shield being driven through the ground is actually a prefabricated section of tunnel lining.

In the parallel drift method a series of parallel, horizontal holes (drifts) are bored using microtunneling machinery (microtunnels are too small for human miners to work inside of) such as augers or small versions of TMBs. These drifts are filled; for example, steel pipes may be driven into them and then the pipes packed with grout. The filled drifts form a protective arch around the tunnel path. Excavation machinery is used to remove the soil from inside the arch.

The final method is the tunnel boring machine method. The types and arrangement of cutting devices on the face of the TBM are determined by the geology at the tunnel site. The face slowly rotates and grinds away the rock and soil in front of it (e.g., the TBMs used to build the Channel Tunnel could rotate up to 12 revolutions per minute in optimal soil). The TBM is constantly pushed forward to keep the face in contact with its target. Forward pressure may be exerted by jacks at the rear of the TBM pushing against the most recently installed section of tunnel lining. Alternatively, gripper arms may extend outward from the sides of the TBM and push against rocky tunnel walls to hold the machine in place while the face is pushed forward. Muck is passed through holes in the face and carried by conveyor belt to the rear of the TBM, where it drops into carts that transport it out of the tunnel. Bentonite may be pumped through the TBM face to make the soil surface more workable and to carry away the muck. Some TBMs are equipped at the rear with robotic arms that position and attach segments of tunnel lining as soon as the machine has moved for-

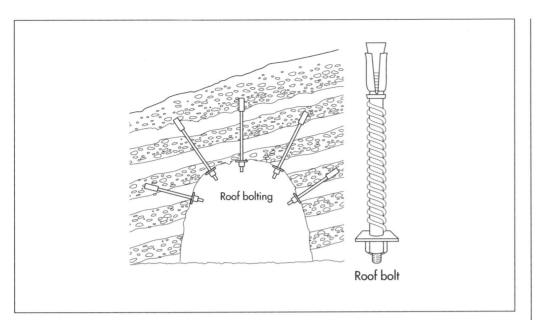

Roof bolting

Roof bolt

ward a sufficient distance. In other cases, the NATM is used to create a preliminary lining as the TBM progresses.

Especially in cases where two TBMs dig toward each other from opposite ends of a tunnel, it may be too difficult or expensive to remove them when the digging is completed. As it nears the end of its mission, the TBM may be steered away from the tunnel's path to dig a short spur in which it is permanently sealed.

Final lining

4 In some cases, the final lining is placed during the excavation process. Two examples are TBMs that install lining segments and prefabricated tunnels that are jacked into place. In other cases, a final lining must be constructed after the entire tunnel is excavated. One option is to pour a reinforced concrete lining in place. Slipforming is an efficient technique in which a section of form is slowly moved forward as the concrete is poured between it and the tunnel wall; the concrete hardens quickly enough to support itself by the time the form moves on.

A second option is to install segments of preformed concrete or steel lining, much as some TBMs do. Lining segments are constructed so that several of them can be joined to form a complete ring a few feet (a meter or two) wide. Once a ring has been bolted into place, grout is injected between it and the tunnel wall.

A third option is to spray a layer of shotcrete several inches (70 mm or more) thick onto the tunnel walls. One or two layers of wire mesh might be placed first to reinforce the shotcrete, or reinforcing fibers might be added to the concrete mixture to increase its strength.

Byproducts/Waste

Sometimes the earth removed from a tunnel is simply discarded into a landfill. In other cases, however, it becomes raw material for other projects. For example, it may be used to form the base course for an approach roadway or to create roadway embankments for wider shoulders or erosion control.

Quality Control

Besides maintaining ground stability around the tunnel and ensuring structural integrity of the tunnel lining, proper alignment of the excavation path must be achieved. Two valuable tools are global positioning system (GPS) sensors that receive precise locational data via satellite signals and guidance systems that project and detect a laser beam within the tunnel.

The Future

Exploration methods, materials, and machinery are possible areas of improvement. Sound waves transmitted through the earth can now generate a virtual CAT scan of the tunnel path, reducing the need to drill core samples and pilot tunnels. Some examples of materials research involve cutting tools that are more effective and durable, concrete with more precisely controlled hardening rates, and better processes for modifying soil to make it easier to cut, dig, or remove. Recent developments in machine technology include multiple-headed TBMs that can bore two or three parallel tunnels simultaneously and a TBM that can turn a corner up to 90° while cutting. Better remote control capabilities for digging machinery would improve safety by reducing the amount of time people have to be underground during the digging process.

Where to Learn More

Periodicals

Burroughs, Dan, et al. "Depressing Traffic Top-Down." *Civil Engineering* (January 1994): 62.

Campo, David W., and Donald P. Richards. "Tunneling Beneath Cairo." *Civil Engineering* (January 2000): 36.

Iseley, Tom. "Microtunneling MARTA." *Engineering* (December 1991): 50.

O'Connor, Leo. "Tunneling Under the Channel." *Mechanical Engineering* (December 1993): 60.

Other

The Cumberland Gap Tunnel. http://www.efl.fha.dot.gov/cumgap/tunnel.htm (January 2000).

"A Short History of Tunnelling." http://pisces.sbu.ac.uk/BE/CECM/Civ-eng/tunhist.html (January 2000).

"Tunnel Jacking." Central Artery/Tunnel Project. http://www.bigdig.com (January 2001).

—Loretta Hall

Ukulele

Background

The ukulele is a string instrument that originated in Portugal in the second century B.C. With a small, guitar-shaped body that is fitted with four strings, it is considered a member of the chordophone family. Sound is produced through these instruments by plucking and strumming the strings. The strings in turn vibrate and are amplified by the resonating body. The ukulele is manufactured in a similar way as a full size guitar.

History

String instruments date back many centuries and have been developed independently by most ancient cultures. The earliest instruments were single strings tied to bows. Evidence of these primitive instruments has been found in Asia and Africa dating back over 3,000 years. Over time, instrument makers added more strings.

Ukuleles first had their start in Portugal in 139 B.C. in the Lusitani tribe. The development of the ukulele has been influenced by instruments from Spain, South America, and Africa. By the thirteenth, century four-string instruments were being used in Spain. When six string instruments were introduced in the 1700s the popularity of chordophonesexploded. Although ukuleles are most commonlu associated with Hawaii, it wasn't until 1879 that the first ukelele was brought over from Portugal. One of the Portuguese immigrants on the ship *Ravenscrag*, João Fernandez, started playing his four-string Portuguese instrument known as a braghuina. Local residents were intrigued with the instrument, adopted it as their own, and renamed it ukulele which in Hawaiian means "jumping flea." This name reflected

the way the islanders thought the fingers jumped around the fretboard when it was played. Within 10 years of its introduction, the ukulele became the most popular instrument in Hawaii.

The first ukuleles were made by hand, a process that was both painstaking and time consuming. Subsequently, the number of ukuleles in existence was quite low prior to 1910. Eventually, special wood cutting and shaping machines were created to produce ukuleles. The instrument was steadily modified making it look and sound more like the modern day ukulele. Manuel Nunes was one of the most important innovators. He modified the instrument by replacing steel strings with gut strings. He also suggested a different tuning pattern to make chord formation easier. He also began using wood from the koa tree to produce a lighter, more resonant ukulele.

The ukulele was introduced to the United States mainland during the Panama-Pacific Exposition in San Francisco during 1915. Record sales of Hawaiian music grew rapidly and United States guitar manufacturers began selling their own version of the ukulele. By the 1920s and 1930s, the popularity of the ukulele spread throughout North America, and its sound became closely associated with vaudeville music shows. Since then, the ukulele has often been played as a jazz and solo instrument.

The plastic ukulele, called the "TV Pal" was developed by Mario Maccaferri in 1950. He was a well-known guitar maker who became intrigued with plastics. He used his instrument-making skill to produce the plastic ukulele which sold over nine million units between the time it was introduced and

In 1950 Mario Maccaferri produced the plastic ukulele and sold over nine million units between the time it was introduced and 1958.

1958. Its popularity was mainly due to the fact that it was inexpensive, had a good sound, and was tied in with the popular television show "Arthur Godfrey and his Ukulele."

Raw Materials

The body of the ukulele is primarily made from wood, although plastic instruments have also been sold. Woods from all over the world are used including Hawaiian koa, maple, walnut, rosewood, myrtle, brazilian canary, cocobolo, madrone, elm, lacewood, and black limba. The type of wood has a significant impact on the sound, tone, and quality.

For example, mahogany is a "soft" hardwood and it creates a warm, mellow tone. It is thought by many manufacturers to be the finest wood for making ukuleles. It also has excellent aging properties, sounding better as it gets older. Koa wood is the most revered of Hawaiian woods for ukulele manufacturing. These trees have unique grain patterns and colors making every ukulele made from them distinct. Typically, the same type of wood is used for the entire instrument.

Beyond the wood, other materials used in a ukulele's manufacture are nylon, steel, plastic, coatings, and glues. The strings are typically made from nylon although some ukuleles are produced with steel strings. The wood is treated with different lacquers for both protection and decoration. Various types of glue can be used such as superglue, aliphatic or yellow glue, hide glue, and epoxy. For instruments made in the tropics, synthetic adhesives are superior because they are less prone to degradation by fungus.

Design

The ukulele is a portable instrument with a small guitar-like body. It consists of a short neck, a main body, four strings and tuning keys, a bridge, a fretboard, and a sound hole. There are a variety of different types of ukuleles including the soprano, concert, tenor, and baritone. The most common type is the soprano ukulele which is about 21 in (53 cm) long. The strings are tuned to the notes G-C-E-A.

The Manufacturing Process

The various parts of the ukelele are made in separate processes and then put together in a finishing step. The process begins with wood selection which is the most important factor because it will will affect sound quality and instrument appearance.

Bookmatching

1 The front piece of the ukelele is produced by a process called bookmatching. In this method a piece of wood is cut into two equal sheets. This gives a distinct symmetrical grain pattern. At this point, the wood pieces are kiln dried before any more work is done. The wood pieces are then glued together and sanded to the desired thickness. The bottom piece of the ukelele is made in the same way.

2 The wood pieces are then sent to a shaping machine, which cuts the wood into the ukelele shape. For the front piece, the soundhole is also cut out at this point.

Strutting

3 The next step is to glue wood braces on the underside of the front section. This process is called strutting. It serves to reinforce the wood against the pressure created when the strings are plucked. It also helps control the way the instrument vibrates. The bottom piece may also have some strips of wood glued to it to provide more strength to the instrument.

Making the sides

4 The sides of the ukulele are produced by cutting and sanding an appropriate length of wood. The wood is then softened in water and placed in a mold. This mold is designed to cause the wood to take on the curved shape of the ukulele. It is clamped down and held in place for the required amount of time. The two ends of the piece of wood are joined together with glue at the place where the neck of the instrument will go. A small piece of wood, called an endblock, is fastened here and near the bottom of the ukulele so the front, back, and neck can be attached.

5 After the sides are joined and the endblocks are attached, the front and back of the instrument are glued on. The excess

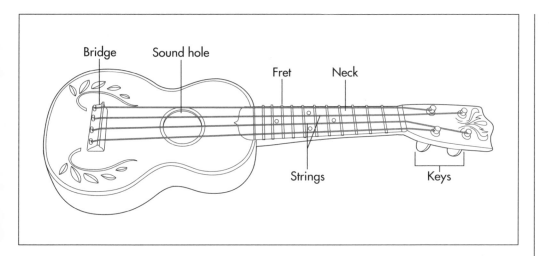

Diagram labels: Bridge, Sound hole, Fret, Neck, Strings, Keys

wood is shaved off and the joints are sanded to make them smooth.

Neck production

6 The ukulele's neck is carved from a single piece of wood. This may be made using a harder wood than the main body because it holds the strings tight and is consequently under greater stress. The whole part is then sanded and a wooden piece is attached to make the fingerboard. On the fingerboard, small grooves are cut across the width of the neck. Thin metal or wood strips are placed in these grooves and glued. These strips are the frets that allow the musician to change the sound of the instrument.

7 The neck is then attached to the main body of the instrument. When the glue dries, the entire instrument is stained or painted for decorative purposes.

Attaching the bridge

8 Next, the bridge and saddle are attached to ukulele just below the sound hole. These pieces are held on to the body of the instrument by tiny screws and wood glue. Another bridge is placed at the top of the instrument's neck. The saddle is the section where one end of the strings are attached. The strings pass over both bridges and this area creates the distinct length where the strings vibrate.

9 At the top of the ukulele, four holes are drilled and pegs, which hold the strings, are inserted. Tuning keys are attached to these pegs and fastened to the neck with screws. These keys have a gear mechanism which rotates the pegs and tighten the strings when turned.

Attaching the strings

10 Strings are then put on the ukulele. They are first tied to the bottom saddle section and strung up to the tuning pegs. Each peg has a hole in it where the string is inserted and tied. The tuning keys are turned to make the strings tight. The instrument is then inspected for flaws and put into the final packaging. Depending on the manufacturer, the whole process of making a ukulele can take weeks.

Quality Control

As of the year 2000, there are only three major ukulele manufacturers in the world. These are small companies and many of the instruments are handmade. This enables workers to inspect the instrument during every step of the manufacturing process to ensure a high quality product. It begins with inspection of the incoming raw materials and parts. The physical appearance and condition of the wood evaluated and rejected if it does not meet specifications. Final inspections are done on the finished product and in this way, most flaws are detected.

The Future

Improvements in the future of ukulele manufacture will focus on better quality, growing sales, and increasing output. The quality of a ukulele is primarily dependent on the type of wood used. Manufacturers are constantly looking for new wood sources and blends that can give a cleaner, more consistent sound quality. Sales growth will be driven largely by promotional efforts. In manufacturing, improvements in string quality, wood consistency and instrument durability can still be realized. Other improvements will focus on automating the production process and increasing production speeds.

Where to Learn More

Books

Beloff, Jim. *Jumpin Jim's Ukulele Tips 'N' Tunes.* Hal Leonnard Publishing Corporation, 1994.

Beloff, Jim. *The Ukulele: A Visual History.* Miller Freeman, 1997.

Brosnac, Donald, ed. *Guitar History.* New York: Bold Strummer Ltd., 1995.

Other

Ukuleles by Kawika. 1626 Kino'ole Street, Hilo, HI 96720- 5021. (808) 969-7751. kawika@ilhawaii.net. http://www.ukuleles. com (January 2001).

Ukulele Hall of Fame Museum. 15 Concord Ave., Cranston, RI 02910. (401) 461-1668. ukeinfo@ukulele.org. http://www.ukulele. org (January 2001).

—*Perry Romanowski*

Vacuum Cleaner

Background

The vacuum cleaner is the appliance that frightens the cat, is chased by the dog, and, perhaps, gives a home the most immediate appearance of being clean. Imagining a home without a vacuum cleaner is next to impossible; yet, like many time- and effort-saving devices, its widespread use is less than a century old.

History

There were no mechanical devices for cleaning rugs or carpeting until the 1840s. Before then, carpet cleaning was the duty of housemaids for the well-to-do and the women of the family for everyone else. Most rugs were made of rags that were woven together or braided in long ropes that were then stitched together as floor coverings. Carpets were woven of finer materials. Rugs and small carpets were taken outside several times a year, hung on heavy clotheslines, and beaten with fan-shaped beaters to drive out the dust. Larger carpets were left in place and brushed; curtains were also cleaned by beating and brushing.

When carpets and rugs were cleaned, the furniture and many ornaments that characterized the fussy Victorian style had to be moved: a time-consuming and inefficient process. Even worse, the beaten- or brushed-out dust quickly resettled on the floors and furniture. This, of course, did nothing to sanitize the house.

Relief from this arduous task was still a long time coming. The vacuum cleaner had three significant ancestors, the first of which was the street-sweeping machine. Public streets collected much of the waste from private homes and were filthy. Joseph Whitworth, an enterprising English gentleman of the 1840s, mounted large coarse-bristled brushes onto a rotating drum inside a horse-drawn van. The turning brushes picked up street dirt and deposited it in the van. The home carpet sweeper was invented in 1858 by H. H. Herrick, but its complexity and inefficiency limited its success.

Carpet sweepers

Finally, in 1876, Melville Reuben Bissell, owner of a china shop in Grand Rapids, Michigan, made the first popular and successful carpet sweeper by putting rotary brushes in a small canister with a push handle. Bissell's invention was spurred by his own need: bits of packing-crate straw became imbedded in his carpet. The Bissell carpet sweeper picked up both straw and dust and contained them in the canister for later disposal. Bissell named his first model the "Grand Rapids" after his home town. It revolutionized home care by making the need for beating carpets less frequent.

On the other side of the Atlantic a British company called Ewbank dominated the market. By 1880, Ewbank sweepers were found in many homes including the palaces of Britain's royal family. Models came in several sizes; with Miniatures for ladies to operate, followed by the larger Standard and the Parlour Queen, which boasted "a very powerful pattern for the thickest piles." The carpet sweeper was dominant through the 1930s; its internal parts were cast of aluminum, making these machines light and easy to use.

Unfortunately, carpet sweepers lacked vacuum suction. They were effective to a certain point, but could not pull dust and dirt from

The first Model O Hoover vacuum was made in 1908 with a grey cheesecloth bag, cleaning tools, and a weight of only 40 lb (18 kg).

deep within carpet pile. Inventor Hubert Cecil Booth saw a demonstration at the Empire Music Hall in London of an American machine that blew compressed air through carpeting; this produced a cloud of dust (proving how much was trapped inside the carpet), but the same dust only settled back into the carpet. The Americans had also experimented with suction devices since about 1859, but only a few factory cleaners reached the marketplace. Booth saw the future in suction. He proved this to friends in two startling demonstrations. In one, he placed a handkerchief on the carpet and sucked on the handkerchief with his mouth. The underside of the kerchief was filled with dirt. Even more startling, Booth was so eager to prove his thinking to friends that he knelt in front of a chair in a restaurant and sucked on the chair covering. Coughing and spluttering, he spat the extracted dirt into a hankie.

Vacuum cleaners

Booth gave the vacuum cleaner its start. His first vacuum cleaner, called "puffin Billy," was made of a piston pump. It did not contain any brushes; all the cleaning was done by suction through long tubes with nozzles on the ends. It was a large machine, mounted in a horse-drawn van that was pulled through the streets. The vans of the British Vacuum Cleaning Company (BVCC) were bright red; uniformed operators would haul hose off the van and route it through the windows of a building to reach all the rooms inside. Booth was harassed by complaints about the noise of his vacuum machines and was even fined for frightening horses. The BVCC's most prestigious engagement was cleaning the carpets in Westminster Abbey in London before the 1901 coronation of King Edward VII and Queen Alexandra.

The coronation cleaning led to a demonstration at Buckingham Palace, which had a system installed after the royal family saw the dirt Booth was able to suction out of the palace. Booth's vacuum system, however, was not suitable for individual homeowners. Some large buildings had Booth's machine installed in the basement with a network of tubes fitted into the walls of the rooms with sockets in the walls. Short lengths of tubing with nozzles were connected to the sockets, and this central cleaning system sucked the dust into a container in the basement. Booth

rented his machines rather than sell them, but in the United States, David T. Kenney built similar equipment and sold it, mostly to office buildings like the Flick Building in New York.

Efforts to make smaller vacuum cleaners were slow to develop. Booth made a smaller version call the Trolley Vac in 1906, but it was very expensive and still weighed 100 lb (45 kg). Other cleaners included the Griffith (also debuting in 1906) and the Davies device, patented in 1909, which required a two-man operating crew—fine for wealthy households but not the average home.

In their drive to produce a single-operator vacuum cleaner, inventors experimented with many types of mechanical suction. Davies's machine had a rotating wheel that used four bellows to create suction. Other early vacuum cleaners used a wide range of suction devices, including rocking chairs to work the bellows, assorted hand pumps connected to nozzles, and reverse-action bicycle pumps. Davies produced a smaller machine in 1912 called the Wizard, and Kirby's patent of 1912 was a pushed machine that moved forward like a caterpillar to open a long suctioning container. K. von Meyenburg's invention consisted of a long hose and nozzle that was attached to a bellow device worn like a back pack.

James Murray Spangler, who, like Bissell, suffered from dust allergy and asthma, constructed an electric-powered vacuum cleaner in Canton, Ohio, in 1907. Spangler made a box of wood and tin with a **broom** handle to push it and a **pillow** case to hold the collected dust. Spangler's innovation was to connect the motor to a fan disc and a rotating brush, combining the best of Bissell's brush sweeper with the suction of a powered vacuum cleaner to pull more dust out of carpets.

Spangler himself did not have the money to promote the cleaner, but his relative, William H. "Boss" Hoover, a maker of leather goods, quickly saw the advantages of Spangler's machine. The first Model O Hoover vacuum was made in 1908 with a grey cheesecloth bag, cleaning tools, and a weight of only 40 lb (18 kg). Hoover found that the machines sold very well door-to-door because housekeepers could see the action on their own carpeting. Hoover quickly built a large retailing operation that spread to Britain by 1913; to this day, vacuum cleaning in England is called "hoovering," a measure of the impact the Spangler/Hoover machine had on everyday life.

Other machines by Eureka and Electrolux soon followed and even copied Hoover's door-to-door sales methods. Hoover added a beater rod to the cylinder in 1926, so the cleaner brushed, beat, and suctioned the carpet. In the 1930s, the Great Depression prevented many from buying such luxury goods; to make the vacuum a necessity, Hoover hired renowned industrial designer Henry Dreyfuss to reconfigure the vacuum cleaner. With a body made of Bakelite instead of tin, a lighter total weight, more efficient operation, a signal showing when the bag was full, and other innovations, the streamlined vacuum cleaner resembled a high-speed locomotive. A canister cleaner followed during World War II. Today, the vacuum cleaner is firmly established as a household essential.

Raw Materials

Most upright vacuum cleaner parts are manufactured as individual parts or subassemblies (groups of parts that fit together) by subcontractors using specifications established by the manufacturer. These are sent to the factory where they're inspected, then stored in bins that can be moved to the assembly line as needed. Companies usually do their own injection molding of large plastic parts, including the exterior housing, the connections that support the bag, handle parts, wheels, and the attachments provided with the vacuum cleaner. Some models that have removable plastic canisters to collect the dust that can be snapped off and emptied; these plastic cylinders are also injection molded in the factory using clear plastic pellets. Rubberized parts, like the hose that channels dust from the fan to the bag and the bumper around the edge of the housing, are also made in the factory. The dust bag is made of fabric and is sometimes lined; this help keep fine particles that escape from the replaceable paper bag from seeping out.

Design

Portable vacuum cleaners are made in many general configurations, providing a range of cleaning actions to meet a broad range of customer requirements. The canister type has a cylindrical body containing the motor, fan, and other operating parts, and a removable, disposable paper dust bag. The canister is pulled over the floor on a set of wheels. The upright model is a push-pull device also mounted on wheels; the motor is mounted in a housing over the fan, beater bar and brushes, and drive belt. An upright handle, extending vertically from the back of the machine, carries both the electrical power cord and brackets to hold the dust bag or plastic dust canister. A simple locking device on the rear of the motor unit allows the handle to be lowered so the operator can maneuver it under tables and around other furniture.

Vacuum cleaner design used to focus exclusively on cleaning effectiveness, ease of operation, and low noise level. Since about 1990, however, almost all major manufacturers have also produced lines to reduce dust and allergens during vacuuming. These units usually have removable plastic canisters to contain the dust and are less to let fine particles escape through the bags and back into the air. Many are also equipped with replaceable filters for very fine particles. The partial vacuum produced by the fan has been improved, with more powerful motors and fans that still operate quietly. Panasonic's models feature a bypass motor that pulls dirt directly into the bag preventing wobble of the fan and possible motor burnout. These models also have onboard attachments; the suction can be transferred directly to the attachments. Lighter materials and a lighter total weight for these units compensate for the weight of the attachments.

The Manufacturing Process

Plastic parts

1 Many of a vacuum cleaner's plastic parts begin with computerized drafting and design systems (CADD). The parts are shaped in a two-part steel mold, called a die that is lowered into the chamber of an injection-molding machine.

2 Tiny plastic pellets stored in a large hopper next to the machine are poured into a

An example of a canister-type vacuum cleaner.

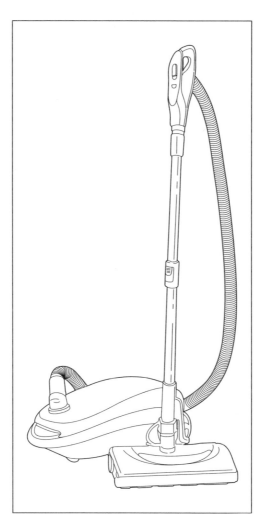

The assembly line

5 Vacuum cleaners are manufactured in an assembly-line process, with workers at assembly stations attaching subassemblies or individual parts to the vacuum as it moves along the line. Assembling an upright vacuum starts with the base, which is made of metal or molded plastic. The steel beater bar with brushes fitted into (a subassembly) is then pulled from a bin and inserted into fitted notches at the front of the base. The beater bar has a locknut on one end and a cap on the other so the owner can open it and replace the brushes when necessary. A rubber drive belt is placed in a guide channel around the beater bar and pulled over a belt guide and motor pulley on the underside of the base.

6 A steel base plate is fitted into notches in the front of the base and latched into place with a cam lock (a turning lever) over the underside of the belt and pulley. The steel base plate is a subassembly that has small rollers on it and openings near its front where the beater bar and brushes will agitate the carpet (to release dirt) during operation.

7 At the rear of the base, an axle is inserted through a tunnel-like opening that passes from one side of the base to the other. A release handle is fitted onto one end of the axle; it is a simple locking lever that allows the vacuum's operator to lower the handle during operation or raise and lock it into place for storage. Wheels are added to both ends of the axle and are locked into place.

8 The fan is bolted onto the base, and the motor assembly is attached to the top side of the base. The electrical connections from the motor to the fan and light, and from the motor to the electrical cord connection are made. A lightbulb is installed in a socket in the front of the base. A plastic housing that forms the top of the vacuum cleaner and fully encloses the motor and fan is snapped into place. It has already had a rubberized bumper wrapped around its sides and front. It also carries a clear plastic panel allowing the light bulb inside to shine through as a "headlight."

9 Plastic fittings that support the bag and handle are attached to the rear of the base. An opening at the back of the base holds a rubberized length of flexible hose

heating vat and melted. The pellets are either purchased in the color desired or colored with pigments as they are melted.

3 The melted plastic, injected under high heat and pressure into the chamber of the injection molding machine, penetrates every part of the mold. The two halves of the mold open enough to let the plastic part fall into a bin. Although the pieces are still hot to the touch, the plastic hardens on contact with the air as the tool opens. The plastic pieces are stored in bins that can be rolled to the assembly line as needed.

4 Many identical plastic parts of the same type are made during the injection process. When the desired number have been made, the tool is removed from the injection molding machine, another one is inserted, and the process repeats as supplies of another part are formed.

that transfers dust from the fan to the bag; this hose is fastened to the base opening and to the plastic fitting leading into the bag support. At the top end of the handle, a plastic unit that holds the top end of the bag is bolted through the handle. Next, electrical connections inside the handle are completed, and the electrical cord that has been attached to the back of the base is tied to the connections within the handle, allowing the machine's operation to be controlled by a switch near the top of the handle. The length of electrical cord leading from the machine to a power outlet is connected. For packing and shipping, this cord is looped and tied with a twist tie; the owner will wrap it around storage supports on the handle.

6 The final touches are added, including attaching the bag, the inner disposable bag, and outer markings (preprinted on decals that list the manufacturer, operating instructions, and information such as the serial number and the power of the motor).

10 The completed machine is taken to the packing department where it is wrapped in a plastic bag and put in a carton. A box of plastic attachments, including nozzles and a hose for upholstery cleaning, is also put in the carton with an information booklet, assembly instructions, and a warranty card. The cartons, which have been preprinted with marketing information, are then closed, sealed, and stored for shipping and distribution.

Byproducts/Waste

The major manufacturers make different styles of vacuum cleaners, but they don't produce true byproducts. They produce or stock replacement parts and supplies (like disposable paper bags) for sale to both customers and retailers.

Imperfect injection-molded parts are remelted and mixed (in controlled quantities) into new batches of plastic. Paper items, such as the bags and shipping materials, are also made by outside suppliers and can be recycled.

Quality Control

Assembly-line workers can reject any imperfect parts or partially assembled machines they find. Supervisors also monitor assembly along the line and can reject parts and partial assemblies. They may periodically remove machines for inspection during the line manufacture. The motors are bench-tested before installation. At the end of the assembly process, each machine is inspected for quality before it is sent to the packing department. Selected machines are also tested for operation before they are packed.

The Future

The vacuum cleaner is an essential part of every home no matter how small. It's typically one of the first small appliances purchased. Many families have several vacuum cleaners for dedicated uses. These specialized uses have helped broaden the lines of vacuums made. Designs have also changed as the importance of minimizing allergens like dust, dust mites, and pet hair has increased. Today's vacuum cleaners are more powerful, versatile, and convenient than their predecessors.

Where to Learn More

Books

Cohen, Daniel. *The Last Hundred Years: Household Technology*. New York: M. Evans and Company, Inc., 1982.

Langone, John. *National Geographic's How Things Work: Everyday Technology Explained*. Washington, D.C., National Geographic Society, 1999.

Rubin, Susan Goldman. *Toilets, Toasters & Telephones: The How and Why of Everyday Objects*. San Diego, CA: Browndeer Press, Harcourt Brace & Company, 1998.

Weaver, Rebecca, and Rodney Dale. *Machines in the Home*. New York: Oxford University Press, Inc., 1992.

Other

Bissell. http://www.bissell.com (January 2001).

Dyson Appliances. http://www.dyson.com/homepage.Asp (January 2001).

Eureka. http://www.eureka.com (January 2001).

Hoover. http://www.hoovercompany.com (January 2001).

—*Gillian S. Holmes*

Vermiculite

The largest vermiculite mining operation in the world is located in the Phalabowra (also sometimes spelled Palabora) district of the Republic of South Africa.

Background

The term vermiculite applies to a group of minerals characterized by their ability to expand into long, worm-like strands when heated. This expansion process is called exfoliation. The name vermiculite is derived from a combination of the Latin word *vermiculare* meaning "to breed worms," and the English suffix -ite, meaning mineral or rock. In its expanded form, vermiculite has a very low density and thermal conductivity, which makes it attractive for use as a soil amendment, lightweight construction aggregate, and thermal insulation filler. Expanded vermiculite also has a very large chemically active surface area, which makes it useful as an absorbent in some chemical processes. When vermiculite is ground into a fine powder, it is used as a filler in inks, paints, plastics, and other materials.

History

Vermiculite and its unique properties were known as early as 1824, when Thomas H. Webb experimented with it in Worcester, Massachusetts. It was Webb who gave the mineral its fanciful name because he thought the long strands looked like a mass of small worms. Vermiculite was regarded as not much more than a scientific curiosity until the early 1900s when more practical uses were sought. The first commercial mining effort occurred in 1915 in Colorado. The material was sold as tung ash, but did not find sufficient buyers, and the venture failed. The first successful vermiculite mine was started by the Zonolite Company in Libby, Montana, in 1923. The mine continued to operate until 1990.

The largest vermiculite mining operation in the world is located in the Phalabowra (also sometimes spelled Palabora) district of the Republic of South Africa. Other countries producing significant amounts of vermiculite include the United States, China, Russia, Brazil, Japan, Zimbabwe, and Australia.

In 1999, there were three active vermiculite mining operations in the United States, two in South Carolina and one in Virginia, which shipped concentrated vermiculite ore to exfoliation plants located throughout the country. In addition to using concentrated vermiculite from domestic mining operations, these plants also imported about 77,000 tons (70,000 metric tons) of concentrated vermiculite from foreign sources—mostly South Africa.

Raw Materials

Technically, vermiculite encompasses a large group of hydrated laminar magnesium-aluminum-iron silicates, which resemble mica. There are two keys to the unique properties of vermiculite. The first is its laminar (or layered) crystalline structure, which provides the hinged plates that make the material expand or unfold in a linear manner, like an accordion. The second is the fact that it contains trapped water, which flashes into steam when heated to force the layers open. There are a great many naturally occurring vermiculite minerals and soils, and their identification often requires sophisticated scientific analysis.

One of the most common forms of vermiculite is generally known as commercial vermiculite. This is the form that is mined and processed for various end uses. It is derived from rocks containing large crystals of the minerals biotite and iron-bearing phlogopite. As these rocks are exposed to the weather, they start to decompose, allowing water to enter and react with the various chemicals present. As the decomposition and chemical reactions proceed, vermiculite is formed.

A typical chemical analysis of commercial vermiculite shows it contains 38-46% silicon oxide (SiO_2), 16-35% magnesium oxide (MgO), 10-16% aluminum oxide (Al_2O_3), 8-16% water, plus lesser amounts of several other chemicals.

When commercial vermiculite flakes are heated and expanded, they undergo a color change that depends on the chemicals present and the temperature of the furnace. The resulting expanded vermiculite granules are usually a gold-brown color with a bulk density of about 4-10 lb/cu ft (64-160 kg/cu m), depending on the size of the granules.

The Manufacturing Process

The manufacturing process used to produce commercial expanded vermiculite consists of two separate operations. The mining and concentrating operations that produce raw vermiculite flakes are conducted at one location. The exfoliation and classifying operations that produce various sizes of lightweight, expanded vermiculite granules for use in other products are conducted in another location. Sometimes these two locations can be half a world apart.

There are many different methods used in both of these operations. The exact methods vary from mine to mine and plant to plant. Here is a typical manufacturing process used to produce commercial expanded vermiculite.

Mining

1 Rocks containing vermiculite are dug from a huge open pit in the ground. The soil on top of the rocks, called the overburden, is removed with power shovels or earth scrapers. The exposed rock layers are then drilled with large pneumatic or hydraulic drills, and the holes are filled with explosive charges. When all personnel and equipment have been moved out of the area, the explosive charges are detonated.

2 The resulting heap of loose rocks are scooped up with power shovels and dumped into trucks or train cars, which carry the rocks to a nearby processing plant.

Concentrating

3 The rocks are fed through a series of crushers and screens to reduce their size. The vermiculite is separated from the surrounding rocks and dirt using various wet or dry techniques depending on the particular mining operation and local environmental regulations. These techniques may include froth flotation, gravity separations, winnowing, or electrostatic separation. In each of these techniques, either the vermiculite itself or the other materials are trapped and separated from each other until the resulting vermiculite flakes are about 90% pure by weight.

4 The vermiculite flakes extracted from various sections of the mine may be blended together before further processing to ensure uniformity of the product.

Grading

5 The separated vermiculite flakes are sorted by size. This may be done with a series of screens or it may be done in a long enclosed wind **tunnel**. In the wind tunnel, the flakes are fed into the upstream end of the tunnel and are carried along the length of the tunnel by the flow of air. The larger flakes, being heavier, fall out of the air stream first and are caught in a hopper at the bottom of the tunnel. This separation by weight continues down the length of the tunnel until all the flakes are caught in hoppers. By controlling the length of each hopper opening and the velocity of the air, the flakes can be sorted into various sizes, or grades, ranging from about 0.63 in (16 mm) down to about 0.02 in (0.8 mm) in diameter. If the particular vermiculite being mined tends to form a high percentage of large flakes, the flakes may be slightly crushed to delaminate them and reduce their size. This process is called debooking and allows the flakes to be quickly heated during the exfoliation process.

6 The graded vermiculite flakes are dumped into large plastic bags or other containers for shipping to various exfolia-

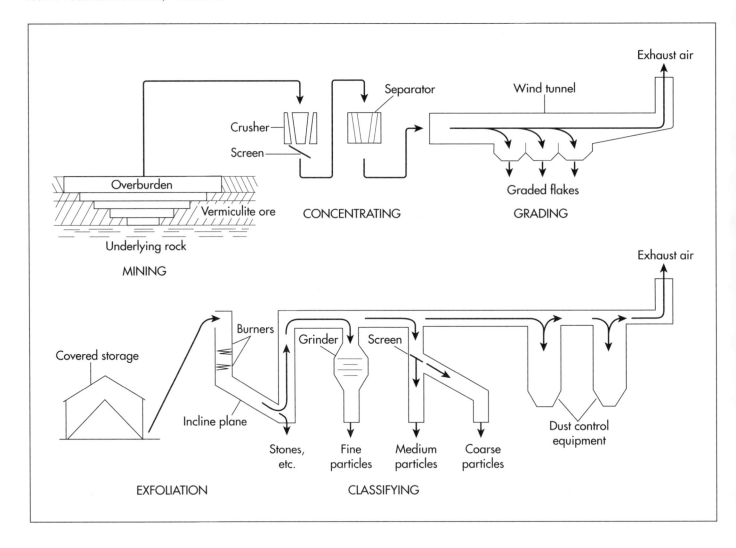

Exhaust air

Separator

Wind tunnel

Crusher

Screen

Overburden

Vermiculite ore

CONCENTRATING

Graded flakes

GRADING

Underlying rock

MINING

Exhaust air

Covered storage

Burners

Grinder

Screen

Incline plane

Dust control equipment

Stones, etc.

Fine particles

Medium particles

Coarse particles

EXFOLIATION

CLASSIFYING

A diagram depicting the processing of vermiculite.

tion plants. If the flakes are to be shipped to plants overseas, they are loaded in bulk into the holds of ships for transport.

Exfoliating

7 The vermiculite flakes are transported by truck or train from the port or mine to the exfoliation plant, where they are offloaded and stored in a covered area to protect them from contaminants and the weather. It is important to prevent the flakes from absorbing moisture. Otherwise, it will take too much energy to heat the flakes to the required temperature to make them expand.

8 The flakes are loaded onto a conveyor belt and lifted to the top of a 20-25 ft (6.1-7.6 m) high vertical furnace lined with ceramic bricks. As the flakes fall down the length of the furnace, they pass through one or more burners fired by **natural gas**. The temperature inside the fur-

nace reaches approximately 1,000-1,500°F (540-810°C), which is sufficient to make the trapped water in the flakes flash to steam and cause the flakes to expand into worm-like particles. At the bottom of the furnace, the particles slide down an inclined plane. This delays the exit of the particles from the furnace and allows the vermiculite to be heated further in order to reach full expansion. Other exfoliation plants may use different furnace configurations, but the general sequence of operations is similar.

Classifying

9 The hot, expanded vermiculite particles are then drawn up a vertical tube by a vacuum. Any small stones or other solid contaminants are too heavy to be carried upward by the gentle flow of air and fall out the bottom of the tube. The air flow also acts to cool the hot vermiculite.

10 If a customer or application requires fine particles, the vermiculite may be ground and screened to produce a specific size or range of sizes before it is packaged for shipping. In some exfoliation plants, the larger particles may also be screened or sorted into various sizes, depending on the final use.

11 The sorted, or classified, vermiculite particles are then deposited into storage hoppers, where they are dispensed into individual 4-6 cu ft (0.10-0.15 cu m) paper or plastic bags for retail sales or placed into larger 50 cu ft (1.3 cu m) bags for use in various commercial applications. The bags are sealed, labeled, and moved to a warehouse for shipping.

Health Aspects

Vermiculite ore deposits may also contain a variety of other materials such as mica, quartz, and feldspar. These deposits vary from one mining location to another. During the manufacturing process, some of these materials may pose potential health hazards to workers. In the United States and many other countries, these hazards are defined in Material Safety Data Sheets (MSDS), which identify the hazard and provide information on the safe handling and disposal of the material.

One of the most common health hazards in processing vermiculite comes from quartz, which is crystalline silica. It is usually only present as larger particles, but when it is ground into finer particles, the dust can be inhaled and cause a lung disease called silicosis. As a result, strict dust control and personal protection measures are incorporated into those areas of the vermiculite-processing operation where the materials are ground, sifted, and bagged. At the consumer level, exposure to silica dust is negligible and does not pose a health hazard.

In some vermiculite ore deposits, there may also be certain amounts of various forms of asbestos. None of the ore bodies currently used by major vermiculite producers pose an asbestos health risk to workers when the material is processed in accordance with the applicable MSDS. In August 2000, the United States Environmental Protection Agency (EPA) issued a report regarding vermiculite sold as a soil amendment. In the report, they concluded there was little or no risk to consumers from asbestos.

The Future

Although there are several other materials that may be used as a substitute for vermiculite, vermiculite's extremely low density and thermal conductivity continue to make it attractive for many applications. In 1999, it was estimated there were approximately 55 million tons (50 million metric tons) of vermiculite reserves in the world.

Where to Learn More

Books

Hornbostel, Caleb. "Vermiculite." In *Construction Materials: Types, Uses, and Applications*. New York: John Wiley & Sons, Inc., 1991.

Other

Grace Construction Products "Grace Specialty Vermiculite History." (2000). http://www.graceconstruction.com/vermiculite/verm_prodhist.html (January 2001).

Hindman, James R. "Vermiculite as an Industrial Mineral" (1997). http://www.mcn.net/~vermiculite/overview.htm (March 22, 2000).

Hindman, James R. "Vermiculite Products and Applications." (1997). http://www.mcn.net/~vermiculite/uses.htm (March 22, 2000).

Potter, Michael J. "Vermiculite." *U.S. Geological Survey, Mineral Commodity Summaries, February 2000*. http://minerals.usgs.gov/minerals/pubs/comodity/vermiculite/index.html (May 18, 2000).

The Schundler Company. "Basic Vermiculite Information and Data." http://www.schundler.com/techverm.htm (May 18, 2000).

The Vermiculite Association. "About Vermiculite." http://www.vermiculite.org (May 18, 2000).

—*Chris Cavette*

Water Gun

Within two years of its introduction in 1990, Larami had sold over 10 million Super Soakers.

Background

Pump-action water guns are a relatively recent addition to the water gun arsenal. Plastic squirt guns have long been a staple summertime toy of American children. These traditional toys eject a relatively weak stream of water, requiring frequent breaks for refilling. The pump-action gun operates on a different principal from its predecessors. The user pumps a handle on the gun, which pressurizes air in a reservoir. When the user then operates a piston-like trigger, the pressurized air ejects water from a separate water tank, resulting in a strong stream of water that can reach as far as 50 ft (15 m). The original pump-action water gun was invented by an American engineer, Lonnie Johnson, and the toys are manufactured under the brand-name Super Soaker by the Larami Corporation of Mount Laurel, New Jersey. Despite the success of the Super Soaker, other companies have taken and expanded on the design.

History

Inventor Lonnie Johnson had a Master's degree in nuclear engineering and was working for the Air Force in 1982 when he came up with the idea of a pressurized squirt gun. He was actually working on a heat pump project that involved the use of water instead of Freon as a refrigeration fluid. While experimenting with some tubing in his bathroom, Johnson shot water through a high-pressure nozzle. He put the refrigeration work aside to make a prototype water gun, which he gave to his six-year-old daughter. The high-powered squirt gun quickly became the hit of the neighborhood, and the inventor began working on a marketable model. Johnson made all the valves and fittings for his water gun himself, and took out several patents to protect his rights. Manufacturing the water gun with his own resources proved to be too expensive, so Johnson marketed his idea to toy companies. It wasn't until 1989 that the inventor met with representatives of the Larami Corporation at that year's American International Toy Fair. Larami had previously marketed a battery-operated squirt gun that was patterned on the Israeli Uzi submachine gun. The battery-operated gun, invented by Alan Amron, went on the market in 1985 and was a big hit in 1986. This gun was powered by an interior motor which pumped out 250 squirts per minute. Due to Larami's water gun experience, it seemed a good candidate to make Johnson's toy. Johnson was wary of letting out the secret of his water gun's design, and at the toy conference he described his model in vague terms. He disclosed enough to get Larami's executives curious, and soon after, Johnson brought his prototype to the company's headquarters. His hand-built model was made of polyvinyl chloride (PVC) tubing, Plexiglas, and plastic soda bottles. To show the executives what the water gun could do, Johnson filled it and shot an immense stream of water all the way across the board room. Larami's president was immediately impressed, and Johnson and the company signed an agreement. The toy was almost instantly popular. Within two years of its introduction in 1990, Larami had sold over 10 million Super Soakers. With the larger models cost up to $40 dollars, it was a significant money-maker.

Controversy erupted over the high-powered water guns in the summer of 1992. In Boston, a water fight escalated into a real gunfight, and one teenager was shot to death, causing Boston's mayor to ask local stores to stop stocking Super Soakers. Shortly after, in New York, teenagers drenched a passerby with a Super Soaker, and the offended victim answered with gunfire, wounding two youths. Politicians and law enforcement personnel faulted the Super Soaker in the wake of these tragedies, though others argued that it was real guns and not water guns that were the problem. Nevertheless, pump-action water guns grew in popularity. By 1998, Larami's Super Soaker was the top-selling summer toy in the world. Retail sales stood at over $200 million. The company had sold more than 300 million Super Soakers over eight years, and produced close to 10 models of varying size and price.

Raw Materials

The prime raw material for water guns is simply plastic. The plastic is mixed with various pigments for colorful effects. Other materials needed are glue and light screws. The dies for the plastic molds are made of steel. Water is used in the manufacturing process, to test parts, as is pressurized air.

Design

The manufacturing process for pump-action water guns is fairly simple, but the design process is lengthy and involved. The selling period for water guns is limited to summer. The design process usually begins in July or August, to create new models for next year's selling season. At that point, marketers usually have a clear idea of what models are selling well for the current year, and they can incorporate sales data and retailers' feedback into their plans for the next summer. Personnel at the water gun company begin their designs with a wish list, noting what features they would like to see improved or changed, and coming up with creative notions for the product. The company might also employ consultants or designers outside the corporation, to make sure it has access to fresh points of view. When the ideas have been hashed out for about six months, the water gun company enlists engineers to make a preliminary model. This first model is known as a "bread box." The bread box has the working features the water gun company

has requested, but the nozzle and pump are fitted into a box looking little like the finished product. The features are evaluated for their performance only at this point; how the gun looks is another question entirely. The bread box may undergo alterations several times. Once it is approved by the company personnel working on the project, the engineers make drawings of the internal layout of the required parts.

Next, designers begin their work, making drawings of how they think the water gun should appear. The designers may make half a dozen to a dozen drawings. The company team reviews the drawings, and may mix and match aspects of different designs. For example they may prefer the handle on one design, and the nozzle on another. Designers make a final drawing based on the company team's preferences.

The designers' drawings are then sent to the engineers, who come up with the parts drawings. In most cases, the engineers need to alter aspects of the design. So the drawings go back to the designers for review, alterations are made, and the engineers make more drawings. Eventually a design is finalized and sent to a model maker.

The model maker makes several models. The first one is not a working model but solid plastic. This allows the company team to see what the gun looks like in three dimensions. Then the team can judge the aesthetics of the new gun better. Next, the model maker produces a working model. Meanwhile, the art department at the company comes up with package designs and colors for the gun. The working model is made of a gray plastic. The model maker produces six of these, and the art department artists paint them to make them look like the toy that the company proposes to sell the next summer. These mock-ups are then distributed to top retail buyers. The buyers give their opinions on what they like or don't like about the product. They also begin to give the company estimates of how many their stores might want to stock.

After the water gun company has received feedback from the retailers, it begins working with its manufacturing contractor to get the new product ready for mass production. Yet another model is created, this time

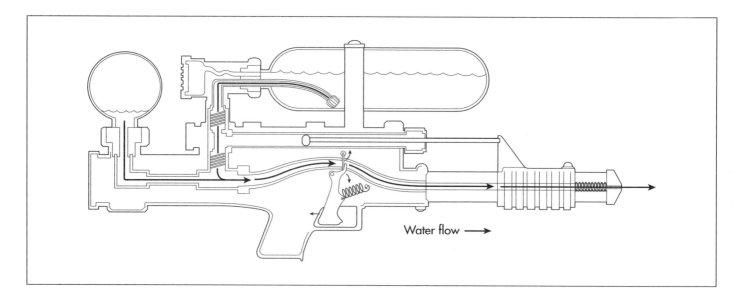

Water flow ⟶

An typical pump-action water gun.

called a tooling model. This lets the experts in the factory see the parts they will need to mold, and they can point out any problems. At this stage, further work is done at the manufacturing plant.

The Manufacturing Process

Making the molds

1 Designs for the individual parts of the water gun are drawn on computer. The drawings are then fed to the computer at the manufacturing facility. This computer controls the cutting of the molds. The molds are cut out of steel, following the computerized design.

Testing the molds

2 Before mass production begins, workers at the manufacturing facility make several test runs. In the first test run, enough uncolored plastic is fed into the machines to make just two or three copies of each part. Engineers study these pieces, called test shots, to make sure there are no problems. Then workers run a slightly larger test. This is called the engineering pilot. Again using uncolored plastic, the molds are filled and a few dozen pieces of each part is made. These parts are then fully assembled, to make two or three dozen model guns. This allows the factory personnel to test the whole production process. They can see how well the pieces are formed and how they fit together. Any problems in assembly

are addressed at this point. The next test run is the color check. The raw plastic used in the molds is naturally light brown, off-white, or gray. Pigments are added to the plastic, and the molds are filled. The colors are checked to make sure they are what the design specified. Finally, workers make a test run of 40-50 pieces, to be set aside for quality control. These are sent to several different labs and tested for both safety and reliability. If everything tests well, then mass production can begin.

Full production

3 Individual pieces of the water guns are made through injection molding or blow molding. Heated plastic is injected or blown into the steel molds under pressure. Workers unmold the parts and take them to an assembly area. The assembly work is done by hand. Some parts simply snap together, others need to be glued or put together with screws. One aspect that makes water gun manufacturing different from the manufacturing of other molded plastic items is that the parts need to be tested for water-tightness. Every pump and every nozzle that comes out of the molds is tested, as are some other vital parts. If possible, air is blown into the part instead of water. If a part is air-tight, then it will certainly hold water. Some parts must be tested with water, however. This means that afterwards they must be dried, or else the guns would leave the assembly room wet. So air is blown through them to dry them. In some cases, because of

the shape of the part, it is not possible to get all the water out. In these instances, the water used is treated with an anti-fungal agent, and the packaging also indicates that some moisture in the gun is to be expected. When running at peak production, the water gun manufacturing plant employs thousands of people and over several months makes millions of guns.

Packaging

4 When the water guns are fully assembled, they are ready for packaging. As the gun itself had a long design process, packaging also has many aspects to be considered for an ideal product. Pump-action water guns are mostly made overseas and shipped long distances. The weight of the packaging is extremely important, since it determines shipping costs. The packaging must also be durable enough to withstand rough handling and attractive enough to grab consumer's attention at the retailer. The overall size of the package is also important. The box may be designed so that several fit together, saving space in the shipping carton. Since the product is packed by hand, the package also needs to be easy to fill, so workers package them quickly.

Quality Control

Quality control starts at the beginning of the design process. As with all toys and products marketed to children in the United States, pump-action water guns must meet safety standards. These standards are set by the American Society for Testing and Materials (ASTM) and concern small parts, sharp points, safety of materials, and other aspects. Manufacturers are also concerned with the reliability of their product. An expensive, top-of-the-line water gun may be designed to last for 10 years, while a less expensive product is generally expected to last a year or two. This reliability is designed into the product, with a testing schedule laid out specifying the number of times the gun is expected to shoot. Just before the main production run, the manufacturer does a test run of 40-50 pieces for quality control purposes. These test pieces are subjected to a variety of trials such as drop tests, to see how well they hold up to both normal use and excessive abuse. If a model is supposed

to last through 2,000 squirts, then the testing facility puts it through 2,000 squirts. Other pieces are tested to make sure no dangerous materials have been used in the making of the product, such as lead or heavy metals, and other safety factors are minutely examined. A large amount of quality control is also built into the manufacturing process. Key parts are tested individually before assembly, either with pressurized air or water. Random samples of fully assembled water guns are also taken. These are tested for function, meaning that they are filled with water and shot.

The Future

Pump-action water guns are being continuously redesigned to cater to the public's demand for bigger, larger water toys. There are many new designs that promise to meet this request. One new water gun will hold 1 gal (3.8 L) of water, have six settings for the nozzle, and three different spray patterns. Another will have 11 nozzle settings and hold 1.3 gal (4.8 L). The amount of soakage per second and length of spray are also a large factors in future design.

Where to Learn More

Periodicals

"Alan Amron's Battery-Powered Big-Squirt Water Guns Have Left the Competition All Washed Up." *People*(September 8, 1986): 89.

Brown, Caryne. "Making Money Making Toys." *Black Enterprise* (November 1993): 68.

Fitzgerald, Kate. "Toy Makers Set to Soak the Market with Water Guns." *Advertising Age* (March 22, 1993): 4.

"Hold Your Water." *People* (June 29, 1992): 89.

Mathews, Jay, with Debra Rosenberg and Nichole Christian. "The Soaking of America." *Newsweek* (June 22, 1992): 58.

Swartz, Mimi. "Child's Play." *New Yorker* (July 13, 1998): 27.

—Angela Woodward

Wood Stain

Iron nails soaked in vinegar render a dark gray or ebony stain and a brown stain may be devised by soaking tobacco in ammonia and water.

Background

Wood pieces are often decorated to add color and appeal. Wood products are often imparted with a wood-tone stain to enhance the natural grain or add depth or tone to the wood. Stain may alter the color and appearance of the wood or hide unattractive grain. Stains are available in a variety of wood tones, including very light, semi-transparent stains to dark, nearly opaque stains.

Stain is a combination of dyes and pigments suspended in a solvent. Soluble dyes dissolve in compatible solvents and provide greater grain clarity, meaning the grain shows through the stain. Insoluble pigments are finely ground coloring materials that disperse but do not dissolve in the solvent. These insoluble pigments tend to cloud the grain. Stains need to be mixed frequently so that the pigments remain evenly dispersed and neither completely reveal or obscure the grain. Stains are generally characterized by the type of solvent that is used in their production. Thus, the most frequently used stains include alcohol (sometimes called non-grain raising stain), water, and oil stains. Each solvent affects the way the stain looks and handles. Today, oil stain is manufactured in the greatest quantity and the most familiar to the amateur woodworker. There are two types of oil stains. These include penetrating oil stain, which sometimes bleeds and fades, and wiping oil stain (sometimes called pigmented stain), which is more consistent and does not streak.

Regardless of solvent, stains generally penetrate only the top layers of the wood. Thus, the stain can be stripped and sanded away, revealing the original color of the wood. Stain must be topcoated or finished, meaning that once it is dry some kind of surface finish is applied to protect the wood surface and stain from moisture, scratches, unwanted stains, dirt, and chemicals. Wood stains are compatible with natural finishes such as varnish or shellac, and synthetic finishes such as **polyurethane** or acrylic.

History

Woodworkers have stained wood for centuries using natural pigments and dyes from plants and minerals. Iron nails soaked in vinegar render a dark gray or ebony stain, brown stain may be devised by soaking tobacco in ammonia and water, and so forth. Many of the earliest stains were essentially thinned paints that rendered opaque color and tone. It is estimated that over 100 years ago stains were first mass-produced, and around 1920 American companies such as Pratt & Lambert not only made a wide variety of oil stains, but were actively advertising and marketing their products.

More recent developments in stains include a wider variety of those with solvent-bases. Water and alcohol stains are considered less environmentally unfriendly. (Mineral spirits essential to oil stains have restricted disposal policies as it may contaminate water supplies.) An interesting array of semi-transparent colors has recently been developed by stain manufacturers to render colorful, non-natural colors sought by some woodworkers. Synthetic pigments have been developed as well, resulting in more consistent coloration than some of the pigments found in the natural world. Gel stains are pigmented stains in a thickened form resembling jelly. Pigments stay mixed evenly and the stain does not drip or splatter as much as a liquid stain.

Raw Materials

The raw materials essential to the production of wood stain vary by type. Water stains use water as the solvent and include water-soluble aniline (chemically derived) dyes to impart color. Non-grain-raising stains, sometimes referred to as alcohol stains, are manufactured using alcohol or glycol as the solvent with alcohol-soluble aniline dyes used in their production. Because alcohol dries almost instantly, this dye is not able to be manipulated much and essentially the stain is set as it is applied.

Oil stains utilize mineral spirits for the solvent. Mineral spirits help the product's viscosity and ease of application and are the volatile ingredient in stains (rags soaked with stains have been known to instantaneously combust and must be carefully disposed). Oil stains also generally use linseed oil as the resin or binder that has been treated with special acids so that it will not penetrate too deeply into the surface of the wood. Pigments come in 50-lb (23-kg) bags and are generally iron oxide pigments (although this may vary). Metallic salts are important ingredients as they help the product oxidize and permit the oil stain to dry. Finally, a thickening agent that also helps control penetration into the wood is needed. These thickeners are often proprietary and may not be discussed by the manufacturer.

The Manufacturing Process

There are many different solvent-based stains. Oil stain is one of the most produced and sold in greatest quantity.

1 First, components must be mixed together in order to begin the process. The main component—linseed oil used as a binder within the stain—is pumped into a tank. Only about half of linseed oil that is needed to make stain is added to the tank in this stage. Next, the solvent, generally mineral spirits, is pumped in. Finally, the pigments are added. The pigments are mixed in powdered form, pre-measured carefully elsewhere and dropped in by hand. This amount is carefully monitored in order to acquire the depth and tone of stain the consumer is expecting. These powdered pigments also have some oil absorption qualities and help

thicken the mixture. Finally, a dedicated thickening agent (varied, and in some cases, a proprietary ingredient) is pumped into the tank.

2 The ingredients must be thoroughly mixed in a process referred to as "the grind." A high-speed dispenser, essentially a saw-tooth blade that rotates at very high speeds, is lowered into the large vat of chemicals and pigment. This blade agitates the slurry for approximately 20 minutes, ensuring that the powdered pigment is evenly distributed throughout the liquids. As this high-speed dispenser rotates within the chemicals for several minutes, the temperature of the mixture rises.

3 The batch must be cooled down. In order to cool down this thick concoction, the rest of the linseed oil is pumped in along with additional solvents (more mineral spirits) and various metallic salts. The nascent stain is quickly cooled and thinned to nearly the viscosity required for a high-quality wiping stain. This single batch of stain is approximately 250 gal (946 L) in volume.

4 Presuming that the batch of stain under production requires no further adjustments for quality standards, it is hooked up to a filtration system that essentially removes all sediment from the oil stain so that the liquid is without grain or lumps. Some companies put their stain batches through two filtration systems to ensure the undesirable solids are eliminated. From the point at which the tank was initially filled with ingredients to the completed decanting may take as long as 2.5 hours.

5 The decanted stain, currently held in a large vat or tank on the second floor of a factory, is now dropped into a filling machine on a lower level of the factory. Here, the liquid is ready to be individually dispersed into cans and packaged. This filling machine automatically fills each can by shining a beam of light into the can with a label already affixed. If the beam remains unbroken, it indicates that the can needs to be filled, and fills it in a quick stream. When the can is filled to the desired level, the beam is broken and the filling stops. Another can moves into its spot, a beam of light is shone in, and the filling of another can commences until all 250 gal (946 L) are gone

Once mixed thoroughly for 20 minutes, the solution is cooled and more ingredients are added. Quality control tests are conducted before the mixture is filtered and packaged for sale as wood stain.

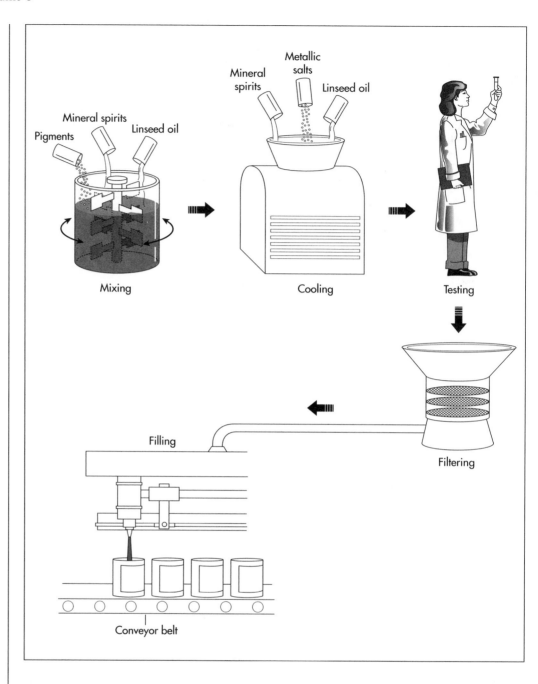

Pigments
Mineral spirits
Linseed oil
Mixing

Mineral spirits
Metallic salts
Linseed oil
Cooling

Testing

Filtering

Filling

Conveyor belt

from the machine. The cans are moved away from the filling machine and they are packed in cartons and readied for shipment.

Quality Control

The creation of oil wood stain is a carefully controlled cooking process. The ingredients are very carefully measured as they are pumped into the mixing tank, and pigments are hand weighed according to proscribed recipes that render the tone desired. The quality of the raw materials—particularly the proprietary thickener, the linseed oil, and the pigments—are essential to producing a quality product. Machinery must be working properly in dispersing the pigments and filtering out undesirable particles. Finished batches are checked for proper viscosity, weight, and color.

Byproducts/Waste

Oil stains generally utilize mineral spirits, a combustible material with a high flash point. Most unused solvents are easily re-used and re-mixed into the stain manufacturing process so that the solvent is generally not a hazard. If for some reason the solvent is contaminated and may not be reused in the product, then the mineral spirits are considered hazardous waste and must be disposed according to federal regulations that pertain to such wastes. These contaminated solvents are shipped to a hazardous waste facility.

The Future

Because oil stains are made with solvents considered hazardous, many woodworkers are turning to the water stains because they are environmentally friendly. Water stains move deeper into the wood than oil. But they don't always have the depth of tone or color that oil stain imparts on the first coat. It may require a few coats to get the desired color. Also, water stains tend to raise the grain, considered undesirable if one wants a smooth, even surface when the piece is topcoated. The future of all paints and oil stains made with linseed oil and mineral spirits is in question as the disposition of the used and contaminated products are becoming an issue.

Where to Learn More

Books

Umstattd, William. *Modern Cabinetmaking*. South Holland, IL: The Goodheart-Wilcox Company, Inc., 1990.

Other

American Furniture Design Company. http://www.americanfurnituredsgn.com (January 2001).

Antiques Resources.Com. http://www.antiqueresources.com (January 2001).

Lowe's Home Improvement Warehouse. http://www.lowes.com (January 2001).

—*Nancy E.V. Bryk*

Wrapping Paper

Background

There are many kinds of wrapping papers manufactured specifically for the types of products they are intended to wrap. For example, wrapping paper is made for bread for sanitary and aesthetic purposes. Originally, waxed paper and cellophane were used, but now polyethylene, polypropylene, and laminates that are a combination of these are preferred. Foil overwraps are also common.

History

Gift giving is associated with many holidays around the world and occasions such as birthdays. Like many other Christmas customs, exchanging gifts originated in ancient, pagan celebrations. The Roman festivals of Saturnalia and Kalends were celebrations of the harvest and winter solstice included gifts of small figurines, food, jewelry, and candles. The Roman New Year featured gifts of candies, cakes, honey, and fruit because sweet gifts foretold a sweet year.

The tradition of giving gifts at Christmas time is associated with the gifts of gold, frankincense, and myrrh delivered to the Christ child by the Three Wise Men. Despite this link, gift-giving was shunned by the early Christians who associated it with the pagan traditions they wanted to leave behind. It was not until the Middle Ages that gift-giving became a popular custom associated with Christmas and the church's abhorrence of gift giving was essentially overwhelmed by popular practice. A whole phalanx of gift givers followed the Three Kings in entering popular culture. Saint Nicholas and Santa Claus (in many forms) are the best known of these; but Italy has the Befana, Russia has both the white-robed girl Kolyada and the ancient Babushka (grandmother), and Scandinavia calls on goats and dwarfs. In England, Boxing Day also arose as a gift-giving day; this day after Christmas was both a church holiday in which alms were given to the poor and a day when families gave boxed presents to their servants and to tradesmen.

The Victorians made an art form out of many holiday customs. Elaborate gift giving was a privilege of the rich, but the Victorians developed the Christmas card as a small gift that friends could give each other to express sentiments without spending a lot of money. The first Christmas card was designed by John Calcott Horsley in 1843 on commission from Sir Henry Cole, a businessman who had fallen behind in writing personal Christmas letters and wanted to send holiday greetings to those awaiting his letters. Before this printed card, people had decorated their calling cards with scraps of colored paper and this practice continued until about 1860 when the Christmas card began to grow into an industry. Developments in the printing industry and the creation of postage stamps helped to fire interest in mass mailing of cards.

The art of designing Victorian Christmas cards led to development of artistic gift wraps. Victorian Christmas papers were intricately printed and ornamented with lace and ribbon. Decorated boxes, loose bags, and coronets bore cutout illustrations of Father Christmas, robins, angels, holly boughs and other seasonal decorations. Often, the gift-wrapping papers matched cards in design, and the association of the two has carried through to today. Developments in printing presses allowed endless sheets of wrapping paper to be printed with consistent quality. The flexography process patented in England in 1890 combined very fluid inks with rubber plates wrapped around the print cylinder to make a printing process ideally suited to coarse or stiff papers that were durable enough for wrapping. The rotary system prints exceptional lengths of printed paper that are rolled on cardboard rolls or cut into smaller sheets.

In the United States, card-making companies expanded their product lines during the period from 1910 to 1925 by making printed or decorated gift wrap. Tissue paper had previously been available in green, white, and red; but sturdier paper was plain brown wrapping paper. Soon, gift wrap was accompanied by gift enclosure cards, and, in the 1980s, gift bags in a wide variety of sizes and shapes became the trend in disguising gifts, even though bags had originally been popular wrappers in Victorian times. Occasion appropriate papers and all the accessories to match are widely available for every gift-giving opportunity.

Raw Materials

Wrapping paper begins with paper that is produced in special mills from wood pulp. The pulp is usually made from trees classified as softwoods; for gift wrap, the pulp is bleached, but other papers like the material called kraft wrapping (familiar as grocery store bags) is made of unbleached pulp. Ink is made from natural and synthetic dyes. The emphasis on protecting the environment by using recyclable papers has been felt by both the paper-making and ink-making industries, which choose bleach and pigments that are easily recycled.

The strength of gift-wrap manufacturers may be in design over actual paper and ink production. The gift-wrap companies buy paper and ink from vendors that specialize in high-quality production of two weights of wrapping paper (called "giftcote" in the industry), heavier kraft paper, and tissue paper. The finished surfaces of certain wrapping papers, like foils, are impressed on the paper during the printing process.

Design

Design is the key to eye-catching gift wrap that will sell well. The design department of a major gift wrap manufacturer includes teams of designers divided into line designers, graphic designers, and artists. Line designers create the initial ideas for particular lines of gift wrap and all the related specialty items. The specialties include everything except greeting cards, although line designers often work with greeting card designers to producing matching or complementary products. Line designers include tissues, bows, ribbon, gift enclosures, wrapping papers, stickers or seals, and gift bags in their concepts.

Graphic designers also participate in design planning, but their primary task is to create artwork that will be used to make a template for the printer or paper mill to repeat over the surface of the gift wrap. Graphic designers may direct an idea to an artist in the gift-wrap firm's own design studio or an outside studio. If the wrap maker acquires the rights to use a familiar image like a cartoon character, the designer works with the artist or firm providing the image to develop a series of designs using that image in keeping with copyright laws. The artists may use any medium to create a gift-wrap design; many are hand-drawn or painted and lettered. The graphic and line designers work with the artist to select the colors, layout of the template, paper weight and style, and paper finish to best enhance the design. Completed designs are reviewed and approved then scanned onto computer records that can be distributed to the paper mill, printer, and others responsible for manufacture so that identical, error-free designs can be reproduced.

Meanwhile, the line designers direct completion of designs for other parts of the line. Gift tags are based heavily on gift wrap designs to so they will coordinate. The tags are complex items to design, however, because the tiny space restricts design possibilities and many special finishes or die-cut shapes are expensive to produce. Gift bags are also a large part of the gift wrap market with unique requirements. Again, designs may be based on gift paper plans or lines of cards, but individual lines of gift bags are also conceived. The bag itself is three-dimensional, so the designer must create designs for the

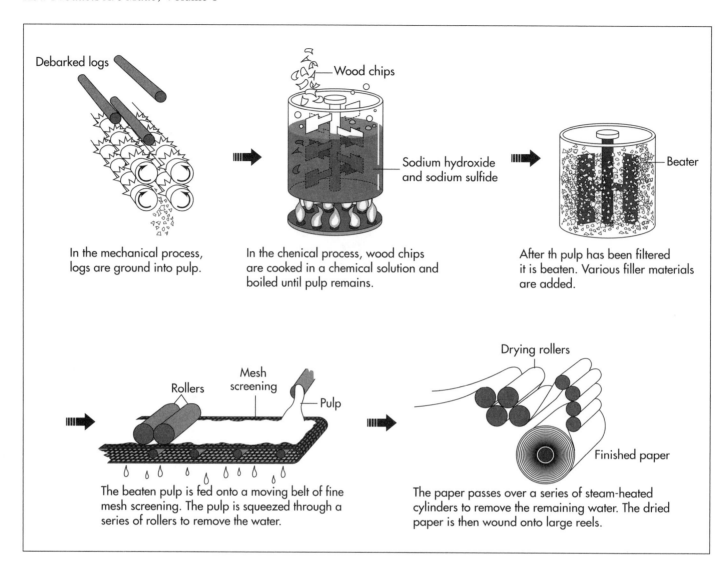

Debarked logs

In the mechanical process, logs are ground into pulp.

Wood chips

Sodium hydroxide and sodium sulfide

In the chenical process, wood chips are cooked in a chemical solution and boiled until pulp remains.

Beater

After th pulp has been filtered it is beaten. Various filler materials are added.

Mesh screening

Rollers

Pulp

The beaten pulp is fed onto a moving belt of fine mesh screening. The pulp is squeezed through a series of rollers to remove the water.

Drying rollers

Finished paper

The paper passes over a series of steam-heated cylinders to remove the remaining water. The dried paper is then wound onto large reels.

The process of making paper.

front, back, sides, bottom, and even the inside. Sometimes different art work is needed for each of these sections, even though they may have coordinated design elements. Finishes on the bags and the design of the handle are important details; the handle can be a solid color, several colors twisted together, and a variety of materials.

Line and graphics designers usually have private work spaces in the office, but they share larger planning rooms where designs, sample papers, references, and other materials for creating the current line can be spread out and shared. Design ideas must also be developed and discussed. Sources may include magazines, movies, traditional images, fashion and home fashion trends, as well as popular topics of conversation. Line designers may work on two to ten products within a single line at any time.

The Manufacturing Process

Manufacturing steps for gift wrap are relatively straightforward after the development and approval of a design is completed.

1 Paper, ink, special finishes, and other items in the design line are purchased from paper and ink vendors. These materials are received and inventoried in the manufacturer's printing plant. A computer file with the digitized artwork from the design team is read by a machine that engraves the image onto a printing cylinder.

2 The printing cylinder is carefully inspected and fitted on a printing press. The presses use rotogravure or flexography processes. Rotogravure requires etched cylinders, while flexography uses rubber plates fitted to a rotating cylinder. The

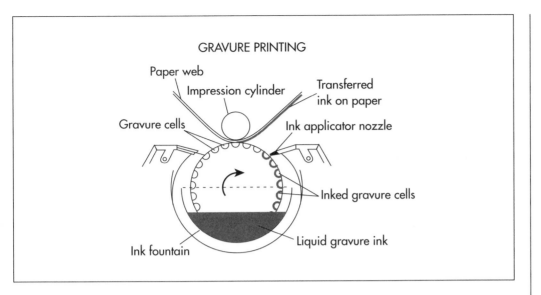

GRAVURE PRINTING

Paper web

Impression cylinder

Transferred ink on paper

Gravure cells

Ink applicator nozzle

Inked gravure cells

Liquid gravure ink

Ink fountain

Gravure printing is used to decorate wrapping paper.

processes require inks unique to each process. Gift-wrap makers have state-of-the-art printing equipment that can apply up to six different colors simultaneously and add special finishes like foil, iridescent, pearlescent, and flocked finishes. As the paper emerges from the press, it is rolled onto large rolls and transferred to another part of the factory. Machines cut and wrap the paper in much smaller rolls for sale or format it for folding and sale in flat packages.

3 Rolls of gift wrap are shrink-wrapped immediately with preprinted clear wrap bearing the manufacturer's information and price. Flat packages are also wrapped and sealed. Both types of gift wrap are bulk-packed in cartons for shipment to card shops, department stores, and other retail outlets.

Quality Control

Because the manufacture of gift wrap is so design-intensive, much of the product's quality is factored in before it ever reaches the manufacturing plant. All workers are responsible for product quality while the product is being processed in their area. In addition, quality control experts are part of the review and approval process during design and are on the floor of the plant throughout production.

Byproducts/Waste

Gift-wrap manufacturers generate lines of products for birthdays, weddings, Christmas, and many other holidays and occasions. Each line includes all the appropriate accessories customers may need for the particular occasion, and lines must be designed to appeal to a wide range of tastes.

Paper is the largest single waste item, and all waste paper is sold to recyclers who grind it up, remix it with appropriate types of pulp, and remanufacture paper from it. Ink waste results from leftover ink that is not fully used by printing of a particular line. The inks are chemically reformulated into new colors and reused. Some manufacturers such as American Greetings produce all of their own inks to make the most of this reformulation process as well as to produce unique qualities and colors of inks. Other makers who purchase ink from subcontractors recycle leftover inks through those manufacturers.

Where to Learn More

Books

Chalmers, Irena. *The Great American Christmas Almanac.* New York: Viking Studio Books, 1988.

Clements, Linda. *The Spirit of Christmas Past: Evocative Memories of Years Gone By.* New York: Smithmark Publishers, 1996.

Coffin, Tristram Potter. *The Book of Christmas Folklore.* New York: Seabury Press, 1973.

Del Re, Gerard, and Patricia Del Re. *The Christmas Almanack.* Garden City, NY: Doubleday & Company, Inc., 1979.

Rogers, Barbara Radcliffe. *The Whole Christmas Catalogue.* Los Angeles: Price Stern Sloan, 1988.

Sansom, William. *A Book of Christmas.* New York: McGraw-Hill Book Company, 1968.

Periodicals

"The Cover of Money." *Time* 136, no. 25 (December 10, 1990):73.

Other

American Greetings. http://www.american-greetings.com (July 18, 2000).

CPS Corporation. http://www.cpscorp.com (July 18, 2000).

Gibson Greetings. http://www.gibsongreetings.com. (July 18, 2000).

Hallmark. http://www.hallmark.com (July 18, 2000).

—*Gillian S. Holmes*

Xylophone

Background

The xylophone is a component of the percussion section of an orchestra and many instrumental groups. Its unique sound, relative rarity, and appearance make it fascinating to the listener. The xylophone has a close cousin called the marimba. Both instruments consist of wooden keys mounted on a wooden frame over a series of metal tubes called resonators. Hammering on the wooden keys causes the impact to resonate through the tubes. The xylophone has a brittle, metallic sound, while the marimba is somewhat more mellow or wooden to the listener.

The xylophone and marimba differ in range. Depending on the model, a xylophone encompasses two to four octaves. Its highest note is the same as C-88 on the piano. The marimba covers two-and-a-half to four-and-a-half octaves with C-76 the highest note. This means that the marimba is one octave lower than the xylophone in range. Music is written for the xylophone as an effects instrument. It rarely is used to play solos with an orchestra or ensemble. The marimba's large resonators make it sound more like an organ. Composers write more music for the marimba as a solo instrument, and its sound range is so wide that it can make music like a full orchestra.

The sound produced from the xylophone depends heavily on the skill of the player. The player stands to play the xylophone and faces the center of the instrument. He or she must stand erect, hold the mallets (hammers or beaters) between the thumb and first joint of the first finger with one mallet per hand. The wrists are used to move the mallets smoothly up and down; the palms face out. The arms are held down near the keyboard and do not move. The xylophonist plays the

lower register by taking one step to the left and the upper register by making one step to the right. The player always returns to center. Notes are struck in the centers of the bars or keys. Flats and sharps are struck along the edges of the bars but not the part of the bar that rests directly on the frame. The lowest end of the xylophone is the widest, and the highest notes are at the narrow end.

The mallets are also important to the sound produced. The instrumentalist must choose the right mallets to either blend in or project above the other instruments, depending on the volume needed and the character of the music. Xylophone players typically use rubber mallets made either of medium, hard, or extra hard rubber. Marimba players use mallets of soft rubber or medium soft woven yarn.

Mallet grip is critical to the proper technique for playing any of the mallet instruments. The player must stay relaxed but completely controlled; ease of movement or flow is very important to the sound produced. Both hands hold the mallets the same way, which is called a "matched grip." The point where each mallet is held between the thumb and the first joint of the first finger is called the pivot point. The other fingers curve around the stick portion of the mallet in a relaxed curl. Any pinching will constrict the sound and tire the player. The pivot point allows the mallet to rebound naturally, and force is provided by the combined movement of the finger, wrist, and forearm. The player will learn to place the pivot point at the point of balance between the ball of the mallet and the end of the stick or handle. The grip is almost the same as the right hand grip for playing the snare drum.

The bars on the keyboard of the xylophone look much like the black and white keys of a

Wooden bars were originally seated on a series of hollow gourds, and the gourds generated the resonating notes that are produced on modern instruments by metal tubes.

piano. The best sound comes from striking the middle of each bar, although very fast passages are played at the ends of the bars. The place where the bar passes over the chord or frame of the xylophone produces a dead sound, so this is avoided. The xylophone is not pounded with the mallets; instead, the correct rebound of the mallets pulls more rounded tones out of the bars. Beginning players learn to strike the centers of the bars to develop their feel for the reach from bar to bar. With increasing skill in getting the right tones from the bars, students can expand the parts of the bars they use to vary the sound and volume.

History

The xylophone is an ancient instrument that originated independently in Africa and Asia. Wooden bars were originally seated on a series of hollow gourds, and the gourds generated the resonating notes that are produced on modern instruments by metal tubes. For centuries, xylophone makers struggled with methods of tuning the wooden bars. Old methods consisted of arranging the bars on tied bundles of straw, and, as still practiced today, placing the bars adjacent to each other in a ladder-like layout. Ancient mallets were made of willow wood with spoon-like bowls on the beaten ends.

African xylophonists had the widest variety of instruments, including some that were plucked instead of hammered and lightweight instruments that were suspended on a rope around the player's neck. They used wooden boxes for resonators as well as clay pots in Nigeria and pits in the ground in Kenya and Central and West Africa. They inserted membranes between the bars and resonators to give the instrument a buzzing sound; these membranes were made of spider cocoons or cigarette papers. In southeastern Africa, the Chopi people play xylophones in groups of as many as six instruments of different sizes and ranges.

In the seventeenth century, African instrumentalists took the xylophone with them to Central America where it was modified and became known as the marimba. The marimba remains popular throughout Mexico and Central America and is considered the national instrument of Guatemala. The Africans who were responsible for the instrument's migration also developed an effective method of tuning it. They carved a gentle arch on the underside of each bar and simply continued carving until the bar was tuned accurately. This arch is called an "arcuate notch" and is the key to the tunefulness of the xylophone, marimba, and all other members of the xylophone family.

Another type of xylophone, the trough xylophone, is characteristic of the ancient instrument invented in Indonesia and Southeast Asia, and is still played today, especially in Java. The trough xylophone has its bars set across a wooden box with an open top and a bottom that slopes downward toward the bass end. Different ranges of bars from alto to bass can be removed and inserted in the box, so its range can be changed to suit the music. The trough xylophone is a favorite teaching instrument.

Early music for the xylophone was traditional and passed down from teacher to student. A European form of the xylophone first known around the fifteenth century and was developed in Central and Eastern Europe; was probably more closely related to the dulcimer than the African and Asian xylophones. In the nineteenth century, this folk instrument was modified by adding extra rows of bars; four rows became standard. Western composers did not "discover" the xylophone or begin writing classical music for it until the mid-1800s. Hans Christian Lumbye entered the history books as the first western composer to write a score for the xylophone in his 1873 "Traumbilder." The French composer Camille Saint-Saens (1835–1921) incorporated the xylophone in his 1874 "Danse Macabre." Spanish composer Manuel de Falla (1876–1946) used the xylophone for some percussion in his dances from "The Three Cornered Hat". The Russian composers Aram Ilyich Khachaturian (1903–1978) and Igor Fydorovich Stravinsky (1882–1971) experimented with many percussive types in their pioneering forays into modern Russian compositions. Khachaturian's "Sabre Dance" from his ballet called "Gayane Suite" has a challenging xylophone part, and Stravinsky's ballet "Petrouchka" includes his best-known use of this unusual instrument.

Modern musicians returned to the xylophone in the 1960s with another flurry of

interest in writing for the instrument. In 1961, Istv n Lang wrote a concerto for xylophone, and his Japanese counterpart, Toshiro Mayuzumi, composed a xylophone concertino in 1965. Also in the 1960s, a variation on the xylophone called the xylorimba was rediscovered. It had been created early in the twentieth century to give the xylophone greater range. Alban Berg (1885–1935) wrote "Three Pieces for Orchestra" in 1915, which demonstrated the xylorimba's capabilities. Another piece called "Hymnody" featured it in a chamber group written in 1963 by English composer Roberto Gerhard. Gerhard was born in Spain and may have learned of the xylorimba through his Spanish connections. His composition required two players for the instrument.

Raw Materials

The materials needed to make an orchestral quality xylophone begin with rosewood for the bars. Some teaching instruments for schools are made with keys fabricated from synthetic materials, but a true xylophone must have rosewood keys. Resonators are made from aluminum tubing that is also acquired in bulk from a specialty metal fabricator. Cords or pads of felt, synthetic, rubber, wood, or other materials support the keys at the nodal points where they rest on the frame over the resonators.

The frame itself may be constructed of metal or any wood, depending on the preferences of the customer and the manufacturer for the finished appearance of the instrument. Xylophones for high school and college marching

The ranges of various percussion instruments.

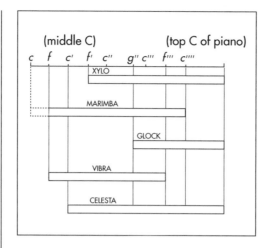

bands take tremendous abuse as they are transported from location to location, so the frame may be constructed of particle board that is easily replaced, patched, or painted if damaged. Instruments used by professional percussionists are usually crafted with frames of fine wood that are finished by skilled craftsmen. The frames, however, are still simple and unornamented, unlike other wooden instruments. Stains and varnishes are used to finish the wood.

The Manufacturing Process

1 Construction of a xylophone begins with a design drawing. Xylophone manufacture is based on traditional designs with little variation, so the design process is limited to selecting the size or range of the xylophone and the finish of the frame.

2 Rosewood for the keys is purchased in bulk. Sections long enough to produce several keys are cut and the wood is stored and aged for approximately two years before the keys are cut. The xylophone bars or keys are cut to lengths ranging from long keys for the low notes to shorter keys for the higher notes. The keys are a standard width of 1.5 in (3.8 cm) and a thickness of 1 in (2.5 cm). Holes are drilled at the support ends where the keys will be fitted to the nodes. The craftsman must then carve out the arcuate notch, the gentle arch on the underside of the key that provides accurate pitch. He does this in a series of cautious steps—carving to rough tune, checking the pitch, carving to truer tune, checking the pitch, carving to final tune, and confirming the pitch. After final tuning, the bar is gently sanded. When all the bars are tuned, they are polished, stained, and varnished. Choice of the color of stain is made by the customer and manufacturer.

3 Aluminum tubular resonators are made from tubing purchased from a metal fabricator. The tubes must be well made with uniform circumferences and smooth finishes that will not detract from the sound quality. The tubes are cut to lengths that depend more on the finished appearance of the xylophone. Usually, the bottom ends of the tubes have a tapered length with longer tubes at the bass end and shorter tubes toward the higher end of the range, or an arched effect from one end to the other. The length of the tubes does not matter for sound production because the tubes are stopped inside like organ pipes. The closing piece is added near the lower end once the tube has been tuned to its bar.

4 The frame is constructed as a separate operation while the keys and tubes are being cut and tuned. The outer perimeter underneath the keys is usually made of quality wood that is finished to match the color of the keys. Legs or supports are made of wood or metal and are bolted to the frame. The feet touching the floor are fitted with castors that can be locked so the xylophone won't move as its being played or with rubber or synthetic feet.

5 When the bars are complete, each one is test-fitted over its resonator and the resonator is tuned by inserting the stop. The bars are then fitted over the nodal points and screwed into place with standard wood screws. The tubes are riveted together and suspended on the frame.

6 Each mallet or beater consists of two parts, the stick and the head. The stick is made of bamboo, rattan, birch, or fiberglass. The spherical heads or ends are made of hard rubber or plastic with an internal core of cord that is wound much like the inside of a golf ball. Heads made for xylophone and marimba mallets are also woven of tightly wrapped wool. Xylophone players use three varieties of mallets constructed with different combinations of stick and head to produce a wider variety of sound. Players choose the beaters based on the music they are playing, the sound of their instrument, and their own wrist strength and technique.

Quality Control

Xylophone makers are skilled craftsmen with woodworking capabilities equivalent to those of cabinet makers. They take professional pride in producing high-quality instruments that live up to or exceed established standards of xylophone making. Because manufacture is a craft, each step is done according to the quality control requirements of the builder. The iterative steps of tuning the bars is considered the single-most important part of xylophone manufacture, and the repetition itself is a quality measure.

Byproducts/Waste

Xylophone makers do generate byproducts. Typically, they offer a line of xylophones ranging from small or piccolo xylophones to bass models for orchestras or individual instrumentalists. Some also make other types of percussion instruments especially those in the xylophone family.

Very little waste results from xylophone manufacture. Rosewood is too valuable a commodity to be used frivolously, and the only wood scrap consists of shavings from tuning the keys and minor end scrap. Aluminum scrap is returned to the supplier for recycling.

The craftsmen handle a limited range of hazardous equipment and almost no hazardous materials. Bench cutters are used to cut the tubular resonators and the wood keys. Hand tools are needed to tune the keys. Safety glasses are worn during all operations. Quantities of stain and varnish are minor; these materials are stored and handled safely, and there are no related disposal or waste hazards.

The Future

The xylophone itself is an established player in an orchestra's percussion array; but its range, repertoire, and opportunities for significant growth are limited by both tradition and possibility. In recent years, its close cousin the marimba has grown considerably in popularity because of the interest in Latin, jazz, and percussive music and a broadening of the repertoire. Music enthusiasts hope the xylophone will also increase in popularity, but it will assuredly be a valued orchestra member because of its unique musical voice.

Where to Learn More

Books

Baines, Anthony. *The Oxford Companion to Musical Instruments.* New York: Oxford University Press, 1992.

Bragard, Roger, and Ferdinand J. De Hen. *Musical Instruments in Art and History.* New York: The Viking Press, 1967.

Dearling, Robert, ed. *The Illustrated Encyclopedia of Musical Instruments.* New York: Schirmer Books, 1996.

Sachs, Curt. *The History of Musical Instruments.* New York: W. W. Norton & Company, Inc., 1940.

Thamm, Duane. *The Complete Xylophone & Marimba Method.* Glenview, IL: Creative Music, 1966.

Other

Mallet Works Music. http://www.malletworks.com (June 26, 2000).

Melbourne Symphony Orchestra, Australia. *The Instruments in an Orchestra.* http://www.mso.com.au/edu/pages/orchestra/percussion (June 26, 2000).

Scott, W. L. *Xylophone.* http://www.musicrafts.com (June 26, 2000).

Vibraphone Web Site. http://www.thevibe.net (June 26, 2000).

—*Gillian S. Holmes*

Index

A

Abiomed, Inc., 6:17

Academy of Sciences (France), 6:43

Accordions, **3:1–5**

Acetylene, **4:1–5**

Acids, in baking powder, 6:44–6:45

Acrylic fingernails, **3:6–10**

Acrylic plastic, **2:1–5**
  electric blankets, 6:155
  patent leather finish, 6:283, 6:286

Action figures, **6:1–4**

Adhesives
  cellophane tape, **1:105–108**
  duct tape, **6:150–153**
  glues, **5:234–237**
  rubber cement, **6:322–324**
  super glues, **1:444–447**

Aerodynamics (Boomerangs), 6:54, 6:57–6:58

Air bags, **1:1–7**

Air conditioners, **3:11–15**

Air-forced needle-free injection systems, 6:275, 6:276–6:277

Air fresheners, **6:5–8**

Aircraft
  airships, **3:16–20**
  business jets, **2:80–85**
  helicopters, **1:223–229**
  hot air balloons, **3:220–224**
  jet engines, **1:230–235**

Airships, **3:16–20**

Akron Candy Company, 6:253

Alcoholic beverages. *See* Beverages

Alcoke, Charles, 6:316–6:317

Alembics (Stills), 6:107

Ali, Muhammad, 6:61

Allen, Herbert, 6:124

Aluminosilicate glass, 6:194, 6:196

Aluminum, **5:1–5**
  baby carriers, 6:25

beverage cans, **2:6–10**
clothes irons, 6:104
ferris wheel components, 6:172
foil, **1:8–13**
gyroscopes, 6:191
hard hats, 6:201
horseshoes, 6:217
sailboat masts, 6:330
skateboard trucks, 6:363
snowshoes, 6:377, 6:378–6:379
xylophone resonators, 6:464

Aluminum beverage cans, **2:6–10**

Aluminum foil, **1:8–13**

Amana Appliances, 6:138, 6:139

Amateur Athletic Union (AAU), 6:48

Ambulances, **5:6–10**

America (Yacht), 6:326, 6:329

American Basketball Association (ABA), 6:49

American Chicle Group, 6:68

American Dental Association, 6:406

American Edwards Laboratories, 6:18

American Eveready Battery Company, 6:176

American Heart Association, 6:13

American Journal of Epidemiology, 6:155

American National Can Company, 6:6

American National Standards Institute (ANSI), 6:200, 6:203

American News Company, 6:116

American Society for Testing and Materials (ASTM), 6:152, 6:175, 6:451

American Water Works Association Research Foundation, 6:165

America's Cup Race, 6:326, 6:329

Ammunition, **2:11–15**

Amusement park rides
  carousels, **4:88–93**
  ferris wheels, **6:171–175**
  roller coasters, **6:316–321**

Angioplasty balloons, **6:9–12**

Index